D0765263

THIN FILM PROCESSES

Contents

Part I

I-1 Introduction

J. L. Vossen

Part II PHYSICAL METHODS OF FILM DEPOSITION

II-1 Glow Discharge Sputter Deposition

J. L. Vossen and J. J. Cuomo

List of Contributors

Numbers in parentheses indicate the pages on which the authors' contributions begin.

VLADIMIR S. BAN (257), RCA Laboratories, Princeton, New Jersey 08540

J. J. CUOMO (11), IBM Thomas J. Watson Research Center, Yorktown Heights, New York 10598

CHERYL A. DECKERT (401), RCA Laboratories, Princeton, New Jersey 08540

DAVID B. FRASER (115), Bell Telephone Laboratories, Murray Hill, New Jersey 07974

JAMES M. E. HARPER, (175), IBM Thomas J. Watson Research Center, Yorktown Heights, New York 10598

J. R. HOLLAHAN (335), Applied Materials Incorporated, Santa Clara, California 95051

WERNER KERN (257, 401), RCA Laboratories, Princeton, New Jersey 08540

FREDERICK A. LOWENHEIM (209), 637 West 7th Street, Plainfield, New Jersey 07060

C. M. MELLIAR-SMITH (497), Bell Telephone Laboratories, Murray Hill, New Jersey 07974

C. J. MOGAB (497), Bell Telephone Laboratories, Murray Hill, New Jersey 07974

ALAN S. PENFOLD (75), Telic Corporation, Santa Monica, California 90404

R. S. ROSLER* (335), Applied Materials Incorporated, Santa Clara, California 95051

* Present address: Motorola, Inc., Phoenix, Arizona 85002

JOHN A. THORNTON (75), Telic Corporation, Santa Monica, California 90404

J. L. VOSSEN (3, 11), RCA Laboratories, Princeton, New Jersey 08540

ROBERT K. WAITS (131), Data Systems Division, Hewlett–Packard Company, Cupertino, California 95014

H. YASUDA (361), Department of Chemical Engineering, University of Missouri–Rolla, Rolla, Missouri 65401

Preface

Remarkable advances have been made in recent years in the science and technology of thin film processes for deposition and etching. It is the purpose of this book to bring together tutorial reviews of selected film deposition and etching processes from a process viewpoint. Emphasis is placed on the practical use of the processes to provide working guidelines for their implementation, a guide to the literature, and an overview of each process.

It was felt that a book of this sort was needed to provide up-to-date reviews of processes which have not been reviewed recently (or at all), and in which significant recent advances have been made. It is hoped that the book will be useful both to those who wish to gain a broad perspective on a process or processes in general and to specialists using these processes.

These reviews were completed in late 1977 and literature published up to that time is included. It is recognized that many of these processes are being investigated actively. We hope that this research and development will shed light on some questions that remain unanswered at this time, leading to more rational selection and use of thin film processes.

Finally, we wish to thank the contributing authors for their considerable efforts in summarizing the voluminous literature of their specialties, RCA Laboratories for the use of their scientific publications facilities, and our wives for patience and understanding during the course of this undertaking.

Part I

I-1

Introduction

J. L. VOSSEN

RCA Laboratories
Princeton, New Jersey

I. THE ORGANIZATION OF THIS BOOK

The processes selected for review fall into four general categories: (1) physical methods of film deposition, (2) chemical methods of film deposition, (3) hybrid (i.e., physical—chemical) methods of film deposition, and (4) patterning techniques.

Part II of the book deals with the mainly physical deposition processes; those selected for review are all sputtering techniques. Chapter II-1 covers conventional sputtering and contains material on sputtering target kinetics that is introductory to the remaining chapters on sputtering. In addition, this chapter provides material on glow discharges and their effects on substrates which applies not only to the sputtering chapters, but also to the glow discharge deposition and plasma etching chapters (IV-1, IV-2, V-2).

Chapters II-2–II-4 deal with various magnetron sputtering configurations. These are highly efficient glow discharge sputtering sources that depend on crossed electric and magnetic fields to produce high ion densities and high deposition rates with minimal substrate bombardment. Chapter II-2 describes cylindrical magnetrons in which the magnetic field is uniformly orthogonal to the electric field. Chapters II-3 and II-4 discuss planar and conical configurations in which there are gradient magnetic fields. The basic physics of uniform and gradient fields are discussed in Chapter II-2, and the process effects of gradient fields are further exposed in Chapters II-3 and II-4.

Chapter II-5 is devoted to ion-beam deposition in which substrates are not immersed in a glow discharge, but are in a relatively good vacuum. Not only are the important consequences of this covered, but the basic physics and technology of ion-gun structures are included. The same gun structures are used in ion-beam etching (ion milling) (Chapter V-2).

The processes that have been categorized as chemical include chemical vapor deposition and deposition of inorganic or metal films from solution by plating, anodization, etc. These processes will be found in Part III.

Hybrid processes (as defined here) are those which use glow discharges to activate chemical deposition processes. Chapter IV-1 discusses plasma deposition—the deposition of inorganic compounds by the glow discharge decomposition of reagents normally used in chemical vapor deposition. Chapter IV-2 discusses the processes involved in polymerization of monomers in a glow discharge.

Part V covers etching techniques for films and substrates. Etching for

Table I

Other Film Deposition Processes

Process	References	Process	References
1. Thermal evaporation from resistance-, electron-beam-, rf-, or laser-heated sources	[1–5]	15. Electrophoresis	[41–44]
		16. Screen printing	[45–49]
		17. Wire flame spraying	[32, 50, 51]
		18. Powder flame spraying	[32, 51, 52]
		19. Plasma flame spraying	[32, 51, 53]
2. Molecular-beam epitaxy	[6, 6a]	20. Arc plasma spraying	[54–57]
		21. Detonation coating	[58]
3. Chemical vapor deposition for epitaxy	[7–11]	22. Diffusion coating	[32, 59, 60]
		23. Pack cementation	[59]
4. Thermal oxidation	[12–17]	24. Cladding	[32, 61]
5. Plasma anodization	[18–20]	25. Explosive cladding	[62]
6. Ion implantation	[21–24]	26. Fluidized-bed coating	[27, 28, 63–70]
7. Electrostatic spraying	[25–30]	27. Vacuum impregnation	[27, 28]
8. Printing	[31]	28. Mechanical (peen) plating	[32]
9. Spraying	[27, 28, 32]		
10. Spin coating	[27, 28, 32, 33]	29. Brush coating	[27, 28]
11. Dip coating	[27, 28]	30. Roller coating	[27, 28]
12. Electric harmonic spraying	[34]	31. Optical lithography	[71–76]
		32. X-ray lithography	[71, 72, 76, 77]
13. Molten-metal dip coating	[32]	33. Electron beam lithography	[71, 72, 75, 76]
14. Liquid-phase epitaxy	[35–40]		

pattern delineation is emphasized. Chapter V-1 details the fundamentals of chemical etching and provides tables of chemical etchants for an extremely wide range of materials. Chapter V-2 deals with four etching processes utilizing vacuum techniques: ion beam etching (milling), rf-sputter etching with inert-gas ions, plasma etching in which reactive-gas ions and excited neutrals form volatile compounds with film surfaces, and reactive-ion etching which is a hybrid process that uses rf-sputter etching equipment with reactive gases similar to those used in plasma etching.

II. OTHER FILM DEPOSITION PROCESSES

Numerous important processes for film deposition are not reviewed in this book for one or more of several reasons. Processes that have been reviewed adequately are not included because an additional review would be redundant. On the other hand, some thin-film processes are not yet mature enough to warrant a review. Also, processes which generally involve very thick films ($>25 \mu$m) have not been included. While these processes are not specifically covered in this book, their importance is nevertheless recognized. Therefore selected references to reviews or other literature on 33 other processes are given in Table I.

REFERENCES

1. R. Glang, *in* "Handbook of Thin Film Technology" (L. I. Maissel and R. Glang, eds.), Ch. 1. McGraw-Hill, New York, 1970.
2. K. W. Raine, *Natl. Phys. Lab. (U.K.), Rep.* No. OP MET 13 (1972).
3. E. B. Draper, *J. Vac. Sci. Technol.* **8**, 333 (1971).
4. H. Schwarz and H. A. Tourtellotte, *J. Vac. Sci. Technol.* **6**, 373 (1969).
5. M. S. Hess and J. F. Milkosky, *J. Appl. Phys.* **43**, 4680 (1972).
6. J. R. Arthur, *in* "Critical Reviews of Solid State Science" (D. E. Schuele and R. W. Hoffman, eds.), Vol. 6, p. 413. CRC Press, Cleveland, 1976.
6a. B. A. Joyce and C. T. Foxon, *in* "Solid State Devices, 1976" (R. Muller and E. Lange, eds.), p. 17. Inst. Phys., London, 1977.
7. G. W. Cullen and C. C. Wang, eds., "Heteroepitaxial Semiconductors for Electronic Devices." Springer-Verlag, Berlin and New York, 1978.
8. J. W. Matthews, ed., "Epitaxial Growth," Part A. Academic Press, New York, 1975.
9. G. W. Cullen, E. Kaldis, R. L. Parker, and C. J. M. Rooymans, eds., "Vapor Growth and Epitaxy." North-Holland Publ., Amsterdam, 1975.
10. H. G. Schneider and V. Ruth, eds., "Advances in Epitaxy and Edotaxy." Elsevier, Berlin and New York, 1976.
11. A. Y. Cho and J. R. Arthur, *Prog. Solid State Chem.* **10**, 157 (1975).
12. K. Hauffe, "Oxidation of Metals." Plenum, New York, 1965.
13. P. Kofstad, "High-Temperature Oxidation of Metals." Wiley, New York, 1966.

14. O. Kubaschewski, "Oxidation of Metals and Alloys," 2nd Ed. Butterworth, London, 1962.
15. E. H. Nicollian, *J. Vac. Sci. Technol.* **14**, 1112 (1977), and references therein.
16. A. M. Goodman and J. M. Breece, *J. Electrochem. Soc.* **117**, 982 (1970).
17. C. W. Wilmsen, *Thin Solid Films* **39**, 105 (1976).
18. C. J. Dell'Oca, D. L. Pulfrey, and L. Young, *Phys. Thin Films* **6**, 1 (1971).
19. J. F. O'Hanlon, *J. Vac. Sci. Technol.* **7**, 330, (1970).
20. J. F. O'Hanlon, *in* "Oxides and Oxide Films" (A. K. Vijh, ed.), Vol. 5, p. 1. Dekker, New York, 1977.
21. P. D. Townsend, J. C. Kelly, and N. E. W. Hartley, "Ion Implantation, Sputtering and Their Applications." Academic Press, New York, 1976.
22. G. Carter and W. A. Grant, "Ion Implantation of Semiconductors." Wiley, New York, 1976.
23. G. Dearnaley, J. H. Freeman, R. S. Nelson, and J. Stephan, "Ion Implantation." North-Holland Publ., Amsterdam, 1973.
24. J. W. Mayer, L. Erikson, and J. A. Davies, "Ion Implantation." Academic Press, New York, 1970.
25. S. Kut, *in* "Science and Technology of Surface Coating" (B. N. Chapman and J. C. Anderson, eds.), p 43. Academic Press, New York, 1974.
26. R. P. Corbett, *in* "Science and Technology of Surface Coating" (B. N. Chapman and J. C. Anderson, eds.), p. 52. Academic Press, New York, 1974.
27. J. J. Licari, "Plastic Coatings for Electronics." McGraw-Hill, New York, 1970.
28. J. J. Licari and E. R. Brands, *in* "Handbook of Materials and Processes for Electronics" (C. A. Harper, ed.), Ch. 5. McGraw-Hill, New York, 1970.
29. M. C. Gourdine, E. L. Collier, G. P. Lewis, H. McCrae, and D. H. Porter, U.S. Patent 3, 613, 993 (1971).
30. L. L. Bromley and J. B. Williams, U.S. Patent 3, 635, 401 (1972).
31. S. Karttunen and P. Oittinen, *in* "Science and Technology of Surface Coating" (B. N. Chapman and J. C. Anderson, eds.), p. 222. Academic Press, New York, 1974.
32. J. A. Murphy, "Surface Preparation and Finishes for Metals," Ch. 6. McGraw-Hill, New York, 1971.
33. B. D. Washo, *IBM J. Res. Dev.* **21**, 190 (1977).
34. S. B. Sample, R. Bollini, D. A. Decker, and J. W. Boarman, *in* "Science and Technology of Surface Coating" (B. N. Chapman and J. C. Anderson, eds.), p. 322. Academic Press, New York, 1974.
35. H. F. Lockwood and M. Ettenberg, *J. Cryst. Growth* **15**, 81 (1972).
36. J. T. Longo, J. S. Harris, E. R. Gertner, and J. C. Chu, *J. Cryst. Growth* **15**, 107 (1972).
37. J. W. Balch and W. W. Anderson, *J. Cryst. Growth* **15**, 204 (1972).
38. I. Crossley and M. B. Small, *J. Cryst. Growth* **15**, 268 (1972).
39. I. Crossley and M. B. Small, *J. Cryst. Growth* **15**, 275 (1972).
40. C. J. Nuese, H. Kressel, and I. Ladany, *J. Vac. Sci. Technol.* **10**, 772 (1973).
41. E. Matejevic, ed., "Surface and Colloid Science," Vol. 7. Wiley, New York, 1974.
42. A. With, *in* "Science and Technology of Surface Coating" (B. N. Chapman and J. C. Anderson, eds.), p. 60. Academic Press, New York, 1974.
43. D. J. Shaw, "Electrophoresis." Academic Press, New York, 1969.
44. H. A. Pohl and W. F. Pickard, eds., "Symposium on Dielectrophoretic and Electrophoretic Deposition." Electrochem. Soc., Princeton, New Jersey, 1969.
45. D. W. Hamer and J. V. Biggers, "Thick Film Hybrid Microcircuit Technology." Wiley, New York, 1972.
46. C. A. Harper, ed., "Handbook of Thick Film Hybrid Microelectronics." McGraw-Hill, New York, 1974.

47. M. L. Topfer, "Thick-Film Microelectronics." Van Nostrand-Reinhold, New York, 1971.
48. G. V. Planer and L. S. Phillips, "Thick Film Circuits." Crane, Russak, New York, 1972.
49. J. I. Biegeleisen and M. A. Cohn, "Silk Screen Techniques." Dover, New York, 1942.
50. H. S. Ingham and A. P. Shepard, "Flame Spray Handbook," Vol. 1: Wire Flame Spraying. Metco, New York, 1965.
51. C. W. Smith, in "Science and Technology of Surface Coating" (B. N. Chapman and J. C. Anderson, eds.), p. 262. Academic Press, New York, 1974.
52. H. S. Ingham and A. P. Shepard, "Flame Spray Handbook," Vol. 2: Powder Flame Spraying. Metco, New York, 1965.
53. H. S. Ingham and A. P. Shepard, "Flame Spray Handbook," Vol. 3: Plasma Flame Process. Metco, New York, 1965.
54. D. H. Harris and R. J. Janowiecki, *Electronics* **43**(3), 108 (1970).
55. J. D. Reardon, *Ind. Res.* **19**(4), 90 (1977).
56. A. R. Moss and W. J. Young, in "Science and Technology of Surface Coating" (B. N. Chapman and J. C. Anderson, eds.), p. 287. Academic Press, New York, 1974.
57. D. R. Marantz, in "Science and Technology of Surface Coating" (B. N. Chapman and J. C. Anderson, eds.), p. 308. Academic Press, New York, 1974.
58. R. G. Smith, in "Science and Technology of Surface Coating" (B. N. Chapman and J. C. Anderson, eds.), p. 271. Academic Press, New York, 1974.
59. R. L. Wachtell, in "Science and Technology of Surface Coating" (B. N. Chapman and J. C. Anderson, eds.), p. 105. Academic Press, New York, 1974.
60. J. C. Gregory, in "Science and Technology of Surface Coating" (B. N. Chapman and J. C. Anderson, eds.), p. 136. Academic Press, New York, 1974.
61. D. R. Gabe, "Principles of Metal Surface Treatment and Protection." Pergamon, Oxford, 1972.
62. R. H. Wittman, *Battelle Tech. Rev.* **16**(7), 17 (1967).
63. W. J. Davis, U.S. Patent 3,004,861 (1961).
64. C. J. Dettling, U.S. Patent 2,974,060 (1961).
65. C. J. Dettling and R. E. Hartline, U.S. Patent 2,987,413 (1961).
66. E. Gemmer, U.S. Patent 2,974,059 (1961).
67. E. Gemmer, U.S. Patent 3,090,696 (1963).
68. C. K. Pettigrew, *Mod. Plast.* **43**(12), 111 (1966).
69. C. K. Pettigrew, *Mod. Plast.* **44**(1), 150 (1966).
70. J. Gaynor, A. H. Robinson, M. Allen, and E. E. Stone, *Mod. Plast.* **43**(5), 133 (1966).
71. A. N. Broers, *Proc. Int. Vac. Congr., 7th, Int. Conf. Solid Surf., 3rd, Vienna* p. 1521 (1977).
72. A. N. Broers, *Proc. Symp. Electron Ion Beam Sci. Technol., 7th, Electrochem. Soc., Princeton, N.J.,* p. 587 (1976).
73. R. J. Ryan, E. B. Davidson, and H. O. Hook, in "Handbook of Materials and Processes for Electronics" (C. Harper, ed.), Ch. 14. McGraw-Hill, New York, 1970.
74. F. H. Dill, *IEEE Trans. Electron Devices* **ED-22**, 440 (1975).
75. L. F. Thompson and R. E. Kerwin, *Annu. Rev. Mater. Sci.* **6**, 267 (1976).
76. E. I. Gordon and D. R. Herriott, *IEEE Trans. Electron Devices* **ED-22**, 371 (1975).
77. H. I. Smith, *Proc. IEEE* **62**, 1361 (1974).

Part II

PHYSICAL METHODS OF FILM DEPOSITION

II-1

Glow Discharge Sputter Deposition

J. L. VOSSEN

RCA Laboratories
Princeton, New Jersey

J. J. CUOMO

IBM Thomas J. Watson Research Center
Yorktown Heights, New York

I. INTRODUCTION

Over the past 20 years or so there have been numerous reviews of sputtering and sputtering processes for film deposition [1–15]. In this chapter we shall take a somewhat different viewpoint than those of most earlier reviewers. We shall attempt to treat this very complex subject from a process viewpoint. That is, we shall discuss the interactions of the process parameters to expose the many permutations and combinations that are available to control the properties of thin films.

Because there are so many interactions among parameters in sputtering systems, it is impossible to separate them completely. Thus, there are necessary, but regrettable references made to later sections throughout the chapter. In an attempt to minimize any confusion that this may cause, we shall give a brief, simple overview of the subject in this section before going to the more detailed discussions.

Figure 1 represents a greatly simplified cross section of a sputtering system. Typically, the target (a plate of the material to be deposited or the material from which a film is to be synthesized) is connected to a negative voltage supply (dc or rf). The substrate holder faces the target. The holder may be grounded, floating, biased, heated, cooled, or some combination of these. A gas is introduced to provide a medium in which a glow discharge can be initiated and maintained. Gas pressures ranging from a few millitorr to about 100 mTorr are used. The most common sputtering gas is argon.

When the glow discharge is started, positive ions strike the target plate and remove mainly neutral target atoms by momentum transfer, and these

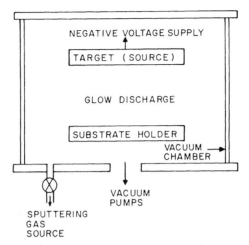

Fig. 1. Simplified cross section of a sputtering system.

condense into thin films. There are, in addition, other particles and radiation produced at the target, all of which may affect film properties (secondary electrons and ions, desorbed gases, x rays, and photons). The electrons and negative ions are accelerated toward the substrate platform and bombard it and the growing film. In some instances, a bias potential (usually negative) is applied to the substrate holder, so that the growing film is subject to positive ion bombardment. This is known variously as *bias sputtering* or *ion plating*. Initially, the term "ion plating" referred to a process in which the deposition source was a thermal evaporation filament instead of a sputtering target and the substrates were connected to a dc sputtering target [16], but it has sometimes been applied to any process in which the substrate is subjected to purposeful ion bombardment during film growth in a glow discharge environment [17].

In some cases, gases or gas mixtures other than Ar are used. Usually this involves some sort of *reactive sputtering* process in which a compound is synthesized by sputtering a metal target (e.g., Ti) in a reactive gas (e.g., O_2 or Ar–O_2 mixtures) to form a compound of the metal and the reactive gas species (e.g., TiO_2). Reactive sputtering is also used to replenish constituents of compound targets lost by dissociation. The reactive version of ion plating is sometimes known as *activated reactive evaporation* [18], but this terminology is more often applied to processes in which an evaporant passes through a glow discharge in transit to an electrically floating or grounded substrate. Reactive sputtering should not be confused with *chemical sputtering* in which the reactive gas (e.g., O_2) reacts with the target surface (e.g., C) to form volatile compounds (e.g.,

CO) that are pumped away [1]. Chemical sputtering is more properly related to ion etching processes (see Chapter V-2).

We shall consider in detail the complex interplay among target kinetics, glow discharge phenomena, substrate conditions, equipment configuration, etc. that bear on the ability to control the properties of thin films.

II. PHYSICAL AND CHEMICAL EFFECTS OF ION BOMBARDMENT ON SURFACES

In sputter deposition, surfaces subject to ion bombardment are usually considered as the source of material from which films are grown. In addition to the neutral (sputtered) material liberated from the bombarded surface which eventually condenses as a film, there are numerous other events that can occur at the target surface which may influence the growth of films profoundly. These include: secondary electron emission, secondary positive and/or negative ion emission, emission of radiation (photons, x rays), reflection of incident particles, heating, chemical dissociation or reaction, bulk diffusion, crystallographic changes, and reflection of some of the emitted particles back to the bombarded surface (backscattering). It should be noted that all of these same phenomena apply to sputter-etching processes (Chapter V-2) in which the workpiece *is* a sputtering target and to *substrates in most glow discharge deposition processes.* (As will be shown later, any material body immersed in a glow discharge acquires a negative potential with respect to its surroundings and must be considered a sputtering target.)

There have been several recent comprehensive reviews of the kinetics involved when a surface is ion bombarded [3, 6, 8, 12]. Therefore, we shall review them only briefly, emphasizing those target effects that can affect the way in which films grow.

A. Emission of Neutral Particles—The Sputtering Yield

The sputtering yield is defined as the number of atoms ejected from a target surface per incident ion. It is the most fundamental parameter of sputtering processes. Yet all of the surface interaction phenomena involved that contribute to the yield of a given surface are not completely understood. Despite this, an impressive body of literature exists showing the yield to be related to momentum transfer from energetic particles to target surface atoms. There is a threshold for sputtering that is approximately equal to the heat of sublimation. In the energy range of practical interest for sputtering processes (10–5000 eV), the yield increases with incident ion energy, and with the mass and d-shell filling of the incident ion [19, 20].

The sputtering yield determines the erosion rate of sputtering targets; and largely, but not completely, determines the deposition rate of sputtered films. Several compilations of experimental sputtering yield and related data have been published [3, 5, 6, 21, 22]. All sputtering yields and related data should be used with caution. In glow discharge systems, bombarding ions are by no means monoenergetic, and it is not necessarily valid to use yield values for pure metals when alloys, compounds, or mixtures are sputtered. As will be shown, the sputtering yield of material A from a matrix of A + B is often very different from the sputtering yield of A from a matrix of A. Also, when sputtering yields of compounds are given, dissociation reactions are often ignored. Despite this, tabulations of sputtering yields are useful, if only to give a rough indication of the deposition or etch rate that might be expected for a given material. Tables I–III give a compilation of sputtering yields and relative film deposition

Table I

Sputtering Yield of Elements at 500 eV

Gas	He	Ne	Ar	Kr	Xe	Reference
Element						
Be	0.24	0.42	0.51	0.48	0.35	[23]
C	0.07	—	0.12	0.13	0.17	[23]
Al	0.16	0.73	1.05	0.96	0.82	[23]
Si	0.13	0.48	0.50	0.50	0.42	[23]
Ti	0.07	0.43	0.51	0.48	0.43	[23]
V	0.06	0.48	0.65	0.62	0.63	[23]
Cr	0.17	0.99	1.18	1.39	1.55	[23]
Mn	—	—	—	1.39	1.43	[23]
Mn	—	—	1.90	—	—	[24]
Bi	—	—	6.64	—	—	[24]
Fe	0.15	0.88	1.10	1.07	1.00	[23]
Fe	—	0.63	0.84	0.77	0.88	[25]
Co	0.13	0.90	1.22	1.08	1.08	[23]
Ni	0.16	1.10	1.45	1.30	1.22	[23]
Ni	—	0.99	1.33	1.06	1.22	[25]
Cu	0.24	1.80	2.35	2.35	2.05	[23]
Cu	—	1.35	2.0	1.91	1.91	[25]
Cu (111)	—	2.1	—	2.50	3.9	[26]
Cu	—	—	1.2	—	—	[27]
Ge	0.08	0.68	1.1	1.12	1.04	[23]
Y	0.05	0.46	0.68	0.66	0.48	[23]
Zr	0.02	0.38	0.65	0.51	0.58	[23]

Table I *(Continued)*

Gas	He	Ne	Ar	Kr	Xe	Reference
Nb	0.03	0.33	0.60	0.55	0.53	[23]
Mo	0.03	0.48	0.80	0.87	0.87	[23]
Mo	—	0.24	0.64	0.59	0.72	[25]
Ru	—	0.57	1.15	1.27	1.20	[23]
Rh	0.06	0.70	1.30	1.43	1.38	[23]
Pd	0.13	1.15	2.08	2.22	2.23	[23]
Ag	0.20	1.77	3.12	3.27	3.32	[23]
Ag	1.0	1.70	2.4	3.1	—	[27]
Ag	—	—	3.06	—	—	[28]
Sm	0.05	0.69	0.80	1.09	1.28	[23]
Gd	0.03	0.48	0.83	1.12	1.20	[23]
Dy	0.03	0.55	0.88	1.15	1.29	[23]
Er	0.03	0.52	0.77	1.07	1.07	[23]
Hf	0.01	0.32	0.70	0.80	—	[23]
Ta	0.01	0.28	0.57	0.87	0.88	[23]
W	0.01	0.28	0.57	0.91	1.01	[23]
Re	0.01	0.37	0.87	1.25	—	[23]
Os	0.01	0.37	0.87	1.27	1.33	[23]
Ir	0.01	0.43	1.01	1.35	1.56	[23]
Pt	0.03	0.63	1.40	1.82	1.93	[23]
Au	0.07	1.08	2.40	3.06	3.01	[23]
Au	0.10	1.3	2.5	—	7.7	[29]
Pb	1.1	—	2.7	—	—	[27]
Th	0.0	0.28	0.62	0.96	1.05	[23]
U	—	0.45	0.85	1.30	0.81	[23]
Sb	—	—	2.83	—	—	[24]
Sn (solid)	—	—	1.2	—	—	[30]
Sn (liquid)	—	—	1.4	—	—	[30]

rates. The latter have been normalized to the sputtering yields of pure metals [21]. All target materials are polycrystalline unless otherwise indicated.

B. Emission of Other Particles

1. Secondary Electrons

Since sputtering targets are held at high negative potentials, secondary electrons are accelerated away from the target surface with an initial energy equal to the target potential. As will be shown in Section III.A, these electrons help to sustain the glow discharge by ionization of neutral sput-

Table II

Sputtering Yield of Elements at 1 keV

Gas	He	N	Ne	N_2	Ar	Kr	Xe	Reference
Element								
Fe	—	—	0.85	—	1.33	1.42	1.82	[25]
Fe	—	0.55	—	0.78	—	—	—	[3]
Ni	—	—	1.22	—	2.21	1.76	2.26	[25]
Ni	—	0.74	—	1.05	—	—	—	[3]
Ni	—	—	—	—	2.0	2.0	2.0	[26]
Cu	—	—	1.88	—	2.85	3.42	3.6	[25]
Cu	—	1.5	—	—	—	—	—	[3]
Cu	—	—	—	—	3.2	2.5	—	[27]
Cu (111)	—	—	2.75	—	4.5	4.65	6.05	[26]
Cu	—	—	—	1.95	—	—	—	[31]
Mo	—	—	0.49	—	1.13	1.27	1.60	[25]
Mo	—	0.16	—	0.3	—	—	—	[3]
Ag	1.8	—	2.4	—	3.8	4.7	—	[27]
Sn	—	—	—	—	0.8	—	—	[32]
W	—	0.18	—	0.2	—	—	—	[3]
Au	—	—	—	—	1.0	—	—	[32]
Au	0.3	—	2.1	—	4.9	—	—	[29]
Pb	1.5	—	—	—	3.0	—	—	[27]
Sn (liquid)	—	—	—	—	1.7	—	—	[30]
Au (111)	—	—	—	—	3.7	—	—	[33]
Au (100)	—	—	—	—	3.0	—	—	[33]
Au (110)	—	—	—	—	2.0	—	—	[33]
Au	—	—	—	—	3.6	—	—	[33]
Al (111)	—	—	—	—	1.0	—	—	[33]

tering gas atoms which in turn bombard the target and release more secondary electrons in an avalanche process. Upon arrival at the substrate, such energy as they retain after collisions in the gas is liberated in the form of heat [38–43]. Many of the secondary electrons are thermalized by collisions in the gas, but even at high gas pressures, a substantial number of electrons retain full target potential upon impact at the substrates [41, 42].

2. Secondary Ions

Most of the data on ion emission from solids due to primary ion bombardment is to be found in the literature of secondary ion mass spectroscopy (SIMS). Most of this literature deals with the formation and emission of positive ions. However, in glow discharge sputtering, it is highly

Table III

Miscellaneous Sputtering Yields

Material	Gas	Energy (keV)	Yield	Reference
Fe	Ar	10	1.0	[32]
Cu	Ne	10	3.2	[34, 35]
Cu	Ar	10	6.25	[34, 35]
Cu	Kr	10	8.0	[35]
Cu	Xe	10	10.2	[35]
Ag	Ar	10	10.4	[36]
Ag	Kr	10	14.8	[36]
Ag	Xe	10	15.5	[35, 36]
Au	Ne	10	3.7	[35]
Au	Ar	10	8.5	[32, 35]
Au	Kr	10	14.6	[35]
Au	Xe	10	20.3	[35]
Sn	Ar	10	2.1	[32]
Mg (0001)	Ar	5	3.13	[37]
Mg (10$\bar{1}$0)	Ar	5	2.70	[37]
Mg (11$\bar{2}$0)	Ar	5	1.64	[37]
Zn (0001)	Ar	3	11.22	[37]
Zn (0001)	Ar	4	11.43	[37]
Zn (0001)	Ar	5	12.65	[37]
Zn (11$\bar{2}$0)	Ar	3	9.19	[37]
Zn (11$\bar{2}$0)	Ar	4	10.07	[37]
Zn (11$\bar{2}$0)	Ar	5	9.35	[37]
Zr (0001)	Ar	5	1.12	[37]
Zr (10$\bar{1}$0)	Ar	5	1.56	[37]
Zr (11$\bar{2}$0)	Ar	4	0.74	[37]
Zr (11$\bar{2}$0)	Ar	5	0.64	[37]
Zr (11$\bar{2}$0)	Ar	10	0.65	[37]
Cd (0001)	Ar	4	19.01	[37]
Cd (0001)	Ar	5	21.25	[37]
Cd (10$\bar{1}$0)	Ar	4	15.86	[37]
Cd (10$\bar{1}$0)	Ar	5	17.86	[37]
Cd (11$\bar{2}$0)	Ar	5	13.58	[37]
PbTe (111)	Ar	0.5	1.4	[28]
GaAs (110)	Ar	0.5	0.9	[28]
GaP (111)	Ar	0.5	0.95	[28]
CdS (10$\bar{1}$0)	Ar	0.5	1.12	[28]
SiC (0001)	Ar	0.5	0.41	[28]
InSb	Ar	0.5	0.55	[28]
TaC	Ar	0.6	0.13	[21]
Mo_2C	Ar	0.6	0.15	[21]

improbable that a positive ion generated at the target surface could escape the negative target field, so these are of little interest. Negative ions have been studied by SIMS to a lesser extent [44–53]. Negative ions result mainly from the sputtering of the anionic species of compounds [48–51] and high-electron-affinity constituents of alloys [53a,b]. Virtually no secondary negative ions are produced during inert gas ion bombardment of pure metal surfaces. Negative ions, like electrons, are accelerated away from the target toward the substrates, thus representing another source of substrate bombardment. Experimental evidence [54] shows that these ions actually arrive at the substrates as energetic neutrals, having suffered electron-stripping collisions in transit through the glow discharge.

3. Reflected Incident Particles

Some of the primary bombarding particles are reflected from the target surface. The literature of ion scattering spectroscopy (ISS) [55–62] contains most of the pertinent literature related to this effect. These reflected particles are neutralized and reflected as atoms, not ions. The amount of reflection is an inverse function of primary bombarding energy because this effect competes with ion implantation. As low primary energies, reflection fractions as high as 0.4 have been observed [56], whereas at high primary bombarding energies (>1000 eV) typical reflection fractions are of the order of 0.05. These particles represent still another source of substrate bombardment during film growth [43].

4. Desorption of Gases

Desorption of gases occurs even for very dense target materials. Initially, adsorbed gas layers on target surfaces are sputtered away, and then, depending on the nature of the target material, chemisorbed gases, occluded gases, and gases generated by decomposition of target compounds are released in that order [63–66]. For adsorbed gases there are marked peaks in the desorption rate at low energies (generally <200 eV) which have been attributed to local heating of the lattice in a radius of the order of 10 Å. Chemisorbed gases are truly sputtered [66]. It is these gases that usually leave the target initially as negative ions as noted above. Occluded gases are both sputtered and thermally desorbed. This is especially a problem with hot-pressed targets [67] and powder targets [68]. The effect of massive gas desorption from porous targets is to contaminate the sputtering gas and, hence, the deposited film. In addition, depending on the nature of the gas, it will have an effect on the film deposition rate (Section VIII.B).

C. Emission of Radiation

In this section, we consider the emission of radiation due to the sputtering process alone at the target surface—not to that contributed by the glow discharge.

1. Photons

Both ultraviolet (UV) and visible radiation are emitted from sputtering targets [69–76]. Sputtered metal and elemental semiconductor atoms leaving the target in an excited or ionized state undergo resonance- and Auger-type electronic transitions with subsequent photon emission characteristic of the metal being sputtered. Electronic transitions do not apply to compounds or glasses. In these cases the radiative process is simply due to the excited species of the sputtered atoms. Only semiquantitative data on this phenomenon have been reported, but the heirarchy of emission proceeds from low to high in the sequence: elemental targets, binary compounds, ternary compounds, etc., glasses. In the context of sputter deposition, this effect is mainly related to radiation damage to surface-sensitive substrates by energetic UV photons. However, it has been used to advantage to monitor sputter deposition and etching rates (Section IV.C).

2. X rays

X rays characteristic of the target material are emitted from the target surface at energies up to that of the primary bombarding ions. These x rays also can damage surface-sensitive substrates. Also energetic secondary electrons originating at the target which arrive at substrates can generate x rays at the substrate surface. Tables of x-ray energies and excitation potentials [77] can sometimes be used to estimate the degree of damage that may be introduced by this effect.

D. Ion Implantation

Primary bombarding particles can embed into the target surface, become neutralized and trapped. A large body of literature exists on ion-surface interactions particularly those dealing with first-wall problems in fusion reactors. Even at relatively low bombarding energies [78–81] substantial implantation of the primary bombarding ions occurs (Fig. 2). When these ions lose their energy, they contribute to target heating and do not give rise to sputtering.

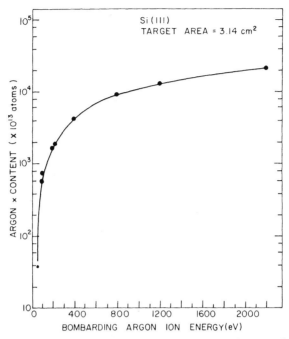

Fig. 2. Argon content of Si versus incident ion energy (after Comas and Carosella [78], with permission of The Electrochemical Society).

E. Altered Surface Layers and Diffusion

1. Multicomponent Targets

The bombardment of a multicomponent solid surface with ions or neutral atoms alters the chemical composition of the surface due to the difference in the sputter yield of the constituents [82]. The region of change is known as the altered layer. Upon initial bombardment of the surface the constituent with highest sputter yield is preferentially removed from the surface, enriching the surface layer in the lower sputter yield material until a steady state is reached. At steady state, the sputtered composition is that of the bulk target composition and the altered layer will recede uniformly with continued sputtering as long as the steady state conditions are maintained. Data on altered layers come from several areas: sputter deposition [83, 84], surface analysis [85, 86], and ion beam alteration of surface properties [86]. Empirical data were generated in attempts to quantify surface analysis techniques [86–91].

2. Yield

The surface composition alteration generally assumes yields of the components as in the elemental form [88], although it is has been shown [87] that some alloys have yields that are greater than those of either element by itself. For example, the yield of Cu from $\langle 100 \rangle$ Cu_3Au was found to be higher than from $\langle 100 \rangle$ single crystal Cu.

3. Steady State

The thickness of the altered layer has been explored in some detail for a variety of targets [83–85, 92–94]. Estimates for metal alloys have been a few tens of angstroms. For oxides the layer is much thicker (~ 1000 Å) [94].

4. Diffusion

Diffusion severely affects the altered layer thickness and the time to reach steady state. These can be due to ion-induced diffusion [87, 91, 94, 95], the ion mean range [93, 94], temperature driven surface diffusion [88], and enhanced diffusion due to the surface implantation of sputter-gas species [93, 95]. As noted in the previous section, surface diffusion will be affected by the phases present, grain size, and the nature of the species in the target. The temperature dependence of diffusion of the two phase Ag–Cu system was examined at 80 and 270°C [96]. Forty minutes sputtering time was required to reach steady state for the 80°C target while the 270°C target required about 200 min to reach steady state. In the 270°C experiment the target showed a definite reddish cast, the altered layer thickness was estimated to be about 1 μm, and the films deposited were enriched in Ag. The difference in sputter yield ratios of Ni/Cu of 1.9 at room temperature [88] to that of 1.6 [96] at 300°C was attributed to diffusion in the target at 300°C.

5. Models

There have been numerous phenomenological models for the altered layer [83–85, 92, 93]. Due to the relationship between multicomponent sputter deposition and altered layer formation [96a] these will be considered in the section on multicomponent sputtering.

F. Dissociation Processes

1. Sublimation

Most of the energy of the incident ions at a sputtering target is transferred to the target surface as heat. When one or more of the target constituents is volatile this can lead to sublimation from the target [97–99]. Even though the back surface of the target is directly cooled, substantial thermal gradients can arise with materials of low thermal diffusivity. This effect can lead to stoichiometric differences between the target and the deposited film. For fixed sputtering conditions, it is possible to compensate for this effect by purposely enriching the target in the volatile material [98].

2. Chemical Dissociation

Since most compounds have dissociation energies in the 10–100 eV range, it is not surprising that sputtering with keV ions results in chemical dissociation. Numerous studies in sputter-deposition systems have shown this by indirection [21, 97–104], but there is also a significant body of literature related to electron spectroscopy for chemical analysis (ESCA) that gives direct, quantitative evidence of this effect [105–107]. What is less obvious is the relationship between film stoichiometry and target bombarding potential. For example, in the case of binary oxide targets, films are, in general, less deficient in oxygen if the target potential is high than if it is low [11, 108–110]. This is because higher target potentials sputter more secondary oxygen ions that are accelerated toward the substrate where they can recombine to form the original compound. At low target potentials, these secondary ions have insufficient energy to survive the collisions encountered between the target and the substrate, and are lost to the vacuum pumps. If truly stoichiometric oxides, nitrides, sulfides, etc. are desired in the film, it is virtually always required to add O_2, N_2, H_2S, etc. to the sputtering gas to ensure stoichiometry by reactive sputtering. In some cases, sputtering with 100% reactive gas will still not result in stoichiometric films [100] because it is probable with weakly bonded compounds that reactive gas bombardment of the substrate will lead to preferential sputtering of the reactive gas rather than incorporation of it in a film [102, 106].

G. Chemical Sputtering

Chemical sputtering involves the reaction of an excited neutral or ionized gas with a surface to form volatile compounds [1, 111, 112]. This technique is mainly used for plasma treatment of organic surfaces and for etching in plasmas (Chapter V-2). However, it is sometimes a factor in film deposition [113]. When targets containing reactive anions (e.g., F^-, Cl^-) are sputtered, some of these anions are sputtered as secondary ions and accelerated toward the substrates where chemical etching reactions can occur. Etching of glass substrates, rather than film deposition, has been observed during "sputter deposition" from targets of TbF_3 and $TbCl_3$ [113].

III. GLOW DISCHARGES

In this section, we give a brief account of glow discharge phenomena relevant to diode sputtering. The treatment of magnetron glow discharges is not included here; (see Chapters II-2–II-4). The effects of these glow discharge phenomena on substrates are described in Section VII.

A. DC Glow Discharges

Figure 3 illustrates the manner in which a glow discharge is formed in a low-pressure gas with a high-impedence dc power supply. When a voltage is first applied, a very small current flows. This is due to the presence of a small number of ions and electrons resulting from a variety of sources (e.g., cosmic radiation). Initially, the current is nearly constant, because all of the charge present is moving. As the voltage is increased, sufficient energy is imparted to the charged particles so that they produce more charged particles by collisions with the electrodes (secondary electron emission) and with neutral gas atoms. As more charge is created the current increases steadily, but the voltage is limited by the output impedence of the power supply. This region is known as the Townsend discharge.

Eventually, an avalanche occurs. Ions strike the cathode, release secondary electrons which form more ions by collision with neutral gas atoms. These ions then return to cathode, produce more electrons that, in turn produce more ions. When the number of electrons generated is just sufficient to produce enough ions to regenerate the same number of electrons, the discharge is self-sustaining. The gas begins to glow, the voltage drops, and the current rises abruptly. At this point, the mode is called the "normal glow." Since the secondary electron emission ratio of most materials is of the order of 0.1, more than one ion must strike a given area of the cathode to produce another secondary electron. The bombardment of the cathode in the normal glow region self-adjusts in area to accomplish this. Initially, the bombardment is not uniform, but is concentrated near the edges of the cathode or at other irregularities on the surface. As more power is supplied, the bombardment increasingly covers the cathode surface until a nearly uniform current density is achieved. (This region of the glow discharge is used for voltage regulator tubes.)

After the bombardment covers the whole cathode surface, further increases in power produce both increased voltage and current density in

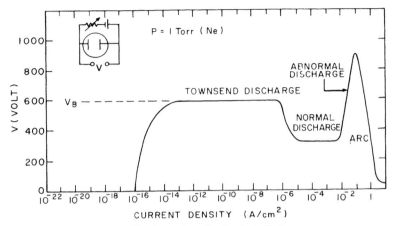

Fig. 3. The formation of a dc glow discharge.

the discharge. This region, the "abnormal glow," is the mode used in sputtering and virtually all other glow discharge processes. If the cathode is not cooled, when the current density reaches about 0.1 A/cm^2, thermionic electrons are emitted in addition to secondary electrons, followed by a further avalanche. The output impedance of the power supply limits the voltage, and the low-voltage high-current arc discharge forms.

The foregoing represents a qualitative description of the various dc discharge modes. Several more quantitative reviews have appeared [114–117].

Crucial to the formation of an abnormal glow is the breakdown voltage V_B (Fig. 3). This voltage is mainly dependent upon the mean-free-path of secondary electrons and the distance between the anode and cathode. Each secondary electron must produce about 10–20 ions for the original avalanche to occur. If the gas pressure is too low or the cathode–anode separation too small, the secondaries cannot undergo a sufficient number of ionizing collisions before they strike the anode. If the pressure and/or separation are too large, ions generated in the gas are slowed by inelastic collisions so that they strike the cathode with insufficient energy to produce secondary electrons. This is a qualitative statement of Paschen's law which relates V_B to the product of gas pressure and electrode separation (Fig. 4). In most sputtering glow discharges, the pressure-separation product is well to the left of the minimum, thus requiring relatively high discharge starting voltages. In close-spaced electrode configurations it is often necessary to increase the gas pressure momentarily to start the discharge. Alternatively, an external ionizing source may be used (e.g., a Tesla coil connected into the chamber by a high-voltage feedthrough).

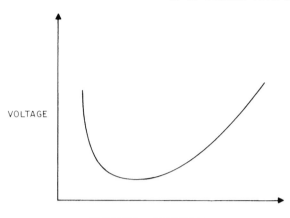

PRESSURE × SEPARATION

Fig. 4. Paschen's Law.

Figure 5 illustrates the luminous regions of a dc glow discharge, the voltage distribution, and the net space charge as a function of distance from cathode to anode. Adjacent to the cathode, there is a brilliant luminous layer known as the cathode glow. This is the region in which incoming discharge ions and positive ions produced at the cathode are neutralized by a variety of processes. This is also the region in which secondary electrons begin to accelerate away from the cathode. The light emitted is characteristic of both the cathode material and the incident ion.

Secondary electrons are repelled at high velocity from the cathode and start to make collisions with neutral gas atoms at a distance away from the cathode corresponding to their mean free path. This leaves a dark space which is very well defined. Since the electrons rapidly lose their energy by collisions, nearly all of the applied voltage appears across this dark space. The dark space is also the region in which positive ions are accelerated toward the cathode. Since the mobility of ions is very much less than that of electrons, the predominant species in the dark space are ions [118]. Acceleration of secondary electrons from the cathode results in ionizing collisions in the negative glow region.

The Faraday dark space and positive column are nearly field-free regions whose sole function is to connect electrically the negative glow to the anode. They are not at all essential to the operation of a glow discharge. In most sputtering systems, the anode is located in the negative glow and these other regions do not exist. The length of an unobstructed negative glow is exactly equal to the range of electrons that have been accelerated from the cathode [119]. When the negative glow is truncated,

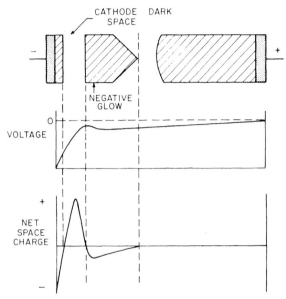

Fig. 5. Luminous regions, voltage and net space charge versus position in a dc glow discharge.

higher voltages must be applied to make up for the ions that would have been generated in the part of it that is blocked by the anode. In general, for uniform cathode bombardment, the anode should be located at least 3–4 times the thickness of the dark space away from the cathode. This distance, of course, is inversely related to gas pressure.

B. Low-Frequency AC Glow Discharges

Low-frequency ac glow discharges are not often used for sputtering. At frequencies up to about 50 kHz, ions are mobile enough so that there is ample time for a complete dc discharge to form on each electrode on each half-cycle. Thus, the discharge is basically the same as a dc discharge, except that both electrodes are alternatively cathode and anode (i.e., both are sputtered).

C. RF Glow Discharges

As the frequency of an applied ac signal is increased above 50 kHz, two important effects occur. First electrons oscillating in the glow space acquire sufficient energy to cause ionizing collisions, thus reducing the dependence of the discharge on secondary electrons and lowering the breakdown voltage [120]. Second, the electrodes are no longer required to be electrical conductors since rf voltages can be coupled through any kind of impedance. Thus, it is literally possible to sputter anything. However, this does not imply that the films deposited will necessarily resemble the target (Section VII).

At typical rf frequencies used for sputtering (5–30 MHz), most ions are sufficiently immobile that one would expect negligible ion bombardment of the electrodes. In fact this is not the case. If one or both of the electrodes is coupled to the rf generator through a series capacitor, a pulsating, negative voltage will be developed on the electrode [121]. Owing to the difference in mobility between electrons and ions, the I–V characteristics of a glow discharge resemble those of a leaky rectifier (Fig. 6). Upon application of an rf voltage through the capacitor, a high initial electron current flows to the electrode. On the second half of the cycle, only a relatively small ion current can flow. Since no charge can be transferred *through* the capacitor, the voltage on the electrode surface must self-bias negatively until the net current (averaged over each cycle) is zero. This results in the pulsating negative potential shown in Fig. 6 [122]. The average dc value of this potential (V_s) is nearly equal to the peak voltage applied [123].

To obtain sputtering from only one electrode in an rf system, it has been shown [120] that the electrode which is to be the sputtering target must be an insulator or must be capacitively coupled to the rf generator and that the area of that electrode must be small compared to that of the directly coupled electrode. Also the ratio of the voltage between the glow space and the small capacitively coupled electrode (V_c) to the voltage between the glow space and the large directly coupled electrode (V_d) is

$$V_c/V_d = (A_d/A_c)^4, \tag{1}$$

where A_d and A_c are the areas of the directly and capacitively coupled electrodes, respectively. In practice, the directly coupled electrode is the system ground, including baseplates, walls, etc., and is quite large with respect to A_c. Thus, the average sheath potential (V_s) varies between the target electrode and ground as shown in Fig. 7. Clearly, to minimize bombardment of grounded fixtures, the area of all grounded parts should be very large by comparison to that of the target. We shall expand on this point further in Section VII.A.

Recently, Logan *et al.* [124] have described a total model for an rf sputtering system in which externally measured parameters can be put into a computer program to predict ionization levels, sputtering, material transport, and other bombardment effects in the system.

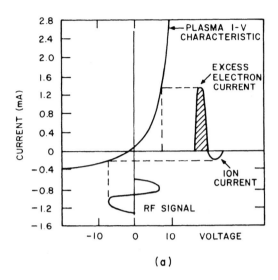

(a)

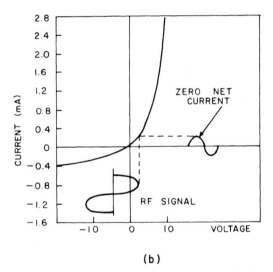

(b)

Fig. 6. The formation of a pulsating negative sheath on a capacitively coupled surface in an rf glow discharge (after Butler and Kino [121]).

D. Discharge Supporting Modes

Glow discharges are relatively inefficient ion sources. Only a few percent of the gas atoms in a glow discharge are ionized. Several techniques have been developed for increasing the ionization efficiency somewhat.

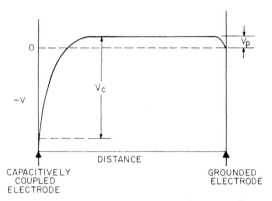

Fig. 7. Distribution of voltage in an rf glow discharge from a small, capacitively coupled electrode (target) to a large, directly coupled electrode.

These include axial and transverse magnetic fields, thermionic additions of electrons, and rf coils. The transverse magnetic field devices are described in Chapters II-2–II-4.

1. Axial Magnetic Field

A magnetic field normal to the target surface constrains secondary electrons to follow a helical, rather than straight-line path, the radius of which is given by

$$r = (mv \sin \theta)/eB, \tag{2}$$

where m is the electron mass, v the velocity, θ the angle of emission, e the electron charge, and B the magnetic flux density. The effect is to give electrons a longer path length for a fixed mean free path, thus increasing the probability of ionizing collisions before the electron reaches the anode. The magnetic field pinches the discharge in toward the center of the target, resulting in nonuniformity of film thickness. For relatively small targets ($<$20-cm diameter) the judicious use of low magnetic fields (10–50 G) can be used to improve uniformity by compensating for excessive edge bombardment (Section VII.F). For targets $>$20 cm in diameter, axial magnetic fields should be avoided.

2. Thermionic Support (Triode Sputtering)

In this technique, ions are generated in a low-voltage ($\sim$50 V), high-current (5–20 A) arc discharge between a thermionic filament and a main anode. The sputtering target is located in this main discharge and ions are extracted from it toward the target which can be powered either by a dc or capacitively coupled rf source. This technique produces a very high ion density, allows operation at much lower pressures than a two-terminal glow discharge, and allows control of the target current density independent of the target voltage. However, it is not necessarily good practice to operate a sputtering system at low gas pressures (Section VI.A). The use of triode sputtering is mainly advantageous for sputter etching (Chapter V-2).

The main limitation of this technique is that it is difficult to scale up to very large sizes because of the difficulties involved in producing a large, uniform thermionic arc discharge. Numerous configurations have been described in the literature [125–132] and a complete analysis of such systems has been published by Tisone and co-workers [133–137].

3. RF Coil

Radio-frequency coils are sometimes employed in ion plating systems to increase the ionization efficiency at low gas pressure [138]. The evaporant passes through the coil in transit to substrates attached to a dc sputtering target. This increases the level of ionization of both the evaporant and the sputtering gas.

IV. EQUIPMENT CONFIGURATION

A. Target Assemblies

Electrodes to which the target material is attached and the counter electrodes to which substrates are attached have been designed in a variety of configurations [5, 9, 10, 16, 39, 97, 139–141]. For bias sputtering and ion plating, the counter electrode is a sputtering target assembly similar to the type used as deposition sources. There are four essential considerations in the design of target assemblies: heat dissipation, electrical isolation and contact, ground shielding, and materials of construction. A versatile assembly suitable for research is shown in Fig. 8. Several multi-target arrangements have been described which are useful in sequential deposition of several materials (see, e.g., Cambey [141]).

1. Heat Dissipation

It is estimated [8, 9] that 1% of the energy incident on a target surface goes into ejection of sputtered particles, 75% into heating of the target, and the remainder is dissipated by secondary electrons that bombard and heat the substrates. The heat generated is usually removed by water cool-

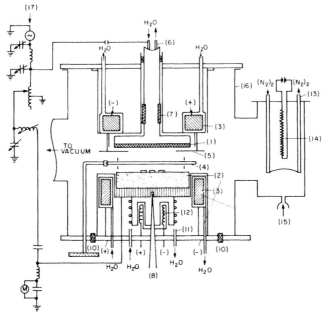

Fig. 8. Schematic of an rf sputtering apparatus. The components are listed as follows: (1) cathode target, (2) anode substrate holder, (3) cathode and anode magnets (water cooled), (4) shutter (SS), (5) cathode shield (SS), (6) cathode water cooling, (7) cathode isolation insulator, (8) substrate thermocouple, (9) substrates, (10) anode isolation insulator, (11) substrate cooling, (12) substrate heating, (13) liquid nitrogen cooled SS shroud, (14) Ti sublimation filaments, (15) sputter gas, (16) SS vacuum chamber, and (17) rf power supply matching network and substrate bias supply.

ing the backing plates to which the target or substrates are attached. Since targets are often soldered or epoxy bonded to the backing plates, the water-cooling channels should be designed so that water is forced through convolutions machined in the material. In this way, no hot spots can develop.

Substrates merely resting on a water-cooled plate are not in good thermal contact, and will be heated almost as much as if the plate were not cooled at all. In cases where substrate heating must be minimized, the substrates must be bonded to the cooled plate (Section IV.D) [142].

2. Electrical Isolation and Contacts

Isolation of the target assembly from grounded parts of the system usually involves a ceramic-to-metal seal with a sufficiently thick ceramic to minimize capacitive losses. Water lines must be isolated from ground by putting several meters of insulating tubing in series with both inlet and outlet connections. Electrical connection from the power supply or matching network is made conveniently by bolting a strap to the water line on the back of the target assembly.

3. Ground Shields

To prevent sputtering of the target assembly itself, a ground shield is contoured around all surfaces at a distance less than that of the cathode dark space. In some cases, shields are placed over the outer rim of the target surface (e.g., to prevent clamping screws from sputtering). In these cases, focusing effects near the ground shield result in excessive erosion of the target [9].

4. Materials of Construction

Many materials have been used to construct sputtering target assemblies, the most common being stainless steel and copper. Stainless steel is highly corrosion resistant, but has very poor thermal conductivity. Copper usually must be gold plated for corrosion protection and is difficult to machine, but it is a good choice. Aluminum, unless it is perfectly protected from corrosion in the presence of water, is a very poor choice.

B. Power Supplies

1. DC

There are many requirements imposed on power supplies used for glow discharge sputtering. The output rating is determined by the size of the target and should be capable of delivering up to 10 W/cm^2 at voltages up to about 5 kV. (Note that this does not apply to magnetron sputtering which generally requires higher output power and current but lower voltage. See Chapters II-2, II-3, and II-5.) The actual power drawn from the supply depends on the process operating parameters (voltage, gas, and gas pressure) and on the secondary electron yield of the target surface. For most metals, the secondary electron yield is initially high and then decreases as surface oxides and other contaminants are sputtered away. Insulating surface contaminants give rise to arcing due to local dielectric breakdown. The power supply must be capable of withstanding these arcs without shutting down, but must be able to distinguish between these small arcs and catastrophic ones such as may happen if a flake of metal shorts the target to ground.

For relatively small targets (up to about 300 cm^2), the most commonly used type of dc supply involves an autotransformer-controlled high-voltage transformer that is magnetically shunted followed by a full-wave bridge rectifier. No smoothing filters are employed, but rf chokes should be used in series with the output to prevent high-frequency spikes (due to arcing) from damaging the bridge rectifier. Magnetic shunting of the power transformer leads to a high output impedance. In the event of a serious arc, the short circuit output current is limited by circulating the current in the transformer and the output voltage drops to zero [143]. This type of supply is limited in that the maximum usable voltage decreases rapidly as the rated output current is approached (Fig. 9).

For larger targets, saturable reactor power supplies are preferred because these are tolerant of arcs, even when operated with very high output currents [143]. They are, however, unstable when the output load is small (Fig. 10).

2. Low-Frequency AC

Power line frequencies are generally used only in processes in which both sides of a substrate are to be coated simultaneously by two targets facing each other. The power supply problems are basically the same as those with dc, except that no rectifiers are used and it is not necessary to use rf chokes in the output circuit.

3. RF (Crystal Controlled)

Most rf generators used for sputtering are crystal controlled to one of the "Industrial, Scientific, and Medical Equipment" (ISM) frequencies allotted by international agreement for unlimited radiation [144]. The fre-

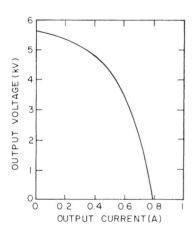

Fig. 9. Typical maximum output voltage and current for a magnetically shunted dc power supply rated at 2 kW.

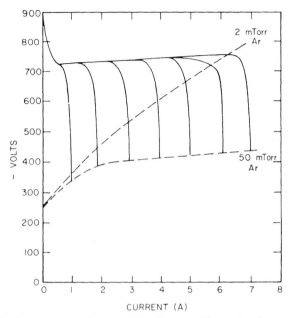

Fig. 10. Typical operating characteristics of a saturable-reactor dc power supply.

quencies and their tolerances most often used are 13.56 ± 0.00678, 27.12 ± 0.0160, and 40.68 ± 0.020 MHz.

Since the output impedance of these generators must be held constant (usually 50 Ω), and since the glow discharge impedance is much higher than this, a matching network must be situated physically close to the target assembly. A typical matching network for a single-ended sputtering system includes a variable-shunt capacitor, a variable-series capacitor, and a fixed inductor. The design of such matching networks are the same as for antenna matching to transmitters [145]. When bias sputtering is contemplated, the situation is more complex. Several manual and automatic circuits have been described for splitting the rf power from a single supply to control voltages and phase at two interacting targets in the same glow discharge [124, 146–151]. In some cases, the substrate target is directly driven, while in others substrate power is coupled through the discharge either by substrate tuning (i.e., inserting a matching network between the substrate target and ground) or by adjusting the system geometry and glow discharge to produce a high plasma potential (Section VII.A).

To prevent damage to the power supply (e.g., during an arc), the power output tubes should be triodes rather than pentodes or tetrodes so that reflected power appears on the plate rather than the grids of the tubes.

4. RF (Self-Excited)

While crystal-controlled rf generators are most commonly used, variable-frequency, self-excited oscillators have advantages in some applications and have been used extensively [97, 152–157]. For output power up to about 1.5 kW, the usual circuit is a Hartley class-C oscillator. For higher power and/or for bias sputtering, in which the rf power must be split between the deposition and substrate targets, push–pull circuits are preferred. The output circuit is usually transformer coupled, and tuning is accomplished by varying the frequency with a capacitor across the primary of the transformer. If the output circuit is slightly over-critically coupled, one observes a very broad, flat-topped tuning peak as compared to the very high-Q output of crystal-controlled oscillators [157]. As a result of this, these generators are much more tolerant of arcs and other disturbances in the glow discharge. For bias sputtering, the secondary of the transformer is center tapped to ground and a series capacitor is inserted in both target legs. These capacitors act as voltage dividers and can vary the two target voltages independently and in a very smooth fashion. The major disadvantage of self-excited oscillators relates to the fact that their output impedance is quite high. Thus, flexible coaxial cables cannot be used to make electrical connection. Instead the generator must be physically close to the target(s) and connections must be made using rigid, air-dielectric coaxial pipes or some other kind of high-impedance connection, which often complicates the mechanical design of the system. As with all kinds of rf equipment, precautions such as shielding, power line filters, etc. should be taken to minimize radio frequency interference.

C. Instrumentation and Control

Depending on the application, one must be able to monitor voltage, current (dc), power (rf), and sputtering gas pressure. In rf glow discharges it is often useful to measure both the average dc self-bias voltage, the peak-to-peak voltage, the waveform, and the phase angle. Pressure measurements can be especially difficult. Thermocouple, Pirani, and high-pressure ion gauges can sometimes be used, but they are easily contaminated by reactive gases (leading to erroneous readings and drift), are not tolerant of rf fields if the output impedance of the sensing head is high, and are difficult to use when gas mixtures are employed (e.g., in reactive sputtering). The most suitable gauges for critical work are low-impedance capacitance manometers with temperature-controlled heads [158]. The membranes in these gauge heads suffer from mechanical hysteresis if they must be cycled from atmosphere to vacuum frequently, so it is advisable to include a shutoff valve between the head and the vacuum chamber to keep them under vacuum at all times.

Many processes are affected by gas flow throughput, which determines how fast impurities are flushed out of the system. In such cases, mass-flow meters should be included in the gas feed supply.

Langmuir probes are especially helpful in calibrating a system to determine plasma and floating potentials (Section VII.A), but they perturb and contaminate the discharge [159] so they should not be used for *in situ* monitoring.

Temperature measurements of anything in a vacuum system are difficult at best. If properly attached and of sufficiently low thermal mass, thermocouples or thermisters can be used [160], but they can also contaminate the system, and the pyrometer or potentiometer must be isolated from ground. Infrared pyrometers are quite useful [161]. Their main advantage is that no material must contact the substrates. However, they must be carefully calibrated for optical losses in the light path (windows, etc.), infrared emission from the glow discharge, and the emissivity of the coated and uncoated substrate surface.

If a given sputtering process is stable and relatively slow, deposition rates can be monitored by controlling conditions and deposition time. In-vacuum monitors (e.g., quartz microbalance) simply do not work because resputtering processes are different on the head than on substrates. Also, they are intolerant of rf fields (high impedance). The most successful techniques for rate monitoring have been optical emission and absorption spectroscopies [162–168], but they must be independently calibrated. These same spectroscopies along with mass spectroscopy are also useful for diagnosing glow discharge species of all kinds (ions, neutrals, molecular fragments, etc.) [169–181]. These techniques are especially useful in reactive sputtering.

D. Substrate Heaters

Given the fact that there is heat input to substrates from the glow discharge, control of substrate temperatures above ambient can be quite difficult. Both resistance (Fig. 8) and radiant heating (Fig. 11) have been used. The resistance heater is relatively massive and has a relatively slow thermal response time. The radiant heater can be used either to heat substrates in the holder shown directly or to heat a solid plate which has a

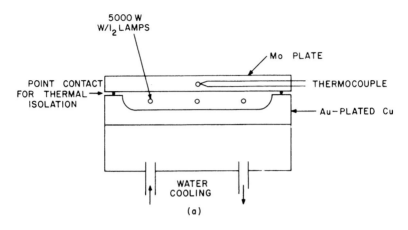

(a)

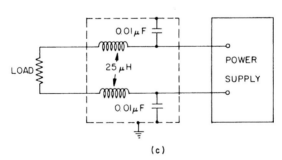

(b)

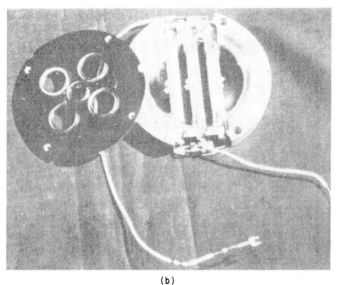

(c)

Fig. 11. Radiant substrate heater. (a) Cross section, (b) photograph, (c) rf decoupling network (courtesy of Silano and Leary [142]).

thermocouple embedded in it. Because of its machinability, high-thermal conductivity and resistance to warping, molybdenum is the preferred material of construction for the heated surface (plate or substrate holder). When rf is used, a decoupling network must be employed to prevent rf interference with the thermocouple controls and heat-input circuits (Fig. 11c).

As with substrate cooling, it is essential to make a good thermal contact between the substrate and heated holder. Ga or In–Ga eutectic can be used up to about 450°C if the substrate holder is made of molybdenum. At higher temperatures there are reactions of Ga or In–Ga with Mo. The Ga or In–Ga is best used at a temperature at which it is molten. Upon solidification the thermal mismatch between it, the substrate and substrate holder sometimes can crack the substrate or break the bond. For lower temperatures a thermally conductive grease has been employed which is a mixture of alumina powder and high vacuum grease [142].

E. Wall Losses

''Wall losses'' is a generic term for any material body in a glow discharge which can act as a point of neutralization for ions. Ions cannot be neutralized in the gas phase because there is no mechanism which can dissipate the heat of the neutralization to conserve both momentum and energy. This can only happen at a surface. Thus, to obtain a uniform ion density across a target surface, all extraneous solid objects (e.g., vacuum chamber walls, support posts, etc.) should be kept well away from the target edges. How far away is far enough depends inversely upon the glow discharge gas pressure. For typical gas pressures (5–50 mTorr) a safe distance is about 10 cm.

F. Shields and Shutters

Dark-space shields around targets and shutters used for preconditioning targets and/or substrates are necessary objects which must be near the target. To minimize their effect on ion neutralization, they should be made as symmetric as possible with respect to the target. As will be shown in Section V.D, the materials from which shields and shutters are constructed are also important to avoid contamination of targets and/or films.

G. Deposition Sources for Bias Sputtering and Ion Plating

Bias sputtering or ion plating are film deposition processes in which substrates are ion bombarded (sputter etched) prior to and/or during deposition from a vapor source. Thus, the substrate holder is a sputtering target assembly which is usually the same type of assembly as a deposition target. However, depending upon the vapor source used, modifications to the substrate target assembly may be required. Vapor sources that can be used include dc or rf sputtering targets, magnetron sputtering targets of all types, resistance-heated filament, rf-heated or electron-beam evaporation sources, or gaseous reagents of the type normally used for chemical vapor deposition. Problems arise mainly with those vapor sources that require that the substrate target be up and the vapor source down in the system. When the reverse is true, the substrates merely rest on an appropriate backing plate, but if the substrates have to be supported against gravity, the substrate holders become a major problem because of contamination (Section V.D), and the nonuniform bombardment that occurs near sharp edges due to fringing fields. This is mainly a problem with thermal evaporation vapor sources in which the molten evaporant must be contained (as, e.g., in the hearth of an electron beam gun). Numerous inverted target assemblies for holding various kinds of substrates have been reported [182–187], all of which probably suffer from these problems to some extent.

H. Scale-Up Problems

Scaling up any deposition process from small laboratory equipment to large production is a very major problem. There are several examples in the literature [188–198a] of laboratory to production scale ups that range from larger bell jars to very large continuous machines. It is almost axiomatic to state that each production application must be treated as a special case, so it would be futile to present a detailed discussion of such systems here.

Most of the problems with large systems arise if they are not designed around the process. That is, if the deposition process is subordinated to a mechanical design, pumping considerations, etc., the process inevitably suffers. The major process-related factor in large sputtering systems is sputtering gas uniformity across large target surfaces to maintain uniform levels of target ion bombardment. Because of implantation of gases into the target and, in some cases, getting of sputtering gas by the target and/or growing films (e.g., in reactive sputtering), sputtering systems must be considered as vacuum pumps. In general, the sputtering gas must be introduced in several locations, and the main vacuum pumps must be located sufficiently far from the active sputtering region that uniform gas flow across large target surfaces is achieved. Particular attention must also be paid to other effects described earlier (wall losses, target shielding, etc.). Systems intended for long operation between cleaning of fixtures must be carefully designed to assure that buildup of film material on system parts will not result in short circuits, arcs (due to flaking material), or jamming of moving parts (e.g., substrate carriers).

It is especially helpful if sputtering targets can be built into one of the walls of vacuum chamber, so that electrical and water-cooling connections can be made at atmospheric pressure. Otherwise one must face the formidable task of ensuring that high-voltage leads and water lines inside the vacuum chamber do not become part of the glow discharge.

V. PRECONDITIONING OF TARGETS, SUBSTRATES, AND SYSTEMS FOR FILM DEPOSITION

As a general rule, the best way to prepare a system for sputter deposition is to run the system first without substrates in exactly the way in which it will be run during the deposition. A preconditioning run of this sort may have one or more of several beneficial effects: (1) the target altered surface layer is established, (2) fixtures are outgassed by bombardment, and (3) fixtures are coated with the material to be deposited, minimizing subsequent contamination.

A. Target Materials

While it is possible to use virtually any kind of material as a sputtering target, for high purity work very dense targets are preferred. Sputtering can and has been done successfully from sintered, hot-pressed, powder, and liquid targets, but the highest purity work has been done with very dense targets (e.g., vacuum-cast or arc-melted materials). Hot-pressed and powder targets have been shown to be nearly limitless sources of gaseous contamination [67]. The major contaminant appears to be oxygen sorbed onto the raw powder surface prior to hot pressing. For oxides this is not a problem, but it can be severe for other materials depending upon the intended application and the tolerance of the desired film properties to oxygen contamination. Even getter sputtering cannot eliminate gaseous contaminants from films if they originate in the target because a large fraction of the contaminant is sputtered as negative ions, accelerated toward the substrate, and implanted in the growing film.

Targets are usually bonded to some kind of water-cooled backing plate. Target bonding must be done with great care to avoid contamination from the bonding material and/or failure of the bond due to nonuniformity and thermal fatigue. In general, epoxy bonds are not recommended because of their poor thermal transfer properties and brittleness. Epoxy bonds usually lead to large amounts of organic contamination. Solder bonds with appropriate plating layers on the back of the target and the front of the backing plate are generally preferred [199].

B. Presputtering of Targets

Presputtering of targets is done to clean and equilibrate target surfaces prior to film deposition with a shutter usually located close to the substrates. For pure metal targets surface oxides are removed, the target surface is brought to thermal equilibrium, and the system is outgassed. The discharge current can be used as a monitor to determine when the system is equilibrated [200]. Initially, oxides with high secondary electron emission ratio are sputtered and background gaseous contaminants (especially H_2O) are being broken down, leading to high discharge current which gradually decreases as the contaminants are removed from the system and/or covered up with film material. When the discharge current falls to a constant value, presputtering can be terminated. Alternatively, glow discharge optical spectroscopy or mass spectroscopy can be used to determine the endpoint (Section IV.C).

When alloy or compound targets are used, one must establish the altered surface layer in addition to the phenomena noted above. The amount of time needed for this must be established empirically. Since there will be inevitably some resputtering of material deposited onto the shutter and subsequent return of some of this material to the target, the shutter must be kept at the same potential as that which the substrate will have when the shutter is removed. This means that shutters should not be grounded, and in some cases they should be biased. If this is not done, there will be a transient in resputtering as the shutter is removed which can change the composition of the first few monolayers of the film [83]. In general, all discharge conditions (voltages, pressure, etc.) should be the same as those used during the subsequent deposition.

C. Sputter Etching of Substrates

Practical methods for cleaning substrates prior to film deposition have been reviewed by Mattox [201, 202]. In this section, we shall restrict our discussion to sputter etching of substrates prior to film deposition, and assume that gross surface contaminants have been removed chemically or otherwise prior to putting the substrates into the vacuum chamber. If it is not done properly, sputter etching can actually produce more surface contamination than was originally present on the substrate.

With glow discharge sputter etching, it is rarely possible to obtain an atomically clean substrate surface. In general, this is only possible in an ultrahigh vacuum system using ion beams followed by high-temperature annealing [203–208].

The first problem in glow discharge sputter etching relates to organic contamination. This may be on the surface as it is put into the vacuum chamber or may be adsorbed into the surface (e.g., from backstreaming vacuum pump fluids). In either case inert gas ion bombardment will polymerize the organic material [209] and render it very difficult to remove [21].

In most cases organic contaminants can be removed by chemical sputtering in O_2 [111, 112, 210]. Clearly, this will not be adequate if there is continuous backstreaming of pump fluids. If the contaminant is a silicone, this will leave a residue of SiO_2. If the substrate is a metal, the O_2 discharge may oxidize the surface [211]. In both cases, a subsequent sputter-etch step in Ar can be used to remove these oxides.

The next problem that must be addressed is backscattering of material emitted from the target surface [11, 21, 200, 212–219]. When the substrate surface is heterogeneous (e.g., an integrated circuit consisting of Si and SiO_2 areas) and/or when the substrate rests upon a target backing plate of a material that is different from that of the substrate, some fraction of all the materials sputtered are returned to the target, forming a new composite surface consisting of all the bombarded materials. Some of the material is returned by simple collisions in the gas phase, and some is returned by being resputtered from the shutter after condensing there. The latter can be largely eliminated by using a "catcher" [213] such as that shown in Fig. 12. The former can only be minimized, but not eliminated in glow discharge sputtering [212]. The amount of material returned to the target and substrates is directly proportional to the sputtering gas pressure and the axial magnetic field applied (if any), and inversely proportional to the target voltage and to the sputtering yield of the materials involved [11, 212, 219]. Even at the lowest gas pressures at which a glow

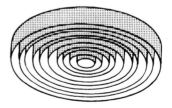

Fig. 12. Catcher (after Maissel *et al.* [213]; copyright 1972 by International Business Machines Corporation, reprinted with permission).

discharge can operate, monolayer quantities of backscattered material are found. Radio-frequency target excitation is much more efficient in minimizing backscattering than is dc, but sputter-etch voltages of 2 kV or more must be used [200, 219]. Reactive gases similar to those used for plasma etching, reactive ion etching, etc. have been shown to produce exceptionally clean substrate surfaces, even at relatively high gas pressures and low rf voltages [220]. For example, various refractory alloys have been cleaned successfully in Ar–HCl mixtures, but cleaning in Ar–CCl$_4$ resulted in C deposits on the substrates. With the Ar–HCl mixtures there may have been some Cl residue remaining after etching.

In some cases, it is desirable to enhance this effect to preform an interface during sputter etching prior to film deposition. For example, ohmic contacts to Si devices have been improved by sputter etching the Si substrate with a Pt backing plate to produce PtSi in the contact regions [221]. This effect has also been used to advantage in forming graded interfacial layers to promote adhesion of the subsequently deposited film [222]. It should be noted, however, that graded interfaces are sometimes detrimental to adhesion [223]. When it is desirable to eliminate this effect, it is necessary to use backing plates made of the same material as the substrate [212]. In the case of oxide dielectric substrates, it has been shown [224] that even this approach is not effective. Sodium ions (a typical contaminant) were found to migrate down from glass and SiO$_2$ surfaces rather than be sputter etched away. Similar effects have been observed in other systems [225].

The structure and chemistry of single crystal Si surfaces after backscattering have been studied in detail using a variety of surface analytic techniques [214, 226, 227]. In addition to backscattered material, significant amounts of Ar were found in the substrate surfaces (2–20 at. %, depending on the bombardment potential), and crystallographic damage was propagated to a depth of 40–110 Å as the sputter-etch voltage increased from 0.5–2.5 kV. The amounts of damage produced and gas incorporated increase with increasing substrate temperature [227].

In some cases, conelike features have been observed on sputter-etched surfaces subject to backscattering [11, 111, 218], or when the substrates contain certain bulk contaminants [228]. An example is shown in Fig. 13. This effect is due to the fact that the sputtering yield of the backscattered metal or other contaminant is lower than that of the substrate or matrix. Thus, the islands of backscattered material or contaminant, act as local etch masks. Once such structures are formed, the sputter-etch rate of the surface drops markedly for geometric reasons [11, 111, 229]. In selected cases, it is possible to add reactive gases that depress the backscat-

Fig. 13. Cones produced on a Si surface sputtered in contact with a Pd backing plate.

tering rate to reduce this effect [218]. The cones continue to sharpen as sputter etching proceeds [230–246]. It should also be noted that such cones can form due to initial microtopography of a substrate surface in the absence of impurities [230–246]. Structures of this sort are not always undesirable (e.g., similar structures prepared by CVD have been used as solar absorbers) [247, 248]. Still another effect occurs if the substrates are nonplanar. Since the angular distribution of material emitted from sputtered surfaces at low voltage is under cosine (i.e., there is more sidewise than normal emission), some of the sputtered material will be deposited on substrate irregularities, and become more difficult to remove from these surfaces as the irregularity becomes more vertical. While this is considered a nuisance when clean surfaces are the desired result [217, 249–251], this effect been put to beneficial use in the coating of the inside walls of through holes in substrates [252], and in enabling one to cover severe substrate topographic features by bias sputtering.

In general, all of these effects and their implications to film–substrate interfaces must be considered carefully in the design of a process involving sputter etching prior to film deposition.

VI. THE SPUTTERING GAS

A. Effects of Gas Species, Pressure, and Flow

The effects of increasing sputtering gas pressure are to increase discharge current, increase backscattering, and to slow energetic particles by inelastic collisions. The first two effects compete and largely determine deposition rates (Section VIII.B). The third effect can be used to maximize or minimize the energy of particles incident on substrates.

In general, it is desirable to use the highest possible gas flow so that impurities are constantly swept out of the sputtering chamber. Numerous workers have found that this improves the properties of thin films (e.g., Fraser and Cook [253]). This implies that the pumping system used must be operated to obtain maximum pumping speed for the gases used.

One must also be careful to ensure that no pressure gradients occur in the vicinity of the sputtering targets. Otherwise such gradients produce nonuniform bombardment and film deposition. Local pressure gradients can occur at the target surface (because of gettering), near the pumping port, or at points of high or low temperature relative to the rest of the system (e.g., near vacuum gauges with hot elements, titanium sublimation pumps, or Meissner traps). To offset these problems it is often necessary to introduce the gas at more than one point in the system. This is especially a problem in reactive sputtering and in very large systems.

As noted in Section II.A, the nature of the sputtering gas largely de-

Table IV

Metastable Neutral Energies and Lifetimes

Species	Metastable energy (eV)	Lifetime (sec)
H	10.20	0.12
H_2	11.86	Long
He	19.82	Very long
He	20.61	Long
N	2.38	6×10^4
N	3.58	13
N_2	6.16	0.9
N_2	8.54	1.7×10^{-4}
O	1.97	110
O	4.17	0.78
O_2	0.98	Very long
Ne	16.62, 16.71	Long
Ar	11.55, 11.72	Long
Kr	9.91, 9.99	—
Xe	8.31, 8.44	—

termines the sputtering yield, and hence, the deposition rate. In addition to that, excited metastable neutrals of the sputtering gas are formed by collisions of electrons with ground-state gas atoms. The potentials required to raise some common sputtering gases from the ground state to the lowest excited state along with the lifetimes of the excited states are given in Table IV [254, 255]. It is well established [172] that sputtered neutral atoms can be ionized by the Penning mechanism (the collision of a metastable neutral with a ground-state neutral to produce an ion and return the metastable neutral to the ground state) when the first ionization potential of the sputtered atom is less than that of the metastable energy. Since the metastable energies of the noble gases are greater than the first ionization potential of most elements, this mechanism can produce a copious supply of ions which can bombard substrate surfaces and/or return to the target to cause self sputtering [212]. It is possible that some sputtered atoms can be doubly ionized by this process [256]. The amount of Penning ionization that occurs increases with increasing gas pressure.

B. Sources of Gas Contamination

The major sources of gaseous impurities in sputtering systems are residual gases (mainly H_2O) left after initial pumpout, wall desorption due to bombardment, the surface of the target (gases adsorbed when the system is vented), occluded gases in the sputtering target, backstreaming of pump fluids, leaks (real or virtual), and impurities in the gas supply itself. Many of these can be eliminated by a combination of good vacuum practice and extensive presputtering before film deposition. For example, to minimize condensation onto the target surface during venting, the cooling water should be shut off or hot water should be valved into the cooling lines to keep the target a few degrees above room temperature. Reactive impurities in inert gas sources can be reduced by passing the gas through hot Ti sponge before introducing it into the process chamber. If the pumping system must be throttled, it is preferable to put the cold trap on the chamber side of the throttling valve so that condensable gases (e.g., H_2O) can be pumped at full efficiency [257].

C. Getter Sputtering

Getter sputtering is a technique sometimes used to reduce the amount of background contamination in the vicinity of sputtering targets and substrates [257a–262]. This technique involves enclosing the target–substrate region of the vacuum system with a rather close-fitting, cooled cylinder. With a shutter closed, the target is presputtered extensively. Films deposited onto the cylindrical enclosure and the shutter getter reactive gases and bury them; and the cylindrical enclosure acts as a conductance-limiting baffle, retarding the entry of more contaminating gases into the active region of the discharge. After all of the reactive gases are gettered, the shutter is opened and deposition proceeds.

This procedure does not eliminate inert gas incorporation in films, but does eliminate most reactive gas incorporation. The main disadvantage of this technique is that the cylindrical enclosure acts as "wall," giving rise to wall losses and excessive nonuniformity in film deposition near the target edges. If the target itself contains dissolved or occluded reactive gases in the bulk, these gases will still be incorporated into the films because they will be sputtered as negative ions and accelerated toward the substrate.

A variant of getter sputtering has been described [263], in which the solid cylindrical container is replaced by a reactive gas (e.g., SiH_4), which decomposes in the diffuse discharge region surrounding the sputtering target. The cation of the reactive gas (e.g., Si) is capable of chemically gettering reactive gases if its free energy of formation of compounds with the reactive gases is more negative than that of the target material. This eliminates the wall problem, but the reactive gas must be introduced far enough from the substrate to exclude the possibility that some of the cations of the gas are incorporated into the growing film.

D. Reactive Sputtering

There have been numerous theoretical treatments and reviews of reactive sputtering processes [264–274]. For purposes of discussion, this topic should be subdivided. In one form of reactive sputtering, the target is a nominally pure metal, alloy, or mixture of species which one desires to synthesize into a compound by sputtering in a pure reactive gas or an inert gas–reactive gas mixture. The reactive gas either is, or contains the ingredient required to synthesize the desired compound. The second type of reactive sputtering involves a compound target that chemically decomposes substantially during inert gas ion bombardment, resulting in a film deficient in one or more constituents of the target. In this case, a reactive gas is added to make up for the lost constituent. The main difference between these two types of reactive sputtering has to do with the deposition rate dependence on partial pressure of the reactive gas.

A large number of reactive gases have been used to synthesize compounds from metal targets or to maintain stoichiometry in the face of decomposition: Air, O_2, or H_2O (oxides), N_2 or NH_3 (nitrides), $O_2 + N_2$ (oxynitrides), H_2S (sulfides), C_2H_2 or CH_4 (carbides), SiH_4 (silicides), HF or

CF$_4$ (fluorides), As (arsenides), etc. There are obvious safety problems with some of these gases.

The question of where compounds are synthesized (at the target, in the gas, or at the substrate) is central to an understanding of reactive sputtering [264]. Reactions in the gas phase can for the most part be ruled out for much the same reasons that ions cannot be neutralized in the gas phase. The heat liberated in the chemical reaction canot be dissipated in a two body collision. Simultaneous conservation of energy and momentum requires the reaction to occur at a surface—either the target or the substrate. At very low reactive gas partial pressure and high target sputtering rate, it is well established that virtually all of the compound synthesis occurs at the substrate and that the stoichiometry of the film depends on the relative rates of arrival at the substrate of metal vapor and reactive gas. Under these conditions, the rate of removal and/or decomposition of compounds at the target surface is far faster than the rate of compound formation at the target surface. However, as the reactive gas partial pressure is increased and/or the target sputtering rate is decreased, a threshold is reached at which the rate of target-compound formation exceeds the removal rate of compounds. For metal targets, this threshold is usually accompanied by a sharp decrease in the sputtering rate. This decrease is due partly to the fact that compounds have generally lower sputtering rates than metals and partly that compounds have higher secondary-electron emission yields than metals. As a result, more of the energy of incoming ions is used to produce and accelerate secondary electrons. With constant-current power supplies, the increased secondary electron current automatically decreases the target voltage for a fixed power setting. A third cause of the drop in sputtering rate is simply due to less efficient sputtering by reactive gas ions than by inert gas ions.

The net effect of all of this is illustrated in Fig. 14. If instead of main-

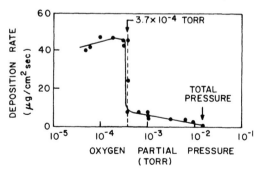

Fig. 14. Deposition rate versus oxygen partial pressure for an iron target in an Ar–O$_2$ mixture (after Heller [269]).

taining constant power one maintains constant voltage, the abrupt decrease in rate is smoothed out considerably. Likewise, for compound targets there is a much more gradual decrease in sputtering rate with increasing partial pressure of reactive gas which is mainly related to the less efficient sputtering ion concentration.

Clearly, the critical partial pressure depends not only on the glow discharge conditions, but on the kinetics of compound formation of the target surface itself. For example, target materials that do not oxidize readily do not show abrupt rate decrease as oxygen partial pressure increases.

It is impossible for us to review in detail all of the reactive sputtering processes that have been described in the literature. Instead, we present a bibliography of reactive sputtering and reactive ion plating processes in Table V. This table includes only processes in which the target (or evaporant in reactive ion plating) is a metal, and the desired compounds are wholly synthesized in the deposition process. Reactive sputtering of compounds to maintain stoichiometry is not included. The table is organized alphabetically by source material with elemental sources first, then binary sources, ternary sources, etc. For multielement sources, the elements are listed according to their abundance in the source.

VII. DEPOSITION WITH SIMULTANEOUS ION BOMBARDMENT OF THE SUBSTRATE AND GROWING FILM

A. Plasma, Floating, and Bias Potentials

The true potentials on substrates in glow discharge sputtering have been studied extensively [124, 147, 454–459]. The luminous region of a glow discharge is not a true plasma in that the concentrations of electrons and ions are not equal, but plasma conditions are roughly approximated. The substrates in a glow discharge may be treated as floating-plane probes. Figure 15 shows the I–V characteristics of such a probe in somewhat idealized fashion. Any material body immersed in a glow discharge, unless it is grounded, will acquire a potential with respect to ground that is slightly negative. This is known as the floating potential V_f. This potential arises because of the higher mobility of electrons, as opposed to ions, in the discharge so that more electrons reach the surface than ions. It can be shown [460] that the floating potential is related to the electron temperature T_e and the masses of the electron m and ion M involved in the discharge:

$$V_f = -(1/2e)kT_e \ln(\pi m/2M), \tag{3}$$

where e is the electron charge and k is Boltzmann's constant.

Table V

Bibliography of Reactive Sputtering and Reactive Ion Plating

Source	Gases	References
Ag	Air	[275]
Ag	Ar + N$_2$	[272]
Ag	Ar + O$_2$	[272, 273, 276, 277]
Al	Ar + N$_2$	[278, 279]
Al	Ar + O$_2$	[18, 261, 273, 280]
Al	O$_2$	[281]
Au	Air	[275]
Au	Ar + O$_2$	[276]
Bi	Ar + O$_2$	[276, 277, 282]
C	N$_2$	[283]
Cd	Ar + H$_2$O	[266]
Cd	Ar + H$_2$S	[266, 284, 285]
Cd	Ar + O$_2$	[266, 276, 277, 286–294]
Cr	Ar + CH$_4$	[295]
Co	Ar + O$_2$	[269, 296, 297]
Cu	Ar + H$_2$S	[284, 285]
Cu	Ar + O$_2$	[264, 298–303]
Cu	Ne + O$_2$	[302]
Fe	Ar + CH$_4$	[295]
Fe	Ar + O$_2$	[269, 273, 276, 304–306]
Ga	Ar + As	[307]
Ge	Ar + N$_2$	[308]
Hf	Ar + C$_2$H$_2$	[18]
Hf	Ar + N$_2$	[309]
Hf	Ar + O$_2$	[309–311]
In	Ar + O$_2$	[266, 286, 288, 312–318]
Mg	Ar + O$_2$	[319]
Mn	Ar + O$_2$	[296, 320, 321]
Mo	Ar + CH$_4$	[295]
Mo	Ar + H$_2$S	[303]
Mo	Ar + N$_2$	[322]
Mo	Ar + O$_2$	[272, 323]
Nb	Ar + CH$_4$	[324]
Nb	Ar + C$_2$H$_2$	[18]
Nb	Ar + N$_2$	[325–327]
Nb	Ar + O$_2$	[310]
Nb	Ar + N$_2$ + CH$_4$	[326]
Nb	Ar + N$_2$ + O$_2$	[328]
Ni	Ar + CH$_4$	[295]
Ni	Ar + O$_2$	[277, 296, 329–331]
Pb	Ar + H$_2$O	[266]
Pb	Ar + H$_2$S	[303]
Pb	Ar + N$_2$	[308]

Table V (*Continued*)

Source	Gases	References
Pb	Ar + O$_2$	[266, 277, 332–334]
Pb	O$_2$	[334]
Pt	Air	[275]
Pt	Ar + O$_2$	[276, 335]
Sb	Ar + O$_2$	[276, 277]
Si	Ar + C$_2$H$_2$	[336]
Si	Ar + N$_2$	[280, 319, 337–350]
Si	Ar + NH$_3$	[268]
Si	Ar + O$_2$	[273, 276, 277, 351–358]
Si	C$_2$H$_2$	[359]
Si	N$_2$	[67, 360, 360a]
Si	NH$_3$	[360]
Si	NH$_3$ + SiH$_4$	[360]
Si	O$_2$	[360b]
Si	Ar + N$_2$ + O$_2$	[357, 361]
Sn	Air	[312]
Sn	Ar + H$_2$S	[303]
Sn	Ar + N$_2$	[308]
Sn	Ar + O$_2$	[276, 277, 286, 288, 362–364]
Ta	Ar + CH$_4$	[295, 365, 366]
Ta	Ar + C$_2$H$_2$	[18, 367]
Ta	Ar + CO	[264]
Ta	Ar + H$_2$	[366]
Ta	Ar + H$_2$O	[266, 368–370]
Ta	Ar + N$_2$	[264, 325, 365, 366, 371–379]
Ta	Ar + O$_2$	[261, 266, 267, 273, 277, 310, 325, 351, 365, 369, 380–395]
Ta	Ar + O$_2$ + N$_2$	[396–400]
Ta	Ar + SiH$_4$	[330]
Te	Ar + O$_2$	[277, 282]
Ti	Ar + CH$_4$	[401]
Ti	Ar + C$_2$H$_2$	[18, 402]
Ti	Ar + H$_2$O	[266]
Ti	Ar + N$_2$	[18, 272, 325, 371, 401, 403]
Ti	Ar + NH$_3$	[18]
Ti	Ar + O$_2$	[266, 270, 272, 274, 276, 310, 319, 323, 325, 351, 383, 401, 404–407]
V	Ar + C$_2$H$_2$	[18]
V	Ar + O$_2$	[408]
W	Ar + CH$_4$	[295]
W	Ar + O$_2$	[276, 277, 310,]
Y	Ar + O$_2$	[18, 408, 409]
Zn	Ar + H$_2$O	[264]

Table V   *(Continued)*

Source	Gases	References
Zn	Ar + H_2S	[284]
Zn	Ar + O_2	[264, 268, 410–414]
Zr	Ar + N_2	[271, 325]
Zr	Ar + O_2	[310, 325, 351, 415, 416]
Al–Si	Ar + O_2	[354, 355]
Au–Ta	Ar + O_2	[417, 418]
Au–W	Ar + O_2	[417]
Ba–Ti	Ar + O_2	[419, 420]
Bi–Ta	Ar + O_2	[421]
Bi–Ti	Ar + O_2	[421]
Cd–Cu	Ar + H_2S	[303]
Cd–Cu	Ar + O_2	[287]
Cd–In	Ar + H_2S	[303]
Cd–In	Ar + O_2	[287]
Cd–Zn	Ar + H_2S	[422]
Cr–Mo	Ar + O_2	[423]
Cu–Fe	Ar + O_2	[424]
Ga–Al	Ar + As	[307]
Gd–Fe	Ar + O_2	[425]
Hf–Ta	Ar + N_2	[426–429]
In–Sn	Ar + O_2	[286, 294, 371, 403, 430–437]
Li–Nb	Ar + O_2	[438]
Nb–C	Ar + N_2	[326]
Ni–Fe	Ar + O_2	[424, 439]
Ni–Ti	Ar + C_2H_2	[440]
Pb–Te	Ar + O_2	[354]
Pb–Ti	Ar + O_2	[415, 441, 442]
Pt–Ta	Ar + O_2	[417]
Pt–W	Ar + O_2	[417]
Si–Al	Ar + N_2	[443]
Sn–In	Ar + O_2	[286, 362]
Sn–Sb	Ar + O_2	[28, 362, 437, 444]
Ta–Si	Ar + O_2	[445]
Ta–Ti	Ar + N_2	[446]
Ti–C	Ar + O_2	[447]
Ti–Ni	C_2H_2	[448]
Zn–Cu	Ar + H_2S	[449]
Cd–Cu–In	Ar + H_2S	[303]
Fe–Cr–Ni	Ar + CH_4	[295, 450]
Mg–Mn–Zn	Ar + O_2	[451]
Ti–Al–V	Ar + C_2H_2	[452]
Pb–Nb–Zr–Fe–Bi–La	Ar + O_2	[453]

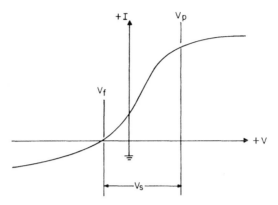

Fig. 15. Idealized I–V characteristics of a Langmuir probe in a glow discharge.

At the point V_p, the probe is at the same potential as the plasma. Ideally, there are no electric fields at this point, and the charged particles migrate to the probe because of their thermal velocities. Again, because electrons are more mobile than ions, what is collected by the probe is predominantly electron current. Under ideal conditions, the plasma potential is given by

$$V_p = kT_i \ln 2/q_i, \qquad (4)$$

where T_i and q_i are the temperature and charge of the ions involved [460].

The assumptions made in the derivations of these equations are by no means strictly valid in glow discharges (especially in the presence of magnetic fields), but they may be used to estimate the electron and ion temperatures in the discharge.

There are three major practical consequences to be drawn from Fig. 15. First, substrates (even if they are grounded) are at a potential that is negative with respect to the plasma. This is shown for a floating substrate in Fig. 15 as V_s. If the substrates are grounded, V_s is reduced by an amount equal to V_f. The second consequence is that *substrates in sputtering systems should rarely be grounded.* If an insulating substrate is attached to a grounded substrate support, the surface of the substrate acquires the potential V_s while its surroundings acquire the lower potential $V_s - V_f$. This leads to nonuniformity of bombardment near the edges of the substrate which, in turn, leads to variable thickness, composition or other properties as a function of distance from the edge of the substrate. The only exception to this rule is in the case of a conducting substrate onto which a conducting film is to be deposited. Finally, since target voltages and substrate voltages (in the case of bias sputtering or ion plating) are measured relative to ground, the existence of the plasma potential introduces a systematic error since the true potential is the sum of the measured voltage and V_p. In some cases the error can be large.

The variation of the floating potential (V_f) with various dc glow discharge parameters is shown in Fig. 16. For rf glow discharges, the behavior is similar, except the magnitudes of the voltages involved are generally

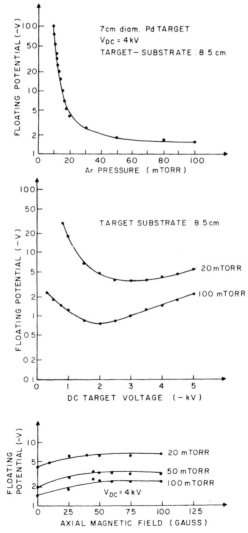

Fig. 16. The variation of the floating potential with Ar pressure, target voltage, and axial magnetic field in a dc sputtering system.

higher because the electron temperatures are higher for otherwise equivalent conditions.

The variation of the plasma potential (V_p) with dc glow discharge conditions is similar to the variation of the floating potential because parameters that increase electron temperature generally increase ion temperatures as well. For example, as the pressure is decreased, V_f and V_p, respectively, become more negative and more positive with respect to ground, leading to higher substrate potentials V_s.

The situation with rf glow discharges is somewhat more complicated because the geometry of the system becomes a major factor. As was pointed out (Eq. (1)), the ratio of the average potential on the capacitively coupled electrode (target) to that of the grounded electrode with respect to the plasma is given by $(A_d/A_c)^4$, where A_c and A_d are the areas of the capacitively coupled and grounded electrodes, respectively. These areas refer to the areas actually in contact with the glow discharge. If the discharge is confined (e.g., by placing a cylinder around the target), a limited portion of the total grounded area is actually in contact with the discharge. This has been done [454], and the plasma potentials have been measured for various confining ratios, R = target area/grounded area versus the sum of the applied target potential to ground and the plasma potential (i.e., the total target potential) as shown in Fig. 17. These effects can be used to impose a negative bias on a substrate without the necessity for making a direct electrical contact, but it is clearly difficult to control.

It should be noted that the potentials described so far are *average* dc potentials. This does not imply that all ions striking substrate surfaces arrive with these potentials. In both dc and rf glow discharges, the thickness of the ion sheath at the substrates is greater than the ionic mean free paths, leading to inelastic collisions which slow many of the ions as they traverse the sheath. For this reason alone, ions reach the surface at all energies from zero up to that corresponding to eV_s [118]. In rf glow discharges, the potential of the ion sheath is time dependent. This leads to modulation of the ion energies depending upon exactly when the ions reach the outer edge of the ion sheath. In this case, incident ions can arrive with potentials ranging from zero to almost $2eV_s$ [123].

Negative substrate bias voltages can be applied from an external power source. If the substrate is electrically conducting, this may be accomplished with a dc power supply [461]. For insulating substrates and/or for the deposition of insulating films on conducting substrates, rf-induced substrate bias is preferred [11, 147, 462–463]. When this is done, the applied bias voltage, V_b, in effect takes the place of the floating potential, and the plasma potential remains unchanged [454]. Thus, the total negative substrate bias is $-(V_p - V_b)$.

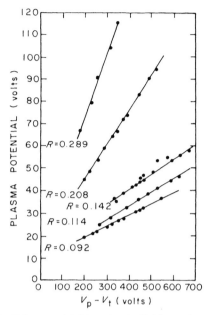

Fig. 17. Plasma potential versus total target potential for various confining ratios, R (see text) in an rf glow discharge (after Coburn and Kay [454]).

If a positive bias is applied to a substrate in a dc glow discharge, the substrate simply becomes a virtual anode, resulting in a very large electron current flow to the substrate. This, in turn, leads to a large amount of substrate heating and a very nonuniform current distribution [456]. For both dc and rf discharges, application of a positive bias to an independent electrode (e.g. a substrate) leads to an increase in the plasma potential to more positive values [454, 456] as shown in Fig. 18. Even for very high positive values of V_b, the effective substrate bias never exceeds the plasma potential. Thus, application of a positive bias is seen to have little effect on the effective substrate potential, but the relatively large increase in V_p leads to rather high levels of bombardment of all grounded surfaces in contact with the discharge. This can lead to rather large amounts of gas desorption from the walls of the chamber. Except in those instances where one might wish to clean up the grounded parts of a sputtering system, the application of positive substrate biases should be avoided.

It can be seen from this discussion that substrates in sputtering systems are nearly always subject to a negative bias (ion bombardment), whether externally applied or not. Consequently, substrates must be treated as sputtering targets, and all of the effects described in Section II

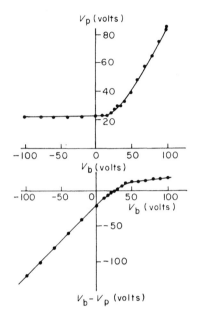

Fig. 18. The plasma potential (V_p) and total substrate potential ($V_b - V_p$) versus substrate bias in an rf glow discharge (after Coburn and Kay [454]).

apply to substrates during film growth as well as to sputtering targets. In some cases, one may wish to maximize substrate bombardment, while in other cases minimum bombardment may be desired. In all cases it should be controlled.

Bombardment of substrates is not limited to charged particles. High-energy neutrals reflected from the target and/or generated in the glow discharge by various excitation processes also bombard the substrate. Clearly these are unaffected by the substrate potential, and bombardment from this source is impossible to eliminate completely in glow discharge sputtering [464]. Recent estimates have been made of the relative contribution to substrate bombardment by ions (10%) and energetic neutrals (90%) in ion plating [465–467].

It should also be pointed out that the phenomena discussed in this section apply to substrates in *any* glow discharge. Thus, they apply not only to sputtering processes, but to all glow-discharge deposition and etching processes.

B. Gas Incorporation and Desorption

During the sputtering process, gas is incorporated in the growing film. The amount depends on target voltage [468–472], substrate bias voltage

[469–471, 473–478], target-to-substrate distance [473, 479], sputter gas pressure [471, 473, 479], inert gas atomic volume [471, 480], substrate temperature, magnetic field, and system geometry. The interplay among these parameters is responsible for the varied results found in the literature. There are two mechanisms by which gas is incorporated: ions can be neutralized at the target surface and reflected at high energy toward the substrate where they may be implanted; and ions can bombard the substrate and be implanted because of the effective substrate bias.

1. Substrate Bias

The effective substrate bias is the most important parameter in determining gas incorporation [468–488]. At very low bias voltages, there is insufficient energy to implant gas, but some thermal desorption can occur. At higher voltages, gas contents increase as V_b^2, where V_b is the substrate bias voltage. At very high bias voltages substrate heating leads to desorption of the implanted gas.

2. Target Voltage

The entrapment of energetic neutral sputter gas atoms reflected from the target surface has been amply demonstrated [468–472]. For example, in the incorporation of He in gold films [468], it was shown that at constant pressure and bias voltage, lower target voltages gave less gas incorporation than higher target voltages. Ar was found to increase in concentration for both triode and diode sputtered W films [478], and the amount of Ar that is incorporated in the film increases with the atomic number of the target material [489].

3. Substrate Temperature

As the substrate temperature is increased the sticking probability of inert sputter gas decreases [468, 473, 483, 487]. However, the rate of reaction increases in reactive sputtering.

4. Pressure

Increasing pressure decreases gas incorporation [471–473]. The increase in gas content with decreasing pressure is due to an increase in the plasma potential.

5. System Geometry

Geometric effects such as target-to-substrate distance have been shown to affect the amount of gas incorporated in films. The greater the distance of the target from the substrate the less gas will be incorporated in the growing films [472–473]. For example, in dc triode sputtering of W films [472] Ar was shown to decrease from about 15 at. % with a target–substrate separation of 9 cm to about 0.1 at. % at a separation of 18 cm.

Another geometric effect found that influences the film gas content is the target size for a fixed system geometry. Here we find seemingly conflicting results. For example, the inert gas concentration in rf sputtered amorphous films with rf bias voltage was found to vary inversely with the target electrode area [471], while in dc triode sputtered films it was found that the Ar concentration increased as the target area increased [473].

C. Stoichiometry of Films

The stoichiometry of films deposited from multicomponent targets when a substrate bias is applied has been addressed by a number of workers [83, 84, 92, 96a, 477, 483, 489–491]. Most of these treatments are theoretical. Rather than review them in detail, we summarize the practical consequences.

To control film stoichiometry, many conditions must be met. The target itself must be homogeneous. This is especially a problem with some alloys and mixtures in which constituents segregate out in patches. The altered surface layer must be established at steady state. There must be no bulk diffusion in the target. If sublimation occurs at the target, the target composition must be enriched in the subliming material. When targets with normally gaseous constituents (e.g., oxides) are used, chemical dissociation at the target must be compensated by reactive sputtering. Preferential resputtering must be compensated by adjusting the target composition and/or by reactive sputtering. Shutters should be biased to the same potential as substrates during deposition to prevent transients in resputtering which change the composition of the first few monolayers of film material.

D. Physical Film Properties

In the absence of problems with film stoichiometry, controlled resputtering in various bias modes often aids in producing films of high quality. Most of the studies of physical film properties as a function of substrate bias have been reviewed elsewhere [9, 11, 13, 147], and they will not be covered in detail here. There have been reports of improved film density, electrical properties, magnetic properties, adhesion, and surface morphology, to name but a few. In addition, variations in grain size, preferred orientation, and crystal polymorphs have been observed in some materials. Ohmic contacts can be made to semiconductors without the need for sintering. Conformal coatings over severe topographic features on substrates can be achieved because of small-angle resputtering. In general, the judicious use of substrate biasing is one of the most powerful tools available in sputtering to tailor the properties of deposited films. To optimize some properties, high bias voltages are required, while to optimize others, substrate bombardment must be minimized.

In summary, there are three basic effects that occur at a substrate during glow discharge sputtering: (1) condensation of energetic vapor, (2) heating, and (3) bombardment by a variety of energetic species. The sum of all of these effects must be carefully controlled, and, since they are all interdependent [11], this is sometimes difficult.

VIII. RATE AND UNIFORMITY OF DEPOSITION

Most deposition rates quoted in the literature should more properly be called "accumulation rates," since they are really the difference between the arrival and resputtering rates of the film material at the substrate. Both the accumulation rate and the uniformity of film deposition have been thoroughly reviewed by Maissel [5, 9], so only a summary is presented here.

For a given target material both rate and uniformity are influenced by system geometry, target voltage, sputtering gas, gas pressure, and power. All other things being equal, rates are linearly proportional to power and decrease with increasing target–substrate separation. The sputtering gas influences deposition rate in the same way as it affects sputtering yields. As the gas pressure is increased the discharge current increases (increasing rate), but return of material to the target by backscattering also increases (decreasing rate). This is further complicated in some cases by increased Penning ionization at higher pressures which increases the rate by self-sputtering. The sum of all of this leads to a gas pressure or a small range of gas pressure at which the rate is a maximum, and this must be determined empirically for each application. The optimum pressure may be anywhere between a few mTorr and several tens of mTorr.

In general, for a given gas pressure there will be an optimum target–substrate separation to produce the best uniformity. For small targets

(<15-cm diameter) this separation is generally small (a few centimeters), while for larger targets, the optimum separation may be considerably larger (10–20 cm). The edges of the target, no matter how the dark-space shield is constructed, represent regions of gross nonuniformity. Under ideal conditions, the best uniformity that one can hope to achieve is ±2– 5% over an area concentric with the deposition target, but ~5 cm smaller in diameter than the target. Often it is much worse than this.

IX. CONCLUSION

In this chapter we have attempted to review the rather complex interplay of the parameters involved in sputter deposition and to introduce some of the glow discharge aspects of related processes.

Unquestionably, the hallmark of the sputtering processes described is versatility, both in terms of materials that can be deposited and process parameters that can be adjusted to tailor the properties of thin films as desired. However, the sheer number of critical process parameters and their complex interrelationships can often make these processes difficult to control. For many applications, the advantages far outweigh the disadvantages, and "diode sputtering" and "ion plating" have been put to use in large-scale production in applications of a wide variety: electronics, optics, abrasion-resistant coatings, lubricating coatings, corrosion protection, and thermal insulation, to name but a few.

In general, these processes are found to be most useful in applications requiring rather thin films (generally <1 μm because of relatively low deposition rates) and/or in cases where the desired material simply cannot be deposited stoichiometrically any other way.

REFERENCES

1. G. K. Wehner, *Adv. Electron. Electron Phys.* **7**, 239 (1955).
2. E. Kay, *Adv. Electron. Electron Phys.* **17**, 245 (1962).
3. M. Kaminsky, "Atomic and Ionic Impact Phenomena on Metal Surfaces." Academic Press, New York, 1965.
4. M. H. Francombe, *in* "Basic Problems in Thin Film Physics" (R. Niedermeyer and H. Mayer, eds.), p. 52. Vanderhoeck & Ruprecht, Göttingen, 1966.
5. L. I. Maissel, *Phys. Thin Films* **3**, 61 (1966).
6. G. Carter and J. S. Colligon, "Ion Bombardment of Solids." Am. Elsevier, New York, 1968.
7. K. L. Chopra, "Thin Film Phenomena." McGraw-Hill, New York, 1969.
8. G. K. Wehner and G. S. Anderson, *in* "Handbook of Thin Film Technology" (L. I. Maissel and R. Glang, eds.), Ch. 3. McGraw-Hill, New York, 1970.
9. L. I. Maissel, *in* "Handbook of Thin Film Technology" (L. I. Maissel and R. Glang, eds.), Ch. 4. McGraw-Hill, New York, 1970.

10. G. N. Jackson, *Thin Solid Films* **5**, 209 (1970).
11. J. L. Vossen, *J. Vac. Sci. Technol.* **8**, S12 (1971).
12. P. D. Townsend, J. C. Kelly, and N. E. W. Hartley, "Ion Implantation, Sputtering and Their Applications," Ch. 6. Academic Press, New York, 1976.
13. W. D. Westwood, *Prog. Surf. Sci.* **7**, 71 (1976).
14. R. J. MacDonald, *Adv. Phys.* **19**, 457 (1970).
15. I. S. T. Tsong and D. J. Barber, *J. Mater. Sci.* **8**, 123 (1973).
16. D. M. Mattox, *Electrochem. Technol.* **2**, 295 (1964).
17. D. M. Mattox, *J. Vac. Sci. Technol.* **10**, 47 (1973).
18. R. E. Bunshah and A. C. Raghuram, *J. Vac. Sci. Technol.* **9**, 1385 (1972).
19. O. Almen and G. Bruce, *Nucl. Instrum. Methods* **11**, 257 (1961).
20. O. Almen and G. Bruce, *Nucl. Instrum. Methods* **11**, 279 (1961).
21. J. L. Vossen and E. B. Davidson, *J. Electrochem. Soc.* **119**, 1708 (1972).
22. P. D. Davidse and L. I. Maissel, *J. Vac. Sci. Technol.* **4**, 33 (1967).
23. G. K. Wehner, Rep. No. 2309. General Mills, Minneapolis (1962).
24. J. L. Vossen, unpublished observations (1974).
25. C. H. Weysenfeld and A. Hoogendoorn, *Proc. Conf. Ion. Phenom. Gases, 5th, Munich* **1**, 124 (1961).
26. D. McKeown and A. Y. Cabezas, *Annu. Rep. Space Sci. Lab.; General Dynamics* July (1962).
27. F. Keywell, *Phys. Rev.* **97**, 1611 (1955).
28. J. Comas and C. B. Cooper, *J. Appl. Phys.* **37**, 2820 (1966).
29. D. McKeown, A. Cabezas, and E. T. Mackenzie, *Annu. Rep. Low Energy Sputtering Stud.; Space Sci. Lab., General Dynamics* July (1961).
30. R. C. Krutenat and C. Panzera, *J. Appl. Phys.* **41**, 4953 (1970).
31. T. W. Snouse, *NASA Tech. Note* **D-2235** (1964).
32. H. Patterson and D. H. Tomlin, *Proc. R. Soc., A* **265**, 474 (1962).
33. M. T. Robinson and A. L. Southern, *J. Appl. Phys.* **38**, 2969 (1967).
34. P. K. Rol, J. M. Fluit, and J. Kistemaker, "Electromagnetic Separation of Radioactive Isotopes," p. 207. Springer-Verlag, Berlin and New York, 1960.
35. O. Almen and G. Bruce, *Trans. Natl. Vac. Symp., 8th, Washington, D.C.*, p. 245 (1961).
36. M. I. Guseva, *Sov. Phys.—Solid State* **1**, 1410 (1959).
37. M. T. Robinson and A. L. Southern, *J. Appl. Phys.* **39**, 3463 (1968).
38. P. A. B. Toombs, *J. Phys. D* **1**, 662 (1968).
39. L. Holland, T. Putner, and G. N. Jackson, *J. Phys. E* **1**, 32 (1968).
40. I. Brodie, L. T. Lamont, and D. O. Myers, *J. Vac. Sci. Technol.* **6**, 124 (1969).
41. D. J. Ball, *J. Appl. Phys.* **43**, 3047 (1972).
42. B. N. Chapman, D. Downer, and L. J. M. Guimaraes, *J. Appl. Phys.* **45**, 2115 (1974).
43. Y. Shintani, K. Nakanishi, T. Takawaki, and O. Tada, *Jpn. J. Appl. Phys.* **14**, 1875 (1975).
44. R. E. Honig, *J. Appl. Phys.* **29**, 549 (1958).
45. R. E. Honig, *Proc. Conf. Ion. Phenom. Gases, 5th, Munich* **1**, 106 (1961).
46. A. K. Ayukhanov and M. K. Abdullaeva, *Bull. Acad. Sci. USSR., Phys. Ser.* **30**, 2083 (1966).
47. Y. A. Fogel, *Sov. Phys.—Usp.* **10**, 17 (1967).
48. A. Benninghoven, *Z. Phys.* **220**, 159 (1969).
49. A. Benninghoven, *Surf. Sci.* **28**, 541 (1971).
50. A. Benninghoven, *Z. Phys.* **230**, 403 (1970).
51. A. Benninghoven, *Surf. Sci.* **35**, 427 (1973).
52. H. W. Werner, *Surf. Sci.* **47**, 301 (1975).

53. V. E. Krohn, *Int. J. Mass Spectrom. Ion Phys.* **22**, 43 (1976).
53a. J. J. Cuomo, R. J. Gambino, J. M. E. Harper, and J. D. Kuptsis, *IBM J. Res. Dev.* **21**, 580 (1977).
53b. J. J. Cuomo, R. J. Gambino, J. M. E. Harper, J. D. Kuptsis, and J. C. Webber, *J. Vac. Sci. Technol.* **15**, 281 (1978).
54. E. Kay, *Colloq. Int. Pulver. Cathod., 2nd, Nice, 1976.*
55. H. D. Hagstrum, *Phys. Rev.* **123**, 758 (1961).
56. E. V. Kornelsen, *Can. J. Phys.* **42**, 364 (1964).
57. E. R. Cawthron, D. L. Cotterell, and M. Oliphant, *Proc. R. Soc., Ser. A* **314**, 53 (1969).
58. J. K. Roberts, *Proc. R. Soc., Ser. A* **135**, 192 (1932).
59. F. O. Goodman, *Surf. Sci.* **24**, 667 (1971).
60. E. P. Suurmeijer and A. L. Boers, *Surf. Sci.* **43**, 309 (1973).
61. F. W. Bingham, *J. Chem. Phys.* **46**, 2003 (1967).
62. H. Niehus and E. Bauer, *Surf. Sci.* **47**, 222 (1975).
63. E. Brown and J. H. Leck, *Br. J. Appl. Phys.* **6**, 161 (1955).
64. K. Erents and G. Carter, *J. Phys. D* **2**, 435 (1969).
65. K. Erents and G. Carter, *J. Phys. D* **2**, 711 (1969).
66. H. F. Winters and P. Sigmund, *J. Appl. Phys.* **45**, 4760 (1974).
67. J. L. Vossen, *J. Vac. Sci. Technol.* **8**, 751 (1971).
68. P. Bacuvier and P. Lavallee, *J. Vac. Sci. Technol.* **13**, 1101 (1976).
69. C. Snoeck and J. Kistemaker, *Adv. Electron. Electron Phys.* **21**, 67 (1965).
70. I. S. T. Tsong, *Phys. Status Solidi A* **7**, 451 (1971).
71. C. B. Kerkdijk and R. Kelly, *Surf. Sci.* **47**, 294 (1975).
72. J. P. Meriaux, R. Goutte, and C. Guilland, *Appl. Phys.* **7**, 313 (1975).
73. C. W. White, D. L. Simms, N. M. Toek, and D. V. Caughan, *Surf. Sci.* **49**, 657 (1975).
74. G. Blaise and M. Bernheim, *Surf. Sci.* **47**, 324 (1975).
75. A. Benninghoven, *Surf. Sci.* **35**, 324 (1973).
76. G. Blaise, *Surf. Sci.* **60**, 65 (1976).
77. E. P. Bertin, "Principles and Practice of X-Ray Spectrometric Analysis," 2nd Ed., pp. 960–963, 968–971. Plenum, New York, 1975.
78. J. Comas and C. A. Carosella, *J. Electrochem. Soc.* **115**, 974 (1968).
79. K. E. Manchester, *J. Electrochem. Soc.* **115**, 656 (1968).
80. J. F. Gibbons, *Proc. IEEE* **56**, 295 (1968).
81. W. Brandt, *Sci. Am.* **218**(3), 90 (1968).
82. E. Gillam, *J. Phys. Chem. Solids* **11**, 55 (1959).
83. D. B. Dove, R. J. Gambino, J. J. Cuomo, and R. J. Kobliska, *J. Vac. Sci. Technol.* **13**, 965 (1976).
84. W. L. Patterson and G. A. Shirn, *J. Vac. Sci. Technol.* **4**, 343 (1967).
85. J. W. Coburn, *J. Vac. Sci. Technol.* **13**, 1037 (1976).
86. H. J. Mathiew and D. Landolt, *Surf. Sci.* **53**, 228 (1975).
87. W. T. Ogar, N. T. Olson, and H. P. Smith, *J. Appl. Phys.* **40**, 4997 (1969).
88. H. Shimizu, M. Ono, and K. Nakayamu, *Surf. Sci.* **36**, 817 (1973).
89. L. A. West, *J. Vac. Sci. Technol.* **13**, 198 (1976).
90. W. Färber, G. Betz, and P. Braun, *Nucl. Instrum. Methods* **132**, 351 (1976).
91. G. Betz, P. Braun, and W. Färber, *J. Appl. Phys.* **48**, 1404 (1977).
92. H. F. Winters and J. W. Coburn, *Appl. Phys. Lett.* **28**, 176 (1976).
93. H. W. Pickering, *J. Vac. Sci. Technol.* **13**, 618 (1976).
94. H. M. Naguib and R. Kelly, *J. Phys. Chem. Solids* **33**, 1751 (1972).
95. D. K. Murti and R. Kelly, *Thin Solid Films* **33**, 149 (1976).
96. G. S. Anderson, *J. Appl. Phys.* **40**, 2884 (1969).
96a. M. L. Tarng and G. K. Wehner, *J. Appl. Phys.* **42**, 2449 (1971).

97. J. L. Vossen and J. J. O'Neill, *RCA Rev.* **29**, 149 (1968).
98. W. J. Takei, N. P. Formigoni, and M. H. Francombe, *J. Vac. Sci. Technol.* **7**, 442 (1970).
99. R. Kelly, *Radiat. Eff.* **32**, 91 (1977).
100. J. L. Vossen, *Proc. Symp. Deposition Thin Films Sputter., 3rd, Univ. Rochester,* p. 80 (1969).
101. H. M. Naguib and R. Kelly, *Radiat. Eff.* **25**, 79 (1975).
102. D. K. Murti and R. Kelly, *Surf. Sci.* **47**, 282 (1975).
103. H. B. Sachse and G. L. Nichols, *J. Appl. Phys.* **41**, 4237 (1970).
104. R. Kelly and J. B. Sanders, *Nucl. Instrum. Methods* **132**, 335 (1976).
105. R. Holm and S. Storp, *Appl. Phys.* **12**, 101 (1977).
106. K. S. Kim and N. Winograd, *Surf. Sci.* **43**, 625 (1974).
107. K. S. Kim, W. E. Baitinger, and N. Winograd, *Surf. Sci.* **55**, 285 (1976).
108. G. V. Jorgenson and G. K. Wehner, *J. Appl. Phys.* **36**, 2672 (1965).
109. R. E. Jones, H. F. Winters, and L. I. Maissel, *J. Vac. Sci. Technol.* **5**, 84 (1968).
110. J. B. Lounsbury, *J. Vac. Sci. Technol.* **6**, 838 (1969).
111. J. L. Vossen, *J. Appl. Phys.* **47**, 544 (1976).
112. L. Holland, *J. Vac. Sci. Technol.* **14**, 5 (1977).
113. J. J. Hanak and J. P. Pellicane, *J. Vac. Sci. Technol.* **13**, 406 (1976).
114. S. C. Brown, "Introduction to Electrical Discharges in Gases." Wiley, New York, 1966.
115. A. von Engel, "Ionized Gases." Oxford Univ. Press, London and New York, 1965.
116. T. Kihara, *Rev. Mod. Phys.* **24**, 45 (1952).
117. R. G. Fowler, *in* "Handbuch der Physik" (S. Flugge, ed.), Vol. 22, p. 59. Springer-Verlag, Berlin and New York, 1956.
118. W. D. Davis and T. A. Vanderslice, *Phys. Rev.* **121**, 219 (1963).
119. A. K. Brewer and J. W. Westhaver, *J. Appl. Phys.* **8**, 779 (1937).
120. H. R. Koenig and L. I. Maissel, *IBM J. Res. Dev.* **14**, 168 (1970).
121. H. S. Butler and G. S. Kino, *Phys. Fluids* **6**, 1346 (1963).
122. G. S. Anderson, W. N. Mayer, and G. K. Wehner, *J. Appl. Phys.* **37**, 574 (1966).
123. R. T. C. Tsui, *Phys. Rev.* **168**, 107 (1968).
124. J. S. Logan, J. H. Keller, and R. G. Simmons, *J. Vac. Sci. Technol.* **14**, 92 (1977).
125. J. W. Nickerson and R. Moseson, *Res/Dev.* **16**(3), 52 (1965).
126. D. Anderson, *SCP Solid State Technol.* **9**(12), 27 (1966).
127. J. Nickerson, *SCP Solid State Technol.* **8**(12), 30 (1965).
128. D. Anderson, *Res/Dev.* **19**(1), 42 (1968).
129. N. Laegreid and R. Moseson, U.S. Patents 3,324,019; 3,297,115; 3,344,054; 3,347,772 (1967).
130. W. N. Huss, *SCP Solid State Technol.* **9**(12), 50 (1966).
131. E. C. Muly and A. J. Aronson, *J. Vac. Sci. Technol.* **6**, 128 (1969).
132. E. C. Muly and A. J. Aronson, *Trans. Natl. Vac. Symp., 13th* p. 145 (1967).
133. T. C. Tisone and J. B. Bindell, *J. Vac. Sci. Technol.* **11**, 519 (1974).
134. R. C. Sun, T. C. Tisone, and P. D. Cruzan, *J. Appl. Phys.* **46**, 112 (1975).
135. R. C. Sun, T. C. Tisone, and P. D. Cruzan, *J. Appl. Phys.* **44**, 1009 (1973).
136. B. E. Nevis and T. C. Tisone, *J. Vac. Sci. Technol.* **11**, 1147 (1974).
137. T. C. Tisone and P. D. Cruzan, *J. Vac. Sci. Technol.* **12**, 677 (1975).
138. Y. Murayama, *Jpn. J. Appl. Phys., Suppl.* **2**, Part 1, 459 (1974).
139. L. I. Maissel and J. H. Vaughn, *Vacuum* **13**, 421 (1963).
140. P. D. Davidse and L. I. Maissel, *J. Appl. Phys.* **37**, 574 (1966).
141. L. A. Cambey, *Proc. Conf. Sputtering Technol. Autom. Prod. Equip.; Materials Research Corp., Orangeburg, N.Y.* p. 129 (1973).

142. P. Silano and P. A. Leary, IBM Res., personal communication (1977).
143. R. Lee, "Electronic Transformers and Circuits," 2nd Ed., Ch. 8. Wiley, New York, 1955.
144. "U.S. Federal Communications Commission Rules and Regulations," Part 18. U.S. Gov. Print. Off., Washington, D.C., 1964.
145. F. E. Terman, "Radio Engineers' Handbook," Sect. 3. McGraw-Hill, New York, 1943.
146. J. S. Logan, *IBM J. Res. Dev.* **14**, 172 (1970).
147. O. Christensen, *Solid State Technol.* **13**(12), 39 (1970).
148. J. S. Logan, U.S. Patent 3,617,459 (1971).
149. N. M. Mazza, *IBM J. Res. Dev.* **14**, 192 (1970).
150. L. J. Kochel, *Rev. Sci. Instrum.* **47**, 1556 (1976).
151. P. Silano, A. Halperin and L. West, *J. Vac. Sci. Technol.* **15**, 116 (1978).
152. P. A. B. Toombs, *J. Phys. D* **1**, 662 (1968).
153. L. Holland, T. Putner, and G. N. Jackson, *J. Phys. E* **1**, 32 (1968).
154. R. B. McDowell, *Solid State Technol.* **12**(2), 23 (1969).
155. R. B. McDowell, U.S. Patent 3,704,219 (1972).
156. J. L. Vossen, U.S. Patent 3,860,507 (1975).
157. J. L. Vossen and J. J. O'Neill, *J. Vac. Sci. Technol.* **12**, 1052 (1975).
158. J. J. Sullivan, *Res/Dev.* **27**(1), 41 (1976).
159. T. L. Thomas and E. L. Battle, *J. Appl. Phys.* **41**, 3428 (1970).
160. A. Amith, *J. Vac. Sci. Technol.* **14**, 803 (1977).
161. C. Misiano, E. Simonetti, and C. Corsi, *Thin Solid Films* **27**, L15 (1975).
162. A. J. Stirling and W. D. Westwood, *J. Appl. Phys.* **41**, 742 (1970).
163. A. J. Stirling and W. D. Westwood, *J. Phys. D* **4**, 246 (1971).
164. A. J. Stirling and W. D. Westwood, *Thin Solid Films* **7**, 1 (1971).
165. A. J. Stirling and W. D. Westwood, *Thin Solid Films* **8**, 199 (1971).
166. H. Ratinen, *J. Appl. Phys.* **44**, 2730 (1973).
167. J. E. Greene and F. Sequeda-Osorio, *J. Vac. Sci. Technol.* **10**, 1144 (1973).
168. H. J. Bauer and E. H. Bogardus, *J. Vac. Sci. Technol.* **11**, 1144 (1974).
169. J. Sosniak, *J. Vac. Sci. Technol.* **4**, 87 (1966).
170. J. B. Lounsbury, *J. Vac. Sci. Technol.* **6**, 838 (1969).
171. J. W. Coburn, *Rev. Sci. Instrum.* **41**, 1219 (1970).
172. J. W. Coburn and E. Kay, *Appl. Phys. Lett.* **18**, 435 (1971).
173. H. Ratinen, *Appl. Phys. Lett.* **21**, 473 (1972).
174. A. B. Arthur and C. B. Cooper, *J. Appl. Phys.* **43**, 863 (1972).
175. J. W. Coburn, E. Taglauer, and E. Kay, *Jpn. J. Appl. Phys., Suppl.* **2**, Part 1, 501 (1974).
176. F. Shinoki and A. Itoh, *Jpn. J. Appl. Phys., Suppl.* **2**, Part 1, 505 (1974).
177. K. Jensen and E. Veje, *Z. Phys.* **269**, 293 (1974).
178. J. W. Coburn, E. W. Eckstein, and E. Kay, *J. Vac. Sci. Technol.* **12**, 151 (1975).
179. E. W. Eckstein, J. W. Coburn, and E. Kay, *Int. J. Mass Spectrom. Ion Phys.* **17**, 129 (1975).
180. J. M. Poitevin, G. Lemperiere, and C. Fourrier, *J. Phys. D* **9**, 1783 (1976).
181. A. J. Purdes, B. F. T. Bolker, J. D. Bucci, and T. C. Tisone, *J. Vac. Sci. Technol.* **14**, 98 (1977).
182. D. M. Mattox, Rep. SC-R-65-997. Sandia Lab., Albuquerque, New Mexico (1966).
183. R. C. Brumfield, J. T. Naff, and A. T. W. Robinson, U.S. Patent 3,514,388 (1970).
184. G. Seeley, P. A. Totta, and G. Wald, U.S. Patent 3,507,248 (1970).
185. D. L. Chambers and D. C. Carmichael, *Res/Dev.* **22**(5), 32 (1971).
186. G. W. White, *Res./Dev.* **24**(7), 43 (1973).

187. G. J. Hale, G. W. White, and D. E. Meyer, *Electron. Packag. Prod.* **15**(5), 48 (1975).
188. S. S. Charschan and H. Westgaard, *Electrochem. Technol.* **2**, 5 (1964).
189. P. Granger and C. R. D. Priestland, *Vacuum* **21**, 309 (1971).
190. C. H. George, *J. Vac. Sci. Technol.* **10**, 393 (1973).
191. H. J. Gläser and H. W. Brandt, U.S. Patent 3,891,536 (1975).
192. D. C. Carmichael, D. L. Chambers, and C. T. Wan, U.S. Patent 3,904,506 (1975).
193. F. H. Gillery, U.S. Patent 3,907,660 (1975).
194. J. J. Bessot, *Thin Solid Films* **32**, 19 (1976).
195. A. W. Morris, *Plat. Surf. Finish.* **63**(10), 42 (1976).
196. C. Altman, *Trans. Natl. Vac. Symp., 9th,* p. 174 (1962).
197. H. Isaak, *Trans. Natl. Vac. Symp., 9th* p. 180 (1962).
198. J. G. Needham, *Trans. Natl. Vac. Symp., 10th, Boston, 1963* p. 402 (1964).
198a. H. J. Gläser, *Proc. Int. Vac. Congr., 7th, Int. Conf. Solid Surf., 3rd, Vienna* p. 1575 (1977).
199. J. van Esdonk and J. F. M. Janssen, *Res/Dev.* **26**(1), 41 (1975).
200. J. E. Houston and R. D. Bland, *J. Appl. Phys.* **44**, 2504 (1973).
201. D. M. Mattox, Rep. SAND 74-0344. Sandia Lab., Albuquerque, New Mexico (1975).
202. D. M. Mattox, "Surface Cleaning in Thin Film Technology." Thin Film Div., Am. Vac. Soc., New York, 1975.
203. H. E. Farnsworth, R. E. Schlier, T. H. George, and R. M. Burger, *J. Appl. Phys.* **29**, 1150 (1958).
204. O. C. Yonts and D. E. Harrison, *J. Appl. Phys.* **31**, 1583 (1960).
205. D. Haneman, *Phys. Rev.* **119**, 563 (1960).
206. R. W. Roberts, *Br. J. Appl. Phys.* **14**, 537 (1963).
207. F. Jona, *J. Phys. Chem. Solids* **28**, 2155 (1967).
208. R. P. H. Gasser, *Q. Rev., Chem. Soc.* **25**, 223 (1971).
209. L. Holland, *Br. J. Appl. Phys.* **9**, 410 (1958).
210. R. B. Gillette, J. R. Hollahan, and G. L. Carlson, *J. Vac. Sci. Technol.* **7**, 534 (1970).
211. J. H. Greiner, *J. Appl. Phys.* **42**, 5151 (1971).
212. J. L. Vossen, J. J. O'Neill, K. M. Finlayson, and L. J. Royer, *RCA Rev.* **31**, 293 (1970).
213. L. I. Maissel, C. L. Standley, and L. V. Gregor, *IBM J. Res. Dev.* **16**, 67 (1972).
214. C. C. Chang, P. Petroff, G. Quintana, and J. Sosniak, *Surf. Sci.* **38**, 341 (1973).
215. H. Dimigen and H. Lüthje, *Thin Solid Films* **27**, 155 (1975).
216. H. Dimigen and H. Lüthje, *Philips Tech. Rev.* **35**, 199 (1975).
217. T. C. Tisone and P. D. Cruzan, *J. Vac. Sci. Technol.* **12**, 677 (1975).
218. C. Misiano and E. Simonetti, *Thin Solid Films* **33**, L15 (1976).
219. G. J. Kominiak and J. E. Uhl, *J. Vac. Sci. Technol.* **13**, 170 (1976).
220. G. J. Kominiak and D. M. Mattox, *Thin Solid Films* **40**, 141 (1977).
221. J. L. Vossen and J. H. Banfield, U.S. Patent 3,640,812 (1972).
222. G. J. Kominiak, *J. Vac. Sci. Technol.* **13**, 1100 (1976).
223. J. L. Vossen, J. J. O'Neill, E. A. James, and O. R. Mesker, *J. Vac. Sci. Techol.* **14**, 85 (1977).
224. D. V. McCaughan and R. A. Kushner, *Thin Solid Films* **22**, 359 (1974).
225. R. R. Hart, H. L. Dunlap, and O. J. Marsh, *J. Appl. Phys.* **46**, 1947 (1975).
226. G. W. Sachse, W. E. Miller, and C. Gross, *Solid-State Electron.* **18**, 431 (1975).
227. J. C. Bean, G. E. Becker, P. M. Petroff, and T. E. Seidel, *J. Appl. Phys.* **48**, 907 (1977).
228. G. K. Wehner, *NASA Spec. Publ.* **SP-5111**, 59 (1972).
229. J. F. Ziegler, J. J. Cuomo, and J. Roth, *Appl. Phys. Lett.* **30**, 268 (1977).
230. M. J. Nobes, J. S. Colligon, and G. Carter, *J. Mater. Sci.* **4**, 730 (1969).
231. A. P. Janssen and J. P. Jones, *J. Phys. D* **4**, 118 (1971).
232. G. Carter, J. S. Colligon, and M. J. Nobes, *J. Mater. Sci.* **6**, 115 (1971).

233. I. H. Wilson and M. W. Kidd, *J. Mater. Sci.* **6**, 1362 (1971).
234. J. Punzel and W. Hauffe, *Phys. Status Solidi A* **14**, K97 (1972).
235. T. Oohashi and S. Yamanaka, *Jpn. J. Appl. Phys.* **11**, 1581 (1972).
236. C. Catana, J. S. Colligon, and G. Carter, *J. Mater. Sci.* **7**, 467 (1972).
237. G. Carter, J. S. Colligon, and M. J. Nobes, *J. Mater. Sci.* **8**, 1473 (1973).
238. R. S. Nelson and D. J. Mazey, *Radiat. Eff.* **18**, 127 (1973).
239. P. Sigmund, *J. Mater. Sci.* **8**, 1545 (1973).
240. M. J. Witcomb, *J. Mater. Sci.* **9**, 1227 (1974).
241. M. J. Witcomb, *J. Mater. Sci.* **9**, 551 (1974).
242. M. J. Witcomb, *Proc. Electron Microsc. Soc. South. Afr., Johannesburg* **4**, 57 (1974).
243. P. G. Glöerson, *J. Vac. Sci. Technol.* **12**, 28 (1975).
244. M. J. Witcomb, *J. Appl. Phys.* **46**, 5053 (1975).
245. G. Carter, *J. Mater. Sci.* **11**, 1091 (1976).
246. M. J. Witcomb, *J. Mater. Sci.* **11**, 859 (1976).
247. J. J. Cuomo, J. F. Ziegler, and J. M. Woodall, *Appl. Phys. Lett.* **26**, 557 (1975).
248. J. J. Cuomo, J. M. Woodall, and T. H. DiStefano, *Proc. Am. Electroplat. Soc. Coat. Sol. Collect., Atlanta, Ga.* p. 133 (1976).
249. R. E. Chapman, *J. Mater. Sci.* **12**, 1125 (1977).
250. H. W. Lehmann, L. Krausbauer, and R. Widmer, *J. Vac. Sci. Technol.* **14**, 281 (1977).
251. B. L. Sopori and W. S. C. Chang, *J. Vac. Sci. Technol.* **14**, 782 (1977).
252. J. L. Vossen, *J. Vac. Sci. Technol.* **11**, 875 (1974).
253. D. B. Fraser and H. D. Cook, *J. Electrochem. Soc.* **119**, 1368 (1972).
254. D. E. Gray, ed., "American Institute of Physics Handbook," 2nd Ed., pp. 7-13–7-15. McGraw-Hill, New York, 1963.
255. E. E. Muschlitz, *Science* **159**, 599 (1968).
256. K. Gerard and H. Hotop, *Chem. Phys. Lett.* **43**, 175 (1976).
257. V. Hoffman, *Electron. Packag. Prod.* **13**(11), 81 (1973).
257a. L. Holland and R. E. L. Cox, *Vacuum* **24**, 107 (1974).
258. H. C. Theuerer and J. J. Hauser, *J. Appl. Phys.* **35**, 554 (1964).
259. H. C. Theuerer and J. J. Hauser, *Trans. Metall. Soc. AIME* **233**, 588 (1965).
260. H. C. Theuerer, E. A. Nesbitt, and D. D. Bacon. *J. Appl. Phys.* **40**, 2994 (1969).
261. F. Vratny, *J. Electrochem. Soc.* **114**, 505 (1967).
262. H. C. Cooke, C. W. Covington, and J. F. Libsch, *Trans. Metall. Soc. AIME* **236**, 314 (1966).
263. J. J. Cuomo and W. W. Molzen, U.S. Patent 3,892,650 (1975).
264. N. Schwartz, *Trans. Natl. Vac. Symp., 10th, Boston, 1963* p. 325 (1964).
265. G. Perny, *Vide* **21**, 106 (1966).
266. J. Pompei, *Proc. Symp. Deposition Thin Films Sputter., 2nd, Univ. Rochester* p. 127 (1967).
267. E. Hollands and D. S. Campbell, *J. Mater. Sci.* **3**, 544 (1968).
268. J. Pompei, *Proc. Symp. Deposition Thin Films Sputter., 3rd, Univ. Rochester* p. 165 (1969).
269. J. Heller, *Thin Solid Films* **17**, 163 (1973).
270. K. G. Geraghty and L. F. Donaghey, *Proc. Conf. Chem. Vap. Dep., 5th, Electrochem. Soc., Princeton, N.J.* p. 219 (1975).
271. F. Shinoki and A. Itoh, *J. Appl. Phys.* **46**, 3381 (1975).
272. T. Abe and T. Yamashina, *Thin Solid Films* **30**, 19 (1975).
273. B. Goranchev, V. Orlinov, and V. Popova, *Thin Solid Films* **33**, 173 (1976).
274. L. F. Donaghey and K. G. Geraghty, *Thin Solid Films* **38**, 271 (1976).
275. T. Suzuki, *Z. Naturforsch., Teil A* **12**, 497 (1957).

276. L. Holland, "Vacuum Deposition of Thin Films," pp. 455–463. Wiley, New York, 1954.
277. M. L. Lieberman and R. C. Medrud, *J. Electrochem. Soc.* **116**, 242 (1969).
278. A. J. Noreika, *J. Vac. Sci Technol.* **6**, 194 (1969).
279. Y. Hirohata, T. Abe, and T. Toshiro, *Oyo Butsuri* **5**, 402 (1976).
280. C. Weissmantel, *Thin Solid Films* **32**, 11 (1976).
281. R. F. Bunshah and R. J. Schramm, *Thin Solid Films* **40**, 211 (1977).
282. M. Lieberman, *J. Appl. Phys.* **40**, 2659 (1969).
283. J. J. Cuomo and W. Reuter, *IBM Tech. Discl. Bull.* **19**, 742 (1976).
284. G. Perny and B. Laville-St.-Martin, *Thin Solid Films* **6**, R25 (1970).
285. T. K. Lakshmanan and J. M. Mitchell, *Trans. Natl. Vac. Symp., 10th, Boston, 1963* p. 335 (1964).
286. J. L. Vossen, *Phys. Thin Films* **9**, 1 (1977).
287. T. K. Lakshmanan, *J. Electrochem. Soc.* **110**, 548 (1963).
288. L. Holland and G. Siddall, *Vacuum* **3**, 375 (1953).
289. J. S. Preston, *Proc. R. Soc., Ser. A* **202**, 449 (1950).
290. G. Helwig, *Z. Phys.* **132**, 621 (1952).
291. H. Dunstädter, *Z. Phys.* **137**, 383 (1954).
292. F. Lappe, *Z. Phys.* **137**, 380 (1954).
293. J. Stuke, *Z. Phys.* **137**, 401 (1954).
294. R. R. Mehta and S. F. Vogel, *J. Electrochem. Soc.* **119**, 752 (1972).
295. G. L. Harding, *J. Vac. Sci. Technol.* **13**, 1070 (1976).
296. J. G. Froemel and M. Sapoff, *Proc. Symp. Deposition Thin Films Sputter., 1st, Univ. Rochester* p. 62 (1966).
297. M. Hecq, A. Hecq, and J. van Cakenberghe, *Thin Solid Films* **42**, 97 (1977).
298. G. Perny, *C. R. Acad. Sci., Ser. B* **256**, 2160 (1963).
299. G. Perny and B. Laville-St.-Martin, *J. Phys. (Paris)* **25**, 5 (1964).
300. G. Perny and B. Laville-St.-Martin, *J. Phys. (Paris)* **25**, 993 (1964).
301. M. Samirant and G. Perny, *C. R. Acad. Sci., Ser. B* **270**, 603 (1970).
302. A. J. Purdes, B. T. F. Bolker, J. D. Bucci, and T. C. Tisone, *J. Vac. Sci. Technol.* **14**, 98 (1977).
303. M. Samirant, B. Laville-St.-Martin, and G. Perny, *Thin Solid Films* **8**, 293 (1971).
304. S. K. Banjeree, *Nature (London)* **202**, 1098 (1964).
305. J. Heller, *IEEE Trans. Magn.* **MAG-12**, 396 (1976).
306. V. I. Orlinov and G. A. Sarov, *Bulg. J. Phys.* **2**, 156 (1975).
307. J. M. Berak and D. J. Quinn, *J. Vac. Sci. Technol.* **13**, 609 (1976).
308. J. J. Hantzpergue, Y. Doucet, Y. Pauleau, and J. C. Remy, *Ann. Chim. (Paris)* **10**, 211 (1975).
309. F. T. J. Smith, *J. Appl. Phys.* **41**, 4227 (1970).
310. C. H. Lane, *Proc. Natl. Electron. Conf.* **20**, 221 (1964).
311. F. Huber, W. Witt, and I. H. Pratt, *Proc. Electron. Components Conf., Washington, D.C.*, p. 66 (1967).
312. J. S. Preston, U.S. Patent 2,769,778 (1956).
313. A. Thelen and H. König, *Naturwissenschaften* **43**, 297 (1956).
314. V. M. Vainshtein and V. I. Fistul, *Sov. Phys.—Semicond.* **1**, 104 (1967).
315. V. I. Fistul and V. M. Vainshtein, *Sov. Phys.—Solid State* **8**, 2769 (1967).
316. H. K. Muller, *Phys. Status Solidi* **27**, 723 (1968).
317. H. K. Muller, *Phys. Status Solidi* **27**, 733 (1968).
318. V. M. Vainshtein, L. Geraisimova, and I. N. Nikolaeva, *Izv. Akad. Nauk SSSR, Neorg. Mater.* **4**, 357 (1968).

319. T. I. Danilina and K. I. Smirnova, *Mikroelektronika* (*Akad. Nauk SSSR*) **5**, 286 (1976).
320. R. M. Valletta and W. A. Pliskin, *J. Electrochem. Soc.* **114**, 944 (1967).
321. L. D. Locker, R. W. Landorf, C. L. Naegele, and F. Vratny, *J. Electrochem. Soc.* **119**, 183 (1972).
322. F. Shoji and S. Nagata, *Jpn. J. Appl. Phys.* **13**, 1072 (1974).
323. T. Abe and T. Yamashima, *Thin Solid Films* **30**, 19 (1975).
324. H. J. Spitzer, *J. Vac. Sci. Technol.* **10**, 20 (1973).
325. D. Gerstenberg, *Bell Lab. Rec.* **42**, 364 (1964).
326. J. Spitz and A. Aubert, *Proc. Conf. Chem. Vap. Dep., 5th, Electrochem. Soc., Princeton, N.J.* p. 258 (1975).
327. H. J. Spitzer, *J. Vac. Sci. Technol.* **9**, 333 (1972).
328. S. J. Ingrey and W. D. Westwood, *Appl. Opt.* **15**, 607 (1976).
329. A. J. Stirling and W. D. Westwood, *Thin Solid Films* **8**, 199 (1971).
330. A. E. Feuersanger, U.S. Patent 3,627,682 (1971).
331. A. E. Feuersanger, M. S. Wasserman, and I. H. Pratt, *Proc. Electron. Components Conf., Washington, D.C.* p. 52 (1970).
332. F. Lappe, *Phys. Chem. Solids* **23**, 1563 (1962).
333. F. Lappe, *Z. Phys.* **137**, 380 (1954).
334. J. R. Clarke, A. K. Weiss, J. L. Donovan, J. E. Greene, and R. E. Klinger, *J. Vac. Sci. Technol.* **14**, 219 (1977).
335. W. D. Westwood and C. D. Bennewitz, *J. Appl. Phys.* **45**, 2313 (1974).
336. K. E. Haq, *Appl. Phys. Lett.* **26**, 255 (1975).
337. P. Pileur and J. C. Viret, *C. R. Colloq. Int. Pulver. Cathod., 1st, Montpellier* p. 123 (1974).
338. C. J. Mogab and E. Lugujjo, *J. Appl. Phys.* **47**, 1302 (1976).
339. W. B. Pennebaker, Br. Patent 1,118,757 (1968).
340. A. W. Stephens, J. L. Vossen, and W. Kern, *J. Electrochem. Soc.* **123**, 303 (1976).
341. G. J. Kominiak, *J. Electrochem. Soc.* **122**, 1271 (1975).
342. C. J. Mogab, P. M. Petroff, and T. T. Sheng, *J. Electrochem. Soc.* **122**, 815 (1975).
343. P. C. Y. Chen, *Thin Solid Films* **21**, 245 (1974).
344. W. Rothemund and C. R. Fritzsche, *Thin Solid Films* **15**, 199 (1973).
345. P. J. Burkhardt and R. F. Marvel, *J. Electrochem. Soc.* **116**, 864 (1969).
346. A. R. Janus and G. A. Shirn, *J. Vac. Sci. Technol.* **4**, 37 (1967).
347. L. F. Cordes, *Appl. Phys. Lett.* **11**, 383 (1967).
348. S. M. Hu, D. R. Kerr, and L. V. Gregor, *Appl. Phys. Lett.* **10**, 97 (1967).
349. S. M. Hu and L. V. Gregor, *J. Electrochem. Soc.* **114**, 826 (1967).
350. S. M. Hu, *J. Electrochem. Soc.* **113**, 693 (1966).
351. E. E. Smith and D. R. Kennedy, *Proc. Inst. Electr. Eng.* **109**, 504 (1962).
352. M. M. Nekrasov, V. I. Dudkin, S. V. Smirnova, I. K. Miklailova, and L. I. Avramenko, *Poluprovodn. Tekh. Mikroelektron.* **3**, 20 (1969).
353. R. S. Clark, *Trans. Metall. Soc. AIME* **233**, 592 (1965).
354. W. R. Sinclair and F. G. Peters, *J. Am. Ceram. Soc.* **46**, 20 (1963).
355. J. C. Williams, W. R. Sinclair, and S. E. Koonce, *J. Am. Ceram. Soc.* **46**, 161 (1963).
356. E. Sawatzky and E. Kay, *J. Appl. Phys.* **39**, 5613 (1968).
357. S. Mirsch and J. Bauer, *Phys. Status Solidi A* **26**, 579 (1974).
358. R. M. Valletta, J. A. Perri, and J. Riseman, *Electrochem. Technol.* **4**, 402 (1966).
359. Y. Murayama and T. Takeo, *Thin Solid Films* **40**, 309 (1977).
360. J. Chin and M. B. Elsner, *J. Vac. Sci. Technol.* **12**, 821 (1975).
360a. M. Fukutomi, M. Kitajima, M. Okada, and R. Wanatabe, *J. Electrochem. Soc.* **124**, 1420 (1977).

360b. D. Kropman, M. Vinnal, and P. Putk, *Vacuum* **27**, 125 (1977).
361. R. I. Frank and W. L. Moberg, *J. Electrochem. Soc.* **117**, 524 (1970).
362. W. R. Sinclair, F. G. Peters, D. W. Stillinger, and S. E. Koonce, *J. Electrochem. Soc.* **112**, 1096 (1965).
363. M. Hecq and E. Portier, *Thin Solid Films* **9**, 341 (1972).
364. S. Yamanaka and T. Oohashi, *Jpn. J. Appl. Phys.* **8**, 1058 (1969).
365. D. Gerstenberg and C. J. Calbick, *J. Appl. Phys.* **35**, 402 (1964).
366. D. Gerstenberg, *J. Electrochem. Soc.* **113**, 542 (1966).
367. W. Grossklauss and R. F. Bunshah, *J. Vac. Sci. Technol.* **12**, 811 (1975).
368. W. D. Westwood and F. C. Livermore, *Thin Solid Films* **5**, 407 (1970).
369. P. R. Stuart, *Vacuum* **19**, 509 (1969).
370. W. D. Westwood, D. J. Willmott, and P. S. Wilcox, *J. Vac. Sci. Technol.* **9**, 987 (1972).
371. Y. Murayama, *J. Vac. Sci. Technol.* **12**, 818 (1975).
372. T. Koikeda and S. Kumagai, *Proc. Symp. Deposition Thin Films Sputter., 3rd, Univ. Rochester* p. 68 (1969).
373. D. Gerstenberg and E. H. Mayer, *Proc. Electron. Components Conf., Washington, D.C.* p. 57 (1962).
374. A. Sato, Y. Oda, and Y. Hishinuma, *Proc. Electron Components Conf., Washington, D.C.* p. 58 (1970).
375. M. R. Wormald, B. Y. Underwood, and K. W. Allen, *Nucl. Instrum. Methods* **107**, 233 (1973).
376. H. J. Coyne and R. N. Tauber, *J. Appl. Phys.* **39**, 5585 (1968).
377. E. A. Buvinger, *Appl. Phys. Lett.* **7**, 14 (1965).
378. J. W. Balde, S. S. Charschan, and J. J. Dineen, *Bell Syst. Tech. J.* **43**, 127 (1964).
379. M. Nakamura, M. Fujimori, and Y. Nishimura, *Jpn. J. Appl. Phys.* **12**, 30 (1973).
380. F. Vratny and D. J. Harrington, *J. Electrochem. Soc.* **112**, 484 (1965).
381. W. J. Pendergast, U.S. Patent 3,258,413 (1966).
382. E. Krikorian and R. J. Sneed, *J. Appl. Phys.* **37**, 3674 (1966).
383. Y. A. Prokhorov, V. V. Shakmanov, L. B. Shelyakin, and V. E. Yurasova, *Fiz. Tverd. Tela (Leningrad)* **9**, 1398 (1967).
384. W. D. Westwood, R. J. Boynton, and S. J. Ingrey, *J. Vac. Sci. Technol.* **11**, 381 (1974).
385. P. Lloyd, *Solid-State Electron.* **3**, 74 (1961).
386. W. R. Hardy and D. Mills, *Thin Solid Films* **18**, 309 (1973).
387. R. S. Clark and C. D. Orr, *IEEE Trans. Parts, Mater. Packag.* **PMP-1**, S-31 (1965).
388. L. I. Maissel and J. H. Vaughan, *Vacuum* **13**, 421 (1963).
389. W. D. Westwood and N. Waterhouse, *J. Appl. Phys.* **42**, 2946 (1971).
390. W. D. Westwood and P. S. Wilcox, *J. Appl. Phys.* **42**, 4055 (1971).
391. F. C. Livermore, P. S. Wilcox, and W. D. Westwood, *J. Vac. Sci. Teachnol.* **8**, 155 (1971).
392. L. G. Feinstein and D. Gerstenberg, *Thin Solid Films* **10**, 79 (1969).
393. J. Harvey and J. Corkhill, *Thin Solid Films* **6**, 277 (1970).
394. N. Waterhouse, P. S. Wilcox, and D. J. Willmott, *J. Appl. Phys.* **42**, 5649 (1971).
395. P. N. Baker, *Thin Solid Films* **6**, R57 (1970).
396. A. Perinati and G. F. Piacenti, *J. Vac. Sci. Technol.* **14**, 169 (1977).
397. S. J. Ingrey, W. D. Westwood, and B. K. Maclaurin, *Thin Solid Films* **30**, 377 (1975).
398. G. L. Parisi, *Proc. Electron. Component Conf., Washington, D.C.* p. 367 (1969).
399. W. R. Hardy, J. Schewchun, D. Kuenzig, and C. Tam, *Thin Solid Films* **8**, 81 (1971).
400. W. D. Westwood, *J. Vac. Sci. Technol.* **11**, 466 (1974).
401. F. Shinoki and A. Itoh, *Jpn. J. Appl Phys., Suppl.* **2**, Part 1, 505 (1974).
402. A. C. Raghuram and R. F. Bunshah, *J. Vac. Sci. Technol.* **9**, 1389 (1972).

72   J. L. VOSSEN AND J. J. CUOMO

403. P. J. Clarke, *J. Vac. Sci. Technol.* **14**, 141 (1977).
404. T. K. Lakshmanan, C. A. Wysocki, and W. J. Slegesky, *IEEE Trans. Component Parts* **CP-11**, 14 (1964).
405. W. Grossklaus and R. F. Bunshah, *J. Vac. Sci. Technol.* **12**, 593 (1975).
406. L. Lavielle, B. Laville-St.-Martin and G. Perny, *Thin Solid Films* **8**, 245 (1971).
407. K. G. Geraghty and L. F. Donaghey, *Thin Solid Films* **40**, 375 (1977).
408. J. Duchene, M. Terraillon, and M. Pailly, *Thin Solid Films* **12**, 231 (1972).
409. R. M. Goldstein and S. C. Wigginton, *Thin Solid Films* **3**, R41 (1969).
410. M. Minakata, N. Chubachi, and Y. Kikuchi, *Jpn. J. Appl. Phys.* **11**, 1852 (1972).
411. N. F. Foster, *J. Vac. Sci. Technol.* **6**, 111 (1969).
412. G. A. Rosgonyi and W. J. Polito, *J. Vac. Sci. Technol.* **6**, 115 (1969).
413. D. L. Raimondi and E. Kay, *J. Vac. Sci. Technol.* **7**, 96 (1970).
414. R. A. Mickelsen and W. D. Kingery, *J. Appl. Phys.* **37**, 3541 (1966).
415. N. F. Jackson, E. Hollands, and D. S. Campbell, *Proc. Jt. IERE-IEE Conf. Appl. Thin Films Electron. Eng., London* p. 13-7 (1967).
416. R. Blickensderfer, R. L. Lincoln, and P. A. Romans, *Thin Solid Films* **37**, L73 (1976).
417. C. H. Lane and J. P. Farrell, *U.S. Gov. Res. Dev. Rep.* **67**, 73 (1967).
418. J. Harvey and J. Corkhill, *Thin Solid Films* **8**, 427 (1971).
419. R. Vu Huy Dat and C. Baumberger, *Phys. Status Solidi* **22**, K67 (1967).
420. W. J. Takei, N. P. Formigoni, and M. H. Francombe, *Appl. Phys. Lett.* **15**, 256 (1969).
421. J. G. Titchmarsh and P. A. B. Toombs, *J. Vac. Sci. Technol.* **7**, 103 (1969).
422. J. Deforges, S. Durand, and P. Bugnet, *Thin Solid Films* **18**, 231 (1973).
423. L. F. Cordes, U.S. Patent 3,703,456 (1972).
424. L. V. Kirenskii, P. Galepov and I. A. Turpanov, *Kristallografiya* **11**, 346 (1966).
425. E. Sawatzsky and E. Kay, *IBM J. Res. Dev.* **13**, 696 (1969).
426. W. R. Knolle and W. C. Ballamy, *J. Electrochem. Soc.* **122**, 561 (1975).
427. R. B. Liebert and T. H. Conklin, U.S. Patent 3,723,278 (1973).
428. L. D. Locker, C. L. Naegele, and F. Vratny, *J. Electrochem. Soc.* **118**, 1856 (1971).
429. L. D. Locker and D. L. Malm, *Rev. Sci. Instrum.* **42**, 1692 (1971).
430. Y. T. Sihvonen and D. R. Boyd, *Rev. Sci. Instrum.* **31**, 992 (1960).
431. D. R. Boyd, Y. T. Sihvonen, and C. D. Woelke, U.S. Patent 3,235,476 (1966).
432. V. A. Williams, *J. Electrochem. Soc.* **113**, 234 (1966).
433. J. M. Pankratz, *J. Electron. Mater.* **1**, 182 (1972).
434. M. Bonnet and M. Marchal, *C. R. Colloq. Int. Pulver. Cathod., 1st, Montpellier* p. 157 (1973).
435. W. W. Molzen, *J. Vac. Sci. Technol.* **12**, 818 (1975).
436. J. A. Thornton and V. L. Hedgcoth, *J. Vac. Sci. Technol.* **13**, 117 (1976).
437. H. W. Lehmann and R. Widmer, *Thin Solid Films* **27**, 359 (1975).
438. N. F. Foster, *J. Appl. Phys.* **40**, 420 (1969).
439. D. C. Lewis, W. D. Westwood, and A. G. Sadler, *J. Can. Ceram. Soc.* **33**, 153 (1964).
440. R. Nimmagadda and R. F. Bunshah, *J. Vac. Sci. Technol.* **12**, 815 (1975).
441. W. P. Beckley and D. S. Campbell, *Vide* **99**, 14 (1962).
442. C. J. Ridge, P. J. Harrop, and D. S. Campbell, *Thin Solid Films* **2**, 413 (1968).
443. J. Chin and N. B. Elsner, *Proc. Conf. Chem. Vap. Dep., 5th, Electrochem. Soc., Princeton, N. J.*, p. 241 (1975).
444. V. M. Vaynshetyn, *Sov. J. Opt. Technol.* **34**, 45 (1967).
445. T. Umezawa and S. Yajima, *Thin Solid Films* **29**, 43 (1975).
446. G. Gabor and E. Hahn, *Proc. Symp. Deposition Thin Films Sputter., 3rd, Univ. Rochester*, p. 68 (1969).
447. W. D. Sproul and M. H. Richman, *Thin Solid Films* **28**, L39 (1975).
448. V. K. Sarin, R. F. Bunshah, and R. Nimmagadda, *Thin Solid Films* **40**, 183 (1977).

449. P. Bugnet, J. Deforges, S. Durand, and G. Battailler, *C. R. Acad. Sci.*, *Ser. B* **269**, 313 (1969).
450. G. L. Harding, D. R. McKenzie, and B. Window, *J. Vac. Sci. Tech·ol.* **13**, 1073 (1976).
451. D. S. Lo, G. F. Sauter, and W. J. Simon, *J. Appl. Phys.* **40**, 5402 (1969).
452. R. Nimmagadda and R. F. Bunshah, *J. Vac. Sci. Technol.* **12**, 585 (1975).
453. S. F. Vogel and I. C. Barlow, *J. Vac. Sci. Technol.* **10**, 381 (1973).
454. J. W. Coburn and E. Kay, *J. Appl Phys.* **43**, 4965 (1972).
455. O. Christensen and M. Brunot, *C. R. Colloq. Int. Pulver. Cathod.*, *1st, Montpellier*, p. 37 (1973).
456. O. Christensen and B. J. Klein, *J. Phys. E* **7**, 261 (1974).
457. O. Christensen, *Thin Solid Films* **27**, 63 (1975).
458. L. Holland, *Thin Solid Films* **27**, 185 (1975).
459. P. C. Huang and P. M. Schaible, *Electrochem. Soc. Extend. Abstr.* **76-2**, 754 (1976).
460. F. F. Chen, *In* "Plasma Diagnostic Techniques" (R. H. Huddlestone and S. L. Leonard, eds.), p. 113. Academic Press, New York, 1965.
461. L. I. Maissel and P. M. Schaible, *J. Appl. Phys.* **36**, 237 (1965).
462. J. L. Vossen and J. J. O'Neill, *RCA Rev.* **29**, 566 (1968).
463. J. L. Vossen and J. J. O'Neill, *RCA Rev.* **31**, 276 (1970).
464. I. Brodie, L. T. Lamont, and R. L. Jepson, *Phys. Rev. Lett.* **21**, 1224 (1968).
465. D. G. Teer, B. L. Delcea, and A. J. Kirkham, *J. Adhes.* **8**, 171 (1976).
466. D. G. Teer, *J. Phys. D* **9**, L187 (1976).
467. D. G. Teer, *J. Adhes.* **8**, 289 (1977).
468. D. M. Mattox and G. J. Kominiak, *J. Vac. Sci. Technol.* **8**, 194 (1971).
469. R. S. Berg, G. J. Kominiak, and D. M. Mattox, *J. Vac. Sci. Technol.* **11**, 52 (1974).
470. I. V. Mitchell and R. C. Maddison, *Vacuum* **21**, 591 (1971).
471. J. J. Cuomo and R. J. Gambino, *J. Vac. Sci. Technol.* **14**, 152 (1977).
472. W. W. Y. Lee and D. W. Oblas, *J. Appl. Phys.* **46**, 1728 (1975).
473. G. C. Schwartz and R. E. Jones, *IBM J. Res. Dev.* **14**, 52 (1970).
474. E. A. Fagen, R. S. Nowicki, and R. W. Saguin, *J. Appl. Phys.* **45**, 50 (1974).
475. A. G. Blachman, *Metall. Trans.* **2**, 699 (1971).
476. A. G. Blachman, *J. Vac. Sci. Technol.* **10**, 299 (1973).
477. J. J. Cuomo and R. J. Gambino, *J. Vac. Sci. Technol.* **12**, 79 (1975).
478. S. Rigo, G. Amsel, and M. Croset, *J. Appl. Phys.* **47**, 2800 (1976).
479. W. W. Y. Lee and D. W. Oblas, *J. Vac. Sci. Technol.* **7**, 129 (1970).
480. P. H. Schmidt, *J. Vac. Sci. Technol.* **10**, 611 (1973).
481. E. Kay, *J. Vac. Sci. Technol.* **7**, 317 (1970).
482. M. A. Frisch and W. Routher, *Annu. Conf. Mass Spectrom. Allied Top.*, *23rd, Houston, Tex., 1975.*
483. H. F. Winters, D. L. Raimondi, and D. E. Horne, *J. Appl. Phys.* **40**, 2996 (1969).
484. B. Navinsek and T. Zabkar, *Thin Solid Films* **36**, 41 (1976).
485. S. Esho and S. Fujiwara, *AIP Conf. Proc.*, *Magn. Magn. Mater.; Am. Inst. Phys.*, New York p. 34 (1976).
486. G. A. N. Connell, *Phys. Rev. B* **13**, 787 (1976).
487. H. F. Winters and E. Kay, *J. Appl. Phys.* **38**, 3928 (1967).
488. W. Hoffmeister and M. Zuegel, *Thin Solid Films* **3**, 35 (1969).
489. P. K. Hoff and Z. E. Switkowski, *Appl. Phys. Lett.* **29**, 549 (1976).
490. J. E. Greene, *Surf. Sci.*, to be published.
491. R. J. Gambino and J. J. Cuomo, *J. Vac. Sci. Technol.* **15**, 296 (1978).

II-2

Cylindrical Magnetron Sputtering

JOHN A. THORNTON AND ALAN S. PENFOLD

Telic Corporation
Santa Monica, California

I. INTRODUCTION

We define magnetrons as diode sputtering sources in which magnetic fields are used in concert with the cathode surface to form electron traps which are so configured that the $\mathbf{E} \times \mathbf{B}$ electron drift currents close on themselves.

The role of magnetrons in sputtering technology can best be understood by reviewing the limitations of conventional planar diodes. Figure 1 shows a typical planar diode (see Chapter II-1, Sections III and IV). A low pressure abnormal negative glow discharge [1, 2] is maintained between the cathode (target) and an adjacent anode (substrate mounting table). Electrons emitted from the cathode by ion bombardment are accelerated to near the full applied potential in the cathode dark space, and enter the negative glow as so-called primary electrons, where they collide with gas atoms and produce the ions required to sustain the discharge. The primary electron mean free path (mfp) increases with increasing electron energy and decreasing pressure. At low pressures, ions are produced far from the cathode where their chances of being lost to the walls are great; and also many primary electrons hit the anode with high energies, causing a loss that is not offset by impact-induced secondary emission. Therefore, ionization efficiencies are low, and self-sustained discharges cannot be maintained in planar diodes at pressures below about 10 mTorr (1.3 Pa) [3]. Currents are limited because an increase in voltage increases the electron mfp and reduces the ionization efficiency. As the pressure is increased at a fixed voltage, the mfp decreases and larger currents are possible, as shown in Fig. 2. However, at high pressures sputtered atom transport is reduced by collisional scattering [4]; improvements in deposition rate cease above about 120 mTorr, as shown in the figure. Typical deposition conditions for Ni sputtered in Ar are: cathode-to-substrate separation 4.5 cm, voltage 3000 V, pressure 75 mTorr (10 Pa), current density ~ 1 mA/cm^2, and deposition rate 360 Å /min.

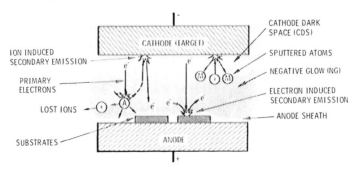

Fig. 1. Schematic representation of the plasma in planar diode sputtering.

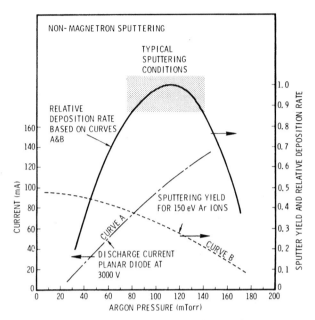

Fig. 2. Influence of working gas pressure on deposition rate for a nonmagnetron sputtering source. The relative deposition rate is calculated for illustrative purposes from the 3000 V discharge current data (curve A) of Kay [3], for a nickel planar diode, and the sputtered atom transport (net yield) data (curve B) of Laegreid and Wehner [4], for a triode with a spherical nickel target.

A magnetic field parallel to the cathode surface can restrain the primary electron motion to the vicinity of the cathode and thereby increase the ionization efficiency. However, the **E × B** motion causes the discharge to be swept to one side [5]. This difficulty can be overcome by using cylindrical cathodes which allow the **E × B** drift currents to close on themselves. This is the general magnetron geometry. Both post [6–12] and hollow [13] cathodes have been investigated for sputtering. However their performance is limited by end losses.

Remarkable performance is achieved when end losses are eliminated. High currents can be obtained, nearly independent of voltage, even at low pressures. This characterizes what we shall define as the *magnetron mode* of operation. This operation may be achieved with electron-reflecting surfaces which result in the formation of an intense uniform plasma sheet over the cathode, as shown in Figs. 3a and b, or with curved magnetic fields which form a series of plasma rings over the cathode, as shown in Figs. 3c and d. These are all cylindrical magnetrons but the hollow cathode types (Figs. 3b and d) are often called inverted magnetrons. We will

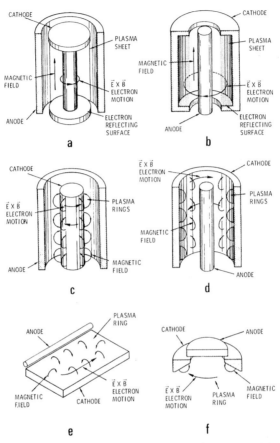

Fig. 3. The configuration of various magnetron sputtering sources: (a)–(d) are cylindrical magnetrons; (b) and (d) are often called inverted magnetrons and sometimes hollow cathodes; (a) is referred to here as a cylindrical-post magnetron and (b) as a cylindrical-hollow magnetron; (e) is the planar magnetron; (f) is often referred to as a sputter gun. In figures vectors are indicated by arrows over the vector quantities.

use the term cylindrical-post magnetron to describe type 3a, and the term cylindrical-hollow magnetron to describe type 3b. The magnetron mode of operation may also be obtained from plasma rings sustained over planar surfaces, as shown in Fig. 3e. These planar magnetrons are discussed in Chapter II-3. When the magnetron mode is obtained from plasma rings sustained on short cylinders or within cylindrical surface cavities, as shown in Fig. 3f, the devices are known as sputter guns. They are discussed in Chapter II-4. The emphasis in this chapter is on cylindrical-post and cylindrical-hollow magnetrons, but the general principles which will be discussed have application to all the devices.

Typical magnetron-mode operating conditions for a cylindrical magnetron are voltage 800 V, magnetic field 150 G, pressure 1 mTorr (0.13 Pa), current density 20 mA/cm^2, cathode erosion rate 12,000 Å/min, and deposition rate 2000 Å/min. The power is generally dc, but rf may be used. Large cathode sizes are possible. This is an important attribute. Furthermore, substrates are not subject to the plasma bombardment common in planar diodes.

As early as 1921 [14], work on the transport of thermionically emitted electrons in cylindrical magnetrons established the importance of plasma oscillations in both the vacuum [15–17] and plasma cases [18–21] and led to their use in microwave oscillators [22–24]. Townsend discharges have been sustained in the geometries shown in Figs. 3a and b to form pressure gauges [25–28], and low pressure negative space charge discharges have been used in geometries 3a–d to form sputter ion pumps [29–31].

The first operation in the magnetron mode for sputtering appears to have been achieved by Penning and Moubis using configurations 3a [32] and 3b [33]. Very active development of cylindrical magnetron technology has occurred over the period from 1969 to the present [34–40a]. Types of apparatus with the general configurations shown in Figs. 3a–d have been operated. Direct current cathodes have ranged in length from 10 to 200 cm. Both cylindrical-post and cylindrical-hollow magnetrons have been operated with rf power. Direct current magnetron sputtering has recently been reported using configurations 3a [41–43], 3c [42, 44], and 3d [44].

II. PRINCIPLES OF OPERATION

This section reviews certain fundamental concepts of plasma physics relevant to plasma magnetrons and describes the principles of operation of the cylindrical magnetron. These same principles apply to the more complex geometries of the planar magnetron and sputter gun described in Chapters II-3 and II-4.

A. Basic Glow Discharge Parameters

A glow discharge plasma can be defined as a region of relatively low temperature gas which is sustained in an ionized state by energetic electrons. The negative glow, shown in Fig. 1, and the positive column (Section II.F) are examples of glow discharge plasmas [1, 2]. The state of a glow discharge plasma can be characterized by the density of neutral atoms (or molecules) n_A, the electron density n_e, and the electron energy distribution, which can often be approximated by an electron temperature T_e (electrons are frequently maintained in a near-thermal velocity distribution by electron–electron collisions and the randomizing effects of plasma oscillations) [44a]. Electron energies and temperatures are often specified in electron volts, where $1\ eV = 11600\ K$ [45]. Average electron energies in glow discharges are typically several electron volts.

Plasmas differ from nonionized gases by their propensity for undergoing collective behavior [45, 46]. Three parameters (derived from n_e, T_e, and n_A) provide useful measures of the tendency toward collective behavior.

The Debye length

$$\lambda_D = 743(T_e/n_e)^{1/2} \quad cm, \tag{1}$$

where T_e is in electron volts and n_e in (cubic centimeters)$^{-1}$, is a measure of the distance over which significant departures from charge neutrality can occur. A plasma cannot exist in a space of size L much less than λ_D.

The plasma frequency,

$$f_p = \omega_p/2\pi = 9000\ (n_e)^{1/2} \quad Hz, \tag{2}$$

provides a measure of the tendency for electrostatic waves to develop. Waves can form if $\omega_p \gg \nu$, where ν is the electron collision frequency and ω_p the plasma angular frequency.

A critical degree of ionization [47] can be written as

$$\alpha_c \sim 1.73 \times 10^{12}\ \sigma_{eA}\ T_e^2, \tag{3}$$

where σ_{eA} (centimeters^2) is the electron–atom collision cross section at the average electron energy. When $n_e/n_A \gg \alpha_c$, long-range Coulomb collisions dominate and the charged particles behave as though they were in a fully ionized gas.

In magnetron plasmas $\lambda_D \ll L$, $\omega_p \gg \nu$, and n_e/n_A is often greater than α_c. These plasmas are much richer in collective behavior than are those in conventional planar diodes.

B. Single Particle Motion

The equation of motion of a particle of charge e, mass m, and velocity $\mathbf{v}$ in an electric field $\mathbf{E}$ and magnetic field $\mathbf{B}$ is [45–46, 48–50]

$$\frac{d\mathbf{v}}{dt} = \frac{e}{m}\ (\mathbf{E} + \mathbf{v} \times \mathbf{B}). \tag{4}$$

Only the electron motion is influenced by magnetic fields of the strengths used in sputtering sources operating in the magnetron mode.

When $\mathbf{B}$ is uniform and $\mathbf{E}$ is zero, the electrons drift along the field lines with a speed $v_\parallel$ which is unaffected by the magnetic field, and orbit the field lines with a gyro or cyclotron frequency

$$\omega_c = eB/m_e = 1.76 \times 10^7 B \quad \text{rad/sec,} \tag{5}$$

and at the gyro or Larmor radius

$$r_g = \frac{m_e}{e} \left(\frac{v_\perp}{B} \right) = 3.37(W_\perp)^{1/2}/B \quad \text{cm,} \tag{6}$$

where B is in Gauss and $W_\perp$ is the energy associated with the electron motion perpendicular to the field in electron volts. The motion is a helix, as shown in Fig. 4a.

When **B** and **E** are uniform and **E** is parallel to **B**, the particles are freely accelerated and the helix pitch increases continuously. When there is an electric field component $E_\perp$ (volts per centimeter) perpendicular to **B**, a drift of speed

$$v_E = 10^8 E_\perp/B \quad \text{cm/sec} \tag{7}$$

develops in a direction perpendicular to both $E_\perp$ and **B** and combines with the orbiting motion as shown in Fig. 4b. This is the **E × B** drift. For a particle created at rest in uniform and perpendicular **E** and **B** fields, the trajectory becomes the cycloid generated by a circle of radius r_g [given by

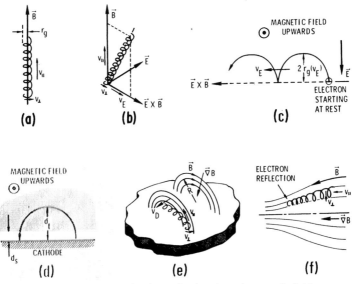

(a)   **(b)**   **(c)**

(d)   **(e)**   **(f)**

Fig. 4. Electron motion in static electric and magnetic fields.

Eq. (6) with $v_\perp = v_E$ from Eq. (7)] moving with velocity v_E as shown in Fig. 4c.

Electrons are emitted from a cold cathode with an initial energy (~ 5 eV) that is small compared to the energy acquired as they are accelerated in the cathode dark space (CDS). Their motion is not exactly cycloidal because the CDS electric field is nonuniform. However, for a planar cathode the distance d_t to the turning point is given by Eq. (6), with $W_\perp$ equal to the total energy which the electrons have gained from the electric field in passing to the turning point [14]. When d_t exceeds the sheath thickness d_s, as is generally the case in sputtering systems, $W_\perp$ is equal to the voltage drop in the CDS (approximately the discharge voltage) and d_t is easily estimated. The trajectory is shown in Fig. 4d. In the case of thin sheaths ($d_s \ll r_g$) over cylindrical cathodes of radius R_0, the turning distance d_t varies from r_g when $R_0 \gg d_t$ to $2\, r_g$ when $R_0 \ll d_t$ [14].

Curved magnetic fields of the type shown in Figs. 3c–f have an inwardly directed gradient ∇B. Such a gradient is largely perpendicular to $\mathbf{B}$. It induces a $\nabla B \times \mathbf{B}$ drift that is perpendicular to both $\mathbf{B}$ and ∇B and proportional to $v_\perp^2$. This combines with an identically directed drift caused by the centrifugal force associated with the speed $v_\parallel$ along the curved field lines [51]. See Fig. 4e. For the circular field lines shown, the total drift can be approximated as

$$v_D \sim (v_\parallel^2 + \tfrac{1}{2}v_\perp^2)/\omega_c R, \tag{8}$$

where R is the field radius.

In the region where the magnetic field lines in the magnetrons shown in Figs. 3c–f enter the cathode, there is a gradient ∇B parallel to $\mathbf{B}$ (see Fig. 8c). Electrons moving in such a field tend to conserve their magnetic moment μ_M [52]:

$$\mu_M \equiv m_e v_\perp^2 / B. \tag{9}$$

Therefore $v_\perp$ must increase as they move in the direction of increasing field strength. Conservation of energy then requires that $v_\parallel$ decrease, and an electron may be reflected, as indicated in Fig. 4f.

C. Mobility and Diffusion

The current density $J_\perp$ which flows across a magnetic field because of a perpendicular electric field $E_\perp$ is

$$J_\perp = en_e\mu_{e\perp}E_\perp + en_i\mu_{i\perp}E_\perp. \tag{10}$$

The electron and ion mobilities are given by [53]

$$\mu_\perp = \frac{\mu}{1 + (\omega_c/\nu)^2}, \tag{11}$$

where $\mu = \mu_\| = e/m\nu$ is the mobility in the absence of a magnetic field or along a field line and ν is the collision frequency for the species in question. The diffusion coefficients, which can be written as $D = \mu kT/e$ (where k is the Boltzmann constant), therefore obey similar relationships [54].

Operation in the positive space charge (PSC) magnetron mode (Section I) requires that $\mu_{e\perp} \gg \mu_{i\perp}$. In the presence of a sufficiently strong magnetic field so that $\omega_c/\nu \gg 1$ for both the electrons and ions, it can be shown by substituting $\nu_i = e/m\,\mu_i$ into Eq. (11) that $\mu_{e\perp}/\mu_{i\perp} = \mu_i/\mu_e$. Thus $\mu_{e\perp} < \mu_{i\perp}$. A typical cylindrical magnetron operates at a pressure of 1 mTorr with a magnetic field strength of 100 G. From Eq. (5), $\omega_{ci} = 2.4 \times 10^4$ rad/sec and $\omega_{ce} = 1.8 \times 10^9$ rad/sec. At 1 mTorr, and an electric field of a few volts per cm, the ion mobility for Ar^+ in Ar is about 10^6 cm²/volt-sec [55, 55a], and $\omega_{ci}/\nu_i = \omega_{ci}m_i\mu_i/e \sim 1$. Thus, from Eq. (11) we see that the ion mobility is only slightly reduced by the magnetic field. Consider electrons with an energy of 10 eV. For degrees of ionization of a few per cent or less, their motion will be dominated by collisions with gas atoms; see Eq. (3). Collision cross sections are shown in Fig. 5. For an electron energy of 10 eV the total cross section is $\sim 2 \times 10^{-15}$ cm² and the collision frequency $\nu_e = n_A\sigma v_e$ is about 10^7 sec⁻¹. Thus the calculated mobility μ_e is $\sim 1.5 \times 10^8$ cm²/volt-sec, which is in good agreement with experiment [58a]. However, $\omega_{ce}/\nu_e \sim 180$; consequently, the magnetic field reduced the electron mobility by a factor of more than 10^4, so that $\mu_{e\perp} \sim 5 \times 10^3$ cm²/volt-sec and $\mu_{e\perp} \ll \mu_{i\perp}$. However, despite this prediction, operation in the PSC magnetron mode still occurs. The transition to a negative space charge mode in cylindrical discharge tubes with central cathodes is not observed until fields of several thousand gauss are used (see Section II.G and [6–12, 59–60]). We ascribe this sustained PSC operation to collective behavior, in the form of plasma oscillations, which fortunately allow the low energy electrons to migrate effectively across fields strong enough to restrain the motion of the primary electrons. Similar anomalous electron transport observed in Penning discharge tubes is also attributed to plasma instabilities [61–64].

D. Plasma Sheaths

Because of their different masses, electrons and ions tend to pass from a plasma to an adjacent surface at different rates. Thus a space charge region, in which one species is largely excluded, forms adjacent to such surfaces. The potential variation between the surface and the plasma is largely confined to this layer, called a sheath [65]. The nature of the sheath depends on the current density passing across it. Except for cases

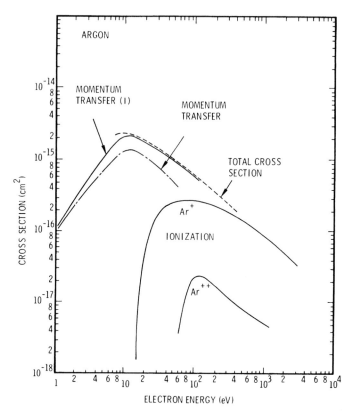

Fig. 5. Collision cross sections for electrons in argon gas. The data are total cross section, Delcroix [47]; momentum transfer (1), Christophorou [56]; momentum transfer (2), Huxley and Crompton [57]; ionization, Kieffer and Dunn [58].

involving very high current densities to anodes, the space charge region will contain primarily particles of the low-mobility species. Thus, for typical magnetron plasmas where $\mu_{e\perp} \gg \mu_{i\perp}$ (Section II.C), electrically isolated surfaces within the plasma develop a positive space charge sheath and float at a negative potential relative to the plasma. Substrates surrounding a cylindrical-post magnetron may assume a positive potential of 1 or 2 V, probably because of the relatively low electron mobility outside of the intense plasma region.

The cathode dark space is a positive space charge sheath. Such a sheath is shown schematically in Fig. 6. The thickness d_s is taken to be that of the region where the electron density is negligible and where the potential drop V_s occurs. The ion current density J_i is related to d_s and V_s by the Child–Langmuir law [1, 45]

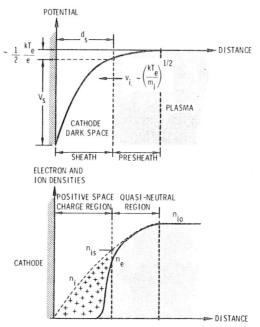

Fig. 6. Schematic representation of the positive space charge sheath that develops over a cathode.

$$J_i = 27.3(40/M)^{1/2}(V_s^{3/2}/d_s^2) \quad \text{mA/cm}^2, \tag{12}$$

where V_s is in kilovolts and d_s is in millimeters. M is the ion molecular weight. It is difficult to relate J_i to the density n_{i0} of ions in the bulk plasma, because there is a quasi-neutral presheath region where a potential drop V_p of the order of $\frac{1}{2}(kT_e/e)$ occurs [45, 66]. For estimates, the presheath density can be assumed to obey a Boltzmann distribution so that $n_{is}/n_{i0} = \exp(-eV_p/kT_e)$ and

$$J_i \sim 0.6en_{i0}(kT_e/m_i)^{1/2}. \tag{13}$$

E. Plasma Waves and Stability

The plasma state is rich in wave phenomena, particularly when a magnetic field is present [45, 46, 67, 68]. The electrons in a magnetron plasma can be pictured as drifting through a sea of relatively stationary atoms and ions in response to the various electric and magnetic field interactions described in Section II.B. Severe departures from charge neutrality, capable of generating waves, can occur in the form of charge bunching and separation over distances of the order of the Debye length. Thus, for example, charge separation associated with the azimuthal electron drift in a cylindrical magnetron can give rise to plasma oscillations and an azimuthal perturbation electric field which interacts with the axial magnetic field to produce radial $E_\theta B_z$ electron drift waves [69–72]. A similar effect (diocotron instability) can occur if the radial variation in electron density has a strong maximum within the plasma region [64, 73, 74]. Electron electrostatic waves generated by such charge inhomogeneities can propagate along field lines with frequencies of the order of ω_p, or across field lines with a frequency of the order of $(\omega_p^2 + \omega_c^2)^{1/2}$. Ion electrostatic waves can propagate in all directions at a velocity of the order of $(kT_e/m_i)^{1/2}$. If feedback mechanisms are present, oscillations of the type described above can grow into macroscopic instabilities capable of enhancing electron transport across the magnetic field, as discussed in Section II.C.

Plasma instabilities can also be important in permitting a collisionless exchange of energy between particles within the plasma. Examples are the two-stream instability and Landau damping [45, 46]. Evidence of such energy transfer has been seen between the primary electrons and the negative glow over thermionic cathodes [75, 76]. Electric field oscillations can negate the static field conservation relations that were used, for example, in developing the turning radius criteria discussed in Section II.B. Such oscillations are believed to play an important role in the primary electron energy exchange and in the operation of plasma magnetrons (see Section II.G).

F. Cold Cathode Discharges

A low pressure cold cathode discharge is one which is maintained primarily by secondary electrons emitted from the cathode because of the bombardment by ions from the plasma. These secondary electrons are accelerated in the cathode dark space and enter the negative glow as shown in Fig. 1, where, known as primary electrons, they must produce sufficient ions to release one further electron from the cathode [2]. This requirement can be expressed by the following relationship for the minimum potential to sustain such a discharge [77]:

$$V_{min} = \mathscr{E}_0/\gamma_i\epsilon_i\epsilon_c, \tag{14}$$

where γ_i is the number of electrons released per incident ion [the secondary electron coefficient: typically 0.1 for low-energy Ar^+ incident on metals [78]), $\mathscr{E}_0$ the average energy required for producing ions [about 30 eV for Ar^+ [79]], ϵ_i the ion collection efficiency, and ϵ_c the fraction of the full complement of ions $(V/\mathscr{E}_0)$ that is made by the average primary electron before it is lost from the system. In conventional sputtering systems

$V_{\min}$ is large because ϵ_i and ϵ_e are small. In magnetron sources, ϵ_i and ϵ_e are near unity but γ_i is believed to be reduced to an effective value of about $\frac{1}{2}\gamma_i$ by electron recapture at the cathode (Section II.G). As a result, $V \approx 2\,V_{\min} \approx 20\mathscr{E}_0$.

The negative glow (NG) is the energy exchange region for the primary electrons, and its extent corresponds to the extent of their travel [1, 2]. The electron distribution in the NG consists of primary electrons, ultimate electrons (primaries that have transferred their energy), and much larger numbers of low energy ionization products. In the classical glow discharge a positive column (PC) extends from the NG to the anode. The electric field in the PC is just sufficient to transport the low energy electrons from the NG to the anode and to produce sufficient ionization to make up for wall losses. In conventional sputtering sources the substrates generally intercept the NG, as shown in Fig. 1, and there is no PC. In magnetrons the magnetic field restricts the primary electrons and the NG to the vicinity of the cathode, and a PC-like region can be identified (Section II.G).

A positive or negative space charge sheath will form at the anode, depending on whether the electron thermal flux is greater or less than the discharge current [65]. Thus the anode sheath potential drop, which determines the plasma potential relative to the anode, is strongly dependent on the placement of the anode within the magnetic field.

G. Discharge between Coaxial Cylinders in an Axial Magnetic Field

Consider a cold cathode discharge between a central cathode and a surrounding anode separated by distance D, as shown in Fig. 7, with a magnetic field of such strength that $d_s < d_t < D$ and $\omega_c \gg \nu$. Cylindrical coordinates are used. We neglect end effects (they are considered in the next section) and consider particle motion in the $r-\theta$ plane.

The primary electrons are emitted from the cathode with energies of a few electron volts [78] and move in cycloidlike orbits. There is a strong probability of recapture at the cathode. (Electron bombardment-induced secondary emission is generally negligible, since most electrons returning to the cathode have insufficient energy.) It is believed that the recapture probability is reduced by interactions with plasma oscillations [18, 24]. On a random phase basis about one-half will escape recapture and enter the negative glow. One defines an effective secondary emission coefficient Γ_i equal to γ_i times a factor which accounts for the probability of recapture [61, 80, 81]. Thus Γ_i is perhaps close to $\frac{1}{2}\gamma_i$. Secondary emission from the end reflectors (which is unaffected by the magnetic field) is of little signifi-

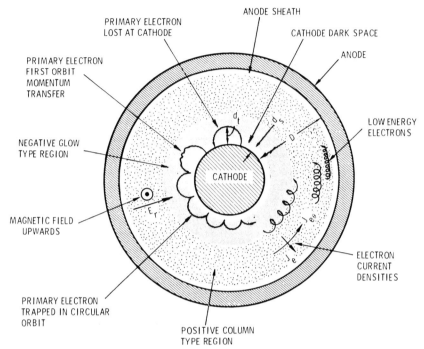

Fig. 7. Schematic representation of electron transport processes in a cold cathode discharge between a cylindrical cathode and a coaxial anode in a uniform axial magnetic field.

cance in most geometries because of the relatively small end reflector area in contact with the intense plasma.

Primary electrons which enter the NG are believed to become trapped on similar cycloidlike paths which orbit, but do not reach, the cathode, as shown in Fig. 7. Collisions in the radial E field beyond the sheath [(Eq. (10)] and plasma oscillations cause the electrons to migrate to the anode, but with decreasing azimuthal speed because of a weakening electric field [Eq. (7)] and with reduced gyro radii [Eq. (5)] because of the concurrent energy loss. The azimuthal drifts produce an azimuthal (Hall) current density J_θ equal to $(\omega_c/\nu)J_r$ [82]. The total Hall current is typical an order of magnitude larger than the discharge current and constitutes a solenoidal current that can cause a 10% reduction in magnetic field strength in the vicinity of the cathode.

Such discharges exhibit a bright glow extending a distance of about $2d_t$ from the cathode and a dimmer glow from there to the anode. These regions can be identified as having NG- and PC-like characteristics, respectively.

Essentially identical discharges can be formed in cylindrical-hollow magnetrons. The primary differences are the stronger radial E field in the PC adjacent to the anode (in configurations with axial anodes) and the formation of a negative space charge (NSC) anode sheath if the anode diameter is small.

If the magnetic field is made very strong, the electron mobility can be decreased to the point where a strong electric field and considerable ionization occurs in the vicinity of the anode [6, 8, 9, 59, 60]. A mode of this form (NSC) is particularly pronounced in inverted magnetrons with axial anodes [13]. These conditions have been suggested for sputtering [6, 8, 9], but the magnetron mode (Section I) is vastly superior. Numerous plasma instabilities have been reported when using strong magnetic fields [60, 83–85].

H. Formation of Electron Traps

Effective sputtering requires that the primary electrons be used efficiently ($\epsilon_e \sim 1$) to produce a copious supply of ions in the vicinity of the cathode ($\epsilon_i \sim 1$) [see Eq. (14)]. The ionization collision cross sections are small for the primary electrons (Fig. 5), so that these electrons must be trapped in orbits that permit long travel distances adjacent to the cathode. For cylindrical magnetrons this requirement can be satisfied in the $r-\theta$ plane by a magnetic field parallel to the cathode surface. Electrons can thus cross field lines and pass from the cathode only by making the desired energy exchange. However, the electrons can move freely along field lines (Fig. 4a). End losses can be eliminated by installing reflecting surface wings maintained at cathode potential (Figs. 3a,b) or by causing the magnetic field lines to intersect the cathode (Figs. 3c–f). Primary and ultimate electron trajectories for several trap configurations are shown in Fig. 8.

I. Anode Considerations

The low-energy electrons must be removed from the trap to complete the electrical circuit. The anode should be placed within the magnetic field so that it terminates the trap at a point where the primary electrons have dissipated their energy. Plasma oscillations are believed to assist in the migration of low energy electrons across field lines to the anode. Anode placement is further aided by the fact that $\mu_{e\parallel} > \mu_{e\perp}$. Thus in the configuration shown in Fig. 8a, all electrons reaching the radius R are swept into the annular anode located at the end of the cathode; i.e., a virtual anode that does not disrupt the sputtered flux separates the substrates

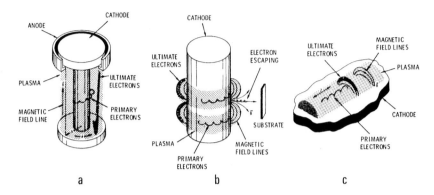

a   b   c

Fig. 8. Schematic representation of primary and ultimate electron trajectories for several electron trap configurations. (Ultimate electrons are defined as primary electrons which have given up most of their energy. Low energy electrons which are the products of the ionization will have identical trajectories.)

from the plasma and removes plasma substrate heating (ions are electrostatically forced to stay in the vicinity of the electrons). Heating can result at substrate positions that intersect field lines, as shown in Fig. 8b.

An anode of insufficient size or with poor placement can cause a significant anode voltage drop. Poor placement can also improperly terminate the trap and permit electrons to reach the substrates. Anode design should take into account the poor mobility of the low-energy electrons in the θ direction in Fig. 8a and the z direction in Fig. 8c.

III. CYLINDRICAL MAGNETRON APPARATUS

Engineering design details affect the operational stability, the profile of substrate coating thickness, and the ability to have long-time trouble-free operation [34, 36–38]. Proper anode placement and design can greatly reduce spurious electrical activity. Proper design of the field coil system allows adjustment of the field shape for optimum performance and allows the shape to be held accurately. The shape of the field is very important in this application, particularly when the ratio of cathode length to electron trap diameter is large. In many cases it is essential that the designer have access to some means for determining and plotting the contours of magnetic flux. With axial symmetry, as in this case, these contours show the shape of the field, and the values determine the degree of saturation in the various magnetic materials which may be present.

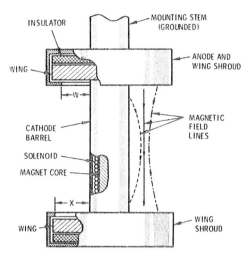

Fig. 9. Representative design of a cylindrical-post magnetron sputtering source.

A. Cylindrical-Post Magnetrons

A cylindrical-post magnetron cathode is a spool formed from a cylindrical barrel and two disklike wings. Figure 9 shows a representative design wherein the dimension of one wing W is smaller than that of the other X. If the magnetic field lines are straight, an intense plasma will form uniformly along the barrel and will extend outward a distance W, whereupon it is terminated abruptly by the anode action of the shroud on the smaller wing, as shown in Fig. 8a. Because electrons in the plasma move with ease along field lines, a virtual anode is projected along the length of the barrel from the smallest wing shroud (Section II.I). The water cooling means for the barrel, the wings, and the anode are not shown in Fig. 9.

Any metal surface which is at cathode potential but does not face the intense plasma between the wings sputters very slowly. Such a surface receives a net coating which may become oxidized, yielding a surface layer with relatively poor electrical conductivity. This layer is susceptible to intermittent spurious electrical activity in the form of small sparks and occasional power arcs. A power arc constricts the entire discharge current into a small cathode spot which moves rapidly over the surface because of the $\mathbf{E} \times \mathbf{B}$ motor action. The discharge voltage falls to a small value like 30 V. Metal will be melted at the attachment spot if the current is too high. This type of melting has been found to occur above about 8 A for Al and above 25 A for higher melting point materials like Cr. The power supply must be capable of handling arc conditions and must prevent uncontrolled current flow. The wing shrouds shield the wing insulators (which may be pyrex) from coating material. Small gaps located between the insulators and their metal surroundings prevent degradation of the insulators by the spurious electrical activity. The wing shrouds also prevent material from the sides and edges of the wings from reaching the substrates.

The smallest wing size W should be at least three times larger than the gyro radius of the primary electrons emitted from the cathode. With this criterion the field strength, wing size, and operating voltage are related to each other in the following way [compare with Eq. (6)]:

$$BW = 10V^{1/2} \quad \text{(G-cm)}. \tag{15}$$

Fields from 30 to 200 G are usually employed.

A solenoid placed coaxially and externally to the cathode may be used as the field generator. Careful attention must be paid to field curvature resulting from end effects. A solenoid, wound on a core of good magnetic material, may be located within the cathode as shown in Fig. 9. If this solenoid is in contact with cooling water, it must be operated at a negative potential with respect to the cathode to avoid corrosion through pinholes in the insulation. A cathode solenoid by itself will generate a field which is strongly bowed outward (dashed line). The plasma will concentrate toward the midplane of the barrel and the sputter erosion will not be uniform. External and internal solenoids can act in concert to produce bowed in, bowed out, or straight field lines as shown. A field which is straight and parallel to the barrel results in uniform barrel erosion up to within a small distance from the wings. The use of an internal solenoid for field shaping is particularly important for cathodes with a large length to wing size ratio.

The power-weight product for a solenoid of copper wire designed to produce a field of B (gauss) over a diameter D (centimeters) and a length L (centimeters) when operating at a temperature T (degrees Celsius) is

$$PW = (BLD/108)^2(1 + (T/234)) \quad \text{(W-g)}. \tag{16}$$

B. Cylindrical-Hollow Magnetrons

The schematic arrangement of a typical cylindrical-hollow magnetron cathode is shown in Fig. 10. Two useful aspects of design are shown. First, the anodes are joined to a tubular backstrap and both are made from magnetic material. This greatly reduces field curvature near the ends and also increases the field strength in the plasma inside the cathode. Second, the solenoid is divided into three or more coils whose current ratios may

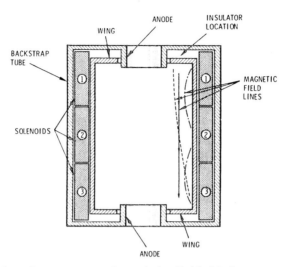

Fig. 10. Schematic arrangement of a typical cylindrical-hollow magnetron sputtering source.

be controlled to provide a wide variety of field shapes including wedges and double traps, as shown. The solenoid sections should be operated in electrical series connection to eliminate changes in field shape caused by unequal heating of the sections. Similar techniques can be used with post cathodes.

C. End Containment

Cylindrical magnetron cathodes may employ end containment which is electrostatic, magnetic, or a combination of the two, depending on how the field is shaped. With straight field lines, electrostatic end containment is used at both ends. Then the cathode sheath thickness at the wings is observed to be much larger than that at the barrel. Since the ion current density is inversely proportional to the square of the sheath thickness [Child–Langmuir law, Eq. (12)], the ion current drawn to the wings is a small fraction of that drawn to the barrel. Variations of field shape cause other forms of containment to occur. A bowed-out shape causes magnetic containment at both ends. A wedge shape (Fig. 10, dashed line) causes magnetic containment at one end and electrostatic at the other. A multiply bowed field creates multiple traps (see Figs. 3c and d). It is also possible to create multiple traps enclosed by a single trap where the latter has electrostatic end containment [39].

D. Magnetic Materials

Magnetic materials can be sputtered using cylindrical magnetrons, provided the cathode cylinder is uniformly saturated with magnetic field. To accomplish this with a cylindrical-post cathode the core of the internal solenoid is provided with flanges which make close mechanical contact with the cathode barrel so as to form a closed magnetic structure. Then the barrel can be saturated and the field lines outside the barrel can be accurately straight and parallel to the surface. Great care must be taken to achieve uniform erosion so that the wall thickness of the barrel remains uniform. Variations of voltage with field shape can be used as a diagnostic tool for achieving the required uniformity.

E. Cathode Fabrication

Cathodes of common engineering metals can be easily fabricated from commercially available tube and rod stock. Cathodes of higher purity metals can be formed by vacuum casting or hot pressing. Soft metals such as In, Pb, Sn, and Cd have been cast over tubular sections that contain the vacuum seals. Electroplated or plasma sprayed cathodes can be used when coatings are thin and purity is not critical. Cathodes can also be formed by wrapping sheet on a tubular support. Compounds may be fabricated in cathode shapes by hot pressing. If the compound is porous it may be mounted on a tubular vacuum shell.

IV. PLASMA DISCHARGE

The current–voltage ($I-V$) characteristic reveals a great deal about the ionization processes in a plasma discharge. Discharges operating in the magnetron mode (Section I) obey an $I-V$ relationship of the form $I \propto V^n$, where n is an index to the performance of the electron trap and is typically in the range 5–9. Thus, with appropriate magnetic field strengths, discharges operate in the magnetron mode at a voltage that is nearly constant and is related to the basic ionization mechanisms by Eq. (14) with $\gamma_i = \Gamma_i$ (see Section II.G). Typical $I-V$ curves for cylindrical-post magnetron operation at various pressures and magnetic field strengths are shown in Fig. 11. If the magnetic field is too weak, the voltage will increase abruptly as shown by A in Fig. 11. Discharges can be stably operated on this part of the characteristic using a constant voltage power supply [41]. However, operation in the magnetron mode using a constant current power supply is generally preferred for sputtering.

At a fixed voltage and magnetic field strength the current increases

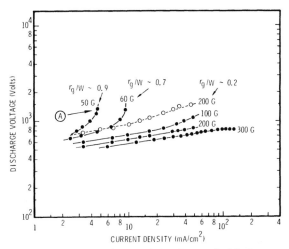

Fig. 11. Typical current–voltage characteristics for a cylindrical-post magnetron sputtering source operating under various conditions of pressure and magnetic field strength. Copper cathode, argon: ●, 10 mTorr; ○, 0.5 mTorr.

with pressure (more collision targets to make ionization). At a fixed voltage and pressure the current increases with the magnetic field strength (higher ϵ_i). With proper choice of magnetic field, magnetron mode operation with a near common $I–V$ characteristic has been achieved for a wide range of apparatus sizes in both the cylindrical-post and cylindrical-hollow magnetron configurations, as shown in Fig. 12.

Figure 13 shows the variation of discharge voltage with pressure at constant currents and magnetic fields. The discharge will fail to operate in the magnetron mode if the primary electrons do not exchange sufficient momentum with the plasma during their first cycloidal orbit and are lost at the cathode (Section II.G), or if after entering the trap, they escape without producing the required ionization. The latter action causes the abrupt voltage increase shown at A in Fig. 11 and at B in Fig. 13. The abrupt extinction of the discharge at C in Fig. 13 is believed to result from the first action.

Figure 14 shows the conditions of pressure and magnetic field strength that permitted stable operation of a cylindrical-post magnetron at a current density of 10 mA/cm². The criterion of Eq. (15) establishes a minimum magnetic field strength that must be exceeded even at relatively high pressures, as shown at D in Fig. 14. High field strengths apparently permit lower pressure operation until the magnetic field decreases the cycloidal path to the point where Γ_i is too low to permit the discharge to sustain itself.

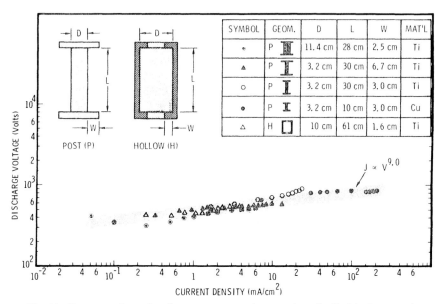

Fig. 12. Current–voltage data for magnetron mode operation of cylindrical-post and cylindrical-hollow magnetron sputtering sources with various geometries.

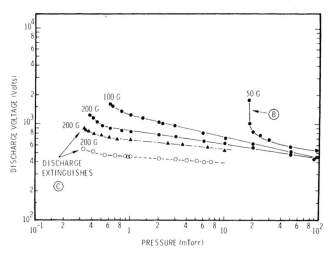

Fig. 13. Operating characteristics of a cylindrical-post magnetron sputtering source, showing the variation of discharge voltage with pressure at constant discharge current and magnetic field strength. Copper cathode (3.2 cm diam. ×30 cm long): ●, current, 3 A, argon; ▲, current, 1 A, argon; ○, current, 1 A, oxygen.

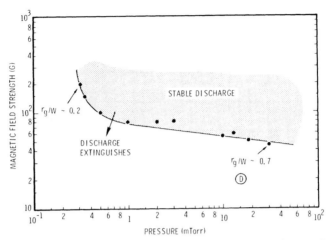

Fig. 14. Conditions of pressure and magnetic field strength for stable operation of a representative cylindrical-post magnetron at fixed current. Copper cathode (3.2-cm diam. × 30 cm long), argon: current density, 10 mA/cm².

Figure 13 shows that operation at the low pressure limit is surprisingly independent of the gas species (compare argon and oxygen data). This behavior is attributed to the role of plasma oscillations in establishing Γ_i (Section II.G). Oscillations in voltage with amplitudes approaching the applied potential are observed at low pressures just prior to extinguishing of the discharge. Oscillations at frequencies in the range 50–500 kHz, with a maximum frequency that varies inversely with the square root of the working gas atomic mass, have been detected using plasma probes in cylindrical magnetron sputtering sources operating with Ar, O_2, and He [86]. This frequency range is consistent with electrostatic waves oscillating in the discharge cavity, and is about equal to the ion–ion collision frequency. The amplitude tended to increase with increasing magnetic field strength and significantly increased with decreasing pressure.

Electrostatic probe ion density measurements indicate that the degree of ionization in the electron trap is a few percent or less. These measurements are generally consistent with estimates based on the cathode current density using Eq. (13). Gas flux calculations indicate that to allow the observed operating conditions, an Ar atom passing through the discharge must have a 1–30% chance of being ionized, depending on the discharge current and the size of the plasma column.

Typical magnetron operation is at a potential of 700 V (Fig. 11). Measurements with an electrostatic probe placed over the anode of a cylindrical magnetron indicate that the primary electrons exchange essentially all of their energy in the discharge, implying that $\epsilon_c \sim 1$ [44a, 86]. Γ_i is believed to be equal to approximately $\frac{1}{20}$ (Section II.G). Letting $\mathscr{E}_0 = 30$ eV/ion [79] and using Eq. (13), we obtain $\epsilon_i \sim 0.9$. This collection efficiency for the ions is generally consistent with electrostatic probe measurements at positions beyond the virtual anode.

V. RF OPERATION

Cylindrical magnetrons of various types have been operated successfully as rf devices. Most experience has been with two cylindrical-hollow cathodes placed end to end and driven as a balanced double-ended system. Ion sheaths appear over both cylinders and also over the surfaces of the vacuum envelope. The latter sheath can be eliminated by appropriate electrical adjustment of the system. Operation is best if the magnetic field forms separate traps over each electrode with a common trap joining the two [39].

The voltage on an electrode can rise only slightly above the plasma potential [6, 7, 87–89]. Therefore, on a time-average basis, each electrode is at a negative potential with respect to the plasma. This is the induced bias [87]. The vacuum chamber walls are similarly biased [90]. In a proper system the bias is negligibly small over all surfaces which should not be sputtering.

The action in an rf system is complicated and it is helpful to have a simple model to aid understanding. Suppose the time required for an ion to cross the sheath is large compared to an rf cycle time. This is a reasonable viewpoint for frequencies above a few megahertz. The ions are then unable to follow rf variations of voltage and are accelerated across the sheath by the time-average voltage they experience. This is equal to the negative bias across the sheath. The ions are thus collected at the electrode at a constant rate throughout the cycle. Even though the ion density is independent of time, the sheath thickness is not. It varies as electrons move among the ions. The sheath thickness is maximum when the electrode is most negative with respect to the plasma and is zero when there is no voltage difference. This is a probelike behavior [87]. Thus the electrodes, and the vacuum chamber wall, are cathodes throughout most of the rf cycle and are anodes for relatively brief fractions of the cycle. The total applied voltage is shared among the various sheaths.

Assuming a square wave of line current gives mathematical simplification to the equations of the model and allows solutions to be obtained in closed form [91]. The sheath size is governed by a Child–Langmuir law, just as in the dc case. In the limit of ion current small compared to displacement current, it is found to be

$$JD^2 = 46(40/M)^{1/2}(V/1000)^{3/2}, \qquad (17)$$

where D is the maximum size of the sheath during the cycle (millimeters), J the collected ion current density (milliamperes per (centimeter)^2), M the ion mass in atomic units, and V the sheath bias voltage. The latter decreases as the ratio of ion current to displacement current rises and is always less than $\frac{5}{12}$ of the peak voltage which appears between the plasma and the electrode. The constant 46 is 1.7 times larger than that for a dc Child–Langmuir law [Eq. (12)]. The electrode voltage (with respect to the plasma) is a clipped triangular wave. The effective sheath capacity is the ratio of the maximum ion charge in the sheath to the maximum voltage across the sheath. It is found to be $1.1/D$ Pf/cm^2.

In diode rf sputtering systems without magnetron action, the vacuum chamber forms a giant hollow cathode with electrostatic containment at the walls. This action, along with the fact that the wall area is usually large compared to the target electrode, forms an efficient electron trap at the walls and results in a voltage division which places most of the applied voltage on the target sheath. Thus wall sputtering is effectively suppressed.

The corresponding situation for cylindrical-post magnetrons has not been investigated in detail. However, the foregoing arguments suggest that each magnetic, electric, and geometrical situation must be examined with care to ensure that the action is desirable in all respects. In making judgments the following points may help.

1. Each electrode has a dc bias and tends to act like a dc magnetron sputtering cathode.

2. Ions in front of a given electrode may be produced by primary electrons from some other electrode.

3. An rf probelike behavior is superimposed on the dc picture. This can modify the efficiency of an electron trap during one rf cycle.

4. Magnetic fields offer a strong impedance to the motion of electrons across field lines.

5. Plasma oscillations may play an important role and structured discharge forms may appear [92].

VI. EROSION AND DEPOSITION PROFILES

Agreement is usually obtained between a measured profile of coating thickness and the corresponding calculated one. This gives confidence to predictions made concerning altered configurations. The operating pressure is usually so low that sputtered material is not gas scattered. Thus, a substrate placed a short distance behind a small pinhole in a plate receives an image of the cathode which pictures in detail the sputter emission rate, as shown in Fig. 15 (note: cathode did not have wing shrouds). A small object placed between the cathode and the pinhole is imaged as a sharp shadow. The law of angular emission of sputtered material may not be a cosine law, but the practical effect on the calculation of profiles is negligible. Sputtering from the cathode wings is too small to require consideration in most cases (see Fig. 15).

A. Erosion Profile

An erosion profile which is uniform along the length of the cathode is usually wanted. Figure 16 shows a representative erosion profile for a cylindrical-post magnetron. Variations in erosion rate around the barrel do not occur (so substrates do not have to be rotated around the cathode).

Fig. 15. Pinhole camera image of a chromium cylindrical-post magnetron cathode formed by sputtered chromium landing on a glass plate placed a small distance behind a mask with a small aperture. The cathode did not have wing shrouds.

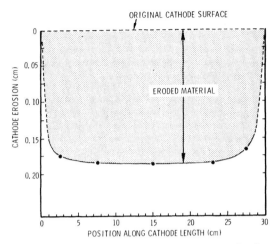

Fig. 16. Sputter erosion of material along the length of a titanium cylindrical-post magnetron cathode (3.2 cm diam. ×30 cm long). (Note the larger scale for the radial direction.)

With a magnetic material like iron, 3.2-mm thick cathode walls have been employed and over 2 mm have been eroded before wall thickness variations gave rise to deleterious changes of field shape. Thick-walled cathodes can be used with nonmagnetic materials of good thermal conductivity. Aluminum barrels with walls 2.5-cm thick have been employed successfully.

B. Deposition Profile

Consider a substrate whose normal is parallel to the midplane of a post cathode and inclined at an angle θ with respect to the radial direction. Let the substrate be a distance X from the cathode axis and a distance Z from one end of the cathode. Let the cathode have a radius R, length L, and erosion rate C_0. If the erosion rate is uniform, and the transport is collisionless, the coating rate can be represented by the following formula:

$$C = C_0(R/X)F(X,Z)\cos\theta. \tag{18}$$

The R/X term describes a coating rate which falls off inversely with distance, as expected for a cylindrical system. The function F gives the departure from this ideal behavior because of end losses. The exact equation for F is quite complicated, so an approximation is usually used instead [93]. The approximation has an error smaller than 2% if X/R is larger than

2. The equation approximating F is

$$F = G(a,b_1) + G(a,b_2), \qquad (19)$$

where $a = R/X$, $b_1 = Z/X$, and $b_2 = (L - Z)/X$. In terms of a general value b, the function G is given by

$$G(a,b) = G_0(b) + aG_1(b)[1 + aG_2(b)], \qquad (20)$$

where

$$G_0(b) = \frac{1}{\pi}\left[\tan^{-1} b + \frac{b}{1 + b^2}\right], \qquad (21)$$

$$G_1(b) = \frac{1}{2}\frac{b}{[1 + b^2]^2}, \qquad (22)$$

and

$$G_2(b) = \frac{2}{\pi}\left[\frac{8}{3(1 + b^2)} - 1\right]. \qquad (23)$$

Figure 17 shows a contour map of the function F for a cathode whose length is ten times its diameter. It displays the end effect correction as a function of radial and axial position near the cathode. An auxiliary contour (the dashed line labeled 80%) shows the contour on which the values of F fall to 80% of their values at the midplane of the cathode. This contour is of minimum axial extent at the radial position $X = \frac{1}{2}L$. Roughly speaking, the cathode acts as a line source when the radial position is less than this, and as a point source when it is more than this.

A representative comparison of a measured profile to one which was calculated is shown in Fig. 18.

A hollow cathode whose wall erosion rate is uniform has a special characteristic which makes it well suited for coating objects of complex shape. The coating rate is independent of the orientation of the substrate and is equal to the erosion rate of the walls. This holds true when two conditions are met. First, the substrate surface must be far enough from the ends so that end losses can be ignored. Second, it must have an unobstructed view of the cathode surface [94].

VII. COATING DEPOSITION

A. Substrate Environment

The environment at substrate locations surrounding a cylindrical-post magnetron is considerably different from that in a conventional planar diode sputtering source (see Chapter II-1). Of particular importance is the

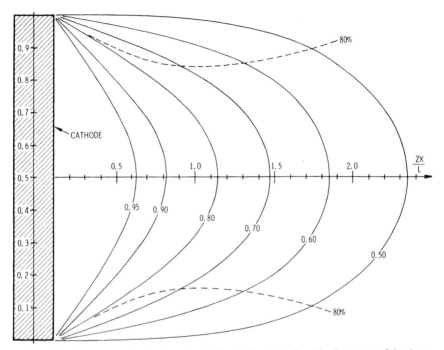

Fig. 17. Contour map for the function F [Eq. (18)] which gives the departure of the deposition rate from ideal inverse-R behavior, due to end losses; based on the assumption of uniform cathode current density, cosine emission, and collisionless transport. Solid lines are contours of constant F. Dashed lines are contours on which F values fall to 80% of their values on the cathode midplane.

fact that the substrates are not in direct contact with the intense plasma (see Fig. 19). Substrates are generally located at radii beyond the virtual anode where the charged particle density due to escaping ions is typically $\frac{1}{50}$ to $\frac{1}{100}$ that in the discharge. In the case of a cylindrical-hollow magnetron the charged particle density on the axis is typically about $\frac{1}{3}$ that adjacent to the target wall. (Similar charge particle densities are found in the region of the virtual anode of a cylindrical-post magnetron.)

A second important difference is the wide range of working gas pressures that are possible with magnetrons. Magnetrons can be effectively operated at such low pressures that the mean free path of both the working gas and the sputtered atoms is of the same order of magnitude as the source-to-substrate distance (see Fig. 15). Thus mechanical masks can be effectively used to define deposition areas. Furthermore, the sputtered atoms pass to the substrates with little loss of kinetic energy, as do energetic neutral atoms (ions that are neutralized and reflected at the target surface [95–99]). Atom reflection (Chapter II-1) is believed to be particu-

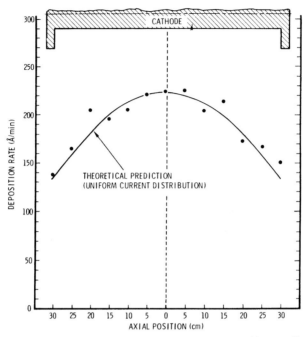

Fig. 18. Comparison between calculated and measured deposition profiles for a chromium cylindrical-post magnetron sputtering source. Argon pressure, 2.5 mTorr; current 8 A; voltage, 830 V; magnetic field, 65 G; and substrate radius, 40 cm.

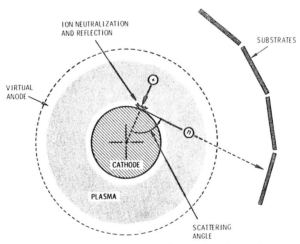

Fig. 19. Schematic representation showing substrate location and atom reflection in a cylindrical-post magnetron sputtering system.

larly important in cylindrical-post magnetrons because atoms undergoing considerably less than 180° reflections pass in the direction of substrates, as indicated in Fig. 19. For binary collisions the cross section and the reflected atom energy increase with decreasing scattering angle [98, 100–102].

The substrate heating is much reduced in magnetrons by the absence of plasma bombardment. Thus plastics, including heat-sensitive membranes, have been effectively coated with metals by using cylindrical-post magnetrons [103]. The heating flux has been found to be proportional to the sputtered flux and to vary from 15 to 25 eV/atom for metals such as Ti, Cr, and Cu to ≥ 50 eV/atom for heavy metals such as Mo, Ta, and W where ion reflection and sputtered atom kinetic energy become more significant and where sputtering yields are moderate [104]. Heating rates can be predicted in most cases. Typical contributions are heat of condensation, 5 eV/atom; kinetic energy of sputtered atom, 10 eV/atom; radiation from plasma discharge, 5 eV/atom; and reflected neutrals, 10 eV/atom. The heating rate per atom increases with decreasing intrinsic sputtering yield. Thus heating rates several times larger are found for dielectric compounds sputtered by rf- or dc-reactive methods. Heating rates per atom are typically a factor of two larger for cylindrical-hollow magnetrons because of the higher plasma density adjacent to the substrates.

Electrically isolated substrates on the axis of a cylindrical-hollow magnetron generally assume a potential of $-20 -- 50$ V relative to the anode. When metals are sputtered, the ion component of the diffusion flux [45, 54] is typically $\frac{1}{5}$ to $\frac{1}{3}$ of the sputtered atom flux. Relatively high potentials (1000–3000 V) are required to achieve current densities effective for bias sputtering because of the low plasma density. In the case of cylindrical-post magnetrons, ion bombardment fluxes are typically only about $\frac{1}{10}$ to $\frac{1}{5}$ of the deposition flux even with an applied negative substrate bias. Substrate sputter cleaning and bias sputtering can often be executed effectively in magnetron devices by designing the substrate holders to form, in concert with the substrates, an effective magnetron electrode that can operate in the existing magnetic field to provide a magnetron-type discharge and a controlled ion flux to the substrate [40].

B. Influence of Operating Parameters on Coating Properties

Figure 20 shows a schematic representation of the influence of substrate temperature and argon working pressure on the structure of metal coatings [105]. The diagram was formulated from thick coating data accumulated using both cylindrical-post and cylindrical-hollow magnetron sources [106]. Zone 1 is associated with coating flux shadowing due to

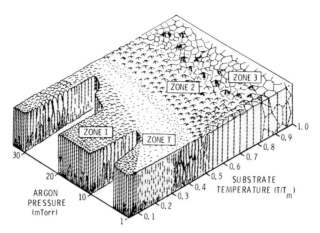

Fig. 20. Schematic representation of the influence of substrate temperature and argon working pressure on the structure of metal coatings deposited by sputtering using cylindrical magnetron sources. T is the substrate temperature and T_m is the melting point of the coating material in absolute degrees (from Thornton [106]).

roughness of the substrate and growing coating, and consists of tapered crystals which are poorly bonded [106, 107]. Zone T is a transition structure consisting of a dense array of poorly defined fibrous grains [106]. Zone 2 is characterized by evolutionary growth due to adatom diffusion and consists of columnar grains separated by distinct dense intercrystalline boundaries [105, 107, 108]. Zone 3 consists of equiaxed grains and results from bulk diffusion processes such as recrystallization [107]. The exact mechanism by which an inert gas promotes the open Zone 1 structure (see Fig. 20) is not known but is believed to involve a reduction in adatom mobility [106]. The oblique component of the sputtered flux in a cylindrical-hollow magnetron also tends to promote Zone 1 characteristics [106]. The Zone 1 structure may be suppressed by ion bombardment yielding a Zone T-like structure [109, 110].

Thin films also exhibit the general characteristics of the zone model. Coatings of a number of metals (Al, Ti, V, Cr, Ni, Cu, Zr, Nb, Mo, Ta, and W) deposited using cylindrical-post magnetrons in the thickness range 500–5000 Å have been found to have high optical reflectance, moderate electrical resistivities, and to be in a state of compression when deposited in the Zone T region [111–113]. Transition pressures have been identified above which the coatings are in tension, and the resistivity and reflectance become strongly dependent on working gas pressure, increasing and decreasing, respectively, with increasing pressure. The transition pressures, although not exactly the same for the different properties, may signal passage (at constant T/T_m) into Zone 1. Of particular interest is the observation that the transition pressure increases with atomic mass, varying from about 1 mTorr for Ti to more than 10 mTorr for Ta and W. This dependence is believed to be related to the kinetic energy of the sputtered atoms and to reflected neutrals. The general occurrence of compressive stresses in low-pressure sputtered deposits may be a fundamental characteristic of sputtered metal film growth. It was first observed in cylindrical magnetrons because they can operate below the transition pressure for even the low-mass metals. A few reports of compressive stresses in heavy elements have been made for other devices capable of operating below the transition pressure for heavy materials [114, 115].

VIII. REACTIVE SPUTTERING

Reactive sputtering (see Chapter II-1, Section VI.D) can be effectively done in both cylindrical-post and cylindrical-hollow magnetrons. Working pressures are generally sufficiently low so that gas phase reactions involving the reactive species are unimportant. However, the high circulating electron currents in the plasma (Section II.G) are believed to play an important role in the process chemistry by causing dissociation, excitation, and ionization of the reactive gas molecules. The nonionized active species pass to the substrates and onto the cathode where they have a high probability of chemisorption [116–119] and of being sputtered by the bombarding ions [120]. Ionized reactive gas species have a high probability of being neutralized and reflected as energetic atoms because their mass is generally low [121]. Such scattering is believed to be particularly important in cylindrical-post magnetrons, as shown in Fig. 19. Thus an intense flux of energetic free radicals is generated at the cathode and accompanies the sputtered metal atoms to the substrate, where they are generally capable of overcoming the activation energies associated with film growth. (The cathode-to-substrate transport of these species may be virtually collisionless at low pressures.) This flux, and the species arriving at the substrates directly from the gas phase, may be operative in forming the final compound. An example is copper oxide, where substrate holder rotation has been found to affect the coating composition.

When a reactive-gas–argon mixture is used, an increase in reactive-gas injection causes an increase in the density of reactive species present on the cathode surface and incorporated within the deposited coatings. At a critical injection rate, for otherwise constant sputtering conditions, a transition in operating conditions is generally observed which, for many metal-reactant combinations, is abrupt and involves a significant decrease in deposition rate (see Fig. 21). Such transitions have also been reported by investigators using a variety of conventional apparatus configurations

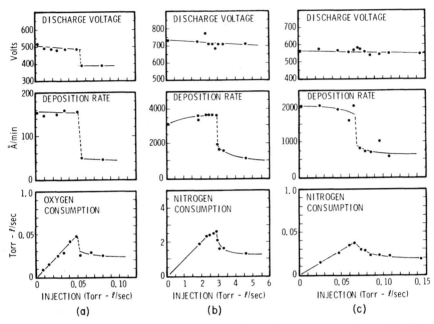

Fig. 21. Discharge voltage, deposition rate, and reactive gas consumption as a function of reactive gas injection rate for reactive sputtering in cylindrical-post and -hollow magnetron sources. The data show abrupt transitions in performance that occur with increasing injection rate at otherwise constant sputtering conditions. Consumption is defined as that injected reactive gas which does not pass into the vacuum pumps. The relatively small consumption at high currents in hollow cathode is due to the small substrate area. (a) cylindrical-post magnetron, chromium–oxygen, current, 1 A, argon pressure, 1 mTorr; (b) cylindrical-post magnetron, niobium–nitrogen, current, 10 A, argon pressure, 1.5 mTorr; (c) cylindrical-hollow magnetron, niobium–nitrogen, current, 10 A, argon pressure, 1.5 mTorr.

[122–127]. The uniformity of the cathode current density and chemistry, and the symmetry, facilitate the interpretation of data from cylindrical-post magnetrons. The transition for a large number of cases has been found to occur when the atomic flux of injected reactive gas is of the same order of magnitude as the total sputtered metal atom flux. Scaling studies for several metals [128] show that over a surprising range of conditions the transition is independent of current density (factor of 10), reactive gas partial pressure (factor of 3), and deposition surface area (factor of 5). The studies imply that the plasma formation of constituents such as radicals is rate limiting under many conditions. The decrease in deposition rate is believed to be associated largely with an increase in the cathode secondary emission coefficient (causing a lowering of voltage) and the formation of cathode surface compounds [124, 129] with low sputtering yields [130].

Similar behavior has been observed in cylindrical-hollow magnetrons (see Fig. 21). An important aspect of these devices is the sputtering of reactive constituents back and forth across the cylinder until they are finally incident on the substrate and incorporated into the coating.

A wide range of materials have been formed using cylindrical-post magnetrons. A few of these include Al_2O_3, AlN (using N_2), TiO_2, TiC (using C_2H_2), Cr_2O_3 [131], Fe_2O_3, Cu_2S (using H_2S), ZrN, NbN [132], CdS (using H_2S), In_2O_3 [133], InP (using PH_3) [134], Ta_2O_5, TaN, and WO_3. Similar materials have been formed using cylindrical-hollow magnetrons.

IX. SUMMARY

An operation defined as the magnetron mode can be achieved in diode sputtering sources configured so that magnetic fields, in concert with the cathode surface, form electron traps such that the $E \times B$ electron drift currents close on themselves. Primary electrons are injected into such traps by the electric field in the cathode dark space and can escape only by undergoing energy exchange processes of the type required to sustain the discharge. Accordingly, the ionization process is very efficient. High currents and sputtering rates can be achieved, at moderate (~ 1000 V) and near constant voltages, even at low pressure.

This chapter has primarily discussed cylindrical magnetrons that are formed with uniform axial magnetic fields and reflecting surfaces to remove end losses. These have been called cylindrical post- and cylindrical-hollow magnetrons. They can be configured in a wide range of sizes, provided certain requirements on magnetic field strength and uniformity, and reflector size, are satisfied. Post devices with lengths from 10 to 200 cm have been operated with dc. Uniform current densities of over 200 mA/cm^2 have been achieved. Both post and hollow devices have been operated with rf power.

Long cathodes can provide uniform coatings over large areas. Hollow cathodes are effective for coating complex shaped objects. Substrate plasma bombardment is much reduced, particularly with cylindrical-post magnetrons. A large inventory of coating material can be stored on a cylindrical cathode and used efficiently because of the uniform current density. For applications involving exotic materials, difficulty may be encountered in fabricating cylindrical sputtering targets. When sputtering magnetic materials, the cathode must be of sufficient size to incorporate internal magnetic field coils. Special magnetic field considerations are generally required for in-line systems where the substrates pass by one or more cathodes.

Considerable experience has been accumulated using cylindrical magnetrons in production situations. For example, a cylindrical-post magnetron 10 cm in diameter and 40 cm long has been used to deposit photomask quality chromium on glass for over seven years with virtually no maintenance. A cathode of similar size has been used to reliably produce iron oxide coatings by reactive sputtering at a current density of 14 mA/cm².

REFERENCES

1. J. D. Cobine, "Gaseous Conductors," pp. 205–248. Dover, New York, 1958.
2. A. von Engel, "Ionized Gases," pp. 217–253. Oxford Univ. Press, London and New York, 1965.
3. E. Kay, *Tech. Met. Res.* **1**, Part 3, 1269 (1968).
4. N. Laegreid and G. K. Wehner, *J. Appl. Phys.* **32**, 365 (1961).
5. L. Maissel, *in* "Handbook of Thin Film Technology" (L. J. Maissel and R. Glang, eds.), p. 4-1. McGraw-Hill, New York, 1970.
6. K. Wasa and S. Hayakawa, *Proc. IEEE* **55**, 2179 (1967).
7. K. Wasa and S. Hayakawa, *Microelectron. Reliab.* **6**, 213 (1967).
8. K. Wasa and S. Hayakawa, *IEEE Trans. Parts, Mater. Packag.* **PMP-3**, 71 (1967).
9. K. Wasa and S. Hayakawa, *Rev. Sci. Instrum.* **40**, 693 (1969).
10. R. F. Tischler, *Soc. Vac. Coaters, Proc. Annu. Tech. Conf. 14th, Cleveland, Ohio* p. 71 (1971).
11. L. Lamont, *NASA Spec. Publ.* **SP-5111**, 139 (1972).
12. S. Aoshima and T. Asamaki, *Jpn. J. Appl. Phys., Suppl.* **2**, Part 1, 253 (1974).
13. W. D. Gill and E. Kay, *Rev. Sci. Instrum.* **36**, 277 (1965).
14. A. W. Hull, *Phys. Rev.* **18**, 31 (1921).
15. R. L. Jepsen, *in* "Cross-Field Microwave Devices" (E. Okress, ed.), Vol. 1, p. 251. Academic Press, New York, 1961.
16. J. M. Osepchuk, *in* "Cross-Field Microwave Devices" (E. Okress, ed.), Vol. 1, p. 275. Academic Press, New York, 1961.
17. K. Mouthaan and C. Süsskind, *J. Appl. Phys.* **37**, 2598 (1966).
18. J. E. Drummond, *in* "Plasma Physics" (J. E. Drummond, ed.), p. 332. McGraw-Hill, New York, 1961.
19. S. Saito, N. Sato, and Y. Hatta, *Appl. Phys. Lett.* **5**, 46 (1964).
20. S. Saito, N. Sato, and Y. Hatta, *J. Phys. Soc. Jpn.* **21**, 2695 (1966).
21. S. Saito and Y. Hatta, *J. Phys. Soc. Jpn.* **26**, 175 (1969).
22. J. B. Fisk, H. D. Hagstrum, and P. L. Hartman, *Bell Syst. Tech. J.* **25**, 167 (1946).
23. G. B. Collins, *in* "Microwave Magnetrons" (G. B. Collins, ed.), p. 1. McGraw-Hill, New York, 1948.
24. R. Latham, A. H. King, and L. Rushforth, "The Magnetron." Chapman & Hall, London, 1952.
25. A. H. Beck and A. D. Brisbane, *Vacuum* **11**, 137 (1952).
26. J. P. Hobson and P. A. Redhead, *Can. J. Phys.* **36**, 271 (1958).
27. P. A. Redhead, *Can. J. Phys.* **37**, 1260 (1959).
28. P. A. Redhead, *in* "Advances in Vacuum Science Technology" (E. Thomas, ed.), p. 410. Pergamon, Oxford, 1960.

29. W. Knauer and E. R. Stack, *Trans. Natl. Vac. Symp., 10th, Boston, 1963* pp. 180–184 (1964).
30. D. Andrew, *Proc. Int. Vac. Congr., 4th, Manchester, England* p. 325 (1968).
31. M. Wutz, *Vacuum* **19**, 1 (1969).
32. F. M. Penning and J. H. A. Moubis, *Proc., K. Ned. Akad. Wet.* **43**, 41 (1940).
33. F. M. Penning, *Physica (Utrecht)* **3**, 873 (1936).
34. A. S. Penfold and J. A. Thornton, U.S. Patent 3,884,793 (1975).
35. A. S. Penfold, U.S. Patent 3,919,678 (1975).
36. A. S. Penfold and J. A. Thornton, U.S. Patent 3,995,187 (1977).
37. A. S. Penfold and J. A. Thornton, U.S. Patent 4,030,996 (1977).
38. A. S. Penfold and J. A. Thornton, U.S. Patent 4,031,424 (1977).
39. A. S. Penfold and J. A. Thornton, U.S. Patent 4,041,353 (1977).
40. J. A. Thornton, U.S. Patent Appl. No. 821, 698 (1977).
40a. J. A. Thornton, *J. Vac. Sci. Technol.* **15**, 171 (1978).
41. K. I. Kirov, N. A. Ivanov, E. D. Atansova, and G. M. Minchev, *Vacuum* **26**, 237 (1976).
42. N. Hosokawa, T. Tsukada, and T. Misumi, *J. Vac. Sci. Technol.* **14**, 143 (1977).
43. F. R. Arcidiacono, *Proc. Electron. Components Conf., 27th; IEEE, New York* p. 232 (1977).
44. U. Heisig, K. Goedicke, and S. Schiller, *Proc. Int. Symp. Electron Ion Beam Sci. Technol., 7th, Washington, D.C.* p. 129 (1976).
44a. J. A. Thornton, *J. Vac. Sci. Technol.* **15**, 188 (1978).
45. F. F. Chen, "Introduction to Plasma Physics." Plenum, New York, 1974.
46. N. A. Krall and A. W. Trivelpiece, "Principles of Plasma Physics." McGraw-Hill, New York, 1973.
47. J. L. Delcroix, "Introduction to the Theory of Ionized Gases," pp. 128–130. Wiley (Interscience), New York, 1960.
48. L. Spitzer, Jr., "Physics of Fully Ionized Gases." Wiley (Interscience), New York, 1956.
49. J. G. Linhart, "Plasma Physics." North-Holland Publ., Amsterdam, 1960.
50. H. Alfvén and C. G. Fälthammar, "Cosmical Electrodynamics." Oxford Univ. Press, London and New York, 1963.
51. Reference 45, pp. 25–26; ref. 46, p. 626.
52. Reference 45, pp. 27–31; ref. 46, pp. 622–623.
53. Reference 46, pp. 328–330.
54. S. Glasstone and R. H. Lovberg, "Controlled Thermonuclear Reactions," pp. 451–469. Van Nostrand, New York, 1960.
55. Reference 2, pp. 114, 124.
55a. E. W. McDaniel and E. A. Mason, "The Mobility of Ions in Gases," p. 275. Wiley, New York, 1973.
56. L. G. Christophorou, "Atomic and Molecular Radiation Physics," p. 283. Wiley (Interscience), New York, 1971.
57. L. G. H. Huxley and R. W. Crompton, "The Diffusion and Drift of Electrons in Gases," p. 611. Wiley, New York, 1974.
58. L. J. Kieffer and G. H. Dunn, *Rev. Mod. Phys.* **38**, 1 (1966).
58a. A. Gilardini, "Low Energy Electron Collisions in Gases," p. 336. Wiley, New York, 1972.
59. S. Hayakawa and K. Wasa, *Electr. Eng. in Jpn (Engl. transl. of Denki Gakkai Zasshi)* **83**, 36 (1963).
60. S. Hayakawa and K. Wasa, *J. Phys. Soc. Jpn.* **20**, 1692 (1965).

61. J. C. Helmer and R. L. Jepson, *Proc. IRE* **49**, 1920 (1961).
62. W. Knauer, *J. Appl. Phys.* **33**, 2093 (1962).
63. E. H. Hirsch, *Br. J. Appl. Phys.* **15**, 1535 (1964).
64. R. F. Mukhamedov, *Sov. Phys.—Tech. Phys.* **18**, 1057 (1974).
65. M. H. Mittleman, *in* "Symposium of Plasma Dynamics" (F. H. Clauser, ed.), p. 54. Addison-Wesley, Reading, Massachusetts, 1960.
66. D. Bohm, E. H. S. Burhop, and H. S. W. Massey, *in* "The Characteristics of Electrical Discharges in Magnetic Fields" (A. Guthrie and R. K. Wakerling, eds.), p. 13. McGraw-Hill, New York, 1949.
67. J. F. Denisse and J. L. Delcroix, "Plasma Waves." Wiley (Interscience), New York, 1963.
68. T. H. Stix, "The Theory of Plasma Waves." McGraw-Hill, New York, 1962.
69. F. C. Hoh, *Phys. Fluids* **6**, 1184 (1963).
70. D. M. Kerr, Jr., *Phys. Fluids* **9**, 2531 (1966).
71. E. B. Hooper, Jr., *Phys. Fluids* **13**, 96 (1970).
72. D. B. Ilic, T. D. Rognlien, S. A. Self, and F. W. Crawford, *Phys. Fluids* **16**, 1042 (1973).
73. R. H. Levy, *Phys. Fluids* **8**, 1288 (1965).
74. W. Knauer and R. L. Poeschel, *Proc. Int. Conf. Phenom. Ion. Gases, 7th, Belgrade,* **2**, 719 (1966).
75. T. K. Allen, R. A. Bailey, and K. G. Emeleus, *Br. J. Appl. Phys.* **6**, 320 (1955).
76. A. B. Cannara and F. W. Crawford, *J. Appl. Phys.* **36**, 3132 (1965).
77. A. S. Penfold and J. A. Thornton, in preparation (1978).
78. E. S. McDaniel, "Collision Phenomena in Ionized Gases," Ch. 13. Wiley, New York, 1964.
79. Reference 56, p. 35.
80. P. A. Redhead, *Can. J. Phys.* **36**, 255 (1958).
81. R. L. Jepsen, *J. Appl. Phys.* **32**, 2619 (1961).
82. G. W. Sutton and A. Sherman, "Engineering Magnetohydrodynamics," p. 394. McGraw-Hill, New York, 1965.
83. S. Hayakawa and K. Wasa, *J. Phys. Soc. Jpn.* **19**, 1990 (1964).
84. K. Wasa and S. Hayakawa, *J. Phys. Soc. Jpn.* **20**, 1219 (1965).
85. K. Wasa and S. Hayakawa, *J. Phys. Soc. Jpn.* **20**, 1732 (1965).
86. J. A. Thornton, in preparation (1978).
87. H. S. Butler and G. S. Kino, *Phys. Fluids* **6**, 1346 (1963).
88. P. D. Davidse and L. J. Maissel, *J. Appl. Phys.* **37**, 574 (1966).
89. R. T. C. Tsui, *Phys. Rev.* **168**, 107 (1968).
90. J. S. Logan, J. H. Keller, and R. G. Simmons, *J. Vac. Sci. Technol.* **14**, 92 (1977).
91. A. S. Penfold and J. A. Thornton, in preparation (1978).
92. L. T. Lamont, Jr. and J. J. Delone, Jr., *J. Vac. Sci. Technol.* **7**, 155 (1970).
93. A. S. Penfold and J. A. Thornton, *Soc. Vac. Coaters, Proc. Annu. Tech. Conf., 19th, Toronto* p. 8 (1976); *Met. Finish.* **75**, 33 (1977).
94. J. A. Thornton and V. L. Hedgcoth, *J. Vac. Sci. Technol.* **12**, 93 (1975).
95. H. F. Winters and E. Kay, *J. Appl. Phys.* **38**, 3928 (1967).
96. I. Brodie, L. T. Lamont, Jr., and R. L. Jepson, *Phys. Rev.* **21**, 1224 (1968).
97. W. W. Lee and D. Oblas, *J. Vac. Sci. Technol.* **7**, 129 (1970).
98. W. W. Y. Lee and D. Oblas, *J. Appl. Phys.* **46**, 1728 (1975).
99. G. Heim and E. Kay, *J. Appl. Phys.* **46**, 4006 (1975).
100. F. W. Bingham, Res. Rep. SC-RR-66-506. Sandia Lab., Albuquerque, New Mexico (1966), available from Clearinghouse Fed. Sci. Tech. Inf., Springfield, Virginia.

101. D. P. Smith, *J. Appl. Phys.* **38**, 340 (1967).
102. G. Carter, *J. Vac. Sci. Technol.* **7**, 31 (1970).
103. J. A. Thornton, *Soc. Vac. Coaters, Proc. Annu. Tech. Conf., 18th, Key Biscayne, Fla.* p. 8 (1975); *Met. Finish.* **74**, 46 (1976).
104. J. A. Thornton, *Thin Solid Films* to be published (1978).
105. J. A. Thornton, *Annu. Rev. Mater. Sci.* **7**, 239 (1977).
106. J. A. Thornton, *J. Vac. Sci. Technol.* **11**, 666 (1974).
107. B. A. Movchan and A. V. Demchishin, *Phys. Met. Metall. (USSR)* **28**, 83 (1969).
108. J. A. Thornton, *J. Vac. Sci. Technol.* **12**, 830 (1975).
109. D. M. Mattox and G. J. Kominiak, *J. Vac. Sci. Technol.* **9**, 528 (1972).
110. J. A. Thornton, *Thin Solid Films* **40**, 335 (1977).
111. D. W. Hoffman and J. A. Thornton, *Thin Solid Films* **40**, 355 (1977).
112. J. A. Thornton and D. W. Hoffman, *J. Vac. Sci. Technol.* **14**, 164 (1977).
113. D. W. Hoffman and J. A. Thornton, *Thin Solid Films* **45**, 387 (1977).
114. P. R. Stuart, *Vacuum* **19**, 507 (1969).
115. R. G. Sun, T. C. Tisone, and P. D. Cruzen, *J. Appl. Phys.* **46**, 112 (1975).
116. W. R. Ott and W. L. Fite, *J. Appl. Phys.* **40**, 3402 (1969).
117. M. Nagasaka and T. Yamashina, *J. Vac. Sci. Technol.* **8**, 605 (1971).
118. R. O. Adams and L. E. Musgrave, *J. Vac. Sci. Technol.* **9**, 539 (1972).
119. M. Nagasaka and T. Yamashina, *J. Vac. Sci. Technol.* **9**, 543 (1972).
120. H. F. Winters, *J. Vac. Sci. Technol.* **8**, 17 (1971).
121. H. F. Winters and D. E. Horne, *Surf. Sci.* **24**, 587 (1971).
122. E. Hollands and D. S. Campbell, *J. Mater. Sci.* **3**, 544 (1968).
123. J. Harvey and J. Corkhill, *Thin Solid Films* **6**, 277 (1970).
124. A. J. Stirling and W. D. Westwood, *Thin Solid Films* **7**, 1 (1971).
125. J. Heller, *Thin Solid Films* **17**, 163 (1973).
126. T. Abe and T. Yamashina, *Thin Solid Films* **30**, 19 (1975).
127. F. Shinoki and A. Itoh, *J. Appl. Phys.* **46**, 3381 (1975).
128. J. A. Thornton, *Soc. Vac. Coaters, Proc. Annu. Tech. Conf., 20th, Atlanta, Ga.* p. 5 (1977).
129. H. F. Winters, *J. Appl. Phys.* **43**, 4809 (1972).
130. O. Almen and G. Bruce, *Trans. Natl. Vac. Symp., 8th* p. 245 (1962).
131. J. A. Thornton, *Proc. Am. Electroplat. Soc. Coat. Sol. Collect. Symp., Atlanta, Ga.* p. 63 (1976).
132. J. A. Thornton and F. M. Kilbane, in preparation (1978).
133. J. A. Thornton and V. L. Hedgcoth, *J. Vac. Sci. Technol.* **13**, 117 (1976).
134. J. A. Thornton and A. D. Jonath, *Conf. Rec., IEEE Photovoltaic Spec. Conf., 12th, Baton Rouge, La.* p. 549 (1976).

II-3

The Sputter and S-Gun Magnetrons*

DAVID B. FRASER

Bell Telephone Laboratories
Murray Hill, New Jersey

I. INTRODUCTION

The Sputter Gun and S-Gun are circular magnetrons invented by Clarke [1]. As in other magnetrons, an intense plasma region is formed near the cathode due to the $E \times B$ field configuration (Fig. 3, Chapter II-2). In common with other magnetrons these sources are high current–low voltage devices operating at relatively low pressures while conventional diode sources tend to be high voltage–low current sources operating at higher pressures. To maximize film deposition rates in conventional sputtering diodes, the cathode is placed close to the substrates which are on the anode, and a maximum potential is applied to the cathode. However, bombardment of the substrate will occur due to the flux of electrons, and thermal degradation of the substrates can result. Magnetrons have high cathode erosion rates and yield useful film deposition rates with substrates supported in planetary or other movable support systems. The cir-

* Sputter Gun® and S-Gun® are registered trademarks of Sloan Technology and Varian Associates, respectively.

cular magnetrons have proven to be very useful for a variety of film deposition requirements.

II. DESCRIPTION

Both of these magnetrons utilize circular cathodes and concentric central anodes. Cross-sectional views are shown in Figs. 1 and 2. The cathodes are clamped without bonding into water cooled housings and in each, contact to the cooled metal surface is achieved through thermal expansion of the cathode during sputtering operation. The anodes are also water cooled and insulated, permitting them to be electrically biased. The cathodes are shown schematically in Figs. 3 and 4 with the **B** field and the plasma ring. Permanent magnets are used to obtain the magnetic field and values of 0.015 T are typical of the flux density near the cathode. In the S-Gun power densities of about 50 W/cm² are utilized in the erosion zone of the cathode when depositing Al or its alloys with Ar discharges powered at 700 V and 10 A. The dc power supplies are high, constant current devices capable of shutting off and restarting when arcs occur. In sputtering, arcs may occur during start up due to an oxide film or particles on the target surface and also, during operation when a flake of detached film material enters the plasma zone. Conventional dc high current supplies are generally inadequate as a power source for the magnetrons since they lack the necessary arc-suppression characteristics. The effect of power supply output characteristics has to be carefully considered during design in order to achieve efficient operation of the magnetron sputter source. This will be pointed out when specific examples of source operation are discussed.

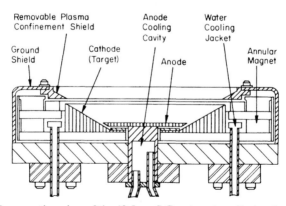

Fig. 1. Cross-section view of the 12.5 cm S-Gun (courtesy Varian Associates).

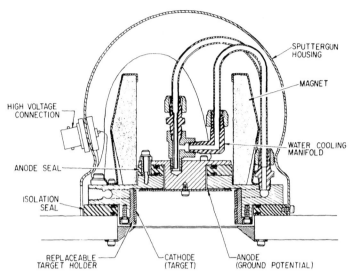

Fig. 2. Cross-section view of the 7.5 cm Sputter-Gun (courtesy of Sloan Technology).

Operation of the circular magnetrons may be accomplished with rf generators as energy sources. The S-Gun may be obtained in an rf model (also capable of being operated in the dc mode) which has shield and insulator differences compared to the dc model. Operation with rf substrate bias is also possible, permitting sputter etching of the substrates *in situ* prior to film deposition.

III. OPERATIONAL CHARACTERISTICS

The current–voltage relationship for a magnetron is dictated by the electron trap [2] formed by the configurations of the electrodes and the magnetic field. In Fig. 5 a typical set of three I–V curves is shown for an

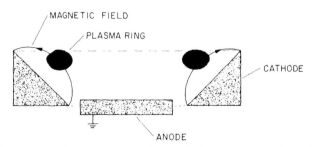

Fig. 3. Cross-section view of the 12.5 cm S-Gun electrodes with **B** field and plasma ring.

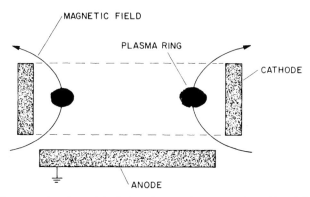

Fig. 4. Cross-section view of the 7.5 cm Sputter-Gun electrodes with **B** field and plasma ring.

Al target with anode grounded and three different Ar pressures. As the pressure is decreased, the nearconstant voltage characteristic of the high pressure $I-V$ curve is replaced at low pressure by one with an increasing voltage for increasing current. At low pressure, the discharge current is limited by the ionization determined by the magnetic field and electrode configuration. Note that since the target erodes with use, the magnitude of

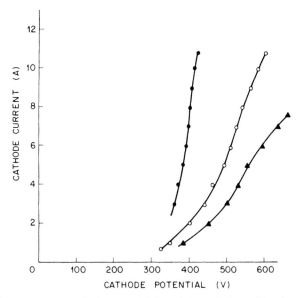

Fig. 5. Current versus cathode potential for 12.5 cm S-Gun with Al cathode, anode grounded and three Ar pressures. ▲, 2.5 mTorr; ○, 4 mTorr; ●, 10 mTorr.

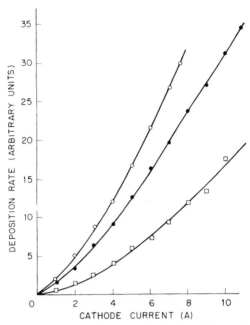

Fig. 6. Deposition rate versus cathode current for 12.5 cm S-Gun with Al cathode, anode grounded, and three Ar pressures. O, 2.5 mTorr; ●, 4 mTorr; □, 10 mTorr.

the magnetic field increases near the target surface. Hence, the $I-V$ characteristics will change with target use and the target is capable of operating at successively lower pressure. It is for this reason that the design of the S-Gun target was changed from its initial uniform conical surface to one that has a step. This step "breaks in" the target and permits immediate operation of the source at normal pressure and voltages. At pressures of 2 mTorr and above, the S-Gun operates with $I \propto V^n$ with n in the range of 6–7.

In Fig. 6 a set of curves relating Al deposition rate to S-Gun cathode current is shown for three Ar pressures. The anode was grounded in each case. At 2.5 mTorr, the Ar appears to be twice as efficient as at 10 mTorr in producing film deposition; e.g., it takes a current of 8 A to produce at 10 mTorr the same deposition rate obtained at 2.5 mTorr with a current of 4 A. Thus, it is of interest to consider the operation of the S-Gun over a range of Ar pressure in two modes: (a) constant current and (b) constant power. In Fig. 7 the deposition rates in a 48 cm planetary system of Al–2% Si are shown for the two modes over a range of Ar pressures. The source was operated with anode biased at +40 V relative to ground. Note that for small pressure changes, the deposition rate change is less in a con-

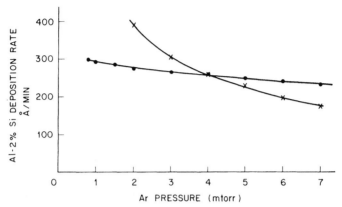

Fig. 7. Deposition rate versus Ar pressure for 12.5 cm S-Gun with Al cathode, anode at +40 V relative to ground and two modes of operation. ×, 5 A constant current; ●, constant power.

stant power mode of operation. As the pressure is decreased, the voltage required to sustain a given current increases and the sputtering efficiency increases [3] yielding a higher deposition rate. Reduced gas scatter also occurs at lower Ar pressures and increases the deposition rate in both modes of operation.

The sputter deposition efficiency is determined by the impedance of the magnetron and the characteristics of the power supply. As an example, an Al–0.5% Cu S-Gun target was used to deposit films in a 48 cm planetary system at different Ar pressures with the power supply set at maximum power output and a +40 V bias applied to the anode. Three variables are shown in Fig. 8 as a function of Ar pressure, deposition rate, input power to the cathode, and deposition efficiency. The latter variable is simply the deposition rate divided by the power input to the cathode. Maximum values of deposition rate and power input occurred at 3 mTorr pressure. However, efficiency increased as pressure decreased due to increased cathode potential [3] and reduced gas scatter. In summary, the current output at high pressure and voltage output at low pressure limit the deposition rate achieved with targets possessing excellent heat transfer.

The use of poor thermal conductors as targets may lead to excessive heating and target degradation. Materials with low melting points such as In or In–Sn alloys will melt and fragile materials such as polycrystalline Si and hot-pressed ceramics, crack under the severe thermal stressing of high power input. An unusual effect has been noted in a 99.999% Al S-Gun target with 2 cm and larger grains. Extended operation of this target

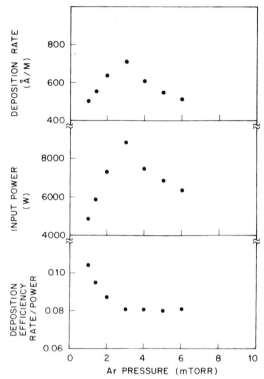

Fig. 8. Deposition rate, input power and deposition efficiency versus Ar pressure for a 12.5 cm S-Gun with Al–0.5% Cu cathode, power supply set at maximum output, and anode at +40 V relative to ground.

at 7 kW (40–50 W/cm²) with adequate cooling water flow led to uniform target shrinkage such that after 60% target use, the diameter had shrunk by 2%. Dilute alloys of Si and Cu in Al have also exhibited shrinkage [4]. Contact with the water cooled jacket may eventually be lost and severe target heating may occur. It is believed that a mass transfer of metal into the bombardment zone occurs, leading to uniform shrinkage of the target. Use of small grained targets appears to reduce the effect.

IV. BIAS OPERATION

Bias may be applied either electrically or magnetically in both types of circular magnetron. Magnetic bias has been used with the 7.5 cm source either to increase or decrease the substrate bombardment during film deposition [5]. Both magnetrons have been biased electrically by use of a pos-

itive potential (relative to ground) applied to the anode. The plasma potential in these sources is a few volts negative with respect to the anode;
thus, the use of a high positive potential on the anode causes the magnetron to act as a source of positive ions as well as neutrals.

The use of an auxiliary bias magnet beneath the stationary substrate is
shown in Fig. 9 for three cases: an opposing magnet, no magnet, and a
magnet in the field aiding configuration. The effect of the bias may be seen
in the plot of temperature at the substrates versus time (Fig. 10). From the
initial temperature versus time curve slopes, the flux of energy to the substrates may be estimated. These values plus electrical probe data are presented in Table I with deposited Cu film resistivity. All films were deposited in 15 min and were 5000 Å thick. The current at -40 V bias of the
electrical probe is an indication of the positive ion flux to the substrate.
The highest current corresponds to the highest film resistivity (no bias
magnet) and may indicate more entrapped Ar in the Cu film. The fact that

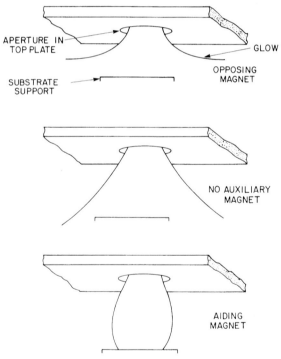

Fig. 9. Influence of a circular auxiliary magnet beneath the substrate support upon the
glow extending from a 7.5 cm Sputter Gun. Three cases are shown: (a) opposing magnet, (b)
no magnet, and (c) aiding magnet (Fraser and Cook [5]).

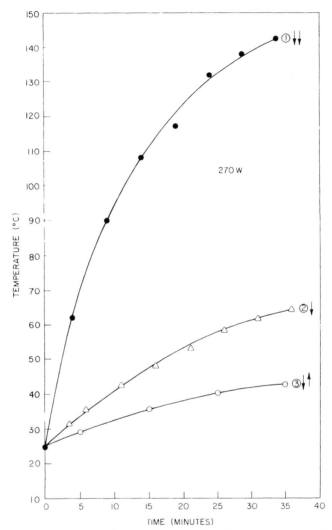

Fig. 10. Temperature versus time as a function of auxiliary magnet orientation for a 7.5 cm Sputter Gun. A power input of 270 W was used in each case (Fraser and Cook [5]).

the Cu film deposition rate was the same in each case indicates that film growth proceeds primarily from the neutral particle flux.

Deposition of indium–tin oxide (ITO) was also performed with and without auxiliary magnets at a power input of 175 W. The results are summarized in Table II. Briefly, the increased energy flux to the substrate with an aiding magnet yields the lowest resistivity while the opposing

Table I

Operating Characteristics and Film Properties for Cu Deposited with Various Magnetic Bias Configurations[a]

Magnets	T_f (°C)	Energy flux (W/cm²)	Probe current (mA/cm²)	Wall potential (V)	Deposition rate (Å/M)	Resistivity (μΩ cm)	Current (at −40 V) (mA/cm²)
↓ ↓	142	1.4×10^{-1}	1.38	−10.3	330	3.5	0.061
↓	64	2.6×10^{-2}	0.75	−7.1	330	38	0.084
↓ ↑	42	5.1×10^{-4}	0.05	−3.6	330	13	0.028

[a] See Fraser and Cook [5].

magnet yields films of only about 3.5 times higher resistivity. The opposing magnet mode has been used to deposit ITO on 6 μm stretched plastic materials, on emulsion photo masks, and on Cr and iron oxide coated photomasks without thermally degrading the substrates. All of the preceeding experiments with the auxiliary bias magnet were performed with the magnetron anode grounded. If the anode is biased positively with respect to ground, the wall potential will rise to a value somewhat lower than the anode potential. For example, when an indium–tin oxide target in 5 mTorr of Ar was biased at − 250 V and the anode at + 150 V (anode current 0.2 A), the wall potential was + 127 V and the grounded probe current was + 0.08 mA/cm² at a distance of 19 cm beneath the magnetron aperture. Auxiliary external electrodes may also be used to help confine the discharge. A ring electrode, 18 cm in diameter, centered on the 7.5 cm magnetron axis and just below the magnetron has been used as a plasma potential-determining anode. The experimental layout is shown schematically in Fig. 11. The current to a 1.5 cm² Pt probe, 14 cm below the bottom edge of the ring electrode (19 cm below the magnetron), is shown in Fig.

Table II

Deposition Rate and Film Resistivity for ITO Films Deposited under Various Magnetic Bias Conditions[a]

Magnets	Deposition rate (Å/M)	Sheet resistance (Ω/□)	Resistivity (μΩ cm)
↓ ↓	573	3.75	215
↓	459	11	505
↓ ↑	495	14	693

[a] See Fraser and Cook [5].

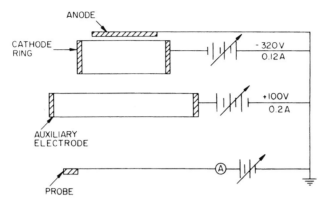

Fig. 11. Schematic view of 7.5 cm Sputter Gun used with an auxiliary external ring electrode.

12 for two positions in the chamber as a function of probe voltage. By adjusting the relative removal and deposition rates, film deposition or etching was obtained at the substrate.

In the 12.5 cm magnetron, the use of positive bias of + 100 V or more on the anode has led to surface roughening of deposited Al films. Also, the

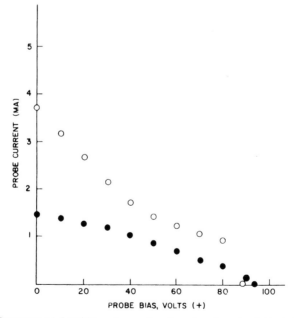

Fig. 12. Current to a 1.5 cm² Pt probe as a function of probe potential (relative to ground) for two positions of the probe. ○, centered 19 cm under Sputter Gun; ●, 5 cm off center.

planets used to support the substrates and the stainless steel bell jar became hot (>60°C) as a result of the energetic bombardment. Thus, bias
sputter deposition may be accomplished with both magnetrons and may
find application in film step coverage problems [6] and in reactive sputter
deposition.

Recently, the application of rf bias to sputter etch planetary-mounted
substrates has become available; this feature may find application in bias
deposition processes as well as sputter etching.

V. FILM DEPOSITION

In common with other magnetrons, the S-Gun and the Sputter Gun
have high target erosion rates and can yield high deposition rates. However, the sources yield a lobe of vapor of sputtered material that requires
substrate motion during deposition. As an example, the distribution from
an S-Gun obtained by Turner [7] is shown in Fig. 13 with the distribution

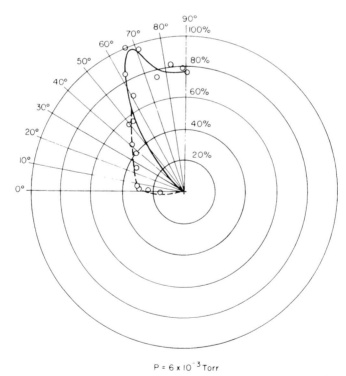

$P = 6 \times 10^{-3}$ Torr

Fig. 13. Angular distribution of sputtered material from a 12.5 cm S-Gun, the solid line is
the calculated flux from a ring source and the broken line is the measured flux (Turner [7]).

calculated from an annular source. For low angles, the flux is over cosine and in the side lobes there is a large flux of material in excess of that calculated for the ring source. This characteristic permits the use of domed rotating planetary fixtures with the S-Gun source to achieve deposited film thickness uniformity of $\pm 6\%$ over all substrates. However, the substrate angle relative to the planet shell may have to be changed from that normally used in film deposition with evaporation sources [7].

Flat rotating planetaries are often used with the Sputter Gun to achieve similar film thickness uniformity. Although, power and time parameters can be used to predict deposited film thickness, quartz crystal thickness monitors may be used conveniently with these magnetrons.

Because of the extended source geometry and gas scattering, the circular magnetrons do yield better step coverage in integrated circuit metallization than in single evaporation sources. A comparison of step coverage of electron beam evaporated Al and of S-Gun sputtered Al over similar oxide steps is shown in Fig. 14. The crack evident in the evaporated film is not present in the sputtered film. In the case of the sputter deposited film, the metal thins to 50% of the flat surface thickness at the step. Both films were deposited without any heating before or during deposition. Extreme step geometries such as those involving reentrant angles may require the use of bias sputter deposition [6] to ensure adequate film coverage of the step.

The ionizing radiation present in the plasma of the magnetrons such as long wavelength UV or x-ray photons can cause radiation damage when

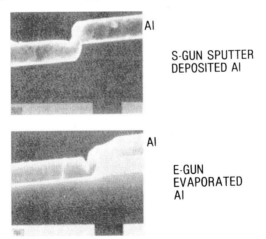

Al

S-GUN SPUTTER
DEPOSITED Al

Al

E-GUN
EVAPORATED
Al

COMPARISON OF STEP COVERAGE

Fig. 14. A comparison of step coverage from Al–0.5% Cu films deposited by electron beam evaporation and by S-Gun sputter deposition onto SiO_2 steps.

MOS integrated circuit silicon wafers are metallized. Test wafers Al metallized by filament evaporation, electron beam evaporation, and S-Gun
magnetron sputter deposition yielded similar $C-V$ characteristics of temperature and voltage stressed MOS capacitors following high temperature
annealing (475°, 30 min) [8]. Similar tests at these laboratories have cor-
roborated the latter results.

In the metallization of integrated circuits, the use of Al–Si alloys to
prevent shallow junction penetration [9, 10] or Al alloys containing Cu,
Mg, or Cr to retard electromigration and suppress hillock growth is well
known [11]. Some of the difficulty in preserving alloy composition in the
deposited film is avoided by using magnetron sputter deposition from a
single target rather than simultaneous electron beam evaporation from
multiple sources. The sputtered films essentially reproduce the composition of source target [8].

Magnetic materials may be deposited by using targets which have
been thinned by machining. The thinning operation permits sufficient
magnetic field strength to be maintained near the target surface such that
magnetron operation is achieved [12].

Reactive sputter deposition of TiN using a Ti target and Ar–N_2 sputtering gas and of ITO using an In–Sn alloy target and Ar–O_2 gas have
been performed using dc magnetron operation [13].

VI. CONCLUSIONS

The S-Gun and Sputter Gun, circular magnetrons provide flexible
sputter deposition capability for a wide variety of film materials. As for
other magnetrons these sources also permit practical film deposition on
thermally sensitive substrates.

ACKNOWLEDGMENTS

The author wishes to thank R. E. Kerwin, H. J. Levinstein, A. K. Sinha, and R. S.
Wagner for their discussions and encouragement and J. L. Fink for his assistance.

REFERENCES

1. P. J. Clarke, U.S. Patent 3,616,450 (1971).
2. J. A. Thornton and A. S. Penfold, Ch. II-2.
3. N. Laegried and G. K. Wehner, *J. Appl. Phys.* **32**, 365 (1961).
4. T. N. Fogarty, D. B. Fraser, and W. J. Valentine, *J. Vac. Sci. Technol.* **15**, 178 (1978).
5. D. B. Fraser and H. D. Cook, *J. Vac. Sci. Technol.* **14**, 147 (1977).
6. J. L. Vossen, *J. Vac. Sci. Technol.* **8**, S12 (1971).

7. F. T. Turner, *Varian Seminar Proc.: Thin Film Coat., Ion Implant. Auger Surf. Anal., Palo Alto, Calif. 1975.*

8. R. W. Wilson and L. E. Terry, *J. Vac. Sci. Technol.* **13,** 157 (1976).

9. J. R. Black, *J. Electrochem. Soc.* **115,** 242C (1968).

10. P. A. Totta and R. P. Sopher, *IBM J. Res. Dev.* **13,** 226 (1969).

11. A. Gangulee and F. M. d'Heurle, *Appl. Phys. Lett.* **19,** 76 (1971).

12. V. Hoffman, Varian Associates, personal communication (1976).

13. P. J. Clarke, *J. Vac. Sci. Technol.* **14,** 141 (1977).

II-4

Planar Magnetron Sputtering

ROBERT K. WAITS

Data Systems Division
Hewlett-Packard Company
Cupertino, California

I. INTRODUCTION

The planar magnetron (PM) has emerged as an elegant embodiment of the long-sought [1, 2] high-rate sputtering source. In essence, it is the classic dc or rf sputtering arrangement consisting of a planar cathode and its

131

surrounding dark-space shield (Chapter II-1) with the essential addition of permanent magnets directly behind the cathode. The magnets are arrayed so that there is at least one region in front of the cathode surface where the locus of magnetic field lines parallel to the cathode surface is a closed path. Although there are many variations in geometry, all have in common a closed path or region in front of a substantially flat cathode surface where the magnetic field is normal to the electric field. Bounding this region, the magnetic field lines enter the cathode surface. Ideally, at the point of entry the field lines are normal to the cathode face. The magnetic field can be supplied by electromagnets, but at the loss of simplicity.

The discharge plasma (ionization region) is constrained to an area adjacent to the cathode surface by one or more endless toroidal electron-trapping regions bounded by a tunnel-shaped magnetic field. The same principle can be applied to tubular and other nonplanar surfaces, but this discussion will be limited to planar structures. The basic principle of all magnetically enhanced sputtering techniques was discovered by Penning [3] and further developed by Kay and others [4–8], ultimately spawning the Sputter Gun (Chapter II-3) and cylindrical magnetron (Chapter II-2) sources. Penning's work had led earlier to the invention of the getter-ion pump [9], the development of which also indirectly contributed to the understanding [10, 11, 11a] and evolution of magnetically enhanced sputtering sources [12].

The planar magnetron structure is an example of an embarassingly "obvious" solution to a technological problem that eluded discovery and implementation for more than 30 years. One of the first descriptions of a planar magnetron device may have been that of Kesaev and Pashkova [13] who, in mercury arc lamp studies, used an electromagnet to stabilize a plasma on the surface of a mercury pool (Fig. 1). Striking photographs of circular and square (!) plasma regions were published. The use of permanent magnets inside a tubular cathode for an ion pump application was reported by Knauer and Stack [11]. Once again, photographs clearly showed multiple toroidal plasma rings surrounding the cylindrical cathode. This work led to a patent [11a] for ion pump structures that included two disk planar magnetron configurations. These devices were either overlooked or ignored by those intent on developing high-rate sputtering.* Some 15 years later the principle was reintroduced in the form of the planar magnetron by Chapin [14]. This work was not overlooked. Within one year PM sputtering was included in a review of deposition technology

---

* In 1969 research at the CVC division of Bendix Company led to a planar magnetron sputtering source in the form of a long rectangle. The plasma was thermionically supported and the magnetic field, supplied by permanent magnets, did not form a closed "tunnel" [13a].

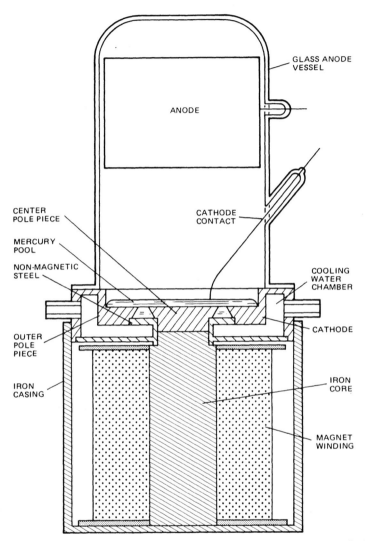

Fig. 1. Cross section of device for electromagnetic (planar magnetron) anchoring of cathode spot in mercury-discharge lamp (after Kesaev and Pashkova [13]).

[15], further developed [16, 17], and independently reported [18, 19]. By early 1975, versions of the PM sputtering source were available commercially from several manufacturers [17, 20]. By 1976, if not earlier, both rf and dc planar magnetron sputtering were being used in daily production. Planar magnetron sputtering has recently been reviewed [21], and was briefly mentioned in a recent review of glow discharge sputtering [22].

For a general discussion of the theory of operation of magnetron sputtering sources, see Chapter II-2, Section II. The applications of PM sputtering are wide ranging and are discussed in Section IV.

Schiller *et al.* [23] have used the term "ring gap plasmatron" to describe PM sputtering sources. However, the words "plasmatron" and "duoplasmatron" have been used previously to describe certain gas discharge tubes and particular classes of ion-beam sources [23a]. "Magnetron" has generally evolved as an inclusive term describing devices operating with crossed electric and magnetic fields (including microwave electron tubes, some ionization vacuum gauges, and sputtering sources).

II. DC PLANAR MAGNETRON SPUTTERING

A. System Configuration

A typical sputtering system would be similar to those discussed in Chapter II-1, Section IV. A planar cathode (sputtering source or target) is parallel to an anode surface, usually grounded, that serves as a substrate holder or carrier. One simple geometry is a disk-shaped cathode having a toroidal plasma ring facing and parallel to a fixed planar substrate holder [14, 23]. An alternative arrangement is a rectangular (or oval) cathode in conjunction with a means whereby the substrates are moved continuously in a direction perpendicular to the cathode width during deposition [16, 17, 24]. The plasma region is in the shape of an elongated ring which, in the case of a long cathode, behaves as two parallel line sources. The two configurations are shown in Fig. 2.

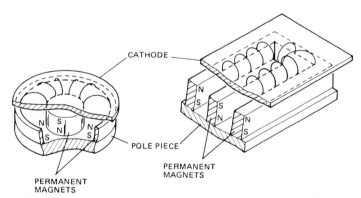

Fig. 2. Circular and rectangular planar magnetron sputtering sources. Curved lines represent magnetic field lines.

1. Cathode and Magnet Geometry

Magnet arrangements can be varied substantially, the only constraints being those of geometry and the requirement that there be at least one closed path where the magnetic field lines are parallel to the cathode surface. For example, the paths may be single or multiple circles or ovals, concentric circles or ovals, or cloverleaf or multilobed [11] patterns. Permanent magnets, electromagnets [14], or a combination of both [25] have been used. Although electromagnets add complexity to the system (e.g., power supply and control, insulation, cooling), their use may be justified if it is desired to alternate easily between conventional and magnetron sputtering.

The maximum transverse component of magnetic field in front of the target is typically in the range 200–500 G (0.02–0.05 T), although the threshold flux density for the magnetron discharge may be as low as 80–90 G [26].

Various permanent magnet materials and pole piece arrangements have been used. Suitable magnet materials are barium (or barium–strontium) ferrites, alnico alloys, and cobalt rare-earth alloys [27]. All of these materials have a high coercive force and residual induction and do not readily demagnetize by being "open circuited," which is the normal situation for a magnetron cathode. Ferrites do not corrode, which allows them to be placed within a water-cooled cathode assembly. Cobalt rare-earth magnets, although expensive, may be of use in sputtering ferromagnetic materials where the target must be magnetically saturated in order to achieve a useful field beyond the cathode face. The choice of magnet material determines the pole piece geometry. Ideally the magnetic field should enter and leave the target normal to the target face to avoid sputtering from the target edges and maximize the transverse field component. It is preferable to have the magnet material adjacent to the target surface rather than to the pole pieces, since there will be greater limb leakage from the pole piece material (soft iron or cold-drawn steel) than from the magnet material [28]. Since ferrite magnets are magnetized along their thickness (generally 2–3 cm), they can be placed behind the cathode and backed by a flat pole piece, as in Fig. 2. Alnico magnets are magnetized along their length and have been used in a radial arrangement behind a round cathode having a center cylinder and an outer ring as pole pieces, as in Fig. 3a [29]. Long alnico magnets can also be used adjacent to the cathode, but more space is required behind the cathode (see Fig. 3b). An oval pattern of U-shaped magnets has also been reported [25]. A slight amount of sputtering occurs along the corners and sides of targets due to fringing magnetic field lines from the limbs of the outer magnets or pole

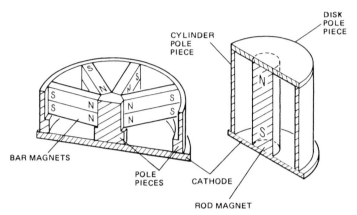

Fig. 3. Disk cathode bar magnet arrangements; (a) radial magnets, (b) central rod magnet.

pieces. This can be minimized—at the cost of less efficient use of target material—by moving the pole faces farther in from the edge of the target. A convex target surface [19] or modified pole faces may minimize the fringe effect. Computer-aided magnetic circuit design should allow the optimization of magnet–pole piece geometry for a given cathode design and magnetic material.

The advantages of permanent magnets over electromagnets can be summarized as no dc power required, no additional heat to dissipate, no field interruption, no danger of magnet insulation failure, and less weight and volume for equivalent magnetic field. The disadvantages are the field cannot readily be varied or turned off and the target will continually attract magnetic "dirt" [28].

2. Cathode Shielding and Insulation

Two basic designs have been used: (1) The cathode and shielding are entirely within the deposition vacuum chamber with power and coolant brought in via vacuum feedthroughs [16, 24]. (2) The cathode and surrounding electrical insulator form part of the vacuum chamber wall so that high-voltage connections and coolant seals are outside the vacuum chamber [14].

The advantages of the first design are that the cathode can be readily retrofitted to an existing vacuum system as, for example, a replacement for an electron-beam deposition source, and it can be placed to deposit in any direction. The disadvantages are that it requires vacuum feedthroughs for coolant and high voltage and may, depending on the de-

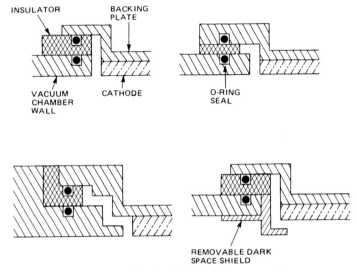

Fig. 4. Examples of cathode isolation geometries.

sign, have vacuum-to-coolant O-ring seals. Slight leakage through such seals may not be detected, and total seal failure can be catastrophic.

The second design has the advantage of not requiring vacuum-to-coolant seals if a cathode backing plate is used. A disadvantage is that some portion of the deposition chamber wall must be designed to receive the cathode assembly.

Cathode isolation details can vary from a simple stacked arrangement of the chamber wall-insulator-cathode to a reentrant or stair-step profile. Dark-space or ground shielding may be flush with the cathode surface or it may extend beyond or in front of the cathode surface. It has been reported that arcing along the cathode surface can be suppressed by a ground shield extending slightly over the outside edge of the cathode [14, 24]. A removable dark-space shield is advantageous when frequent clean up of sputtered deposits is necessary. An alternative would be to make that portion of the vacuum chamber wall surrounding the cathode easily removable. Various insulator–shield arrangements are illustrated in Fig. 4. The concept of the dark-space shield loses some meaning in the context of magnetron sputtering, since outside the magnetic field region at 5–10 mTorr argon pressure the dark-space thickness is several centimeters [30].

The insulating material may be polytetrafluoroethylene (Teflon®), acetal resin (Delrin®),* or a similar insulator having suitable mechanical

---

* Teflon and Delrin are registered trademarks of E. I. duPont de Nemours and Co., Wilmington, Delaware.

and vacuum characteristics, i.e., low permeability, low water absorption, and low outgassing rate [31]. Glass or ceramic may be used in some applications (e.g., small cathodes), but a more flexible material is preferable to brittle ceramic because the mating surfaces may not be absolutely flat or coplanar and a slight amount of "give" may be necessary to effect a vacuum-tight seal. Under such conditions a ceramic may crack. The plastic material used for cathode insulation fabrication should be free of seams that may be gas permeable.

3. Cathode Construction

The cathode assembly consists of the source material, generally 3–10 mm thick, bonded to a backing plate which in turn forms one wall of a coolant plenum. The magnet assembly may be within or external to the coolant channel. Source material and target fabrication techniques are similar to those used for conventional sputtering targets (Chapter II-1, Section V.A). Ferromagnetic targets such as Ni or Co must be thin enough to be saturated by the magnetic field; for ferrite magnets this is 2–3 mm. Targets should be free of voids or bubbles because at the high power densities and erosion rates typical of magnetron sputtering, local melting and cathode "spitting" can occur as these voids are uncovered. Targets should be stress free.

The backing plate material must not be ferromagnetic; it should have good thermal conductivity through its thickness; and the backing plate and coolant plenum must be designed to withstand atmospheric pressure plus coolant pressure without deformation. The total pressure may exceed 3 atm. Both copper and nonmagnetic stainless steel have been used, however, stainless steel is a very poor thermal conductor.

Cathode cooling in PM sputtering is critical because of the high power dissipation at the cathode. In conventional sputtering an estimated 55–70% [30, 32] of the power delivered to the cathode is dissipated in the coolant. For PM operation, cooling limitations will generally determine the maximum operating power of the cathode (Section D.2). In high-power applications, cathode bonding is essential. Disk targets bonded to a stud-mounted backing plate, as sometimes used for conventional rf sputtering targets less than about 30 cm in diameter, may be marginal at high power densities. In lower-power usage with targets that can operate at higher temperatures, it may be feasible to bolt or clamp the target to the backing plate, preferably with a malleable foil intermediate layer [33].

An innovative nonbonded aluminum cathode design has recently been developed [33a]. Four aluminum bars (the source material) approximately 2.5-cm-wide and 2.5-cm-thick, having mitered corners, are arranged as a rectangular frame and held in place by channeled magnetically soft pole pieces within and surrounding the frame. The pole pieces are clamped to a water-cooled copper backing plate by magnets behind the backing plate. The faces of the pole pieces are approximately coplanar with the aluminum target surface.

Target bonding details are rarely given. A relatively simple technique [34] applicable to Cu and many other materials is to use a low melting-point solder such as 50:50 In–Sn (m.p. 117°C, Table I). Ga can be used at room temperature to wet the (precleaned) surfaces to be joined. This allows the bonding operation to be carried out with minimum oxidation of the surfaces to be bonded. The absolute minimum amount must be used because Ga forms a brittle alloy with In–Sn. Ga cannot be used with materials such as Al in which rapid Ga diffusion occurs at room temperature. Solder bonding [38] is preferable to the use of metal-filled organic adhe-

Table I

Low Melting-Point Solders[a]

Composition (wt %)				Temperature (°C)	
Bi[b]	In	Pb[b]	Sn	Liquid	Solid
49	21	18	12	58	58
56	—	22	22	104	95
—	52	—	48	117	117
—	50	—	50	127	117
—	25	37.5	37.5	138	—
58	—	—	42	138	—
—	42	—	58	145	117
—	80	15	—	149	141
		(Ag 5)			
—	99	—	—	153	153
	(Cu 1)				
—	100	—	—	157	157
—	12	18	70	174	150
—	70	30	—	174	160
—	—	37	63	182	—
—	—	30	70	186	183
—	—	40	60	188	183
—	—	50	50	214	183
—	—	60	40	238	183[c]

[a] Data from Rosebury [35], Kohl [36], and Manko [37] and Indium Corporation of America, Utica, New York.
[b] Bi and Pb have vapor pressures $<10^{-8}$ Torr at 300°C; In and Sn have substantially lower vapor pressures.
[c] Most ductile of Sn–Pb solders.

sives [39, 40] where low thermal resistance and minimum outgassing are required. Vacuum brazing of sputtering targets has also been described [38]. Joining processes suitable for vacuum applications including brazing, soldering, and glass– and ceramic–metal sealing have been reviewed previously [36].

Some early planar magnetron designs sealed the target directly to the coolant plenum with no backing plate. This is very risky due to the high differential pressure exerted on the cathode (which may not be a suitable structural material), the possibility of diffusion through the target [41], possible electrolytic corrosion at the O-ring seal, and finally, the likelihood of the target bursting and flooding the vacuum chamber with coolant when the erosion depth reaches some critical point.

Water cooling is the most common method of maintaining a low cathode temperature, although gas flow or even oil cooling [42] have been used. The flow rates required are relatively high. Systems should be designed conservatively for a water-temperature rise of 10°C or less. Table II lists required minimum water flow versus power dissipated for a 10°C ΔT [43]. High flow rates require that the cathode cooling channels offer minimum flow resistance if high water pressures are to be avoided. The thermal characteristics of the cathode are important; they are discussed in Section II.D. The use of chilled water at some temperature above the dew point of the room air will afford an extra margin of safety. The electrical conductivity of the cooling water should be low enough to minimize current leakage to ground when maximum voltage is applied. If 0.5 m or more of 1-cm inner diameter insulating water line (for example, polyvinyl chloride) is inserted between the cathode and ground, a water resistivity of about 10 kΩ cm would reduce leakage to less than 1 mA at 600 V.

One disadvantage of PM sputtering is that the target erodes only in the transverse magnetic field region; as the erosion proceeds, the bombarding ions are focused into an increasingly narrow region leading to a V-shaped

Table II

Water Flow Rates for $\Delta T = 10°C$

Power dissipated (kW)	Water flow rate (liter/min)	(gal/min)	Power dissipated (kW)	Water flow rate (liter/min)	(gal/min)
2	2.8	0.74	15	21	5.6
4	5.6	1.5	20	28	7.4
6	8.4	2.2	25	35	9.3
8	11.2	3.0	30	42	11
10	14	3.7			

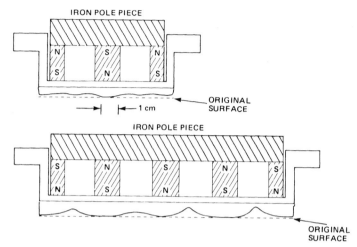

Fig. 5. Relation of cathode erosion profile to magnet position; (a) Al_2O_3 cathode, 10×25 cm, single plasma ring, operated at 7 W/cm² rf, ~1 mTorr Ar pressure, (b) Al–4% Cu cathode, 19×25 cm, dual plasma rings, operated at 5.6 W/cm² dc, ~9 mTorr Ar pressure. Note unequal erosion rates for the two plasma regions due to unequal magnetic field strengths.

erosion profile as shown in Fig. 5. End-of-life for the target occurs as soon as this groove reaches the backing plate. Reported material utilization at this point ranges from 26–45% [14, 23, 24, 43]. Various means have been employed to improve material utilization: increased target thickness in the erosion region, an oscillating electromagnetic field to modulate the position of the permanent magnet field [25], and mechanical movement of the magnet array [44, 45]. Mechanical movement is more amenable to the cylindrical-cathode internal-magnet magnetron [46] than to a planar cathode. Expensive target materials may warrant the complexity necessary to increase the erosion area.

4. Anode Placement

An auxiliary anode has been reported to be effective in preventing electron bombardment of the substrate [14, 17, 23] (see Section II.E). Configurations have included a peripheral anode ring between the cathode and the dark-space shield of a disk target [23]; a water-cooled tube connected to the positive side of the floating-cathode power supply and located slightly in front of and surrounding the circular cathode surface [14]; and anode rod or rods along the sides of a rectangular cathode [16]. The relative effectiveness of these arrangements has not been reported. It may be that the magnetron discharge acts much like a conventional rf or dc plasma in that the position of a positively biased auxiliary electrode is not critical [47]. The placement of the anode with respect to the magnetic field lines may be important, however (see Chapter II-2). The most effective anode placement should be near the regions where the magnetic field enters the cathode, as in the Sputter Gun (Chapter II-3).

It has been suggested [48, 49] that both the annular shield surrounding the target and the center of the target (where the magnetic field is normal to the target surface) must be anodes. If a central anode is not provided and an easily oxidizable material such as Al is used as a source, the center of the target may form an insulating layer that will charge up to the cathode potential. If this potential exceeds the oxide film breakdown potential, arcs can occur that can contaminate the growing film (Section II.F.3). A planar disk cathode source having a grounded central anode (akin to a flattened Sputter Gun) has been compared with a conical cathode Sputter Gun [41] and found to give equivalent deposition rates for Cu and Al targets for a given power input. It was noted that even with a central anode it was more "difficult" to apply high dc power levels to the flat cathode structure than to the conical structure. There was no problem when rf power was used with the flat cathode.

5. Substrate Heating

Substrate heating poses a problem in PM sputtering. Tubular quartz lamps are often used as radiant heaters; the close source–substrate distance typically used in PM sputtering prevents uniform surface heating during deposition by a conventional quartz-lamp radiant heater array. In an in-line system, radiant heaters can be used to preheat the substrates prior to entering the deposition zone. Deposition will then occur on a cooling substrate. This may or may not be a disadvantage. The effect would be less for a planar or cylindrical rotating substrate carrier system (see Section II.B.2). If constant uniform heat is necessary, radiant surface preheating combined with some type of backside substrate carrier heating to minimize thermal conductive losses may be an effective technique. In any substrate heating arrangement it is difficult to avoid heating extraneous portions of the deposition chamber and fixturing, resulting in desorption of residual gases and ubiquitous low-vapor-pressure contaminants (e.g., alkali metals). This is especially true for radiant heating. The determination of actual substrate surface temperature for calibration or temperature control purposes is not trivial and is discussed in Chapter II-1.

B. Thickness Uniformity Control

A planar magnetron is inherently a nonuniform deposition source because the sputtered material originates from the regions of high plasma density. Since these regions are dependent on magnet location and field strength, the PM technique allows adjustment of the effective source geometry by modifying the magnet arrangement and the resultant erosion pattern on the target. Substrate motion appropriate to the source geometry, with or without shaped shields, is another method that can improve film thickness uniformity.

Actual sputtering thickness distributions usually agree with calculations based on a cosine emission of sputtered material from the erosion area [50]. However, such calculations may be inaccurate at high deposition rates because of considerable self scattering of emitted material in the high source–material flux region above the erosion area. This may result in a "virtual source" in front of the cathode surface and an "under-cosine" distribution [51, 52]. It should be recognized that thickness distribution can change as the erosion depth increases and the effective emission width narrows [34].

1. Magnet Pattern

a. Disk Source. The thickness distribution from the planar disk magnetron has been reported to agree with that calculated for a ring source having a cosine emission characteristic [14, 23]. For such a source, best film uniformity is obtained when the source inner radius is about 0.7 of the source-to-substrate distance h and the outer ring radius is about $0.8h$ [53]. If we assume a plasma ring width of 2 cm and a minimum source–substrate spacing of 5 cm,* the optimum plasma ring outside diameter would be ~9.5 cm. Larger single-ring disk sources would have to be proportionately farther away from the substrates. A better magnet geometry for a large disk source would be n concentric rings [54, 55] of opposite magnetic polarity resulting in n-1 toroidal plasma regions. In any source utilizing multiple independent closed plasma regions, it is important to have equal magnet strengths and field intensities in each individual plasma region to equalize cathode erosion rates for maximum material utilization. Figure 5b illustrates the effect of unequal magnetic field strengths for two independent oval patterns on a 19 × 25 cm cathode. Concentric-ring cathodes up to 61 cm in diameter with six equally spaced plasma rings have been constructed and 80% target utilization realized [56]. Figure 6 illustrates calculated thickness distributions for a single-ring source and for concentric-ring sources having from two to six rings, based on the analysis of Glang [53]. A source–substrate distance of 5 cm was chosen, and each ring was assumed to have uniform cosine emission from an effective width of 2 cm. The ring radius in centimeters at midwidth was $4n$,

---

* Generally the minimum source–substrate distance for magnetron sputtering is 5–6 cm in order to be outside the plasma region.

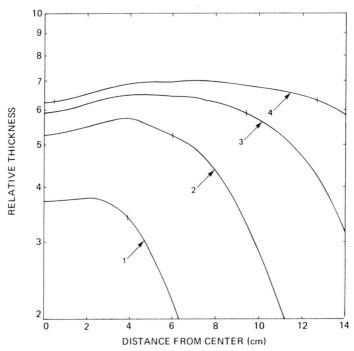

Fig. 6a. Calculated thickness distribution for multiple concentric ring sources. Relative thickness is plotted versus distance from center for (a) one to four rings and (b) four to six rings. The source–substrate spacing was taken as 5 cm. The effective ring width was assumed to be 2 cm; ring spacing is 2 cm. Short vertical lines indicate approximate boundaries for ± 5% uniformity. Based on ring source analysis of Glang [53].

where n is the ring number. This results in an array of 2-cm-wide rings with 2 cm spacing between rings. As Table III indicates, uniformity efficiency over a circular region improves for three or more concentric rings.

b. Rectangle Source. A line source can be approximated by either a rectangular or oval cathode and a magnet geometry that provides an elongated ring plasma pattern. Good deposition uniformity requires substrate motion perpendicular to the cathode width. For a cathode length/width ratio of about 2.5 and a source–substrate distance one-half the target width, the deposition profile is thick in the center [16, 43]. Figure 7 shows two examples of magnet arrangements for improved uniformity from a rectangular cathode [57]. Increasing the length/width ration to ≥3 can also improve uniformity of a rectangular magnet arrangement. Obviously, a long narrow cathode having the end semicircular plasma regions beyond the substrate region will give uniformity distribution approximating an in-

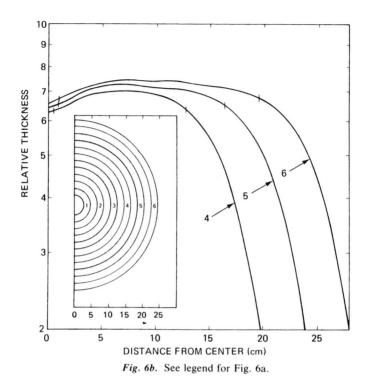

Fig. 6b. See legend for Fig. 6a.

finite line source. Cathodes as long as 150 cm and 20 cm wide have been reported [24, 58].

Other than magnet rearrangement, the plasma intensity can be modified by varying the spacing between the magnets and the cathode surface [45], by modifying the magnets or pole faces with gaps, or by varying pole face widths or angles. Special closed-curve geometries can be devised to suit the specific deposition profile requirements, whether uniform or nonuniform.

Table III

Single and Concentric Ring Source Efficiencies[a]

No. of rings	Target radius (cm)	Efficiency %	No. of rings	Target radius (cm)	Efficiency %
1	6	42	4	18	50
2	10	36	5	22	55
3	14	45	6	26	57

[a] Efficiency is the substrate area for ±5% uniformity divided by target area (see Fig. 6).

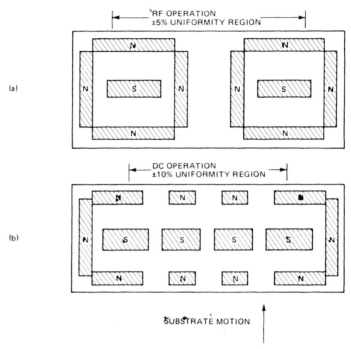

Fig. 7. Magnet arrangements for improved deposition thickness distribution from rectangular cathodes: (a) dual rings, (b) magnet gaps. Experimentally determined uniformities are indicated (Muto *et al.* [57]).

2. Substrate Motion

Various types of substrate motion relative to the sputtering source are shown in Fig. 8. Sputter sources can be mounted to sputter in any direction. When used with planetary motion (Fig. 8a), as in replacing an existing electron-beam source, the source–substrate distance is large and the effective deposition rate is lowered. Planetary motion implies multiple rotation of the substrates and continuously varying angles of incidence, instantaneous deposition rates, and source–substrate distance. If a disk source is used with planar rotation (Fig. 8b), the source–substrate distance can be less but an aperture shield [59] is required for good deposition uniformity. However, modification of the cathode and magnet arrangement to get a wedge-shaped source distribution may be possible.*
Figure 8c shows substrates located on the interior of a drum-shaped substrate holder rotating around a planar cathode. A cylindrical cathode

* For example, the Perkin–Elmer Ultek "delta source" [59a].

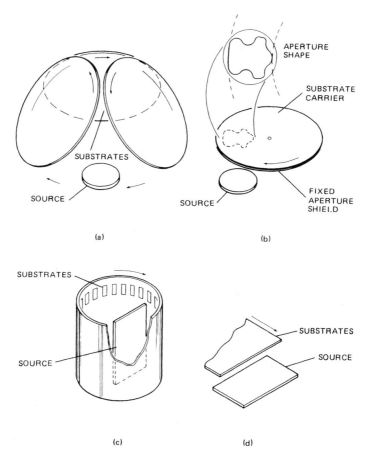

Fig. 8. Planar magnetron deposition with substrate motion; (a) planetary motion, (b) planar rotation with aperture shield, (c) drum rotation, (d) linear motion.

(Chapter II-2) would be more appropriate unless multiple sources (different materials) are required [60]. Substrates can also be placed on the exterior of a drum-shaped carrier and single or multiple PM sources located outside the drum. Planar rotary motion, as in Fig. 8b or c, results in a nearly constant source–substrate spacing; but if multiple passes are required, the deposition rate varies. The linear motion arrangement in Fig. 8d is, in effect, similar to Fig. 8c but is easily adapted to continuous substrate motion from and to a magazine chamber or air-to-vacuum load lock.

The substrate exposure time for linear substrate motion from an extended oval plasma ring having semicircular end regions has been calculated [45]; the optimum arrangement was a wide plasma region with a minimum end radius.

Linear motion makes optimum use of the high-rate capability of the rectangular magnetron source by allowing minimum source–substrate spacing and maximum throughput. All substrate motion, however, results in time-varying instantaneous deposition rates and incident angles. Such variation can affect the structure and properties of the deposited films.

3. Cathode Shielding for Uniformity Control

Shaped aperture shields have been used to achieve uniform deposition from a magnetron disk source with planar substrate rotation (Perkin–Elmer Ultek, Inc. [59] and Fig. 8b) and from a rectangular source with linear motion [17]. Because the emission from the source is nonuniform, the shield profiles were empirically determined. Such shields have the disadvantages of (1) collecting deposited material that must be periodically removed and (2) reducing the effective deposition rate for a given cathode erosion rate. In most cases, modification of magnet geometry is preferable to aperture shielding. An exception would be for rate or angle-of-incidence control; for example, single or stacked aperture shields can be placed as close to the substrate as is practical in order to intercept deposition at low rates and grazing incident angles [61].

C. Voltage, Current, and Pressure Relationships

Planar magnetron sources usually are operated in argon at a pressure of 1–10 mTorr and at cathode potentials of 300–700 V. Under these conditions, current densities can vary from 4 to 60 mA/cm^2; power densities are in the range 1–36 W/cm^2.

Typical voltage–current characteristics are shown in Fig. 9a for various pressures.* For an optimum magnetic field shape and intensity, these curves follow the relation:

$$I = kV^n, \tag{1}$$

where I is the cathode current (or current density) and V is the cathode potential (Chapter II-2, Section IV). The more efficient the electron trapping in the plasma, the higher the exponent n. Thus, for a given cathode material and configuration the magnetron operates at some characteristic, nearly constant voltage.

Figure 9b displays the same data as Fig. 9a in terms of the voltage re-

* The current density in the erosion area of the target may be up to four times the average current densities given.

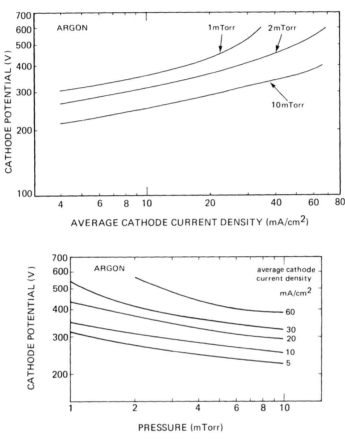

Fig. 9. (a) Current–voltage characteristics for rectangular planar magnetron cathode at various pressures. (b) Cathode voltage versus pressure for constant average cathode current density. Target size 37.5 × 12.1 cm, oval ring plasma geometry (data from Van Vorous [24] and Materials Research Corp. [62]).

quired to maintain a constant average current density at various pressures. Almost no data have been published concerning the effect of parameters such as magnetic field geometry, target material, and pressure on the constants of Eq. (1). Since the power (P) is the product of cathode potential and current, Eq. (1) becomes

$$P = kV^{n+1}, \tag{2}$$

where k and n are the same constants as in Eq. (1). For the 10 mTorr curve of Fig. 9a, $n = 4.6$ and $k = 9.0 \times 10^{-13}$ for average power density in W/cm^2 and cathode potential in volts.

For a given cathode configuration, a family of current–voltage curves versus pressure can be obtained that enable a suitable operating point to be chosen for a given power input. Similar or argon-calibrated gauges must be used when transferring a process from one deposition system to another. Sputtering gas pressure as measured with thermal-conductivity gauges, ionization gauges, and capacitance manometers are not directly comparable because these gauges are usually calibrated in terms of nitrogen-equivalent pressure and all except the capacitance manometer have differing calibration factors or response curves for various gases [63]. Pressures given in the literature may be nitrogen- or argon-equivalent readings. Since the distinction is rarely made, direct comparison of data is uncertain.

D. Deposition Rate

1. Factors Determining Deposition Rate

Sputtering rates are determined primarily by the ion–current density at the target and secondarily by the ion energy (voltage). In practice, for magnetron sputtering, deposition rates have been found to be directly proportional to the power delivered to the target. The factors determining deposition rates are power density in the erosion area, size of the erosion area, source–substrate distance, source material, and pressure. Other factors such as sputtering gas composition will not be considered here. The above parameters are listed in approximate order of importance but some are interrelated, such as pressure, power density, and size of the erosion area. Also, as will be discussed later, the thermal and mechanical properties of the target may limit the maximum deposition rate attainable.

For maximum deposition rate, the substrate should be as close as possible to the source while remaining outside the plasma region. Typical minimum spacings are 5–7 cm. Source–substrate distance and deposition rate will vary with time when substrate motion is used to improve uniformity or increase capacity.

Substrate motion complicates the definition of deposition rate. For planetary or other rotary motion of substrates, the deposition rate reported is usually the average rate determined by dividing thickness by deposition time. For static deposition with PM sources, the thickness is usually nonuniform, so a thickness average (or even maximum thickness) is used to calculate reported deposition rates.

For linear motion of substrates under a rectangular target, an average deposition rate (R) can be defined by

$$R = tS/W, \qquad (3)$$

where t is the average film thickness for one pass of the substrate at speed S under a target of width W measured in the direction of substrate motion. The rate, R, will have the unit thickness/time.

A deposition rate efficiency (R_E) can be defined as the slope of the linear deposition rate versus power density curve, or

$$R_E = R \; (\text{Å}/\text{min})/P \; (\text{W}/\text{cm}^2). \qquad (4)$$

Deposition rate efficiency is a useful concept for optimizing deposition conditions, magnet arrangement, source–substrate spacing, etc. Table IV summarizes some deposition rates and calculated rate efficiencies reported for dc PM sputtering. Average power densities are 10–30 W/cm^2 and rate efficiencies are 800–1200 Å-cm^2/W-min for copper and 200–650 Å-cm^2/W-min for aluminum. For a constant cathode size, magnet configuration, and power density, the deposition efficiencies vary with material. For nonferromagnetic materials, this is partly due to differences in sput-

Table IV

DC PM Sputtering—Deposition Rates and Efficiencies

Cathode material	Reported rate $\left(\dfrac{\text{kÅ}}{\text{min}}\right)$	Calculated efficiency $\left(\dfrac{\text{Å}}{\text{min}}\dfrac{\text{cm}^2}{\text{W}}\right)$	Av. power density $\left(\dfrac{\text{W}}{\text{cm}^2}\right)$	Source–substrate distance (substrate motion) (cm)	Argon pressure (mTorr)	Cathode dimensions (cm)	Reference
Cu	10.7	1020					
Al	5.5	520	10.5	5	4	9 × 21	[16]
Ti	3.9	370		(static)			
Au	20	1250					
Cu	20	1250		not			
Cr	10	630	16	reported	not reported	15 × 41	[17]
Al	7	440		(static)			
Ti	4	250					
Cu	25	780		5		diam.	
Al	11	340	30	(static)	3–5	14 (est.)	[23]
Au	17	1100					
Cu	13	840		not			
Cr	10	650	15.5	reported	not reported	13 × 30	[24]
Al	10	650		(linear)			
Ti	3.3	210					
Al	2	185	10.8	7 (linear)	11	10 × 25	[64]

tering yield. If the deposition rate or efficiency of a given material of yield A is known, then the expected rate for another material of yield B under the same conditions can be estimated by multiplying the rate for A by the ratio B/A. Sputtering yields at 600 V have been tabulated [65].

Sputtering yields increase with voltage in the 300–800 V range used for dc PM sputtering. Thus part of the rate versus power dependence is due to sputtering yield increasing with voltage. The sputtering yield for many metals doubles or triples between 200 and 600 V. Following the reasoning of Lamont and Turner [66], we can show, assuming a linear relation between sputtering yield and voltage between 300 and 800 V, that for PM sputtering the deposition rate should be proportional to power. Deviations occur due to effects such as gas scattering of sputtered material, re-emission from the substrate, and the presence of inefficient sputtering gas species such as hydrogen [67] that contribute to ion current but not to sputtered material flux.

Sputter gas pressure has less influence on deposition rate for dc PM sputtering than for conventional sputtering. As shown in Fig. 9b, as gas pressure decreases, the effective plasma resistance increases, requiring a slightly higher voltage to maintain a given current density. Therefore if the deposition rate for a given current is plotted versus pressure, the rate will increase with decreasing pressure. If power is held constant, the deposition rate versus pressure curve will exhibit a maximum (Fig. 10). For the conditions of Fig. 10, the deposition rate changes less than 10% between 3 and 15 mTorr. The decreasing rate at high pressures is caused by gas scattering of the sputtered material; the reduction at lower pressure was suggested to be due to less efficient ion collection by the cathode [23].

The magnetron discharge is a very efficient sputtering source; it has been estimated that the sputter rate per watt of discharge power may exceed 60% of the theoretical limit set by the ratio of sputtering yield to mean ion energy [23]. In comparison to conventional sputtering, planar magnetron sputtering has been reported to result in three times the cathode erosion rate efficiency, e.g., for copper sputtered in argon, 2 mg/sec-kW for PM sputtering versus 0.7 mg/sec-kW for conventional sputtering [23].

2. Rate Limitations

The major factor limiting deposition rate is the maximum power flux that can be applied to the cathode without causing cracking, sublimation, or melting (if a molten target is used, then in most cases we are dealing more with thermal evaporation than with sputtering).

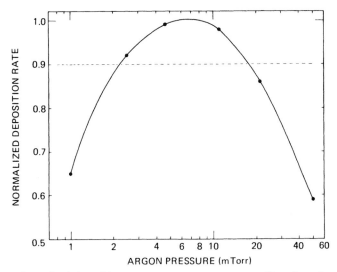

Fig. 10. Normalized deposition rate versus argon pressure. Chromium disc cathode; source–substrate distance 5 cm; estimated power density 18 W/cm² (data from Schiller *et al.* [23]).

If we assume that the coolant flow channels are properly sized so that the necessary minimum flow rates are possible (see Table II), the cooling rate of the target depends on the thermal conductances of the backing plate, the target backing-plate interface, and the target itself. These conductances depend on the thermal resistance (thickness × thermal conductivity) of the materials involved.

Table V lists calculated maximum power densities that can be applied to various materials for a 100°C temperature gradient through the material thickness for a thickness of 1.0 cm. Assumptions in the calculations are *only* conductive heat transfer, uniform distribution of power flux at the target surface, and no thermal resistance at the target backing–plate coolant interfaces—obviously an ideal situation. These values are equal numerically to the thermal conductivity of the material in the units W/m K, i.e., watts per meter per degree Kelvin. (See Rosebury [69] for a tabulation of thermal conductivities of many metals and alloys.)

For good thermal conductors, the allowable power density can easily exceed the ability of water cooling to remove the heat from the back of the target. If we assume 1 liter/sec (16 gal/min) as an upper limit for cooling water flow and 30°C as the maximum tolerable water-temperature rise, then the power limit for a water-cooled cathode is 125 kW [70]. For an

Table V

Power Density (W/cm²) for a 100°C Temperature Gradient through a 1 cm Cathode Thickness^a

		Power density
Dielectrics	Glasses	
	vitreous silica (fused quartz)	2–4
	and most borosilicate, aluminosilicate, and soda-lime glasses	
	80:20 PbO–SiO$_2$	0.6
	Oxides (polycrystalline)	
	BeO	51
	Al$_2$O$_3$	30
	MgO	8.7
	Oxides (single crystal)	
	sapphire (Al$_2$O$_3$)	7.7–8.3
	quartz (SiO$_2$)	1.4–2.6
	stainless steel (300-series)	13.4–15.5
Alloys	In–Sn 50:50	70 (est)
	Pb–Sn 60:40	47
	Pb–Sn 50:50	43
Conducting compounds	molybdenum disilicide	31
	Ta, Ti, and Zr carbides	21

Metals/semiconductors	Power density
Ag	427
Cu	398
Au	315
Al	237
W	178
Si	149
Mo	138
Cr	94
Ni	91
In	82
Fe	80
Ge	76
Pt	73
Pd	72
Sn	67
Ta	58
Pb	35
Ti	22

^a Data from Weast [68] and Rosebury [69].

154

aluminum cathode having an area of 360 cm^2 (a 21-cm-diameter disk or a 12 × 30 cm rectangle), the power limit for a 100°C cathode temperature gradient is 85 kW and the required water flow is 40 liter/min (11 gal/min) for a 30°C water-temperature rise.

For poor thermal conductors, for example, glass or quartz, the maximum power density (total power divided by erosion area) would be a safer number to use than the average power density (total power divided by target area). A stress-free low thermal expansion material such as fused quartz can support a much higher temperature gradient than a target having internal stresses or a high thermal expansion coefficient. For example, successful operation of an SiO$_2$ target in a magnetically enhanced, high-rate rf mode at 72 W/cm^2 has been reported [71]. At this power density the target surface was near its melting point and the authors concluded that evaporation as well as sputtering was occurring. In contrast, Table VI lists some results for rf sputtering of aluminum oxide targets.

The data of Table V also allow determination of the conditions when a poor thermal conductor such as stainless steel can be used as a backing plate. In all cases, we have neglected the target backing-plate interface (Section II.A.3). For solder-bonded targets, this is not a significant thermal barrier unless voids or discontinuities are present [38].

3. Rate and Thickness Control

If all the factors affecting deposition rate can be held constant, then the rate will be constant and reproducible. The power supply should be

Table VI

Aluminum Oxide Targets—Power Density Comparison

Target type	Thickness (cm)	Av Power Density (W/cm^2)	Av Power Density for $\Delta T = 100°C$[a] (W/cm^2)	Time to Failure (min)	Reference(s)
Polycrystalline[b]	~0.5	72	16	5	[71]
Sintered	~0.5	120	60	15	[71]
Sintered[c]	0.64	7[d]	50	>10^4	[72, 73]

[a] Gradient through target thickness for uniform power density (Table IV).
[b] General Electric Co. Lucalox®.
[c] Western Gold and Platinum Co. AL-995.
[d] rf PM; maximum power density in erosion area ~20 W/cm^2.

capable of maintaining a constant power. For PM sputtering, the power supply is usually operated in a constant-current mode with voltage limiting. Since the plasma erosion area depends on the magnetic field intensity and shape, this field should be constant or varied in a reproducible manner. Electromagnets require a high-current power supply—another source of variation.

Pressure can be held constant by servo-controlling a leak valve with an ionization gauge or a capacitance manometer [74]. In practice, this is rarely necessary if the diffusion pump throttle-valve mechanism is capable of reproducing its orifice area (conductance) in the throttle position. Turbomolecular pumps do not require a throttle valve if a sufficiently high-capacity backing pump is used. To take advantage of the pumping speed of the liquid-nitrogen trap, the throttle orifice should be located between the diffusion pump and the liquid-nitrogen trap [75]. Usually a micrometer-type needle valve can be used to set the argon flow rate for a foreline pressure below the diffusion-pump forepressure tolerance or at the foreline pressure for maximum throughput [76]. The throttle mechanism then is adjusted to give the approximate sputtering pressure desired. Lastly, the argon needle valve is readjusted for the exact pressure required. If both throughput *and* pressure are to be set accurately, then a vernier control on the throttle is helpful. An ion gauge designed for the pressure range 1–10 mTorr is preferable to a thermal-conductivity vacuum gauge for reproducing the pressure setting. As indicated by Fig. 10, precision pressure reproducibility is not required unless unusually accurate rate control is needed. However, sputtering gas composition and purity should be controlled (see Chapter II-1, Section VI).

For reproducible film thickness, all that remains is to control deposition time. In a linear-motion system, the substrate-carrier speed determines deposition time.

An effect peculiar to magnetron sputtering is the nonuniform erosion of the cathode surface leading to V-shaped grooves. As these grooves deepen, the discharge impedance changes and the voltage required for constant current may change slightly. If power level is critical, a power-feedback control could be used. In most cases, deposition thickness or rate monitors are not necessary for rate–thickness control during sputter deposition. Rate monitors are useful, however, for rapidly determining the effects of sputtering parameters, e.g., pressure, on deposition rate. Care must be taken to distinguish between changes in rate and changes in distribution pattern. Quartz-crystal oscillator monitors, shielded from rf, can be used.

For reactive sputtering, gas composition and throughput must be reproducible (see Chapter II-1, Section VI.D).

E. Substrate Effects

1. Electron and Ion Bombardment

In conventional rf or dc sputtering, secondary electrons emitted by the cathode are accelerated across the dark space and, unless the substrate is shielded or negatively biased in some manner, will bombard the substrate with almost the full energy of the dc cathode potential [30]. It has been estimated that for conventional rf sputtering, of the 5–10% of the cathode power dissipated at the substrate about 60% is due to electron bombardment [77]. For conventional dc sputtering, of an estimated 40% of the applied power dissipated at the substrate, virtually all was transmitted by secondary electrons [30].

In a planar magnetron the secondary electrons are accelerated to the potential difference between the cathode and the plasma, but those electrons under the influence of the transverse magnetic field will be given a curved trajectory of the order of 1-cm radius so that they cannot bombard the substrate directly (see Chapter II-2). By ionization and collision processes, the plasma becomes populated with more or less thermalized electrons having energies of about 1 eV [23]. These electrons are readily collected by an anode adjacent to the region where the magnetic field lines enter the cathode surface (Section II.A.4). Figure 11 shows how an escaping electron can follow magnetic field lines emerging from the cathode.* These electrons can impinge directly on the substrate.

It has been reported [23] that an anode surrounding the cathode at a positive potential of 20–35 V will reduce the current to a grounded substrate to near zero for a cathode potential of − 450 V. Without an anode, a total current equal to the discharge current will be about equally divided between a grounded substrate and adjacent grounded chamber walls [23]. An ungrounded or nonconducting substrate will quickly charge up to a potential sufficient to prevent further electron bombardment; ion bombardment will then occur until some equilibrium "floating" substrate potential is reached.

If a negative bias is applied to the substrate, then ions can be extracted from the plasma and large substrate ion currents can be produced.† The identity of these ions is not known; presumably the majority are discharge gas ions but a significant fraction may be ionized cathode material [46] and residual gas ions [78, 79]. Schiller *et al.* [23] have measured substrate cur-

---

* If an rf glow discharge (positive column) is present in the same vacuum chamber with a nonoperating magnetron target, the discharge glow will graphically delineate the magnetic field pattern in front of the magnetron cathode.

† This is in contrast to cylindrical magnetrons; Chapter II-2, Section VII.A.

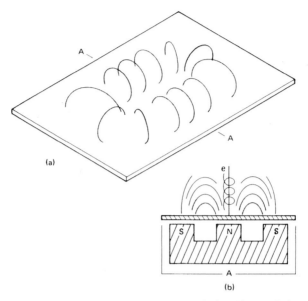

Fig. 11. Rectangular magnetron cathode; (a) magnetic field lines and plasma region, (b) cross section through section A-A showing field lines and escape path for electrons.

rents for substrate bias voltages from 0 to -400 V for disk planar magnetron cathode potentials of 350–485 V, $+35$ V on a concentric-ring anode, and a 5 cm cathode–substrate separation. They found that at 5 mTorr argon pressure for substrate bias potentials between -100 and -400 V, the substrate current was nearly independent of bias potential, especially at the lower cathode potentials. The substrate current saturated at about 8% of the target current. They concluded that for a given pressure, the energy of the bombarding ions could therefore be varied without substantially altering the total ion current or the ratio of ions to sputtered atoms at the substrate. For a given substrate bias, the ion current to the substrate decreased with increasing pressure. The highest ion/sputtered atom ratios at the substrate occurred at pressures slightly below 1 mTorr. The resistivity of Cr films sputtered at 5 mTorr in commercial grade Ar (about 0.2% O_2) at zero bias was 80 $\mu\Omega$ cm. At -400 V bias voltage the resistivity was lowered to nearly 30 $\mu\Omega$ cm. The deposition conditions were 485 V cathode potential and 6 A target current, resulting in about 500 mA substrate current. The substrate area was not given, but if it were equal to the cathode area (estimated to be 150 cm^2) then the ion current density at the substrate would have been about 3 mA/cm^2. This is several times times greater than cathode current densities during conventional dc sputtering. As Vossen [80] has pointed out, substrates can indeed be treated as sputtering targets. In this case, considerable resputtering of the depositing film must have occurred.

Since relatively small bias potentials can lead to large ion bombardment currents, the control of substrate bias is important in many applications of PM sputter deposition. Insulating substrates pose a particular problem in this regard. No data have been reported to date on PM magnetron sputtering with rf bias on insulating substrates.

2. Neutral Bombarding Species

As previously indicated, there is a significant ionization of sputtered species in a magnetron discharge [46]. Many of these ions may be neutralized and reflected from the target surface as atoms. At energies less than 1000 eV, reflection fractions can be as high as 0.4 and the particles can retain a substantial amount of their original energy [80]. Thus, in PM sputtering we might expect a significant flux of reflected high-energy sputtered atom bombardment of the substrate in addition to atoms arriving directly from the target. These same processes can cause high-energy neutral atom bombardment by the sputtering gas and residual gases. Ions of the sputtered species can also drift to the substrate and recombine with electrons at the substrate surface, thus releasing energy. These mechanisms may lead to higher surface mobility for atoms deposited by PM sputtering. Measurements of the energy distribution of PM sputtered atoms have not been reported.

3. Temperature

Manufacturers of PM sputtering equipment have claimed that for a given deposition rate the substrate temperature for PM sputtering is significantly lower than for rf sputtering. Details of the temperature-measurement technique are usually not given, but one example quotes for Cu deposition at 2000 Å/min a peak substrate temperature of $\sim 90°C$ for PM sputtering versus $\sim 400°C$ for conventional rf sputtering [24]. The temperature reached by a substrate depends on the heat capacity of the substrate, as well as the balance between the energy carried to it by the processes listed in Table VII and the energy lost by radiation and conduction. The total energy dissipated at the substrate for conventional sputtering has been variously estimated to be 4–5% of the total power delivered by the power supply for rf deposition [32] or 40% for dc sputtering [30]. These effects are discussed in Chapter II-1.

Table VII

Substrate Heating Processes—Conventional Sputtering from Water-Cooled Cathode

Mechanism	Approximate contribution (%) to substrate heating	
	rf[a]	dc[b]
Secondary electron bombardment	58	98
Argon ion bombardment		
Ion–electron recombination at substrate	22	negligible
Neutral and metastable atom bombardment		
Sputtered atom kinetic energy	15	1.5
Heat of condensation	5	0.5

[a] Lau *et al.* [32] and Lamont and Lang [77].
[b] Ball [30].

Lower substrate heating during PM sputtering is thought to be caused by reduced high-energy electron bombardment of the substrate. Even if all charged-particle bombardment and radiation input were eliminated, it may still be possible to get significant substrate heating by the condensing atoms—the heat of condensation plus their kinetic energy [49, 81] (see Section IV.B). The kinetic energy of PM sputtered atoms is not known, but the average energy may be higher than that generated by conventional dc sputtering due to the high percentage of ionization of the sputtered species [46]. In addition, the lower pressures and short source–substrate distances employed in PM sputtering will result in less kinetic energy loss by collision with sputtering gas atoms.

F. Power Supplies

1. Requirements

In conventional dc sputtering, power supplies are usually operated in a constant-voltage mode at 1000–3000 V and seldom operate at cathode current densities greater than a few milliamperes per square centimeter. In contrast, PM sputtering requires power supplies to operate in a constant-current mode at 300–700 V and at total currents to 25 A or greater. The output may be filtered or unfiltered dc; the latter may be advantageous in some applications [24]. The PM power supply must be able to survive high-voltage (several kilovolts) rf transients generated by the magnetron plasma, as well as by major and minor arcing along the cathode face and from cathode to ground. Power supplies developed for cylindrical or conical magnetrons are also suitable for PM sputtering but are generally not rated above 10 kW. Higher power ratings are required for large-area PM sources.

2. Power Supply Design

A typical magnetron power supply [14, 43] employs a silicon-controlled rectifier dc drive for the control winding of a three-phase saturable reactor. A feedback signal derived from the voltage drop is detected in the dc output circuit to provide essentially constant current output. The power supply must also be provided with adequate short-circuit and rf high-voltage transient protection. In addition, a practical power supply should have the following features on a remote control panel: interlocks with status indicators for cathode water flow, cathode temperature (or output water temperature), vacuum, and others as required for operator protection; variable power output control, e.g., voltage control until an adjustable upper limit is reached—then constant-current control; and meters for output voltage and current. Useful optional features would include: panel control settings for maximum (controlled) power and maximum available voltage, automatic ramping to preset power level, digital power-output meter or digital voltage and current meters, adjustable anode voltage output (0–50 V dc), plasma ignition capability, e.g., tesla coil, and floating (nongrounded) output connections.

3. Arcing

Even though the operating voltage for PM dc sputtering is lower than that for conventional sputtering, the magnetron source is more susceptible to various types of arcing. This is probably caused by the electron current to various adjacent shields and anodes and by the considerable rf component in the discharge plasma (Chapter II-2, Section II.E). Present PM shield arrangements are borrowed from conventional sputtering; improvements related to the PM configuration should reduce cathode-shield arcing problems. Arcing between points on the face of the cathode, which usually occurs with a new cathode or one that has been exposed to air, is believed to be caused by charge build-up on locally insulating areas [14, 48, 49]. Easily oxidizable metals such as Al are prone to surface arcing. By slowly increasing the power on new or air-exposed targets, this arcing can be controlled until the sputtered area cathode surface is clean and the unsputtered region is covered by a backscattered conducting layer. Extreme cathode surface arcing ("racetrack arcs") can be extinguished by sensing a low-voltage, normal-current condition and interrupting the power source for 100 msec or so. A similar technique is used for arc suppression in electron-beam evaporation source power supplies (Hill [43], pp. 27–35).

G. Miscellaneous Techniques

1. Reactive Sputtering

Reactive dc PM sputtering of tantalum nitride films has been investigated [82]. Compared to conventional dc sputtering of Ta in Ar/N_2, PM sputtering required a higher nitrogen concentration for equivalent film resistivities and temperature coefficients of resistance. Similar increased reactive gas concentrations (>25%) were reported necessary for conical-magnetron dc reactive sputtering of titanium nitride and indium–tin oxide [83]. Reactive dc PM sputtering of TiO_2 from a Ti target in an 85:15 Ar–O_2 mixture has been used to deposit the high refractive-index layer of a multilayer dielectric coating on an Al mirror [60]. Optical dispersion of the films was measured for the 0.3–1.1 μm wavelength range. The deposition rate was 240 Å/min at 4.8 cm source–substrate spacing and a pressure of 2 mTorr. Deposition power was not given. Generally, rf is preferable to dc for PM reactive sputter deposition of nonconducting films because of problems caused by the formation of an insulating film on the target, e.g., igniting the discharge.

2. Ion Plating

The combination of substrate ion bombardment and deposition by evaporation in rapid repetitive sequence, termed "alternating ion plating," has been described [84]. A PM arrangement was used to ion bombard a negatively biased moving strip-steel substrate. Nickel was evaporated onto the substrate as it moved through the plasma region. A multipass planar rotating substrate arrangement was also described, again employing a PM plasma at the substrate and electron-beam evaporation. The resulting films were equivalent—in terms of porosity—to those deposited by conventional ion plating.

The use of a PM source for plasma plating, that is, simultaneous ion-beam thermal evaporation and sputtering from a molten PM source combined with a negatively biased substrate, has also been reported [85]. At 5 kW the thermal evaporation rate for Cu was 8 g/min. It is possible that such a source, once started and if sufficiently ionized, could be self-sustaining without an inert gas ambient [86].

3. Composite Targets

Ta/Al resistive films have been deposited by PM sputtering from a 9.5-mm-thick, 10-cm-wide, and 25-cm-long Al target inlaid with 5-mm-thick, 5.2-mm-wide Ta strips spaced 3.2 mm apart across the target width [64]. An oval ring plasma pattern was used. It was found that this Ta to Al area ratio (approximately 1.68:1 for the total cathode area) resulted in 53 at. % Ta in films deposited on substrates moving parallel to the Ta strips at 2.5 cm/min. Assuming a 2:1 Al:Ta sputtering yield ratio, a 59.5 at. % Ta film would be expected. Film resistivity and temperature coefficient of resistance uniformity was better than ±10% across a 20-cm width, indicating good uniformity of composition. The deposition rate at 300 W (280 V) was about 640 Å/min at 11 mTorr argon pressure.

4. Sputter Cleaning

Schiller *et al.* [45] have investigated the efficiency of sputter-cleaning sheet steel in a continuous linear-feed system employing a PM discharge in Ar. In this application, an electromagnet array was located above the grounded moving substrate. A water-cooled anode and surrounding ground shield were located below the substrate and facing the surface to be cleaned. It was found that even with ferromagnetic substrates up to 2-mm thick, a sufficient magnetic field could be generated adjacent to the substrate surface. By switching off the electromagnets, conventional glow discharge plasma cleaning could be compared with PM discharge cleaning. A 5-μm-thick Ni film was evaporated onto the moving substrate subsequent to discharge cleaning. Planar magnetron discharge cleaning was found to be superior to glow discharge cleaning in terms of energy consumption for equal Ni-film adherence. Methods of obtaining uniformity of discharge exposure by varying the magnet arrangement and substrate–magnet spacing were evaluated in terms of the temperature uniformity achieved. The best uniformity obtained was 3% over the 13-cm width of the steel strip.

It is possible that a similar positive anode, grounded substrate PM arrangement could be used for sputter etching or reactive ion etching of conducting materials.

III. RF PLANAR MAGNETRON SPUTTERING

Chapin [14] described a disk cathode rf PM source employing an electromagnet. No operating data were given other than that the discharge had a low impedance and that insulators could be sputtered at high rates.

Subsequently, rf PM deposition of Al_2O_3 [72] and SiO_2 [29, 60] were reported.

A. Comparison of RF and DC PM Sputtering

Planar magnetron sputtering with an rf potential on the cathode (usually 13.56 MHz or 27 MHz) enables the use of nonconducting materials as targets and allows direct or reactive deposition of dielectric films in the same manner as conventional rf sputtering. However, because of other considerations, such as the additional complexity of the rf power supply and impedance matching network, extraneous plasma generation within the deposition chamber [29], the difficulty of ensuring an efficient rf ground, and the possibility of rf radiation and rf electromagnetic interference (RFI/EMI), dc PM sputtering remains the better choice for deposition of electrically conducting materials.

Planar magnetron rf sputtering is somewhat similar to PM dc sputtering in that, because of low plasma impedance, high power densities are possible at low applied rf potentials. For example, at a power density of 2 W/cm^2 the dc self-bias on the target is about 360 V for PM rf sputtering compared to about 3500 V for conventional rf sputtering [66].

As Fig. 5 shows, the erosion areas of an rf target differ from those of a dc PM target. Instead of a single erosion groove approximately midway between the magnet faces of opposite polarity, there are two regions of deepest erosion near the edges of the magnet pole faces with a raised area approximately where the groove was for dc sputtering. The cause of this peculiar erosion pattern is not known. The rf target was isostatically pressed and sintered 99.5% Al_2O_3; the dc target was vacuum-cast Al–4% Cu.

B. Power Density, DC-Self-Bias, and Pressure Relationships

Figure 12 shows the relation between dc self-bias potential [66] and average power density for PM operation of a 6-mm-thick Al_2O_3 cathode in argon in the pressure range 10-20 mTorr. (Operation at 1 mTorr follows the same curve within the measurement error.) For comparison, the curve for dc PM operation of a 9-mm-thick aluminum cathode is also shown. The targets were identical in size and magnet configuration. The rf curve is approximately linear above 2 W/cm^2 and follows the relation of Eq. (2), Section II.C, with $n = 0.9$ and $k = 2.8 \times 10^{-5}$. For a conventional rf plasma, $n = \frac{3}{2}$ [66]. For the dc curve, $n = 4.6$ and $k = 1.9 \times 10^{-14}$. As previously mentioned, the larger the value of the voltage exponent n, the more efficient the plasma electron trapping (see Chapter II-2, Section IV).

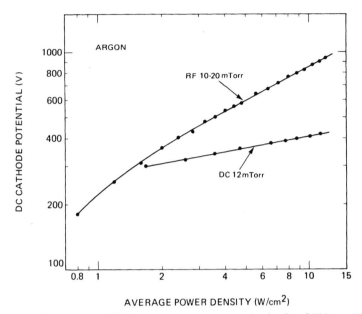

AVERAGE POWER DENSITY (W/cm^2)

Fig. 12. Direct current self-bias *versus* average power density for rf PM operation of 10 × 25 cm Al_2O_3 cathode, argon pressure 10–20 mTorr. Cathode potential versus power density for dc PM operation of similar aluminum target having same magnet configuration shown for comparison.

The rf PM discharge is not as sensitive to pressure as is the dc PM magnetron. The deposition rate also appears to be constant (± 10%) for a given power independent of pressure from 1–20 mTorr.

C. Rate Limitations and Rate Control

As discussed in Section II.D, the low thermal conductivity or high-internal stresses of many targets, e.g., glass, Si, and sintered metallic oxides, will limit the maximum power density and hence the deposition rate. Targets having a relatively high coefficient of expansion combined with poor thermal conductivity may crack because of localized ion-bombardment heating. In non-PM high-rate sputtering of Al_2O_3 (a fairly good heat conductor), power densities of 120 W/cm^2 were attained but target life was limited to 15 min (Table VI). For targets having a high dielectric constant or high rf dissipation factor (dielectric loss tangent), rf bulk heating may be a consideration [80]. Reactive sputtering from a metallic target may be a way of circumventing these power limitations to achieve high rates.

Deposition rates for rf PM sputtering have been reported to be proportional to power [29, 72], so it is reasonable to assume that PM rf rates can be controlled in the same manner as in conventional rf deposition. The dc self-bias at the target may be useful for rate indication and control [66].

D. Power Supplies and Matching Networks

In contrast to dc PM sputtering, rf power supplies developed for conventional 13.56 MHz rf sputtering are also suitable for rf PM sputtering. Commercial rf sputtering supplies can tolerate the occasional arcs and transient high voltage pulses of rf PM sputtering. Conventional rf power supply/discharge plasma matching networks can also be used if the range of the variable components (either capacitors or capacitor and inductor) is sufficient to compensate for the lower PM plasma impedance. Matching network design for rf sputtering has been described previously [66, 87–90a].

Proper rf grounding of the power supply, vacuum chamber, and internal shielding is important to prevent unwanted substrate–plasma interaction and RFI/EMI. In many applications the substrates must be electrically isolated so that they can be biased or electrically ''floated'' to control charged-particle bombardment [80].

E. Deposition Rates

For a given average power density (power divided by cathode area), the deposition rates for rf PM sputtering are comparable to conventional rf sputtering, i.e., the slopes of the deposition rate versus power density curves are similar. Table VIII compares reported rf deposition rates for

Table VIII

RF Sputtering—Deposition Efficiency Comparison

Target	Sputtering method	Power density range (W/cm^2)	Rate / Power density $\left(\dfrac{\text{Å}}{\text{min}} \dfrac{\text{cm}^2}{\text{W}}\right)$	Source– substrate spacing (cm)	Reference
Al$_2$O$_3$	conventional	1.2–2.4	50	3	[91]
Al$_2$O$_3$	PM[a]	3–8	51	7	[72]
Al$_2$O$_3$	high rate	20–120	52	2.5	[71]
SiO$_2$	PM	1–9	110	9	[29]
SiO$_2$	PM[b]	26	93	4.8	[60]
SiO$_2$	high rate	<20	~100	3.3	[71]

[a] Linear substrate motion.
[b] Drum rotation of substrates.

Al_2O_3 and SiO_2. These rates are lower than those for metals due to lower sputtering yields [65]. The addition of O_2 to the Ar sputtering gas has been reported to reduce deposition rates for PM rf sputtering of Al_2O_3 [72] and SiO_2 [29, 60], as is the case for conventional rf sputtering of oxides. Since the oxygen effect is partial-pressure rather than total-pressure dependent [92, 93], and PM rf sputtering pressures are typically lower than those for conventional rf sputtering, the percentage of O_2 required to reduce the sputtering rate is greater.

IV. APPLICATIONS

A. Industrial Coatings

Planar magnetron deposition offers obvious advantages in coating large surface areas and could replace or supplement electron-beam deposition for plastic film metallization and for antireflection and thermal-barrier coating of architectural glass [42, 43, 94, 94a]. Planar magnetron sources are advantageous in such applications because cathode size (usually length) can be scaled up, cathodes can be run side-by-side or sequentially in continuous coating machines, and they can be placed to deposit in any direction [24]. Other applications include decorative coatings on plastic [94b], functional coatings on plastic (antireflective lens coatings, video disk metallization), and anticorrosion or abrasion-resistant coatings on metals.

B. Thin Film Electronics

1. Integrated Circuits

a. Al metallization. The evolution and status of Al and Al-alloy metallization have recently been reviewed [95], and magnetron sputtering was characterized as a promising deposition technique that is amenable to process automation. An acceptable integrated circuit metallization process must deposit rapidly about 1 μm of Al alloy, usually containing 3–4% Cu and/or 1–2% Si with a reproducibility and uniformity of ±5% or better. The resulting film must form a continuous layer over the various Si, SiO_2, and Si_3N_4 steps that abound on large-scale integrated circuits. In addition, the deposition process must not cause radiation damage to the underlying interfaces or dielectric layers by x-ray, electron, or ion bombardment, nor contaminate the metal–dielectric interface with mobile alkali-metal ions [95]. Typically, substrate heating ($\sim300°C$) is used during electron-beam deposition to promote surface diffusion of the Al in order to achieve adequate step coverage; heating, however, increases the likelihood of mobile ion contamination. Electron-beam evaporators have been retrofitted with single or multiple conical-magnetron sputtering sources that simplify Al–Cu–Si or Al–Si alloy deposition, but substrate heating is still required for vertical step coverage [96].

A theoretical analysis of deposition from an extended sputtering source with translational motion suggests that such motion "will not drastically change the propensity for microcrack formation" [97]. The applicability of this analysis to magnetron sputtering seemed to be confirmed by the results of aluminum deposition with a PM source with linear substrate motion, since substrate heating was required for adequate step coverage [61]. However, other work has indicated that excellent step coverage can be achieved with PM sputtering without substrate heating [98]. The substrates were 7 cm from a 19 × 25 cm cathode with transverse substrate motion perpendicular to the long dimension of the cathode.

The reason for conflicting results with PM sputtering is not understood. It may be simply that the step contours tested were different; but it is tempting to speculate whether substrate heating, substrate bias, or both, may play a role. It is easy to show [49, 81] that the heat flux due to condensation of Al atoms having an average kinetic energy of 6 eV will be 0.142 W/cm^2 at a deposition rate of 1 $\mu m/min$. If we assume that the average kinetic energy of PM sputtered aluminum atoms is 12 eV (or that 12 eV represents the average energy imparted to the substrate per Al atom by the first five processes listed in Table VII), then the heat flux will be 0.242 W/cm^2. For a thermally isolated 300-μm-thick Si wafer, this represents a temperature rise rate of either 2.8°C/sec (6 eV) or 4.7°C/sec (12 eV). Thus, starting at 25°C the substrate will reach ∼ 190°C in 1 min in the former case and over 300°C in the latter case. The critical point for microcrack formation occurs when self-shadowing starts. This depends on step geometry, but one can conclude that heat flux from the impinging atoms may not get the substrate hot enough, soon enough, to heal a microcrack unless the substrates are ∼ > 100°C at the start of the deposition.

A negative substrate bias of the proper magnitude can promote microcrack healing by resputtering of the growing film [80, 99, 100]. If, in addition, the incoming Al is ionized, then conformal coating would be enhanced by self-sputtering and ion-plating effects. Any such unintentional bias effect would be sensitive to the degree of electrical isolation of the substrate, the position (and motion) of the substrate in relation to the cathode [72], the cathode power [23, 72], and the source–substrate spacing. An auxiliary anode would reduce or eliminate electron bombardment of the substrate and thus influence the floating bias. If bias effects are indeed responsible for enhanced step coverage during dc PM Al deposition, then variance in the reported experimental results would be expected. A comparison of electron-beam deposited pure Al with PM sputtered Al indicated that the PM deposited Al, having smaller more randomly oriented grains, was more resistant to grain growth during heat treatment but the median time to failure due to electromigration was less [101].

b. Passivation. A low-temperature hermetic passivation process employing Al_2O_3 or Si_3N_4 is desirable [95]. Previously mentioned studies indicate that good step coverage can be achieved on aluminum metallization by rf PM sputtering of Al_2O_3 [72] or SiO_2 [29]. It has been shown that for conventional rf sputtering, an rf substrate bias of -60 V is sufficient for vertical-wall step coverage with SiO_2 and a -100 V bias enables Al_2O_3 to cover a reentrant step [100]. Thus, PM rf sputtering offers an alternative low-temperature deposition technique.

2. Thin-film Thermal Printhead Fabrication

Thermal printheads employing a matrix of thin-film resistor elements to develop an image on thermally sensitive paper [102–104] are being used on computers and portable and hand-held printing calculators. The printing element requires an abrasion- and chemically-resistant overlayer to prevent premature wearout. A 6-μm-thick Al_2O_3 layer deposited by the PM sputtering process previously described [72] has provided an effective printhead wear-resistant layer. The resistive element (Ta–Al) and the low-resistance interconnections (Al–Cu) are also deposited sequentially from dual independent cathodes in an in-line dc PM sputtering system [104].

C. Miscellaneous Applications

Table IX lists recent PM sputtering applications. In many applications, PM sputtering has the potential for continuous automated processing [105, 107].

On a smaller scale, dc sputter deposition from 5–7-cm diameter disk PM sources has been found advantageous for coating heat-sensitive biological samples with Au, Au–Pd, or C prior to examination by scanning electron microscopy. For example, not until soil fungi samples prepared by PM sputtering of Au were observed were the obliterative effects of filament evaporation coating noted. The conclusion was that vacuum evaporation was an unreliable coating technique for retention of microorganism surface structure detail [108].

Table IX

PM Sputtering Applications

Application	Materials	Purpose	Reference(s)
Sputter deposition	Al–1% Si	Integrated circuit metallization	[43]
	Al	Integrated circuit metallization	[97, 101, 101a]
	Al–SiO$_2{}^a$–TiO$_x$	Coated Al mirror	[60]
	Cr–Au, Cr–Cu	Ceramic metallization	[106, 107]
	Ta–Al	Resistor film	[102, 64]
	TaN	Resistor film	[82]
	Al$_2$O$_3{}^a$	Evaluation of film properties	[72]
	SiO$_2{}^a$	Evaluation of film properties	[29]
	W	Evaluation of film properties	[101b]
	Cr alloy	Plastic metallization	[94b]
Sputter cleaning	Sheet steel	Clean prior to Ni evaporation	[45]
Ion plating	Ni	PM cathode at substrate during Ni evaporation	[84]
	Cu	PM as sputter/evaporation source	[85]

^{a}rf PM.

V. CONCLUSIONS

Planar magnetron sputtering can deposit metallic films over large areas at rates comparable to electron-beam evaporation without the degree of radiation heating typical of thermal sources. The deposition rates for dielectric films are higher than those for conventional rf sputtering but are lower than the rates achievable by chemical vapor deposition; however, a wider range of materials may be deposited.

At the present stage of development of PM sputtering there are more questions than answers. For example, what is the energy distribution of the sputtered atoms? What percentage is ionized? How can the magnetic field pattern be optimized? What is the effect of controlling the substrate bias? In rf sputtering, what is the effect of an anode? Can direct reactive sputtering of dielectrics produce high-quality films at high rates?

The details of competitively advantageous production processes are often proprietary and this is becoming true for planar magnetron technology. Developments may be several years old or even obsolete before being published. Nevertheless, dissemination of information inevitably occurs to the general benefit of industry.

REFERENCES

1. L. I. Maissel, in "Handbook of Thin Film Technology" (L. I. Maissel and R. Glang, eds.), p. 4-1. McGraw-Hill, New York, 1970.

2. S. D. Dahlgren and E. D. McClanahan, *Proc. Symp. Deposition Thin Films, 3rd, Univ. Rochester* p. 20 (1969); S. D. Dahlgren, E. D. McClanahan, J. W. Johnston, and A. G. Graybeal, *J. Vac. Sci. Technol.* **7**, 398 (1970).

3. F. M. Penning, *Physica (Utrecht)* **3**, 873 (1936); U.S. Patent 2,146,025 (1939).

4. E. Kay, *J. Appl. Phys.* **34**, 760 (1963).

5. W. D. Gill and E. Kay, *Rev. Sci. Instrum.* **36**, 277 (1965).

6. E. Kay and A. P. Poenisch, U.S. Patent 2,282,815 (1966).

6a. P. D. Davidse and L. I. Maissel, U.S. Patent 3,369,991 (1968).

7. J. R. Mullaly, *Res/Dev.* **22**(2), 40 (1971).

8. K. Wasa and S. Hayakawa, *Rev. Sci. Instrum.* **40**, 693 (1969); U.S. Patent 3,528,902 (1970).

9. A. M. Gurewitsch and W. F. Westendorf, *Rev. Sci. Instrum.* **25**, 389 (1954); L. D. Hall, *Rev. Sci. Instrum.* **29**, 367 (1958); *Science* **128**, 279 (1958).

10. J. C. Helmer and R. L. Jepsen, *Proc. IRE* **49**, 1920 (1961).

11. W. Knauer and E. R. Stack, *Trans. Natl. Vac. Symp., 10th, Boston, 1963* p. 180 (1964).

11a. W. Knauer, U.S. Patent 3,216,652 (1965).

12. K. M. Welch, U.S. Patent 4,006,073 (1977).

13. I. G. Kesaev and V. V. Pashkova, *Sov. Phys.—Tech. Phys.* **4**, 254 (1959).

13a. J. P. Argana, CVC Products, Inc., Rochester, New York (private communication).

14. J. S. Chapin, *Res./Dev.* **25**(1), 37 (1974); U.S. Patent Appl. 438,482 (1974).

15. H. R. Smith, *Proc. Electron Beam Process. Semin., 3rd, Stratford, England; Universal Technology Corp., Dayton, Ohio* p. 2cl (1974).

16. R. L. Cormia, P. S. McLeod, and N. K. Tsujimoto, *Proc. Int. Conf. Electron Ion Beam Technol., 6th, Electrochem. Soc., Princeton, N.J.* p. 248 (1974).

17. S. Hurwitt, *Trans. Conf. Prod. Sputter., Materials Research Corp., Orangeburg, N. Y.* p. 3.1 (1974).

18. A. M. Dorodnov, *in* "Fisika i Primenenie Plasmennich Uskoritelej" (A. I. Morosov, ed.), p. 330. Nauka Tehnika, Minsk, 1974.

19. J. F. Corbani, U.S. Patent 3,878,085 (1975).

20. T. Van Vorous, *Circuits Manuf.* **15**(6), 6 (1975).

21. G. N. Jackson, *ElectroComponent Sci. Technol.* **3**, 254 (1977). Abstr.

22. W. D. Westwood, *Prog. Surf. Sci.* **7**, 71 (1976).

23. S. Schiller, U. Heisig, and K. Goedicke, *Thin Solid Films* **40**, 327 (1977).

23a. J. Markus, "Electronics and Nucleonics Dictionary," 3rd Ed. McGraw-Hill, New York, 1966.

24. T. Van Vorous, *Solid State Technol.* **19**(12), 62 (1976).

25. P. S. McLeod, U.S. Patent 3,956,093 (1976).

26. F. A. Green and B. N. Chapman, *J. Vac. Sci. Technol.* **13**, 165 (1976).

27. E. A. Nesbitt and J. H. Wernick, "Rare Earth Permanent Magnets." Academic Press, New York, 1973.

28. R. J. Parker and R. J. Studders, "Permanent Magnets and Their Application." Wiley, New York, 1962.

29. K. Urbanek, *Solid State Technol.* **20**(4), 87 (1977).

30. D. J. Ball, *J. Appl. Phys.* **43**, 3047 (1972).

31. W. G. Perkins, *J. Vac. Sci. Technol.* **10**, 543 (1973).

32. S. S. Lau, R. H. Mills, and D. G. Muth, *J. Vac. Sci. Technol.* **9**, 1196 (1972).

33. L. I. Maissel and J. H. Vaughn, *Vacuum* **13**, 421 (1963).

33a. Materials Research Corp., Data Sheet 1057-AMD, Orangeburg, New York (1978). Unpublished.

34. S. Muto, Hewlett-Packard Co., personal communication (1975).

35. F. Rosebury, "Handbook of Electron Tube and Vacuum Techniques." Addison-Wesley, Reading, Massachusetts, 1965.

36. W. H. Kohl, "Handbook of Materials and Techniques for Vacuum Devices," p. 333. Reinhold, New York, 1967.
37. H. H. Manko, "Solders and Soldering." McGraw-Hill, New York, 1964.
38. J. van Esdonk and F. M. Janssen, Res./Dev. 26(1), 41 (1975).
39. J. A. Seitchik and B. F. Stein, Rev. Sci. Instrum. 39, 1062 (1968).
40. A. K. Gupta, K. V. Kurup, J. Smanthanam, and P. Vijendran, Vacuum 27, 61 (1977).
41. P. J. Clarke, Solid State Technol. 19(12), 77 (1976).
42. N. Veigel, U.S. Patent 4,009,090 (1977).
43. "Physical Vapor Deposition" (R. J. Hill, ed.), pp. 114–149. Airco-Temescal, Berkeley, California, 1976.
44. E. Soxman, personal communication (1976); Anonymous, Res./Dev. 27(9), 63 (1976).
45. S. Schiller, U. Heisig, and K. Steinfelder, Thin Solid Films 33, 331 (1976).
46. N. Hosokawa, T. Tsukada, and T. Misumi, J. Vac. Sci. Technol. 14, 143 (1977).
47. J. W. Coburn and E. Kay, J. Appl. Phys. 43, 4965 (1972).
48. L. T. Lamont, Jr., Varian Vac. Views 9(3), 2 (1975).
49. L. T. Lamont, Jr., J. Vac. Sci. Technol. 14, 122 (1977) (Abstr).
50. G. K. Wehner and G. S. Anderson, in "Handbook of Thin Film Technology" (L. I. Maissel and R. Glang, eds.), p. 3–1. McGraw-Hill, New York, 1970.
51. K. H. Behrndt, J. Vac. Sci. Technol. 9, 995 (1972).
52. E. B. Graper, J. Vac. Sci. Technol. 10, 100 (1973).
53. R. Glang, in "Handbook of Thin Film Technology" (L. I. Maissel and R. Glang, eds.), p. 1–3. McGraw-Hill, New York, 1970.
54. L. Holland, "Vacuum Deposition of Thin Films," p. 152. Wiley, New York, 1956.
55. J. Strong, H. V. Neher, A. E. Whitford, C. H. Cartwright, and R. Hayward, "Procedures in Experimental Physics," p. 177. Prentice-Hall, Englewood Cliffs, New Jersey, 1938.
56. Hugh R. Smith, Industrial Vacuum Engineering, personal communication (1976).
57. S. Muto, W. Ebert, and R. S. Nowicki, Hewlett-Packard Co., unpublished observations (1976).
58. Reference 43, photograph following p. 102.
59. Perkin–Elmer Ultek, Inc., Sales Inf. Bull. No. 700-12. Palo Alto, California (1976). Unpublished.
59a. Perkin–Elmer Ultek, Inc., Engineering Note EN 1007. Palo Alto, California (1977). Unpublished.
60. L. D. Hartsough and P. S. McLeod, J. Vac. Sci. Technol. 14, 123 (1977).
61. P. S. McLeod and L. D. Hartsough, J. Vac. Sci. Technol. 14, 263 (1977).
62. Materials Research Corp., Data Sheet 1002-PED. Orangeburg, New York (1976). Unpublished.
63. S. Dushman and J. M. Lafferty, "Scientific Foundations of Vacuum Technique," 2nd Ed., Ch. 5. Wiley, New York, 1962.
64. W. Sperry and R. K. Waits, unpublished observations (1975).
65. Reference 1, p. 4-40.
66. L. T. Lamont, Jr. and F. T. Turner, J. Vac. Sci. Technol. 11, 47 (1974).
67. G. Stern and H. L. Caswell, J. Vac. Sci. Technol. 4, 128 (1967).
68. R. C. Weast, ed., "Handbook of Chemistry and Physics," 57th Ed. CRC Press, Cleveland, Ohio, 1976.
69. Reference 35, pp. 521–525.
70. Reference 43, p. 121.
71. D. H. Grantham, E. L. Paradis, and D. J. Quinn, J. Vac. Sci. Technol. 7, 343 (1970).
72. R. S. Nowicki, J. Vac. Sci. Technol. 14, 127 (1977).
73. R. K. Waits, unpublished observations (1976).

74. R. J. Ferran and J. J. Sullivan, *J. Vac. Sci. Technol.* **12**, 560 (1975).
75. V. Hoffman, *Electron. Packag. Prod.* **13**(11), 81 (1973).
76. M. Hablanian and P. Forant, *Varian Vac. Views* **11**(1), 4 (1977); **11**(2), 3 (1977).
77. L. T. Lamont, Jr. and A. Lang, *J. Vac. Sci. Technol.* **7**, 198 (1970).
78. J. W. Coburn, *Rev. Sci. Instrum.* **41**, 1219 (1970).
79. J. W. Coburn and E. Kay, *Solid State Technol.* **14**(12), 49 (1971); *Appl. Phys. Lett.* **18**, 435 (1971).
80. J. L. Vossen, *J. Vac. Sci. Technol.* **8**(5), S12 (1972).
81. G. Gafner, *Philos. Mag.* **58**, 1041 (1960).
82. J. Joly and J. B. Ranger, *C. R. Int. Colloq. Pulver. Cathod. Appl. 2nd, Nice* p. 34 (1976).
83. P. J. Clarke, *J. Vac. Sci. Technol.* **14**, 141 (1977).
84. S. Schiller, U. Heisig, and K. Goedicke, *J. Vac. Sci. Technol.* **12**, 858 (1975).
85. S. Schiller, U. Heisig, and K. Goedicke, *J. Vac. Sci. Technol.* **14**, 815 (1977).
86. R. C. Krutenat and W. R. Gesick, *J. Vac. Sci. Technol.* **7**(6), S40 (1970).
87. D. J. Healy and M. Lauriente, *Proc. Natl. Vac. Symp., 13th, San Francisco, Calif.* p. 55 (1966).
88. J. S. Logan, N. M. Mazza, and P. D. Davidse, *J. Vac. Sci. Technol.* **6**, 120 (1969).
89. N. M. Mazza, *IBM J. Res. Dev.* **14**, 192 (1970).
90. E. C. Rock and C. W. Smith, *J. Vac. Sci. Technol.* **12**, 943 (1975).
90a. A. Halperin, P. Silano, and L. West, *J. Vac. Sci. Technol.* **15**, 116 (1978).
91. C. A. T. Salama, *J. Electrochem. Soc.* **117**, 913 (1970).
92. R. E. Jones, H. F. Winters, and L. I. Maissel, *J. Vac. Sci. Technol.* **5**, 84 (1968).
93. J. B. Lounsbury, *J. Vac. Sci. Technol.* **6**, 838 (1969).
94. G. Kienel and H. Walter, *Res./Dev.* **24**(11), 49 (1973).
94a. T. Van Vorous, *Opt. Spectra* **11**(11), 30 (1977).
94b. L. Hughes, R. Lucariello, and P. Blum, *Soc. Vac. Coaters, Proc. Annu. Tech. Conf., 20th Atlanta, Ga.* p. 15 (1977).
95. A. J. Learn, *J. Electrochem. Soc.* **123**, 894 (1976).
96. V. Hoffman, *Solid State Technol.* **19**(12), 57 (1976).
97. J. B. Bindell and T. C. Tisone, *Thin Solid Films* **23**, 31 (1974).
98. D. Chun, Hewlett-Packard Co., personal communication (1977).
99. J. L. Vossen and J. J. O'Neill, Jr., *RCA Rev.* **31**, 276 (1970).
100. T. N. Kennedy, *J. Vac. Sci. Technol.* **13**, 1135 (1976).
101. K. Kauchi, H. Maeda, S. Kishi, and Y. Haneta, *Electrochem. Soc. Extend. Abstr.* **77-1**, 316 (1977).
101a. C. Ladas, *Trans. Conf. Sputter. Electron. Components; Materials Research Corp., Orangeburg, N.Y.* p. 6-1 (1976).
101b. R. Kossowsky, *Electrochem. Soc. Extend. Abstr.* **77-2**, 429 (1977).
102. F. Ura, U.S. Patent 4,007,352 (1976).
103. S. Shibata, K. Murasugi, and K. Kaminishi, *IEEE Trans. Parts, Hybrids Packag.* **PHP-12**, 233 (1976).
104. R. B. Taggart and B. E. Musch, *Hewlett-Packard J.* **28**(3), 9 (1976).
105. M. Hutt, *Solid State Technol.* **19**(12), 74 (1976).
106. A. J. Aronson, *C. R. Int. Colloq. Pulver. Cathod. Appl. 2nd, Nice* p. 175 (1976).
107. A. Aronson and S. Weinig, *Vacuum* **27**, 151 (1977).
108. S. Draggan, *Appl. Environ. Microbiol.* **31**, 313 (1976).

II-5

Ion Beam Deposition

JAMES M. E. HARPER

IBM Thomas J. Watson Research Center
Yorktown Heights, New York

I. INTRODUCTION

Ion beams are used in thin film deposition in two basic configurations. In primary ion beam deposition (Fig. 1a) the ion beam consists of the desired film material and is deposited at low energy (around 100 eV) directly onto a substrate. In secondary ion beam deposition (Fig. 1b), or ion beam sputter deposition, the ion beam is usually an inert or reactive gas at

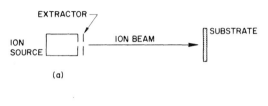

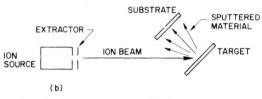

Fig. 1. Ion beam deposition configurations. (a) Primary ion beam deposition, with ion beam deposited directly onto substrate, (b) secondary ion beam deposition, with inert or reactive ion beam sputtering target material onto substrate (also called ion beam sputter deposition).

higher energy (hundreds to thousands of electron volts). The beam is directed at a target of the desired material, which is sputtered and collected on a nearby substrate.

Ion beam deposition allows greater isolation of the substrate from the ion generation process than is found in the conventional diode sputtering configuration. This enables control over the substrate temperature, gas pressure, angle of deposition, and the type of particle bombardment of the growing film, as well as independent control over the ion beam current and energy. This flexibility characterizes ion beam deposition as a useful technique in the study of thin film growth processes and has led to unique film properties not obtained in conventional deposition methods.

In this chapter the two techniques of ion beam deposition will be described, with the emphasis on the major controlling parameters which affect the deposition process. First an introduction will be given to ion source types and the basic processes of ion beam formation and control.

II. ION BEAM GENERATION

Two types of ion source will be described which are used in practical thin film deposition, the Kaufman source and the duoplasmatron. The Penning ion source will also be described to introduce some of the concepts of ion generation. Several other ion sources will be included which have particular characteristics which may be useful in future deposition applications. For further details on ion sources and ion beam transport the reader is referred to the book by Wilson and Brewer [1] and its references, and to conference proceedings [2].

A. Production of Ions

The ionization mechanism used in most ion sources is electron impact ionization in a low voltage gas discharge. The cross section for electron impact ionization has a maximum at low electron energy, about 70 eV for argon [3], therefore high electron energies are not needed for efficient ionization. More important are the electron supply rate and gas pressure, since the mean free path for ionization is inversely proportional to gas pressure (about 0.1 cm at 1 Pa) [4].

Almost any species can be introduced into a gas discharge as a vapor or by sputtering from the solid. The discharge may consist primarily of the desired species, for example argon for secondary ion beam deposition, or it may use an inert gas to support the discharge while the desired material is introduced to be ionized, as in primary ion beam deposition of a metal. The types of discharge have been described elsewhere [5]. Some discharge parameters will be mentioned here for orientation, not as a description of a particular configuration. Source pressures range from 10^{+2} to 10^{-2} Pa, giving mean-free-paths for atomic collisions of 0.01–100 cm. The neutral atom density is several times the ion density [4], but since extraction is more efficient for ions, the ionization efficiency can be high, up to 50–90%. Multiple ionization is usually a few percent of single ionization [4, 6]. The mean-free-path for electron–electron collisions is longer than the size of the source, so the electron and atom populations are usually not in thermal equilibrium [4]. Other mechanisms of ion production used in ion sources are thermal ionization [7], surface contact ionization [8], and field ionization [9, 10].

B. Ion Sources

1. Penning Ion Source

The configuration of a Penning ion source [11, 12] is shown in Fig. 2, consisting of a cylindrical anode with two cathodes forming the end plates. A magnetic field of several hundred gauss is applied parallel to the cylinder axis. With an argon pressure of about 15 Pa and a discharge voltage of several hundred volts, a plasma fills the chamber. Electrons repelled from the cathodes are attracted to the anode but the magnetic field constrains the electrons to follow helical trajectories, causing them to oscillate between the cathodes and increasing their path length, enhancing the ionization efficiency. Ions formed by collisions between fast electrons and gas atoms are attracted to the cathodes, where they bombard the surface, generating more electrons by secondary electron emission. As the discharge voltage is raised, the discharge current increases approximately

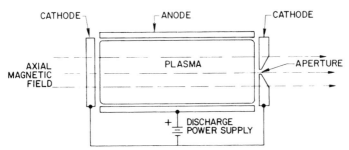

Fig. 2. Penning ion source with aperture in cathode.

linearly at first, indicating that the rate of secondary electron emission is not changing with ion energy. However, the ions bombarding the cathodes heat the surfaces, and if the discharge voltage is further increased, the cathodes become hot enough to emit electrons thermionically. At this point the electron emission increases rapidly, and the discharge changes to a negative resistance characteristic [12], as shown in Fig. 3.

Two points are illustrated here. First, the discharge has different modes of operation depending on the rate at which electrons are supplied to the plasma, as secondary electrons or by thermionic emission. Electron supply rate is therefore a primary control over the ion density. In a Pen-

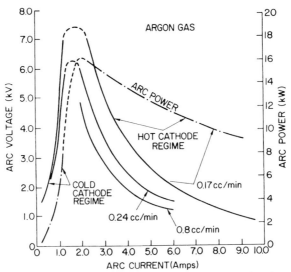

Fig. 3. Current–voltage characteristic of a Penning ion source showing transition from cold cathode regime to hot cathode regime. Argon flow rate is given in cm³/min (from Bennett [12]).

ning discharge, the discharge current is not independent of the discharge voltage, limiting the degree of control, and this is one reason why many ion sources use a separately heated thermionic cathode to supply electrons at a rate essentially independent of the discharge voltage. The thermionic cathode supports the discharge at lower discharge voltages, minimizing heating of the source and sputtering of cathode material into the discharge. The second point is the role of the magnetic field in constraining the electron orbits. The cyclotron radius of a 1000 eV electron in a magnetic field of 10^{-2} T (100 G) is about 1 cm [13]. Thus fields on the order of hundreds of gauss are sufficient to constrain the electron trajectories in a Penning source and allow operation at lower pressures than with no magnetic field.

The ion and electron density in the plasma is high enough to shield electric fields over distances typically less than 0.01 cm (the Debye length [4, 14]). This shielding isolates the plasma from external fields, thus ions are extracted from the plasma only when they diffuse to the plasma boundary, which may be considered as an ion emitting surface. An aperture in the anode or cathode (Fig. 2) is used to extract ions into a beam. The energy of ions in the beam is determined by the potential at which the ions originate. Since the plasma is highly conductive, it is approximately an equipotential region and may be characterized by a plasma potential, a few volts positive with respect to the anode [4]. The ions originate at close to plasma potential, with an energy spread determined primarily by the discharge voltage. For accurate focusing, mass separation, and energy control it is important that the initial ion energies have a narrow spread. In a cold-cathode Penning source this is in the range 10–50 eV [15]. Thermionically supported Penning sources have an energy spread of a few electron volts [16].

2. Kaufman Ion Source

The Kaufman source is an extension of the Penning source concept to include a thermionic cathode and multiple aperture extraction. This broad-beam ion source was originally developed in 1960 for use in space propulsion as an ion thruster [17]. A review of ion thruster technology is given by Kaufman [4], and sources optimized for uniform large area ion milling are described by Reader and Kaufman [18] and Robinson and Kaufman [19].

The cross section of a Kaufman source is shown in Fig. 4 [20]. The anode forms one end of the cylindrical discharge chamber and the screen grid (at cathode potential) the other end. The thermionic cathode circles the chamber wall, supported by four clamps, and the discharge gas is in-

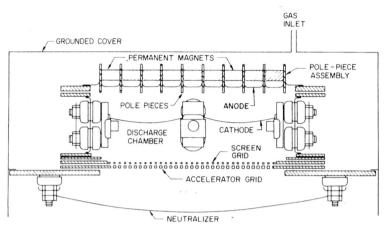

Fig. 4. Cross section of a Kaufman ion source which generates a 10 cm diameter beam (courtesy of Ion Tech, Inc. [20]).

troduced through the anode. Behind the anode is an array of permanent magnets which produce a multipole or "picket fence" field configuration across the anode surface. In some configurations an axial magnetic field is used, with both end surfaces at cathode potential (an example is given in Section III.A). The function of the magnetic field is to keep electrons emitted by the cathode from traveling directly to the anode. The screen grid contains about 1400 holes of 0.21-cm diameter. Aligned with this grid at a separation of 0.1 cm is an extraction or acceleration grid of similar structure [18]. A potential difference between the two grids shapes the plasma boundary into a meniscus at each aperture in the screen grid. The shape and control of this ion emitting surface will be discussed in the section on extraction.

The heated cathode allows operation at lower pressure (0.1 Pa) and discharge voltage (50 V) than a cold cathode Penning source, minimizing multiple ionization and reducing the ion energy spread to 1–10 eV [4, 15]. The discharge current is typically 1–2 A. Primary control over the ion density is through the rate of electron emission, which has two regimes of behavior. For low emission the ion density is proportional to the emission current, which is determined by the filament temperature. At high emission currents the gas pressure determines the ion density. The upper limit on pressure is determined by arcing in the source. An important advantage of independent cathode control is that the beam current may be varied independently of the discharge voltage and beam energy. The extracted argon ion current density from a 10-cm diameter source is in the range 1–2 mA/cm^2 at a beam energy of 500–2000 eV [18]. The ionization efficiency is in the range 50–75% [4].

Various magnetic field configurations have been tested for ionization efficiency and beam uniformity. The multipole configuration [18, 19, 21] (Fig. 4), with about 0.01 T (100 G) at the pole pieces, produces a more uniform beam profile than an axial or diverging field [4].

3. Duoplasmatron

The duoplasmatron, developed in 1956 by von Ardenne [22], is a versatile ion source used to generate high intensity narrow ion beams. The name refers to the dual manner of constricting the plasma in an arc discharge by both electric and magnetic fields to form a small region of high ion density adjacent to the extraction aperture. The configuration is shown in Fig. 5 [23]. With an argon pressure of 1–100 Pa in the source, the hot cathode filament supports a discharge of 1–2 A between the cathode and anode (with 80 V discharge voltage). Between these electrodes is the intermediate or *zwischen* ("in-between") electrode held at a potential intermediate between that of the cathode and anode (40 V). The canal-shaped aperture in the intermediate electrode confines the discharge electrically and an intense magnetic field (0.1–0.3 T) (1000–3000 G) between the intermediate electrode and anode constricts the plasma axially. The anode region is shown in Fig. 6 [23]. The diverging magnetic field region shapes a plasma meniscus next to the anode aperture, which becomes the emitting surface for ion extraction. Argon ion currents of 2 mA [24] and

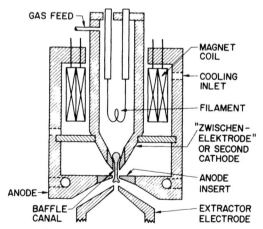

Fig. 5. Cross section of a duoplasmatron ion source showing location of plasma in baffle canal (from Brewer *et al.* [23]).

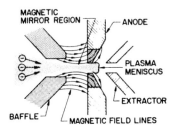

Fig. 6. Anode region of duoplasmatron showing constriction of plasma and plasma meniscus (from Brewer et al. [23]).

higher [15, 25] are obtained at about 20 keV beam energy through an aperture of 0.03-cm diameter with an energy spread of 10–50 eV [15]. The converging region of magnetic field lines acts as a magnetic mirror to reflect electrons back to the cathode region [23], enhancing ionization efficiency.

4. Other Ion Sources

Several additional ion source types will be mentioned, with references, to indicate the variety of approaches to generating ions. These are listed with the hope of stimulating novel applications to thin film deposition processes. The *glow discharge* ion source is simply a glow discharge between two plane electrodes with no magnetic field or thermionic cathode. One variation of this configuration is the *hollow anode* ion source in which the beam is extracted through a hole in the cathode plate [26, 27]. The *hollow cathode* ion source, of the same basic type, may be used in place of a thermionic cathode in applications where lifetime or contamination from the cathode are important [4, 28]. The *twin anode* or *electrostatic* ion source uses the geometrical arrangement of the anode surfaces within a surrounding cathode to produce long oscillatory electron trajectories, gaining enhanced ionization with no magnetic field [29].

The term *arc discharge* ion source applies to several configurations. In one type the desired material is vaporized from a crucible directly into a low voltage thermionically supported arc discharge. With some materials the arc is self sustaining with no support gas [30, 31]. No magnetic field is used. Ions are extracted with a low energy spread of 0.1–1.0 eV [31].

The *electrohydrodynamic* ion source [10] is a liquid-metal field emitting tip which generates a very high brightness, low current ion beam. A strong electric field pulls the liquid into a cusp-shaped tip less than 10^{-4}-cm diameter.

A vapor stream of atoms may be ionized by direct electron bombardment from a hot filament [32]. A variation of this technique, the *ionized cluster* source [33], vaporizes the desired material through a small orifice, inducing cluster formation as the vapor stream expands.

High multiply-charged ion states, for example Ar^{14+}, Kr^{17+}, Xe^{21+}, have been obtained by constraining the ions long enough for sequential ionization [34]. *Negative ion* beams may be generated by sputtering the desired material with a beam of low ionization potential material such as cesium [35], and may also be extracted from an off-axis aperture in a duoplasmatron [36]. *Very high current* hydrogen ion beams (several amperes) have been generated with a combined duoplasmatron and Penning ion source [37].

C. Beam Extraction and Control

1. Ion Extraction

Ions in the interior of the plasma diffuse to the plasma boundary, where they are extracted by an electric field. The rate of ion extraction is determined by Child's law of space charge limited current flow [4, 38]. In a planar geometry the space charge limited current density between two planes a distance d apart with potential difference V is

$$j = (4\epsilon_0/9)(2q/m)^{1/2}(V^{3/2}/d^2),$$

where ϵ_0 is the permeability of free space and q/m is the charge-to-mass ratio of the particles. This relationship determines the upper limit for planar current flow and demonstrates two important controls on the current density. Ion extraction from the plasma increases rapidly with increased extraction voltage and with decreased spacing between the plasma boundary and extraction electrode. For nonplanar geometries only the proportionality constant changes in Child's law, thus a given geometry may be characterized by the ratio $j/V^{3/2}$, the perveance [4].

Acceleration takes place mainly in the extraction region, since this is usually where the greatest potential drop occurs. However, the ion energy at the target is determined only by the potential difference between the target and the point of origin of the ion, which is usually within a few volts of anode potential. The usual arrangement is to have the target at ground potential and raise the entire source chamber to the desired beam voltage, with the extraction electrode at ground potential or lower.

As an example of extraction geometry, Fig. 7 shows the potential distribution through the extraction region of a Kaufman source [4]. The screen electrode is at the source potential (positive) and the accelerator or extraction electrode is negative. The screen voltage is lower than the anode voltage by the discharge voltage, with the difference between plasma potential and anode voltage not indicated. The extraction electrode is held negative, typically −200 V for source voltage of 1000 V, for two reasons. The rate of ion extraction is increased by a larger potential

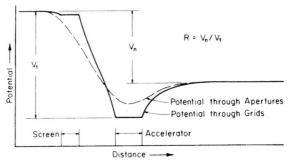

Fig. 7. Potential distribution through the extraction region of a Kaufman source (from Kaufman [4]).

difference between screen and accelerator electrodes, and the extraction electrode serves as a barrier to keep electrons (added to the beam downstream for neutrality) from backstreaming into the source and effectively shorting out the extraction field. The potential distribution is characterized by the net-to-total voltage ratio $R = V_n/V_t$, which is typically between 0.5 and 0.8 [4], the upper limit being determined by electron backstreaming and the lower limit by increased divergence of the beam.

The plasma meniscus adjusts its position such that the ion current density across the plasma boundary satisfies the space charge limited condition for the electric field at that location. In the desired operating condition (Fig. 8a) [4] the ion beam passes through the extraction electrode without impingement. As the ion density is increased, the meniscus moves downstream to a position of higher electric field, closer to the extraction electrode. If the ion density is too high, or extraction voltage too low, the meniscus can move to the position of Fig. 8b in which direct impingement on the extractor occurs, causing sputtering of this electrode. The normally concave shape of the meniscus forms an ideal ion emitting surface which gives the beam an initial convergence to counteract the de-

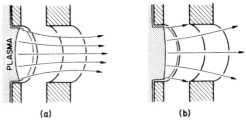

(a)   (b)

Fig. 8. Location of plasma meniscus in one aperture of a Kaufman source. (a) Normal extraction condition, (b) with excessive ion production (from Kaufman [4]).

focusing tendency of the accelerate–decelerate sequence. Thus, the beam may be extracted with a low divergence half-angle, typically 5–10° in a Kaufman source [4]. Beam divergence may also be decreased by increasing the separation between screen and extractor grids or by adding a third grid at ground potential [18, 39, 40]. The tradeoffs between extracted current, grid voltages, beam divergence, and extraction geometry have been studied in detail in the development of ion thrusters [4, 39, 41]. In addition to direct impingement of the beam, which is to be avoided, a second form of ion current to the accelerator grid arises from charge exchange collisions between ions and neutral atoms in the extraction region producing fast neutrals and slow ions. The slow ions are attracted to the accelerator grid and sputter its surfaces. This erosion is visible on the downstream side of the grid [4] but is serious only for very long term operation.

If the aperture diameter and grid spacing are scaled together, the ion current per aperture is independent of aperture size. This enables large beam currents to be extracted through grids containing a large number of closely spaced apertures. The upper limit on current is determined by the difficulty in maintaining a small gap between thin large diameter grids. The ion current density profile is shown in Fig. 9 for several distances from a 10-cm diameter Kaufman source operating at 500 eV [18], indicating uniformity of ± 5% across the inner half of the beam.

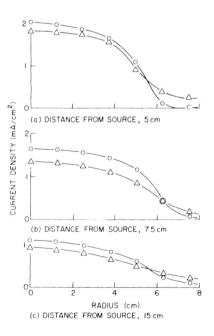

Fig. 9. Current density profile of 500 eV argon ion beam from 10-cm diameter Kaufman source at several distances from source and at two values of chamber gas pressure. ○, 5 × 10⁻⁴ Torr; △, 15 × 10⁻⁴ Torr (from Reader and Kaufman [18]).

The beam profile from a duoplasmatron or other single aperture source is usually close to Gaussian, determined more by the range of transverse velocities and the ion transport optics than by the initial profile at the extractor.

2. Beam Transport

Beam deflection is accomplished by transverse electric fields [42] and mass separation by magnetic fields [43]. If mass separation is desired with a straight trajectory, crossed electric and magnetic fields will select a particular velocity and deflect all others, acting as a mass separator [31, 44].

One type of focusing lens will be mentioned briefly. The Einzel, or unipotential, lens consists of a decelerate section followed by an accelerate section back to the initial beam energy [24]. The field distribution defocuses the beam while it is at high energy and focuses while it is at low energy, resulting in a net focusing of the beam. Examples of using the Einzel lens to focus the beam from a single aperture source are given in Sections III.A and IV.A.

3. Neutralization

The above methods of deflecting and focusing an ion beam apply to a nonneutralized beam. With the broad beam from a Kaufman source, neutralization is necessary to avoid beam spreading by space charge repulsion. This is accomplished by adding electrons from a thermionic filament (Fig. 4) and monitoring the net current to the target or a beam probe. When the net current is zero, the arrival rate of ions equals that of electrons, but the ions are not neutralized by recombination since the mean free path for this process is much larger than the beam diameter [4]. The beam is itself a plasma, in which the electrons rapidly distribute to cancel net charge, thus electron injection does not have to be uniform. The neutralizer filament is sputtered by the beam, and contributes to contamination unless suitable masking is provided. An alternate method of injecting electrons into the beam is from a hollow cathode source (the plasma bridge neutralizer) which may be located outside the beam [4, 44a].

D. System Requirements

1. Materials

Materials in the ion source must be stable at the temperatures involved (several hundred degrees Celsius), have low sputtering yield if subjected to ion bombardment, and low susceptibility to corrosive gases, if used.

Particularly important components are the cathode filaments and electrode apertures. Thermionic cathodes are tantalum or tungsten wire [18], or barium oxide coated mesh [24]. Thermal cycling leads to embrittlement and failure, with exposure to oxygen or more reactive gases shortening filament life greatly. Magnetic fields may be supplied by an external solenoid, usually air cooled, or by permanent magnets, which do not have to be exposed to the plasma. Other surfaces may be molybdenum for thermal stability. In the Kaufman source the screen and accelerator grids span a large diameter, with a small separation, typically 0.1 cm. Thermal expansion leads to distortion, directly affecting extracted ion density and uniformity. These problems are minimized in ion thrusters by using dished molybdenum grids [4], such that thermal distortion occurs uniformly across the grids, but these are difficult to make accurately. Pyrolytic graphite provides the best combination of low thermal expansion, high thermal conductivity, and mechanical stiffness in flat grids [18]. For highest purity films it may be necessary to fabricate parts of the ion source out of materials that are compatible with the desired film [27, 45].

The lifetime of the source is limited by the hot cathode burning out or by buildup of sputtered material in the source, leading to shorting or to insulating coating of electrodes. Flaking of accumulated sputtered material may also cause shorts and insulating supports must be shielded from sputter coating. Lifetimes range from hours to months depending on operating conditions.

2. Vacuum and Gas

The gas pressure in the ion source is determined by the type of source and the ion density needed for the desired ion current. Therefore the pressure at the target is determined by the conductance of the ion source apertures and the pumping speed of the pump, assumed to be in the target region. Background gas in the path of the beam has two effects in addition to the effect of gas pressure on the source itself. Large angle collisions contribute slightly to beam divergence, and charge exchange collisions which result in fast neutrals and slow ions contribute to the sputtering rate at the target without registering as ion current to a probe [18]. The pressure must be below about 1 Pa to sustain a beam without excessive scattering, but is usually maintained around 0.1 Pa or lower to minimize divergence and contamination.

3. Electrical and Other Requirements

The electrical power requirements of the ion sources described above are straightforward. The filaments may be heated with ac or dc current of several amperes at low voltage. The discharge power supply is relatively low voltage, but may need to deliver several amperes in some sources. The extraction voltage power supply is high voltage, low current, and should be protected against arcing or shorting which may occur between the electrodes. Feedback control of the beam current may be applied to the cathode heater power supply.

Heat inputs to the system include the hot cathode in the source, electron and ion bombardment of source surfaces, neutralizer filament, if used, and magnet current if electromagnets are used. The source is cooled by forced air cooling of the magnet coil and chamber, cold water cooling of the source electrodes or surrounding shroud, or more localized cooling on high-power sources, including glycol or liquid-nitrogen cooling of single aperture electrode structures. Multiaperture grids are thin and cannot be effectively cooled, thus they must be stable at their operating temperature of several hundred degrees Celsius. Heat input to the target is primarily from the energy of the ion beam if the target is far enough from the source to neglect radiation. A 1000 eV, 1 mA/cm^2 beam supplies 1 W/cm^2 to the target. Thus the total power input to the target may be 100 W or higher. Water cooling is needed to keep target temperatures to less than about 200°C [46]. One of the advantages of secondary ion beam deposition is that these heat loads are essentially confined to the source and target, and the substrates do not experience a high heat input. In primary ion beam deposition the beam energy and current density are significantly lower, and heat input to the substrate is not a problem.

E. Measurement of Beam Properties

Positive ion current density from a Kaufman source is measured with a probe biased to about -20 V to repel the low-energy electrons in the beam. Secondary electron emission from the probe surface contributes to a high reading, but is a relatively small contribution (a few percent) at low beam energies with refractory metal probes [47]. The secondary electrons may be collected in a Faraday cup with a negatively biased entrance aperture. Beam profile and divergence are measured with an array of Faraday probes [41] or a scanning wire [48]. Ion energy distribution is measured with an electrostatic analyzer [49, 50]. Degree of neutralization is measured with an unbiased probe to sample the net beam current. The beam composition and charge state [6] may be determined by mass separation [31] or measured with a mass spectrometer [50]. The neutral component impinging on the target may be measured by residual gas analysis or an ionizer–mass spectrometer combination [50].

III. SECONDARY ION BEAM DEPOSITION

Several aspects of secondary ion beam deposition (or ion beam sputter deposition, Fig. 1b) differ from other sputtering processes: *low background pressure* gives less gas incorporation and less scattering of sputtered particles on their way to the substrates; *directionality of beam* allows variable angle of incidence of beam on target, and angle of deposition on substrate; *almost monoenergetic beam* with narrow energy spread allows study of the sputter yield and deposition process as a function of ion energy, and enables accurate beam focusing and scanning; *beam independent of target and substrate processes* allows changes in target and substrate materials while maintaining constant beam characteristics; *target and substrates not part of rf circuit* minimizes substrate heating [51]; and *independent control of beam energy and current density.* Some applications of secondary ion beam deposition are reviewed by Weissmantel and Gautherin [51a] and uses of ion beams for device fabrication are discussed by Spencer and Schmidt [51b].

A. Source and Beam Characteristics

Two examples of secondary ion beam deposition systems will be illustrated. A Kaufman source [52] is shown in Fig. 10, equipped with a water-

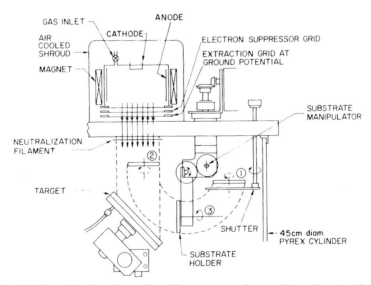

Fig. 10. Secondary ion beam deposition system using an 8-cm diameter Kaufman source, showing substrate positions for (1) target presputter, (2) substrate precleaning, (3) sputter deposition (courtesy of Veeco Instruments, Inc. [52]).

cooled target and rotating substrate holder which can be positioned for substrate precleaning in the beam, film deposition, and target precleaning. The beam diameter is 7 cm, target diameter 13 cm, and substrate holder diameter 6.5 cm. Argon ion energy is variable at 200–2000 eV, with current density up to 1 mA/cm², total beam current about 30 mA. Higher current sources are also available [18].

A duoplasmatron ion source with Einzel lens [53] is shown in Fig. 11. This system produces a 20 keV argon ion beam focused to 0.4-cm diameter at the target with total beam current of 2 mA, corresponding to a current density of about 20 mA/cm². The narrow beam allows deposition of thin films from milligram amounts of target material. Early studies of secondary ion beam deposition were made by Chopra and Randlett [54] using a modified duoplasmatron and by Schmidt *et al.* [27] using a hollow anode source.

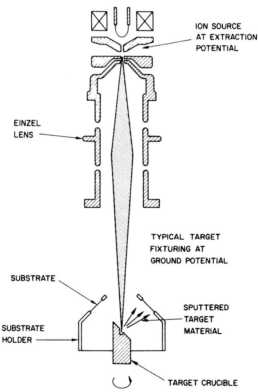

Fig. 11. Secondary ion beam deposition system using a duoplasmatron with Einzel lens to focus the beam (courtesy of General Ionex Corp. [53]).

B. Target Processes

1. Sputter Yield

The sputter yield, s, is the ratio of ejected target atoms to incoming beam ions. Only a summary of the aspects of sputter yield important in ion beam deposition will be given here. More complete descriptions of sputter yield are given by Winters [54a] and Oechsner [54b].

(1) s increases with ion energy E approximately proportional to $E^{1/2}$ from a threshold of a few tens electron volts to a maximum around 10–20 keV, then decreases at higher energy due to implantation [55]. s is of the order of unity in the energy range considered here [56].

(2) s generally increases with mass of ion [54b, 55, 57].

(3) s increases with angle of ion incidence away from normal incidence, to a maximum at an angle which depends on ion and target species and energy, then decreases to zero at grazing incidence [54b, 58].

(4) s depends on the target material, but differences are generally within an order of magnitude [56].

(5) the energy of sputtered particles is in the range of tens of electron volts, with a small fraction at higher energies [59].

The sputtering rate for an ion beam of current density 1.0 mA/cm² with $s = 1$ corresponds to 6×10^{15} atoms/sec/cm² removal rate, which gives about 7 Å/sec for copper, on the order of several monolayers per second. Since the ion density is low enough to treat each ion-target collision independently, the sputtering rate is proportional to current density. The duoplasmatron source in Fig. 11 operates at 20 keV to optimize the sputter yield for high deposition rates. The Kaufman source in Fig. 10 operates at lower energy, 500–2000 eV, to minimize heating of substrates in its complementary use as an ion milling system.

Argon is the commonly used source gas, although some gains may be found by using heavier gases for higher yields. Some dependence of film properties on sputtering gas species has been reported [27, 60].

The increase of sputter yield with angle of ion incidence allows gains by tilting the target. This also directs the sputtered material away from the beam direction, decreasing the rate of accumulation of material in the source. Typically an angle near 45° is used. The maximum sputtered flux is directed normal to the target surface for high ion energy (> 10 keV) and increasingly towards the angle of reflection of the beam for low energy [54b, 60a]. This is because at low ion energy the first ion–target collision becomes dominant as contrasted with the collision cascade at higher energy [54b]. The angular distribution of sputtered material may also be affected by severe erosion and cone formation on the target surface [61].

2. Target Types

In addition to deposition from single element and compound targets, alloy films may be deposited from composite targets with different sectors made of different materials [46]. The relative areas are determined from the relative yields of the materials. In multicomponent targets a difference in sputter yield of target constituents leads to altered layer formation [62] in which the target surface is enriched in the lower sputter yield constituent. After this layer reaches equilibrium, the sputtered flux usually matches the target composition [46, 62a]. A dependence of sputtered film composition on angle of ejection has however been reported [62b] with Ni–Fe and Ni–Cu alloy targets. It is necessary to presputter the target for a period of time before exposing the substrates, to stabilize the ion source and beam characteristics, outgas surfaces by heating, clean the target surface, and establish the altered layer.

Sputtering of insulators requires neutralization of the beam to avoid charging of the target surface, leading to beam deflection and sparking. Neutralization is less of a problem with metals, where secondary electrons from the target help to neutralize the beam. To avoid ion bombardment of unwanted surfaces it may be necessary to collimate the beam with an aperture [45].

3. Effects of Target on Ion Source

The main interaction of the target with the ion source is through the flux of sputtered material which accumulates in the source. This leads eventually to flaking, shorting, or contamination of the target surface. Accumulation of insulating material in the source may lead to difficulty maintaining constant beam conditions or extinguish the source and cause difficulty restarting. These problems are eliminated by removing excess material from the source. The sputtered flux may also modify the discharge slightly by becoming a constituent of the plasma and the beam.

C. Film Properties

Since the substrates in an ion beam sputtering system are essentially isolated from the source and target processes, substrate temperature may be independently controlled, enabling deposition on temperature sensitive materials. Normally the substrate is at ground potential and receives no bombardment by energetic electrons as in diode sputtering. Some bombardment by energetic reflected beam atoms occurs [62c,d]. Substrate size and location are limited somewhat by the conflicting requirements of being close to the target for high deposition rates and also not obstructing the ion beam.

1. Deposition Rate and Uniformity

The primary controls on deposition rate are ion current density, ion energy, and geometrical collection efficiency. The order of magnitude of deposition rate for the systems described above is several hundred angströms per minute. The deposition profile of Ti deposited in the duoplasmatron system shown in Fig. 11 is given in Fig. 12 [24]. The profile is somewhat narrower than the cosine distribution which would be expected from a simple point source. Uniformity of ± 5% is obtainable over a region of ± 15° about the target normal, and may be improved by rotating the substrates. The effects of angle of incidence and ion energy on the deposition profile are discussed by Oechsner [54b].

2. Purity

With an inert gas pressure of 10^{-3} Pa at the substrate, the growing film is bombarded by inert gas at the rate of about 4 monolayers/sec. However, the sticking coefficient of these thermal atoms is very low, leading to little incorporation in the film. Schmidt et al. [27] made a detailed study of the purity of ion beam sputtered films prepared with a hollow anode ion source using various inert gas ion beams at 0.1 mA/cm², 1200 eV. Inert gas incorporation was on the order of 0.1–1.0 at. % in Cr, Mo, Ti, W, and Zr, with less than 100 ppm atomic fraction in Ta and Au. Hinneberg et al. [63] report < 0.1 at. % Ar with traces of Ta, Fe, and W in ion beam sputtered Si. For reactive gases such as oxygen, however, maintaining a low partial pressure is more crucial. With deposition rates of several monolayers per second, oxygen incorporation to several atomic percent is not unlikely with reactive metal films. Getter pumping, liquid-nitrogen trapping, and bakeout of the system help reduce oxygen incorporation. Castellano et al. [46] report 0–1.2 at. % oxygen in Al films, the higher concen-

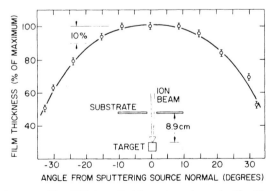

Fig. 12. Thickness profile of Ti deposited in system shown in Fig. 11 using 22 keV, 2.5 mA argon ion beam. The substrate location is shown (from Scaife et al. [24]).

trations found in obliquely deposited films, also up to 2 at. % oxygen in Ni and Au. Schmidt *et al.* [27] found oxidation of Ta films to be a problem. Further oxygen measurements were reported by Westwood and Ingrey [45].

Major metal contaminants can be traced to components of the system subjected to ion bombardment, such as the cathode plate in a hollow anode ion source [27] or a current probe exposed to the beam [46]. Contamination from the source is decreased by operating at a low discharge voltage to minimize energetic particles in the source, optimizing extraction for minimum ion bombardment of extraction apertures, locating the neutralizer out of sight of the target and substrates, using low sputter yield system components, and in critical applications by fabricating source structures from materials compatible with the desired film. Another approach is to mass analyze the beam, but this is usually not feasible in deposition systems which need a reasonably high deposition rate. Schmidt *et al.* [27] found higher purity films sputtered by Xe than by Ar.

3. Structure and Mechanical Properties

Small ion beam deposition systems have been installed in electron microscopes to study the early stages of sputtered film growth [64]. The qualitative features of film growth are found to be similar to evaporated films.

Weissmantel [65] finds Si films to be amorphous when deposited below 500°C and epitaxial on spinel above 750°C. Epitaxial Au and Ag on NaCl were obtained at room temperature by Chopra and Randlett [54]. Crystal structure and orientation are usually the same as obtained by other deposition techniques, although some unusual structures have been observed. Chopra and Randlett [54] obtained fcc modifications of normally bcc Ta, Mo, and W, and normally hcp Re, Hf, and Zr in an intermediate temperature range of substrates using argon ion beam sputtering. The refractory metal films were amorphous when deposited at room temperature. Schmidt *et al.* [60] observed impurity stabilized bcc Zr from an improperly cleaned target. Increases in lattice constant up to several percent were found by Schmidt *et al.* [60] in transition metal films due to gas incorporation. Texture is found to depend on angle of deposition [46], becoming more porous at glancing angles, similar to observations for evaporated films. Some measurements of grain size have been reported [46, 54, 60, 64].

Adhesion of ion beam sputtered films is reported to be superior to evaporated films [60], the improvement attributed to the higher energy of arrival of the sputtered particles. The arriving particle energy (tens of electron volts) helps to clean the substrate by sputtering away adsorbed gases, gives greater mobility to condensing atoms, and provides implantation into the substrate, enhancing chemical bonding and localized rearrangement. Hudson [66] reports improved adhesion of SiO_2 and Al_2O_3 films by depositing at a substrate temperature of 400°C.

Ion beam sputtered films were found by Schmidt *et al.* [27] to have lower internal stress than evaporated films. No curling was observed when removing Au and Ta films from various substrates. Castellano *et al.* [46] found low values of intrinsic stress in Al films. When substrates were exposed to the perimeter of the ion beam, however, subjecting the growing film to ion bombardment, the stress was found to change from tensile to compressive, and correlated with increased porosity and oxygen content of the films.

4. Electrical Properties

Schmidt *et al.* [60] studied the superconducting transition temperature T_c of ion beam deposited transition metals, and found increases over bulk values for Mo, Ti, Zr, and W. Cr was found to be superconducting for the first time. Decreases in T_c were found for Nb, Ru, Ta, and V. The changes in T_c are attributed to lattice expansion, correlating with the atomic size of the sputtering gas used. In all cases films sputtered with Xe gave the highest T_c. Preliminary results in ion beam sputtering of superconducting Nb–Ti films are given by Gautherin *et al.* [62a]. Values of mobility in ion beam deposited Si are discussed by Hinneberg *et al.* [63], and some results on ion beam sputtered GaAs and InSb were reported by Weissmantel *et al.* [65].

D. Reactive Ion Beam Sputtering

Secondary ion beam deposition may also be used to deposit compounds of reactive species, such as oxides and nitrides. Several methods are shown in Fig. 13 for depositing an element (usually a metal) in combination with a gaseous component with which it reacts. Method (a) uses an inert gas ion beam to sputter either the metal or compound target, with reactive gas added to the deposition region. Even in sputtering the compound directly it is usually necessary to add the reactive gas to compensate for the loss of reactive component by dissociation [40]. Method (b) uses an ion beam of the reactive gas itself or a mixture of inert and reactive gas. Usually a mixture is preferable to the pure reactive gas in the beam to maintain a reasonable sputtering rate. Method (c) is a dual ion beam method in which the inert gas ion beam sputters the metal target and a reactive gas ion beam is directed at the substrate surface.

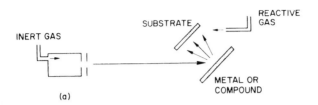

(a)

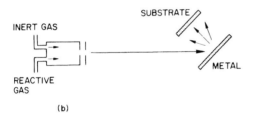

(b)

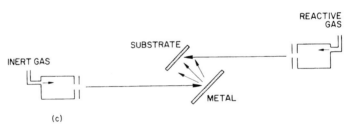

(c)

Fig. 13. Reactive secondary ion beam deposition methods. (a) Inert gas ion beam, reactive gas added to substrate region, (b) reactive and/or inert gas ion beam, (c) dual ion beam method using inert gas ion beam to sputter metal target and reactive gas ion beam to bombard growing film.

In reactive ion beam sputtering the compound formation may take place either at the target or substrate, depending on the technique used. The deposition rate of Ti and Zr as a function of oxygen partial pressure using method (a) is shown in Fig. 14, as measured by Castellano [67]. The rapid drop in deposition rate occurs when the oxide begins to form on the target surface faster than it is sputtered away, and the deposition rate becomes characteristic of the oxide, which has a lower sputter yield than the pure metal. At this point the deposited film becomes a transparent oxide film. The transition pressure is shifted to higher pressure as the beam current density is increased, since the arrival rate of oxygen at the target will

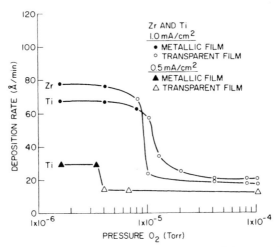

Fig. 14. Deposition rate of Zr and Ti films versus oxygen partial pressure in Kaufman ion source system using 2 keV argon ion beam at 0.5 mA/cm² and 1.0 mA/cm² (from Castellano *et al.* [46], reprinted by permission of Bell Telephone Laboratories, Inc.).

need to increase accordingly. This partial pressure dependence is analogous to reactive glow discharge sputtering. Westwood and Ingrey [45] also find evidence of oxide formation at the target in reactive sputtering of Ta with method (b). Incompletely oxidized films were attributed to the lack of a chemically active substrate region which is found in glow discharge sputtering. Using a single beam of N_2^+ ions to sputter a Si target with method (b), Weissmantel [68] found stoichiometric Si_3N_4 films only at the lowest beam energy and deposition rate. At higher beam energy the films are deficient in nitrogen, and display poor electrical properties compared with glow discharge sputtered films. In this configuration the nitride forms on the target surface.

With the dual beam technique of method (c), Weissmantel [68] used an Ar^+ ion beam to sputter a Si target while the substrate surface was bombarded simultaneously by a N_2^+ beam. Above a minimum level of N_2^+ flux, stoichiometric Si_3N_4 films were always formed. By shadowing a portion of the substrate from the N_2^+ beam it was found that a Si film condensed in the region not exposed to N_2^+ (only 2–3 wt % N in Si). These results show that in this dual beam method the reaction takes place at the substrate. The impact of the reactive ion beam on the substrate dissociates N_2 to atoms enabling reaction with the condensing Si atoms and local rearrangement. Films prepared by the dual beam technique display the same electrical characteristics as rf sputtered films. Thus, the flexibility of ion beam techniques allows unusual conditions to be established at the target and substrate surfaces. The fabrication of efficient photovoltaic heterojunctions of indium–tin oxides on silicon by ion beam sputtering is reported by DuBow et al. [69].

IV. PRIMARY ION BEAM DEPOSITION

Deposition of a charged particle beam (Fig. 1a) has several advantages over neutral particle deposition: control over *deposition energy;* ability to *deflect* and *focus* the beam; ability to *mass analyze* the beam to yield a highly pure deposit. Primary ion beam deposition has been used in two areas of interest. The first is the attempt to fabricate microcircuit elements by maskless deposition. The second is the investigation of the dependence of film properties on deposition energy.

A. Deposition Systems

In 1965, Wolter [70] first investigated the properties of Ag, Cu, and Cr films deposited as ions from a low voltage arc discharge between a thermionic cathode and the molten metal. A versatile arc discharge source is described by Amano et al. [31], in which the arc is self sustaining for Mg and Pb without an inert support gas. This source produces a beam of 10^{-5} A with an extraction voltage of 4–5 kV with an energy spread of several tenths of 1 eV. The beam is subsequently focused through an Einzel lens and decelerated to the substrate in the system shown in Fig. 15 [31].

A different approach is represented by the source used by Aisenberg and Chabot [71] to deposit carbon films of diamondlike structure. In this

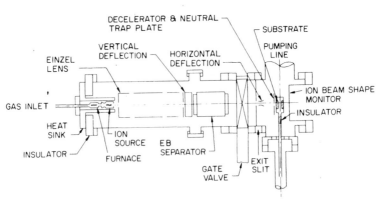

Fig. 15. Primary ion beam deposition system using extraction at high voltage (4 kV) followed by mass separation and deceleration to the substrate (from Amano et al. [31]).

source a Penning discharge arc is maintained between carbon electrodes using argon support gas, and the beam consists of carbon ions and neutrals and argon ions and neutrals. Spencer *et al.* [72] reported further work on diamondlike films with a similar source.

B. Beam Energy Limitations

In primary ion beam deposition the useful energy range of the beam is limited at low energies by low deposition rates and space charge spreading and at high energies by self-sputtering of the growing film or sputtering of the substrate. To determine the beam energy for maximum deposition rate, the dependence of self-sputter yield on energy for the material must be considered. The self-sputter yield s for Al, Cr, Cu, Au, and Ag [73] is shown in Fig. 16. The energy at which $s = 1$, E_1, varies from a few hundred to nearly 1000 eV, depending on the material. Above this energy no material accumulates and the beam sputters the substrate. Hayward and Wolter [73] found that self-sputter yields may be calculated approximately from inert gas yields using the mass conversion factor of Almen and Bruce [73a]. Further discussion of the ion mass dependence of sputter yield is given by Winters [54a], Oechsner [54b], and Sigmund [55]. For maximum deposition rate, self-sputtering yield must be minimized, thus deposition at the lowest energy is required. However, in a deposition system the beam current which can be delivered to a given area of the substrate increases with beam energy approximately linearly [74]. Therefore the combination of factors results in an optimum beam energy of about $\frac{1}{2}E_1$ [74]. Thus the maximum deposition rate occurs when approximately half of the incoming beam ions are being sputtered away. This is a very different environment at the growing film surface compared with most

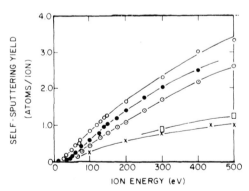

Fig. 16. Self-sputtering yield versus ion energy for several metals. ×, Aluminum; □, chromium; ○, copper; •, gold; ○, silver (from Hayward and Wolter [73]).

other deposition processes, and is approached only by bias sputtering at high substrate bias. Operating the ion beam at an energy greater than E_1 may be used to preclean the substrate [71] or to ion mill.

If an inert support gas is accelerated with the desired material, the inert gas component of the ion beam will also sputter the growing film, changing the deposition rate and modifying the properties of the film [72]. If beam purity is critical, mass separation may be carried out [31]. Neutrals which form in the beam may be removed by deflecting the ion beam prior to deposition [31, 75]. The low deposition energies involved in primary ion beam deposition make it difficult to extract ample current directly from an ion source. For this reason extraction is usually performed at a high voltage with subsequent beam focusing and deceleration [31, 74, 75].

C. Beam Size Limitations

Attempts at maskless deposition have aimed at determining the minimum beam size obtainable with a useful deposition rate. An analysis by Fair [74] identified several factors which limit the beam size in the operation of an indium deposition system. These factors are summarized here.

Space charge repulsion is the most fundamental limit to beam size, because it remains even in a perfect optical system fed by a perfect ion source. Wilson and Brewer [76] discuss the regimes in which space charge repulsion dominates beam size. The effect increases for low ion energy, high current, and high mass. Thus the condition for minimizing space charge repulsion is the opposite of that required for efficient film deposition.

In the system analyzed by Fair, however, the beam size was limited by *chromatic aberration* [74] caused by the energy spread of ions in the source producing a distribution of axial velocities in the beam. An energy spread of 5 eV was sufficient to dominate the attainable spot size. While theoretically capable of depositing a 100 eV indium beam of 0.01-cm diameter at 6×10^{-8} A, the system actually gave 0.075 cm at 1×10^{-8} A, 100 eV. Fair attributed the energy spread to collisions in the source.

Thermal velocity spread perpendicular to the beam axis is due to the temperature of the source [74]. Fair found a crossover between thermal velocity limiting of beam size and space charge limiting in the region of interest. For an indium ion beam of 30–400 eV, current $< 3 \times 10^{-6}$ A and minimum diameter $< 10^{-2}$ cm, thermal velocity dominates. For larger diameter or current, space charge dominates.

Spherical aberration due to nonparallel trajectories is determined primarily by the ratio of beam diameter to lens diameter, and may usually be chosen to be comparable to other limits [74]. The combination of these limits, however, indicates poor performance of a narrow beam maskless deposition system, even with decreased source energy spread and increased brightness [74].

D. Film Properties

1. Deposition Rate and Uniformity

The dependence of deposition rate on beam energy has been discussed. The optimum beam energy is around half of the energy of unity self-sputter yield. Some examples of demonstrated rates will be given without describing each system. Amano et al. [31, 77] deposited 1000–3000 Å of Pb and Mg in 7 hr at 24 eV and 10^{-5} A, with 1.5-cm spot size. The energy E_1 was found to be 200 eV for Pb and 500 eV for Mg. Wolter [70] deposited Cr and Cu at 25 Å/min at 200 eV, 5 × 10^{-6} A, 0.03-cm spot size. Other results are reported by Probyn [78] and by Colligon et al. [75]. Hayward and Wolter [73] studied the transition from deposition to self sputtering for Au, Cu, Ag, Cr, and Al films.

With these low beam currents, uniform coverage is not achieved. The beam current distribution is determined by the ion optics and is usually close to a Gaussian profile. The deposited film, however, may depart from the beam profile depending on the extent of resputtering of the film. The beam current profile and the deposited film profile are compared in Fig. 17 for In deposited by Fair [74] at 5 × 10^{-8} A for 30 min at 300 eV. The flat top

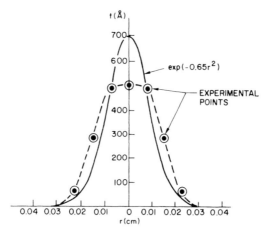

Fig. 17. Ion beam profile (solid line) and deposited film profile (dashed line and points) for primary ion beam deposition of indium at 300 eV, 5 × 10^{-8} A (from Fair [74]).

of the deposit indicates self-sputtering and possibly lateral migration due to the high surface mobility of the film. Scanning the beam improves uniformity, but the scan rates are very low, on the order of 0.01 cm/min for a 200-Å-thick film of In at 100 eV [74]. Thus, a narrow beam maskless deposition system does not look attractive.

2. Purity

Amano et al. [77] found <0.1 at. % heavy mass contamination and <10 at. % light mass by Rutherford backscattering analysis in Pb films deposited at 24–200 eV using a velocity mass filter. Similar purity was obtained for Mg [77]. Thus, high purity may be obtained with a mass analyzed beam. With argon present in the beam, Spencer et al. [72] found evidence of superlattice ordering of Ar in diamond structure C films deposited at 50 eV, with no oxygen or metals detected. Simultaneous sputtering of the film by the inert gas may help clean the surface of loosely bound impurities, similar to bias sputtering.

3. Structure and Mechanical Properties

Amano et al. [77] studied the morphology of Pb and Mg films deposited over a range of energies. With Pb films deposited at 24 eV, coverage of the surface was not achieved until a thickness of 2500 Å. However, films deposited at 48 and 72 eV were continuous at lower thicknesses. Grain size was also found to increase with increased deposition energy from 48 to 120 eV for Pb on C. With Mg, films deposited at 48 eV were smooth, but at 100 and 150 eV craters due to self-sputtering appeared. Colligon et al. [75] found 150 Å Pb films deposited at 50–60 eV to be discontinuous, and continuous when deposited above 600 eV. Wolter [70] found the crystal structure of Cr, Cu, and Ag films to be the same as vacuum evaporated films. Aisenberg [79] reports epitaxial growth of Si on Si by primary ion beam deposition.

Adhesion is also expected to depend on the energy of deposition. Wolter [70] found Ag very adhesive when deposited at 100 eV or higher. Amano and Lawson [77] report that Pb films deposited on graphite at 24–169 eV pass the adhesive tape test. Mg deposited at <100 eV failed the adhesive tape test, while films deposited at >120 eV passed. The adhesion is attributed to a higher degree of implantation at higher energy. Aisenberg and Chabot [71] report diamondlike C films adherent to Si and to stainless steel.

4. Metastable Phases

The demonstration by Aisenberg and Chabot [71] and by Spencer et al. [72] that carbon may be deposited in the diamond phase by primary ion beam deposition shows that under some circumstances metastable phases may be formed. This is a consequence of the highly dynamic state of the growing film surface which experiences a combination of deposition and sputtering. If a range of bond types is available to an incoming ion it may be possible to preferentially resputter those of lower energy and selectively leave a desired bond type, thereby growing a particular phase. Normally one would expect the thermodynamically stable phase to form, but if two or more structures are of competing stability, the possibility arises of preferentially retaining one type. Spencer *et al.* [72] attribute the preferential formation of diamond bonds to the variation of bond energy with crystal orientation, which correlates with the observed (111) orientation perpendicular to the film surface. The charge state of the incoming ion may also affect the growth process and film structure, although this has not been shown to be a controlling factor.

5. Compounds and Reactive Species

Aisenberg and Chabot [80] deposited Al_2O_3 by primary ion beam deposition of Al with oxygen support gas, but the source was hampered by the buildup of insulating coatings on the electrodes. ZnS has been deposited by Van den Berg and Armour [81] by evaporating ZnS powder into an argon ion beam from a duoplasmatron. Hayward and Wolter [73] examined the effects of oxygen on deposition and self-sputtering of Al. Another approach is to create a compound surface layer by ion bombardment with a reactive ion beam, which is more closely related to ion implantation [82].

V. CONCLUDING COMMENTS

While ion beam deposition does not compete with high rate sputtering processes for coverage of large areas, it does offer unique capabilities for studying the sputtering process under controlled conditions. It can be used to shed light on the processes occurring in other deposition methods, for example the study of reactive sputtering by Weissmantel [68]. The ability to control the deposition energy in primary ion beam deposition has been shown to lead to unusual film structures, such as the direct deposition of diamondlike films by Aisenberg and Chabot [71] and by Spencer *et al.* [72]. This capability may be extended into precise control of film growth similar to molecular beam epitaxy.

As larger ion sources are developed [19, 83] for ion milling applications, their application to primary and secondary ion beam deposition will be stimulated. Also the imaginative use of ion beams in creating novel deposition environments appears to be an area only beginning to be explored.

ACKNOWLEDGMENTS

The author thanks R. J. Gambino, H. R. Kaufman, and R. S. Robinson for helpful comments in preparing the manuscript and R. N. Castellano for permission to use his results prior to publication.

REFERENCES

1. R. G. Wilson and G. R. Brewer, "Ion Beams with Applications to Ion Implantation." Wiley, New York, 1973.
2. *Proc. Symp. Ion Sources Form. Ion Beams, BNL 50310 Brookhaven Natl. Lab., Upton, N.Y., 1971; Proc. Int. Conf. Ion Sources, 2nd, Vienna, 1972.* Österreichisthe Studiengesellschaft für Atomenergie, Ges. m.b.h. (SGAE) Vienna, 1972.
3. W. Lotz, *Astrophys. J., Suppl. Ser.* **14**, No. 128, p. 207 (1967).
4. H. R. Kaufman, *Adv. Electron. Electron Phys.* **36**, 265–273 (1974).
5. S. C. Brown, "Introduction to Electrical Discharges in Gases." Wiley, New York, 1966; A. von Engel, "Ionized Gases." Oxford Univ. Press (Clarendon), London and New York, 1965; R. G. Jahn, "Physics of Electric Propulsion." Gordon & Breach, New York, 1970.
6. R. P. Vahrenkamp, *AIAA Pap.* No. 73-1057 (1973).
7. Reference 1, pp. 35–36.
8. Reference 1, pp. 26–32.
9. Reference 1, pp. 36–39.
10. J. F. Mahoney, A. Y. Yahiku, H. L. Daley, R. D. Moore, and J. Perel, *J. Appl. Phys.* **40**, 5101 (1969); V. E. Krohn and G. R. Ringo, *Appl. Phys. Lett.* **27**, 479 (1975).
11. F. M. Penning, *Physica (Utrecht)* **4**, 71 (1937); J. P. Flemming, *J. Vac. Sci. Technol.* **12**, 1369 (1975).
12. J. R. J. Bennett, *IEEE Trans. Nucl. Sci.* **NS-19**, 48 (1972).
13. D. Halliday and R. Resnick, "Physics," Part 2, p. 816. Wiley, New York, 1968.
14. Reference 5, pp. 62–63.
15. Reference 1, p. 100.
16. K. Wittmaack and F. Schulz, *J. Vac. Sci. Technol.* **10**, 918 (1973); C. E. Carlston and G. D. Magnuson, *Rev. Sci. Instrum.* **33**, 905 (1962); B. L. Crowder and N. A. Penebre, *Rev. Sci. Instrum.* **40**, 170 (1969).
17. H. R. Kaufman and P. D. Reader, *Am. Rocket Soc.* [Pap.] No. 1374-60 (1960).
18. P. D. Reader and H. R. Kaufman, *J. Vac. Sci. Technol.* **12**, 1344 (1975).
19. R. S. Robinson and H. R. Kaufman, *AIAA Pap.* No. 76-1016 (1976).
20. Ion Tech, Inc., Fort Collins, Colorado.
21. W. D. Ramsey, *J. Spacecr.* **9**, 318 (1972).
22. M. von Ardenne, "Tabellen der Elektronenphysik, Ionenphysik und Ubermicroskopie," p. 554. Dtsch. Verlag Wiss., Berlin, 1956.
23. G. R. Brewer, M. R. Currie, and R. C. Knechtli, *Proc. IRE* **49**, 1789 (1961).
24. W. A. Scaife, P. R. Hanley, and K. H. Purser, *Proc. Int. Conf. Nucl. Target Dev. Soc., 4th, Argonne, Ill.* p. 75 (1975).
25. B. S. Burton, Jr., in "Electrostatic Propulsion" (D. B. Langmuir, E. Stuhlinger, and J. M. Sellen, Jr., eds.), Progress in Astronautics and Rocketry, Vol. 5, pp. 21–50, Academic Press, New York, 1961; M. G. Betigeri, M. S. Bhatia, and T. P. David, *Nucl. Instrum. Methods* **128**, 29 (1975); N. B. Brooks, P. H. Rose, A. B. Wittkower, and R. P. Bastide, *Rev. Sci. Instrum.* **35**, 894 (1964).

26. R. A. Dugdale, *Proc. Int. Conf. Phenom. Ioniz. Gases, 7th, Belgrade* **3**, 334 (1965).
27. P. H. Schmidt, R. N. Castellano, and E. G. Spencer, *Solid State Technol.* **15**, 27 (1972).
28. R. Schnitzer and F. C. Engesser, *Rev. Sci. Instrum.* **47**, 1219 (1976).
29. R. K. Fitch, T. Mulvey, W. J. Thatcher, and A. H. McIlraith, *J. Phys. D* **3**, 1399 (1970); R. K. Fitch, A. M. Ghander, G. J. Rushton, and R. Singh, *Jpn. J. Appl. Phys., Suppl.* **2**, Part 1, 411 (1974).
30. H. G. Mattes, *Rev. Sci. Instrum.* **45**, 1030 (1974).
31. J. Amano, P. Bryce, and R. P. W. Lawson, *J. Vac. Sci. Technol.* **13**, 591 (1976).
32. Y. Namba and T. Mori, *J. Vac. Sci. Technol.* **13**, 693 (1976); T. Terada, *J. Phys. D* **5**, 756 (1972); Y. Murayama, *J. Vac. Sci. Technol.* **12**, 876 (1975).
33. T. Takagi, I. Yamada, and A. Sasaki, *J. Vac. Sci. Technol.* **12**, 1128 (1975).
34. J. Arianer, E. Baron, M. Brient, A. Cabrespine, A. Liebe, A. Serafini, and T. Thon That, *Nucl. Instrum. Methods* **124**, 157 (1975); E. D. Donets, *IEEE Trans. Nucl. Sci.* **NS-23**, No. 2, 897 (1976); R. Geller, *IEEE Trans. Nucl. Sci.* **NS-23**, No. 2, 904 (1976).
35. R. Middleton and C. T. Adams, *Nucl. Instrum. Methods* **118**, 329 (1974).
36. A. S. Kucherov, L. S. Lebedev, and A. V. Orlovskii, *Prib. Tech. Eksp.* **4**, 21 (1975).
37. C. C. Tsai, W. L. Stirling, and P. M. Ryan, *Rev. Sci. Instrum.* **48**, 651 (1977); R. C. Davis, O. B. Morgan, L. D. Stewart, and W. L. Stirling, *Rev. Sci. Instrum.* **43**, 278 (1972).
38. C. D. Child, *Phys. Rev.* **32**, 492 (1911).
39. H. R. Kaufman, *AIAA Pap.* No. 75-430 (1975).
40. W. Laznovsky, *Res. Dev.* August, p. 47 (1975).
41. G. Aston, *NASA Contract. Rep.* **CR-135034** (1976).
42. Reference 1, pp. 156–158.
43. Reference 1, pp. 207–213.
44. Reference 1, pp. 213–227.
44a. P. D. Reader, D. P. White, and G. C. Isaacson, *Symp. Electron, Ion Photon Beam Technol. 14th, Palo Alto, Calif., 1977.*
45. W. D. Westwood and S. J. Ingrey, *J. Vac. Sci. Technol.* **13**, 104 (1976).
46. R. N. Castellano, M. R. Notis, and G. W. Simmons, *Vacuum* **27**, 109 (1977).
47. L. Maissel, in "Handbook of Thin Film Technology" (L. I. Maissel and R. Glang, eds.), p. 4-3. McGraw-Hill, New York, 1970.
48. J. H. Ormond, *Rev. Sci. Instrum.* **40**, 1247 (1969).
49. G. A. Harrower, *Rev. Sci. Instrum.* **26**, 850 (1955).
50. J. W. Coburn, *Rev. Sci. Instrum.* **41**, 1219 (1970).
51. J. J. Hanak and J. P. Pellicane, *J. Vac. Sci. Technol.* **13**, 406 (1976).
51a. C. Weissmantel and G. Gautherin, "Ion Beam Etching, Sputtering and Plating." Elsevier, Amsterdam, 1978; C. Weissmantel, *Proc. Int. Vac. Congr. 7th, Vienna, 1977.*
51b. E. G. Spencer and P. H. Schmidt, *J. Vac. Sci. Technol.* **8**, S52 (1971).
52. Veeco Instruments, Inc., Plainview, New York.
53. General Ionex Corp., Ipswich, Massachusetts.
54. K. L. Chopra and M. R. Randlett, *Rev. Sci. Instrum.* **38**, 1147 (1967); K. L. Chopra, M. R. Randlett, and R. H. Duff, *Philos. Mag.* **16**, 261 (1967).
54a. H. F. Winters, *Adv. Chem. Ser.* **158**, 1 (1976).
54b. H. Oechsner, *Appl. Phys.* **8**, 185 (1975).
55. P. Sigmund, *Phys. Rev.* **184**, 383 (1969).
56. N. Laegreid and G. K. Wehner, *J. Appl. Phys.* **32**, 365 (1961).
57. H. H. Andersen and H. L. Bay, *Radiat. Eff.* **19**, 139 (1973).
58. H. Oechsner, *Z. Phys.* **261**, 37 (1973).
59. G. K. Wehner and G. S. Anderson, in "Handbook of Thin Film Technology" (L. I. Maissel and R. Glang, eds.), p. 3-23. McGraw-Hill, New York, 1970.

60. P. H. Schmidt, R. N. Castellano, H. Barz, A. S. Cooper, and E. G. Spencer, *J. Appl. Phys.* **44**, 1833 (1973).

60a. B. M. Gurmin, Y. A. Ryzhov, and I. I. Skharban, *Bull. Acad. Sci. USSR, Phys. Ser.* **33**, 752 (1969).

61. B. Navinsek, *Prog. Surf. Sci.* **7**, 49 (1976).

62. E. Gillam, *J. Phys. Chem. Solids* **11**, 55 (1959).

62a. G. Gautherin, C. Schwebel, and C. Weissmantel, *Proc. Int. Vac. Congr. 7th, Vienna, 1977*.

62b. R. R. Olson and G. K. Wehmer, *J. Vac. Sci. Technol.* **14**, 319 (1977).

62c. E. V. Kornelsen, *Can. J. Phys.* **42**, 364 (1964).

62d. H. F. Winters and D. Horne, *Phys. Rev. B* **10**, 55 (1974).

63. H. J. Hinneberg, M. Weidner, G. Hecht, and C. Weissmantel, *Thin Solid Films* **33**, 29 (1976).

64. D. M. Sherman and T. E. Hutchinson, *Rev. Sci. Instrum.* **43**, 1793 (1972); D. M. Sherman, J. S. Maa, and T. E. Hutchinson, *J. Vac. Sci. Technol.* **10**, 155 (1973).

65. C. Weissmantel, O. Fiedler, G. Hecht, and G. Reisse, *Thin Solid Films* **13**, 359 (1972).

66. W. R. Hudson, *NASA Tech. Memo.* **TM X-73511** (1976).

67. R. N. Castellano, *Thin Solid Films* **46**, 213 (1977).

68. C. Weissmantel, *Thin Solid Films* **32**, 11 (1976).

69. J. B. DuBow, D. E. Burk, and J. R. Sites, *Appl. Phys. Lett.* **29**, 494 (1976).

70. A. R. Wolter, *Proc. Microelectron. Symp., 4th, IEEE, St. Louis, Mo.* p. 2A-1 (1965).

71. S. Aisenberg and R. Chabot, *J. Appl. Phys.* **42**, 2953 (1971).

72. E. G. Spencer, P. H. Schmidt, D. C. Joy, and F. J. Sansalone, *Appl. Phys. Lett.* **29**, 118 (1976).

73. W. H. Hayward and A. R. Wolter, *J. Appl. Phys.* **40**, 2911 (1969).

73a. O. Almen and G. Bruce, *Nucl. Instrum. Methods* **11**, 257 (1961).

74. R. B. Fair, *J. Appl. Phys.* **42**, 3176 (1971); R. B. Fair, Ph.D. Thesis, Duke Univ., Durham, North Carolina, 1969.

75. J. S. Colligon, W. A. Grant, J. S. Williams, and R. P. W. Lawson, *Inst. Phys. Conf. Ser.* No. 28, p. 357 (1976).

76. Reference 1, p. 141.

77. J. Amano and R. P. W. Lawson, *J. Vac. Sci. Technol.* **14**, 831 (1977); J. Amano and R. P. W. Lawson, *J. Vac. Sci. Technol.* **14**, 836 (1977).

78. B. A. Probyn, *J. Phys. D* **1**, 457 (1968).

79. S. Aisenberg, *J. Vac. Sci. Technol.* **5**, 172 (1968).

80. S. Aisenberg and R. W. Chabot, AFCRL-TR-73-0176. Air Force Cambridge Res. Lab., Bedford, Massachusetts (1973).

81. J. A. Van den Berg and D. G. Armour, *Vacuum*, **27**, 27 (1977).

82. M. Watanabe and A. Tooi, *Jpn. J. Appl. Phys.* **5**, 737 (1966); V. P. Astakhov, T. B. Karashev, and R. M. Aranovich, *Sov. Phys.—Semicond.* **4**, 1826 (1971); B. Stritzker and W. Buckel, *Z. Phys.* **257**, 1 (1972).

83. J. S. Sovey, *NASA Tech. Memo.* **TM X-73509** (1976).

Part III

CHEMICAL METHODS OF FILM DEPOSITION

III-1

Deposition of Inorganic Films from Solution

FREDERICK A. LOWENHEIM

Consultant
Plainfield, New Jersey

I. INTRODUCTION

Films may be deposited on either metallic or nonmetallic substrates by chemical or electrochemical means from solutions, usually aqueous. Methods may be roughly divided into chemical and electrochemical reactions, although this differentiation is by no means clear cut, since some reactions that appear to be purely chemical in nature involve electrochemical pathways. Perhaps a better distinction would be between those methods which do, and those which do not, require the application of an outside source of electric current. For example, the immersion deposition of a film of, say, tin on a copper substrate involves no application of current, but there is no question that the mechanism involves an electrochemical reaction, with the electrons being provided by cathodic areas on the substrate, and being taken up by the ions of the metal in solution.

Many of the methods to be briefly discussed are by no means limited to the production of thin films, arbitrarily defined as those less than 10 μm thick. Electrodeposition and autocatalytic ("electroless") plating can provide films with virtually no upper limit in thickness.

II. DEPOSITION BY CHEMICAL REACTIONS

These techniques may be classified as (1) deposition by homogeneous chemical reactions, usually reductions of a metal ion in solution by a reducing agent; (2) autocatalytic ("electroless") deposition, similar to (1) except that the reaction takes place only upon certain specific surfaces, called catalytic, whereas in (1) the reduction takes place throughout the solution and the nature of the substrate is of little or no importance; and (3) conversion coatings, in which a reagent in solution reacts with the substrate, a metal, to form a compound of the metal having certain favorable properties.

A. Homogeneous Chemical Reduction

These methods, which have been used to deposit thin films of Ag, Cu, Ni, Au, and PbS among others, rely on the mixing, right before use, of two or more solutions. One contains the metal to be deposited, the other a reducing agent which reacts with the metal ions to reduce them to the metal. Silvering is perhaps the most widely used of these techniques. Since these reactions take place throughout the solution, and not just on the surface to be coated, they are usually applied by a two-nozzle gun aimed at the surface so that the solutions mix just in time to react and deposit the film where it is wanted.

1. Silver

Silver mirrors are deposited on glass and synthetic plastics by many methods; the art is very old and unnumerable formulas are reported; these were summarized by Wein [1]. In spite of minor differences, they all involve a solution of a silver salt (usually $AgNO_3$) as one solution, and a reducing agent dissolved in a second solution; the two solutions are mixed just before using, or are sprayed from a two-nozzle gun on the surface to be silvered. The Brashear formula has been a standard for many years [2–4]. The surface generally should be "sensitized," usually by dipping in a solution of stannous chloride acidified with hydrochloric acid; the concentration of this solution is not critical.

Prominent among the reducing agents used are sucrose, Rochelle salt, formaldehyde, and hydrazine; the last appears to be favored at present, although the others are still in use.

The Brashear process involves two solutions. Silver nitrate solution: $AgNO_3$, 20 g; KOH, 10 g; H_2O, 400 ml. The precipitate formed is just dissolved by the addition of NH_4OH (about 50 ml). Alternatively, the $AgNO_3$ and KOH may be mixed and stored, and made ammoniacal just before use; after the addition of NH_4OH the solution should not be allowed to dry out nor kept for any length of time, since explosive fulminates may be formed.

Reducing solution: sucrose (cane sugar), 2 g; HNO_3, 4 ml; H_2O 1 liter; boil and cool before using. Immediately before using, mix one part of reducer with four parts of Ag solution at a temperature of about 20°C. Another popular process uses Rochelle salt. Silver solution: $AgNO_3$, 100 g/liter, rendered ammoniacal. Reducing solution: $AgNO_3$, 2 g/liter; Rochelle salt, 1.7 g/liter. Use equal volumes of the two solutions.

The formaldehyde process is also used. Silver solution: $AgNO_3$, 20 g/liter; rendered ammoniacal as above. Reducing solution: formaldehyde, 40% solution, 200 ml/liter. Mix one volume of reducing solution with five volumes of the ammoniacal $AgNO_3$. Observe the above precaution concerning ammoniacal $AgNO_3$ solutions in all cases.

An ASTM method [5] for silvering mandrels for electroforming is similar. The silvering solution is 55–60 g/liter $AgNO_3$ plus 56–70 ml/liter 28% NH_4OH, filtering out any precipitate. The usual precaution concerning ammoniacal $AgNO_3$ solutions must be observed. The reducing solution is dextrose 100 g/liter, formaldehyde (40 vol %) 65 ml/liter. A mixing nozzle is used to spray the silvering and reducing solutions simultaneously until the mandrel is completely silvered; a ratio of one part reducer to four parts silvering solution is recommended, at temperatures of 18 to 22°C. The surface must be sensitized with stannous chloride solution as usual; 2.5 g/liter $SnCl_2$ + 40 ml/liter concentrated HCl.

2. Copper

Copper may be deposited on glass and plastic films for much the same purposes as Ag, and is of course less expensive. Copper films are also used in printed circuitry. Wein [6] reviewed the subject and offers many formulas using various reducing agents. Most of these have been superseded by "electroless copper," which is considered in a later section.

3. Gold

Gold films are used for decorative purposes, as well as for optical filters for making transparent mirrors to permit the making of two negatives simultaneously of equal photographic densities, and as transparent electrically conductive films [7].

The processes used are similar to those for silvering: cleaning the surface, sensitizing with a solution of $SnCl_2$ which is then rinsed off, and adding the Au solution. The accepted process is to add to an aqueous solution of $AuCl_3$ first an alkali, then a reducing agent, which may be an alcohol, sugar, glucose, citric acid, or formaldehyde. Wein [7] offers many formulas.

4. Lead Sulfide

Although films of PbS have had other uses, its main application is for coloring metals such as brass, copper, steel, and nickel. Wein [8] reviewed the subject. A typical formula [9] for blue on brass is: lead acetate 30 g/liter; sodium thiosulfate 60 g/liter; acetic acid 30 ml/liter; (temperature 70–95°C.) Deposits must usually be lacquered to retain their original color.

B. Autocatalytic Chemical Reduction (Electroless Plating)*

Autocatalytic plating (often called "electroless") depends on the action of a chemical reducing agent in solution to reduce metallic ions to the metal; but it differs from homogeneous chemical reduction, discussed previously, in that the process takes place only on "catalytic" surfaces rather than throughout the solution (although if the process is not carefully controlled, this reduction can take place throughout the solution, possibly on particles of dust or of catalytic metals, with catastrophic results; in fact this is known as catastrophic decomposition).

Brenner and Riddell invented "electrodeless" Ni plating in 1946, rather accidentally when they observed that the additive NaH_2PO_2 caused apparent cathode efficiencies of more than 100% in a Ni electroplating bath. This led them to the correct conclusion that some chemical reduction was involved; further research resulted in the development of the

* See Pearlstein [10].

original process that the inventors named "electrodeless" plating. The name soon lost the "-de" and later the name autocatalytic was formally adopted, although "electroless" is still widely used. The words are synonymous.

Autocatalytic plating is defined as "deposition of a metallic coating by a controlled chemical reduction that is catalyzed by the metal or alloy being deposited" [11].

The process yields a continuous buildup of a metal or alloy coating on suitable substrates simply by immersion in the electroless solution. A chemical reducing agent in the solution provides the electrons for the reduction $M^{n+} + ne \rightarrow M^0$, but the reaction takes place only on "catalytic" surfaces. Since this is so, once deposition is initiated, the metal deposited must itself be catalytic if plating is to continue. Not all metal substrates are spontaneously plated in electroless plating solutions, but steps can be taken to initiate deposition. Autocatalytic plating takes place linearly with time, in a manner similar to electroplating at constant current density, and there is, at least theoretically, no limit to the deposit thickness.

The chemical reducing agents used in autocatalytic plating are more expensive than electric current, so that where conventional electroplating is applicable, autocatalytic plating cannot complete. Some metals that are easily electroplated are not capable of autocatalytic reduction, at least at present (e.g., Sn and Cr).

To counter these obvious disadvantages, electroless plating possesses several characteristics not shared by other techniques and that account for its growing popularity. Its throwing power is essentially perfect, at least on any surface to which the solution has access, with no excessive buildup on edges and projections (Fig. 1). Deposits may be less porous than electroplates, and hence have better corrosion resistance. Power supplies, electrical contacts, and the other apparatus necessary for electroplating are not required. The process is an integral and necessary step in plating on nonconductors such as synthetic plastics. This process is particularly important in the printed circuit industry. Some electroless deposits have unusual, or even unique magnetic properties.

Autocatalytic plating has been used to yield deposits of Ni, Co, Pd, Pt, Cu, Au, Ag, and some alloys containing these metals plus P or B. Electroless Cr deposition has been claimed but there is serious doubt about its validity.

Chemical reducing agents have included NaH_2PO_2 (the one originally used by the inventors for Ni and Co deposition and still perhaps the most important and most widely investigated), formaldehyde (especially for Cu), hydrazine, borohydrides, amine boranes, and some of their derivatives.

ELECTROLESS   ELECTROLYTIC

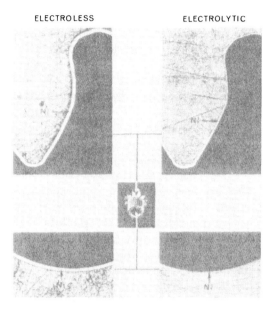

Fig. 1. Uniformity of electroless versus electrolytic nickel deposition. Specimen shown (center) is about 9-mm long and 7-mm diameter across the teeth (after Pearlstein [10, p. 25]).

1. Nickel–Phosphorus

Ni was the first metal so deposited and remains the most important; deposits from hypophosphite baths contain 3–15% P, and those from borohydrides or amine boranes about 0.3–10% B. Hydrazine can deposit almost pure Ni but the process is not widely used. The essential ingredients of a Ni autocatalytic bath are a Ni salt, a reducing agent such as NaH_2PO_2, and a complexing agent for Ni which also serves as a buffer. Also usually included are compounds that increase the rate of deposition (exaltants), and some that increase the stability of the bath. The latter are important, in that unstablized baths may have a tendency to decompose spontaneously, depositing metal not only on the desired surface but throughout the bath, thus essentially ruining it. Exaltants have the property of increasing deposition rates. Many also function as the complexing agents used to keep the Ni in solution: lactate, hydroxyacetate (or glycolate), and succinate; others include propionate, acetate, glycine, malonate, and fluoride. The mechanism by which they speed up deposition has not been explained completely.

Although electroless Ni baths have a fairly long life, a time comes when catalytic nuclei form throughout the bath and cause what has been already mentioned as catastrophic or spontaneous decomposition. Since this is presumably caused by excess of catalytic surface in the bath, certain catalytic poisons may overcome the tendency for this to happen. Organic and inorganic thio compounds and some heavy-metal cations are useful (e.g., 1–5 mg/liter of thiosulfate, 5–10 mg/liter of ethyl xanthate, 1–5 mg/liter of lead or stannous ions may be used). Several other compounds have been recommended. Since these stabilizers act in an opposite way to exaltants, tending to inhibit deposition, their concentrations must be carefully controlled.

According to the usually accepted stoichiometry, 1 g-atom of Ni may be produced for each mole of H_2 gas evolved. The reaction may be formulated as

$$2H_2PO_2 + 2H_2O + Ni^{2+} \rightarrow Ni^0 + H_2 + 4H^+ + 2HPO_3^{2-}. \tag{1}$$

The mechanism of this reaction has been extensively studied, and some disagreement exists; but for the present the most likely route seems to be Lukes' [12] hydride mechanism, in which the effective reducing agent is a hydride ion formed by catalytic decomposition of the hypophosphite:

$$HPO_2 + H_2O \xrightarrow{\text{catalytic surface}} HPO_3^- + 2H^+ + H^-. \tag{2}$$

The hydride ion H^- then reacts either with Ni^{2+} or a hydrolyzed form on the surface, such as $NiOH^+$

$$2H^- + Ni^{2+} \rightarrow Ni^0 + H_2. \tag{3}$$

The efficiency of hypophosphite usage may be decreased by reaction of a hydrogen ion with a hydride ion to produce hydrogen:

$$H^+ + H^- \rightarrow H_2. \tag{4}$$

According to Eq. (1), 1 g-atom of Ni should be deposited for each 2 mole of NaH_2PO_2; actually only about 0.7 g-atom of Ni is produced; since NaH_2PO_2 is expensive, efficiency in its use is important. Alkaline baths exhibit somewhat higher efficiency than acid baths.

Nickel deposits from hypophosphite baths contain 3 to about 15% P, depending on bath composition and operating conditions. After heat treatment the P is present as Ni_3P. The reactions for the formation of elemental P have not been entirely explained.

For bath compositions and operating conditions, texts on electroplating should be consulted [10].

As mentioned, autocatalytic deposition takes place only on catalytic surfaces. These include Ni, Co, Rh, Pd, and steel. Some electronegative metals are also receptive, probably because an initial immersion deposit is formed.

Noncatalytic metals such as Cu, brass, and Ag can be made to receive a deposit by touching them with a part that is actively plating, or by momentarily making them cathodic using a source of dc power. They may be immersed for a short time in a dilute acidic solution of $PdCl_2$, followed by thorough rinsing; the thin immersion deposit of Pd is sufficient to initiate the Ni deposition.

Electroless Ni is plated on steel; coated parts are usually baked to improve adhesion and eliminate possible hydrogen embrittlement. Autocatalytic Ni confers a hard and corrosion resistant surface on Al and Mg. Ti alloys are plated to reduce wear and galling. It has been claimed that electroless Ni–P is superior to hard Cr plate for some moving parts such as cam shafts [13].

Electroless Ni is widely used for plating on nonconductors such as synthetic plastics, discussed below. Only enough is deposited to render the surface conductive, after which the requisite thickness is built up by conventional electroplating.

Vessels for containing electroless Ni baths must, of course, be of a noncatalytic material; stainless steel has been used, and polyethylene bags are used as disposable tank linings.

2. Nickel–Boron

When borohydrides or amine boranes (especially dimethylamine borane or DMAB) are used as the chemical reducing agent in electroless Ni baths, the deposits contain 0.3–10% B. The deposits are harder as plated, and more oxidation resistant than Ni–P deposits; they can be heat treated to yield even greater hardness and wear resistance.

Borohydrides are powerful reducing agents, and in theory one mole of $NaBH_4$ can reduce four equivalents of metal ions, as against one for NaH_2PO_2. Solutions must be highly alkaline to prevent decomposition of the borohydride, and baths are usually kept at pH 11 or above. Complexing agents must, of course, be present to keep Ni in solution at these pH's: tartrates, citrates, ammonia, ethylenediamine, triethylenetetramine, EDTA, and succinates have been recommended. Stabilizers, used for the same purpose as in hypophosphite baths, include organic divalent sulfur compounds such as thiodiglyolic acid and acetylenedithiosalicylic acid, as well as lead salts. The reaction is probably

$$2Ni^{2+} + NaBH_4 + 2H_2O \rightarrow 2Ni^0 + 2H_2 + 4H^+ + NaBO_2. \tag{5}$$

For producing nickel boride, the reaction may be

$$4Ni^{2+} + 2NaBH_4 + 6NaOH \rightarrow 2Ni_2B + 6H_2O + 8Na^+ + H_2. \tag{6}$$

These baths are usually used batchwise, replenishment being limited to the point where about 60 to 80% of the Ni has been deposited. Then the baths must be discarded, since they contain so much $NaBO_2$ that deposition rates are unacceptably low.

Amine borane baths yield deposits similar to those from the borohydride type; at present they are somewhat more widely used than the latter, but since both enjoy much more limited use than the hypophosphite type, this situation is fluid.

These boron compounds are considerably more expensive than NaH_2PO_2, but the desirable properties of the deposits obtained from them may promise increased commercial acceptance.

Hydrazine is capable of producing practically pure (97–99%) Ni; the balance is some O and N, plus trace elements. These baths are not widely used.

3. Other Metals

Although Ni autocatalytic plating has received the most attention, several other metals are capable of being deposited by autocatalytic methods. Cu is widely used for part of the sequence in plating on nonconductors such as synthetic plastics, e.g., as a step in the production of printed circuit boards [14]. There is some controversy about whether electroless copper or electroless nickel is superior for the purpose.

Many proprietary formulations are offered. Most published baths use formaldehyde as the reducing agent. Since the reducing power of formaldehyde increases with alkalinity, most baths are operated at pH 11 or higher. NaOH is the alkali most commonly employed. Other ingredients include a Cu(II) salt, usually $CuSO_4$, a complexing agent to keep the Cu in solution; stabilizers; and perhaps other additives. Tartrate is the most common complexing agent, but EDTA, amines and amine derivatives, and glycolic acid are also used. The reaction may be formulated as

$$Cu^{2+} + 2HCHO + 4OH^- \rightarrow Cu^0 + H_2 + 2H_2O + 2HCO_2^-. \tag{7}$$

In practice more formaldehyde and alkali are consumed than would be indicated by this equation, showing that side reactions occur; one of these is probably the disproportionation of formaldehyde with alkali to form methanol and formate ion:

$$2HCHO + OH^- \rightarrow CH_3OH + HCO_2^-. \tag{8}$$

The mechanism of Cu deposition is believed to be similar to that for Ni, involving a hydride ion originating from the formaldehyde.

Electroless Cu baths are less stable than electroless Ni baths. One of the most troublesome early problems was spontaneous decomposition, and most baths had a short operating life. This decomposition may arise from a competing reaction that is not catalytic (i.e., does not require a catalytic surface) but takes place in the bulk of the solution:

$$2Cu^{2+} + HCOH + 2OH^- \rightarrow Cu_2O + HCO_2^- + 3H_2O. \qquad (9)$$

The Cu_2O particles may disproportionate:

$$Cu_2O \rightarrow Cu^0 + CuO, \qquad (10)$$

thus forming Cu powder in the solution which triggers general bath decomposition. Several ways of inhibiting this undesirable side reaction have been proposed; bubbling air or O_2 through the bath, or vigorous agitation, which has the same effect in saturating the bath with O_2. Small amounts of oxidizing agents are also effective.

As with Ni electroless baths, catalytic poisons also tend to stabilize the Cu baths, but of course they must be present in sufficiently small amounts so that they do not unduly inhibit the deposition. Many stabilizers have been proposed, including mercaptobenzothiazole (MBT), thiourea, cyanide ion, vanadium pentoxide, methyl butynol, and Se compounds. Particulate matter—dust, etc.—from the air can also serve as nuclei for decomposition. Continuous filtration is therefore recommended. Saubestre [15] has reviewed the various methods for stabilizing these baths.

Other metals that can be electrolessly plated, but are not so widely used, include Co from baths essentially similar to electroless Ni baths; either hypophosphite (in alkaline solutions only) or amine boranes can be used, and the deposits are Co–P or Co–B alloys. Electroless Co deposits are utilized primarily for their useful magnetic properties [10]. Variations in solution composition or operating conditions can provide deposits ranging from high coercivity (for high-density recordings) to low coercivity (for high-speed switching devices).

Palladium can be plated autocatalytically using either hydrazine or sodium hypophosphite. These deposits have been used for electrical contacts and connectors. It may be substituted for Au in some applications, and is considerably less expensive on both a volume and a mass basis.

Gold "electroless" deposits are usually misnamed, since they are really immersion deposits; but one truly autocatalytic process using $NaBH_4$ as a reducing agent has been developed. Silver can be plated autocatalytically using DMAB [10].

Reports of the autocatalytic plating of several other metals have been published, but most such processes are really immersion deposits, or they have not been verifiable by other workers. Except for some other metals of the Pt group, it is believed that those mentioned exhaust the possibilities of autocatalytic plating at present.

C. Autocatalytic Chemical Reduction on Nonconductors

Autocatalytic plating has proved to be particularly valuable in the relatively modern technique of plating on nonconductors, especially synthetic plastics and printed circuit boards [14, 16]. Plating on nonconductors has been practiced about as long as electroplating itself; in fact, "galvanoplasty," the reproduction of articles by electrodeposition techniques, was perhaps the earliest application of electroplating. Until about 1960 these techniques relied on making the nonconductor electrically conductive by means of metallic paints, rubbing with graphite, or "silvering." Production was on a piece-by-piece basis, and adhesion of the electroplate was minimal. The part had to be completely encapsulated with the metallic coating if it was to adhere. Such techniques are by no means obsolete, but in the production of electroplated plastic parts they have been entirely superseded by the chemical methods described briefly below.

The producers of ABS and other resins developed copolymer compositions and methods of molding adapted to the subsequent conditioning and surface treatments required.

Chemical conditioning methods were developed to replace the mechanical treatments that had been needed to roughen the substrate so that the subsequent plating would adhere, essentially by a "locking" mechanism. These chemical conditioners not only roughen the surface, but change its chemical nature so that the adhesion of the subsequent electroless plate is at least partly chemical. There is some controversy as to whether the bond is chemical or mechanical. Probably it is a little of both. Regardless of the mechanism, the bond is sufficiently strong so that complete encapsulation of the part by the electroplate is not required. The adhesion is not truly comparable to that obtained in electroplating on metals, but it is sufficient for many purposes.

The process of autocatalytic plating made it possible to deposit a conducting surface to which conventional electroplates could be applied, building up the thickness to any degree required. Many of these processes, especially adapted to plating on plastics, are proprietary. Both electroless Ni and Cu baths are used.

The development of bright leveling acid Cu plating baths (also proprietary) made it possible to deposit a ductile, leveling, and bright Cu plate. They enabled the use of thinner Cu deposits and eliminated the need for mechanical buffing of the Cu.

Although by no means all synthetic plastics are adaptable to the techniques outlined, the list is growing as manufacturers of the plastics realize the importance of being able to plate on them, and research is actively under way to lengthen the list. The principal plated plastics are ABS (acrylonitrile–butadiene–styrene) and polypropylene; but polysulfones, polyamides, and polycarbonates have been plated, with varying success, and new developments appear from time to time in the current literature. The shape and size of the parts also affects their platability. As the size and complexity of the parts increase, so do the problems connected with plating them; sharp variations in cross section, ribs, corners, and holes all pose problems.

The molding process is also important. Use of improper mold-release agents, introduction of strains and stresses into the molded part, and other molding problems can cause later difficulties.

Cleaning is necessary to remove mold-release agents, fingerprints, or other minor soils. It is not usually as vital a step as in plating on metals. If the plastic is not readily wetted by the subsequent conditioning treatment, a solvent treatment may be required. The solvent should have some chemical action on the polymer, so as to render it hydrophilic, but not enough to cause degradation of its basic properties. Acrylonitrile–butadiene–styrene and polypropylene usually do not require this step.

Conditioning both roughens the parts, providing some mechanical "interlocking" of the subsequent plate, and alters the surface chemically by oxidation, providing sites for chemical bonding of the plate. Conditioners for ABS and polypropylene are generally mixtures of chromic and sulfuric acid, sometimes with other additives; and as with the entire field, many proprietary formulations are marketed.

The critical step in the process is variously known as sensitization and nucleation, and a certain amount of confusion exists as to the meaning of the terms. The definitions adopted by ASTM [11] are as follows: sensitization: the adsorption of a reducing agent, often a stannous compound, on the surface; nucleation: the preplating step in which a catalytic material, often a Pd (but sometimes Au) compound, is adsorbed on the surface; the catalyst is not necessarily in its final form; postnucleation: the step where, if necessary, the catalyst is converted to its final form. This is the final step before electroless plating.

Sensitizing/nucleating is often a two-step process. In the first (sensitizing) step, a reducing agent, usually $SnCl_2$, is adsorbed on the surface at the plastic. Following this, rinsing must be thorough so that the subsequent nucleating solution will not be decomposed. The next step (nucleation) is commonly done in dilute solutions of $PdCl_2$. The absorbed sensitizer readily reduces the $PdCl_2$ to Pd, which is deposited on the surface at discrete sites.

An alternative one-step process, in which the sensitizing and nucleating steps have been combined into one solution, containing both the $SnCl_2$ and $PdCl_2$, plus HCl, is often used. The $PdCl_2$ is believed to be reduced to colloidal Pd, which is stabilized by the Sn(IV) hydrolysis products formed by the oxidation of Sn(II). Sodium stannate is sometimes added as further stabilizer.

The Pd particles absorbed on the surface are now surrounded by Sn(IV) ions, or colloidal particles of $SnO_2 \cdot x\,H_2O$. These and any Sn(II) left on the surface must be removed in an "accelerator" which dissolves the Sn compounds, leaving the surface covered with Pd particles as in the two-step process.

The part is now ready for autocatalytic plating with either Cu or Ni. Since the usually invisible Pd layer is very thin, its conductivity is not comparable to that of metallic surfaces, and the electroless layer is also kept fairly thin, so that electroplating must be started at low current densities to avoid burning at contact points; otherwise electroplating is conventional.

More recently, the substitution of Cu–Sn complexes for the much more expensive Pd has been suggested [17].

D. Conversion Coatings

A conversion coating is a coating produced by chemical or electrochemical treatment of a metallic surface that gives a superficial layer containing a compound of the metal, for example, chromate coatings on zinc and cadmium or oxide coatings on steel [11].

The most widely used conversion coatings are chromates, phosphates, and oxides. Most of these coatings are applied by proprietary processes. Formulas and operating instructions must usually be obtained from vendors.

1. Oxides

Oxide coatings are used for blackening ferrous alloys and copper and its alloys. Aluminum and some other metals can be blackened by conversion coatings as well. Some of these black finishes have been suggested as solar energy collectors because of their good absorbance [18].

Many military components require a black finish to avoid detection by reflected light, or to avoid stray reflections that would interfere with the use of optical instruments [19]. Several chemical and electrochemical processes are available. The most commonly used method for blackening steels, stainless steels, and copper alloys is immersion of the metal in a hot, highly alkaline solution containing oxidizing agents to convert the surface to the black oxides Fe_3O_4 or CuO. The oxide film is quite thin (less than 2 μm).

Solutions are usually used at the boiling point. MIL-C13924B covers caustic blackening of several types of iron and steel. After the black oxide is applied and rinsed, a final chromic acid rinse (1 min in 0.6 g/liter CrO_3, pH 2.5, 75°C) is applied, followed by drying in warm air. Black oxide coatings, though improved by this treatment, do not withstand corrosion well in high humidity or salt environments. If not precluded by the intended use, oil or wax films should be applied to these coatings.

Most practical solutions for black oxide finishes on steel are proprietary, although they are all based on similar principles. A published formula is 600 g/liter NaOH, 207 g/liter $NaNO_2$, and 150 g/liter KNO_3 [9, 20].

As with formulations for blackening steel, most processes for immersion black coatings on Cu and brass are proprietary. A published formula is 10 ml/liter 20% $(NH_4)_2S_5$ used at 25°C for about 2 min. Brass must usually be "activated" by removal of Zn from the surface before blackening. A published formula is 150 g/liter $CuCO_3 \cdot Cu(OH)_2$ plus 315 ml/liter 28% NH_4OH. Supplemental oil or wax treatments are normally applied [9, 21]. Corrosion resistance of the treated metals is not improved by the treatment; however the reflectance is greatly diminished.

Aluminum can be blackened by several processes, both proprietary and published. Some formulas are given by Hall [9]. The "Alrok" process uses immersion in a solution of 20 g/liter Na_2CO_3, 1 g/liter Na_2CrO_4, at 65°C for 20 min. Similar coatings have also been promoted as solar energy collectors, as have some anodized coatings described later.

Although they are not all conversion coatings, several other oxide coatings of miscellaneous types may be mentioned here.

Lead dioxide, PbO_2, is formed when Pb or Pb alloy anodes are used in Cr electroplating; in fact the presence of the oxide coating is necessary for proper operation of the bath [22]. Lead dioxide coatings have been suggested as anodes for both the electrowinning of metals and various electrosynthetic processes [23–26]. In one method [23], a Ti anode is used and a film of PbO_2 is deposited on it from a bath containing 100 g/liter HNO_3, 320 g/liter $Pb(NO_3)_2$, 0.3 g/liter $Cu(NO_3)_2$, and 5 g/liter 325-mesh ceramic beads, at 400 A/m². A firmly adherent layer of PbO_2, 1.2-mm thick, is formed in 8 hr. It is said to be stable and useful for electrowinning of Cu and Zn from H_2SO_4 solutions. Similar processes have been reported [24, 25]. Among the many black coatings being considered for solar energy conversion, PbO_2 must also be included [18].

Rubidium dioxide as a coating on Ti has been the subject of several reports. The films are produced in various ways [26–30]. In one process, a Ti strip is etched by immersion in hot oxalic acid; a solution of $SnCl_4$ and

$(NH_4)_2MoO_4$ is brushed on, dried, and heated at 400°C. The anode, now coated with SnO_2, is then coated with RuO_2 by a similar method, using $RuCl_3$ [28]. In another method [30], $RuCl_3$ and $Ti(OC_4H_9)_4$ are dissolved in butanol; the solution is painted onto Ti sheets and air dried, and finally heated to form a layer of RuO_2 on the Ti. These anodes have been of interest principally in the electrolysis of brines, for production of caustic-chlorine, or chlorates.

Manganese dioxide has also been considered as an anode for electrolytic processes [26]; and for solid electrolytic and thin film capacitors [30]. The substrate is first immersed in a 0.1 M solution of $KMnO_4$ at pH 6.7, then in a 0.05 M solution of $Mn(C_2H_3O_2)_2$ and 1 M $NH_4C_2H_3O_2$ at pH 5, at a temperature of 25°C.

2. Chromate Conversion Coatings*

These coatings can be applied to Zn, Cd, Ag, Cu, Sn, and brass (usually electrodeposited coatings of these metals), as well as to Al, Mg, and Zn die castings, and hot-dip galvanized parts, usually by simple immersion in aqueous solutions, although some processes employing electric current are in use.

The primary purpose of chromate coatings on Zn and Cd is to enhance corrosion resistance in various environments (e.g., water, organic vapors, or fingerprints). They also serve as excellent bases for paints because they retard corrosion even if the paint is scratched or porous.

Chromate coatings contain oxides of the basis (substrate) metal plus trivalent and hexavalent Cr in varying proportions. The solutions used to produce them contain hexavalent Cr compounds, along with one or more anions that act as activators or film formers.

Essentially, the coating is formed by reaction between the metal and hexavalent Cr; the latter oxidizes the metal and is itself reduced to the trivalent state. The coating consists of oxides, chromates and the metal, and some salts of trivalent Cr. The reaction causes the pH at the metal/liquid interface to rise, precipitating some trivalent Cr compounds on the surface. This forms a gel, which entraps some hexavalent Cr. This Cr^{6+} is necessary for the desirable properties of the film, since it must leach out slowly to act as a corrosion inhibitor. Therefore, chromate films that have been heated to too high a temperature will not perform their function because the hexavalent Cr is destroyed or insolubilized.

The composition of a typical film on zinc is Cr^{6+}, 8.7%; Cr^{3+}, 23.2%, S

* National Association of Corrosion Engineers [19], Hirata *et al.* [31], Lowenheim [32], and Eppensteiner and Jenkins [33].

(as sulfate), 3.3%; Zn^{2+}, 2.1%; Na^+, 0.3%; H_2O (at 110°C), 19.3%; unaccounted for balance (assumed O_2), 38.1%.

Chromate coatings on Zn and Cd can be clear, iridescent, or olive drab; and can be dyed almost any color. The colored coatings produce the best corrosion resistance and are generally required in military specifications.

The quality of the Zn or Cd plate to which the chromate film is applied is most important in determining its desirable properties. A smooth and fine-grained deposit, relatively free of contaminant metals is required if the chromate film is to perform most effectively.

The process of chromating removes some metal, so that the deposit to be chromated must be thick enough to allow for this. This minimum thickness, for zinc, is about 4 μm, since as much as 1.3 μm may be removed by chromating.

Chromate coatings on Zn, Cd, and Ag are the subject of ASTM and MIL specifications. They include ASTM B 449 and B201; and MIL specs for Zn: QQ-Z-325a; for Cd: QQ-P-416h; for Ag: QQ-S-365a; and for general coatings: MIL-STD-171B.

ASTM B 449 describes recommended practices for chromate coatings on Al but essentially it suggests that the vendor's instructions be followed. These coatings are covered by MIL-C-1706 and MIL-C-5541. There are two classes: IA for maximum corrosion resistance, and 3 for both corrosion resistance plus low electrical resistance. Many proprietary products can meet these specifications: Be can be chromated in essentially the same way as Al, Mg is chromated according to MIL-M-3171.

3. Phosphating*

Phosphate conversion coatings are used primarily as a base for subsequent painting, since they provide a slightly rough surface to which paints adhere better than to a clean metal surface. They are also used to hold lubricants during drawing and shaping, and provide some corrosion protection, especially when a wax or inhibited oil is applied. Like chromate coatings, almost all phosphating is done from proprietary solutions, but the mechanisms are fairly clear.

When iron or steel is exposed to an appropriate acid solution, it dissolves; if insoluble corrosion products are formed, the latter may deposit on the metal surface in the form of a coating. Phosphoric acid has the desired properties. Amorphous or crystalline iron phosphates formed in the reaction are deposited on the iron or steel surface. These tend to protect

* National Association of Corrosion Engineers [19], Lowenheim [32] and Maher and Pradel [34].

the steel from further attack, and, if properly formed, serve as a good bonding surface for organic coatings and paints. If the H_3PO_4 contains other metal salts, such as those of Zn or Mn, the coatings will be mixtures of iron phosphates with phosphates of these metals.

Three types of phosphate coatings are in general use: Fe, Zn, and Mn. The simplest is Fe, in which the substrate itself supplies the metal cations for formation of the phosphate film. Iron phosphating is used almost entirely as a base for paints; the corrosion resistance it provides is small. Zinc phosphating is used for the same purposes, but in addition provides some corrosion protection of its own. Manganese phosphating is used on frictional and bearing surfaces as a protection against wear and seizing of the parts during breaking in. The coating prevents metal-to-metal contact and serves as a reservoir for oil films. These coatings are not used as paint bases. Heavy manganese phosphate coatings are used as wear-resistant finishes. Frictional and bearing surfaces such as piston rings, cam shafts, and gears are typical of such uses. Military specifications MIL-P-16232D, MIL-L-3150, and MIL-L-46010A apply to these coatings; according to the first, the coating weight must be 16.14 g/m^2 minimum [19].

Iron phosphating solutions are essentially alkali or ammonium acid phosphates, plus additives for accelerating the reaction: oxidizing agents such as nitrites or chlorates. The reaction (simplified) is

$$12Fe + 8NaH_2PO_4 + 10H_2O + 7O_2 \rightarrow 2Fe_3(PO_4)_2 \cdot 8H_2O$$
$$+ 2Fe_3O_4 + 4Na_2HPO_4. \quad (11)$$

The mild acid (NaH_2PO_4) dissolves the Fe, forming the soluble ferrous phosphate $Fe(H_2PO_4)_2$. Formation of the coating is a secondary reaction. Even in the presence of oxidizing agents, the Fe is present as Fe(II) because of the H_2 evolution formed by the action of the acid on the metal.

Competing reactions also occur, including the formation of highly insoluble $FePO_4$, which precipitates as a waste product.

In Zn phosphating, the reactions are more complicated. Zinc phosphate is present in the solution ab initio; the bath consists of a complex aqueous mixture of primary zinc phosphate, $Zn(H_2PO_4)_2$, free H_3PO_4, and oxidizing agents. The reactions (also simplified) are

$$Fe + 2H_3PO_4 \rightarrow Fe(H_2PO_4)_2 + H_2, \quad (12)$$

$$3Zn(H_2PO_4)_2 \rightleftarrows Zn_3(PO_4)_2 + H_3PO_4, \quad (13)$$

$$2Zn(H_2PO_4)_2 + Fe(H_2PO_4)_2 + 4H_2O \rightarrow Zn_2Fe(PO_4)_2 \cdot 4H_2O + 4H_3PO_4, \quad (14)$$

$$3Zn(H_2PO_4)_2 + 4H_2O \rightarrow Zn_3(PO_4)_2 \cdot 4H_2O + 4H_3PO_4. \quad (15)$$

When Fe is immersed in a Zn phosphating solution, the H ions (from the H_3PO_4) pickle the surface, and Fe dissolves with H_2 evolution. The increase in pH shifts the equilibrium to the right; the solubility product of $Zn_3(PO_4)_2$ is exceeded and it begins to precipitate on the metal surface. The zinc phosphate coating consists of a thin layer of $Zn_2Fe(PO_4)_3 \cdot 4H_2O$ next to the metal, with an overlay of hopeite, $Zn_3(PO_4)_2 \cdot 4H_2O$.

If phosphate coatings are to perform satisfactorily, they must consist of dense, tightly packed, fine crystals; the coating must be uniform and thin, free of powder, flexible, and must possess satisfactory corrosion resistance. Proper formulation and operating conditions will ensure these characteristics.

Paint-base phosphate coatings can be considerably improved in their protective properties by a final rinse in 0.5 g/liter CrO_3 at pH 2.5–4. [19].

A heavy zinc phosphate coating, defined by MIL-P-16232D as having a minimum coating weight of 1.076 g/m² is often applied as a final finish. Though it has only a short life in salt spray, this can be improved considerably by impregnation with a preservative oil. Waxes and inhibited oils, dry film lubricants, and other materials may be used similarly. Zinc phosphate coatings used with a lubricant (soap) facilitate severe deformation of steel during drawing, stamping, and extruding.

E. Displacement Deposition*

Displacement deposition is also called immersion plating, galvanic replacement, and (when used for recovery of a metal rather than plating it) cementation. It is the deposition of a metal on a substrate by chemical replacement from a solution of a salt of the coating metal, without the use of an outside source of current. It differs from autocatalytic plating in not requiring a chemical reducing agent in the solution; the reducing agent is the substrate itself.

The familiar electromotive force (emf) series (Table I) is often used as a rough guide to the possibilities of displacement plating. Usually, but not always, any metal more negative, or less noble, will displace from solution any metal that is more noble. Thus, Fe replaces Cu from solution; Zn replaces many metals, etc. Often the replacement reaction is so fast, because of large differences in potential, that the deposits are formed above their limiting current density (see electroplating section) and are powdery and nonadherent. Also, the emf series sometimes results in false deductions, because not only potentials but kinetic factors enter the reaction. Also, the potentials in the emf series are determined under highly specific conditions. The relative potentials of metals may reverse under other con-

* Lowenheim [32].

Table I

The Electromotive Force (emf) Series

Electrode	Potential (V)	Electrode	Potential (V)
Li $\rightleftharpoons$ Li$^+$	-3.045	Co $\rightleftharpoons$ Co^{2+}	-0.277
Rb $\rightleftharpoons$ Rb$^+$	-2.93	Ni $\rightleftharpoons$ Ni^{2+}	-0.250
K $\rightleftharpoons$ K$^+$	-2.924	Sn $\rightleftharpoons$ Sn^{2+}	-0.136
Ba $\rightleftharpoons$ Ba^{2+}	-2.90	Pb $\rightleftharpoons$ Pb^{2+}	-0.126
Sr $\rightleftharpoons$ Sr^{2+}	-2.90	Fe $\rightleftharpoons$ Fe^{3+}	-0.04
Ca $\rightleftharpoons$ Ca^{2+}	-2.87	Pt/H$_2$ $\rightleftharpoons$ H$^+$	0.0000
Na $\rightleftharpoons$ Na$^+$	-2.715	Sb $\rightleftharpoons$ Sb^{3+}	$+0.15$
Mg $\rightleftharpoons$ Mg^{2+}	-2.37	Bi $\rightleftharpoons$ Bi^{3+}	$+0.2$
Al $\rightleftharpoons$ Al^{3+}	-1.67	As $\rightleftharpoons$ As^{3+}	$+0.3$
Mn $\rightleftharpoons$ Mn^{2+}	-1.18	Cu $\rightleftharpoons$ Cu^{2+}	$+0.34$
Zn $\rightleftharpoons$ Zn$^+$	-0.762	Pt/OH $\rightleftharpoons$ O$_2$	$+0.40$
Cr $\rightleftharpoons$ Cr^{3+}	-0.74	Cu $\rightleftharpoons$ Cu$^+$	$+0.52$
Cr $\rightleftharpoons$ Cr^{2+}	-0.56	Hg $\rightleftharpoons$ Hg$_2^{2+}$	$+0.789$
Fe $\rightleftharpoons$ Fe^{2+}	-0.441	Ag $\rightleftharpoons$ Ag$^+$	$+0.799$
Cd $\rightleftharpoons$ Cd^{2+}	-0.402	Pd $\rightleftharpoons$ Pd^{2+}	$+0.987$
In $\rightleftharpoons$ In^{3+}	-0.34	Au $\rightleftharpoons$ Au^{3+}	$+1.50$
Tl $\rightleftharpoons$ Tl$^+$	-0.336	Au $\rightleftharpoons$ Au$^+$	$+1.68$

ditions. The possibilities of displacement plating are much more limited than the emf series would suggest.

The deposits produced by displacement deposition are usually powdery, nonadherent, and not useful as coatings. However, in a few cases the process can be made to yield a coherent and adherent coating. Its advantages are obvious: no current is required; the apparatus is simple; and deposition in recesses or on the inside of tubing can be accomplished easily. On the other hand, the thickness of coating that can be attained is strictly limited. Once the substrate has been covered by the coating, action ceases or slows down to the rate at which the substrate metal can diffuse through the coating, or at which the solution can reach the substrate through pores or discontinuities.

Immersion plating has been applied chiefly to Sn films, although processes are available for a few other metals.

1. Tin

Sn is plated by displacement on Cu and its alloys, on steel, and some aluminum alloys. Coating thicknesses are not comparable to those obtainable by electroplating. Typically, for Sn on Cu, the thickness is about 3 μm, obtained in 1 hr.

Copper is more noble than tin, so that the reaction

$$Cu + Sn^{2+} \rightarrow Sn + Cu^{2+} \tag{16}$$

would not normally take place. The electrode potential of Cu, however, can be made much more negative by the incorporation in the plating solution of a complexing agent for Cu. Alkaline stannate solutions containing cyanide have been used, as well as mildly acid solutions containing thiourea. Several proprietary formulations are offered.

Steel wire is immersion plated with Sn or Sn–Cu alloys in a process known as "liquor finishing." The purpose of this coating is twofold: as a coloring mechanism for such items as bobby pins, paper clips, and the like, and as a lubricant in wire drawing. The solution is basically of $Sn(SO_4)_2$ and H_2SO_4, with additions of $CuSO_4$ as needed to obtain the color required.

Aluminum alloy pistons for internal combustion engines are immersion plated with Sn from stannate solutions. In the case of most aluminum alloys, the plate is nonadherent. In fact, the H_2 evolution is often sufficient to blow the Sn deposit off the substrate; but with the particular alloys used for pistons, the deposit is adherent, and somewhat thicker than usual immersion deposits. Tin acts as a lubricant during the running-in period, preventing scoring of the cylinder walls by the very abrasive Al_2O_3 surface layer on the alloy.

2. Gold

Au can be displacement plated on brass and some other substrates; the Zn and Cu in the brass dissolve, precipitating the Au as a thin layer. The coatings are suitable only for imparting a gold color to articles and have no functional value. Many formulas have been published, since the art is a very old one. A typical formulation is: 2.4 g/liter AuCN, 2.1 g/liter KCN, at a temperature of about 80°C.

3. Other Methods

A "nickel dip" is used in porcelain enameling to enhance adhesion of the ground coat on steel. A dilute solution (7–15 g/liter) of $NiSO_4$ is used at pH 3–4 and 70°C.

In plating on Al, it is necessary to replace the natural oxide film on the Al by a film that is platable; this may be either Zn (the "zincate" process) or Sn (a proprietary process known as "Alstan"). Although these processes are indeed immersion plating, the coatings are used only as a base for subsequent electroplating.

III. DEPOSITION BY ELECTROCHEMICAL REACTIONS

A. Electrodeposition

Electroplating [32, 35] is defined as the electrodeposition of an adherent metallic coating upon an electrode for the purpose of securing a surface with properties or dimensions different from those of the basis metal [11]. It is, therefore, a branch of the broader subject of electrodeposition, which is the process of depositing a substance upon an electrode by electrolysis (the production of chemical changes by the passage of current through an electrolyte) [11]. An electrolyte is a conducting medium in which the flow of current is accompanied by movement of matter. Electrolyte also has the meaning of a substance which when dissolved in an appropriate solvent yields an electrolyte of the first meaning. The context will almost always tell which is meant. An electrode is a conductor through which current leaves or enters an electrolytic cell, at which there is a change from conduction by electrons to conduction by ions or vice versa [11]. Whenever current enters or leaves an electrode, and the manner of conduction changes from metallic to electrolytic or vice versa, a chemical change always takes place at the electrode. The electrode at which reduction takes place is the cathode; that at which the reactions are oxidations is the anode. In an electrolytic cell, the cathode is the negative electrode, the anode the positive electrode.

The ions that have a positive charge, and are therefore attracted to the cathode, are cations; those having a negative charge, attracted to the anode, are anions. The solution immediately adjacent to the cathode is the catholyte; that adjacent to the anode is the anolyte.

Metals are characterized by having in their structure electrons not specifically associated with any particular atom, and therefore free to move under the influence of an applied potential without the transfer of matter. Electrolytic conductors are ionic solutions (usually solutions of acids, bases, salts, fused salts, or some gases) in which ions are moved by the potential applied. Electronic or metallic conductors are far better conductors than electrolytes; the conductivity of copper is about 50×10^6 times that of a 1% solution of NaCl.

Ions in solution are usually hydrated or complexed in some other ways. Metals may be in solution in either the cationic or the anionic form; in the latter case, the metal is complexed with a ligand or coordinating agent that gives the total ion a negative charge (e.g., $Zn^{2+} + 4CN^- \rightarrow Zn(CN)_4^{2-}$). Under the influence of an electric charge the ions migrate toward the electrodes. Each ion moves with a velocity characteristic of that ion; the velocity per unit electric field is its mobility. The total conductivity of a solution would be proportional to the sum of the mobilities of its ions, except that electrical forces between the ions (interionic attraction) render this simple assumption untenable unless the solution is extremely dilute. The mobilities of various ions differ considerably, with the result that in a given solution it is likely that more current is carried by the cations than by the anions, or vice versa. The proportion of the total current carried by a given ion is its transport or transference number.

The ions having the highest mobilities, and therefore the best conductivity in aqueous solution are hydrogen and hydroxyl ions; these have mobilities about 5 and 2.5 times that of most other ions, respectively. Therefore, solutions of acids and bases are the best electrolytic conductors. Most plating solutions are formulated to be either acidic or basic. Although migration of ions accounts for the overall conductivity of electrolytic solutions, this migration has little to do with actually bringing the ion to be discharged to the face of the cathode. Diffusion usually plays a large role.

B. Faraday's Laws

Michael Faraday stated his laws of electrolysis in 1833, and these laws have remained perfectly valid ever since. They may be stated as follows.

1.   The amount of chemical change produced by an electric current is proportional to the quantity of electricity that passes.

2.   The amounts of different substances liberated by a given quantity of electricity are proportional to their chemical equivalent weights.

Therefore by measuring the quantity of electricity (in coulombs) that passes, one derives a measure of the amount of chemical change that will be produced. If this chemical change is solely the deposition of metal, it enables calculation of the mass of metal deposited. If one knows the chemical (or electrochemical) equivalent weight of a substance, the amount of that substance that will be liberated by a given number of coulombs can be calculated.

Apparent deviations from these laws can always be accounted for by the failure to take into account all the reactions involved, such as the simultaneous deposition of hydrogen at a cathode on which metal is being deposited or the evolution of oxygen at an anode which is electrolytically dissolving.

The electrochemical equivalent weight of an element or compound is its atomic (or molecular) weight divided by the valence change or number of electrons involved in the reaction. An element has only one atomic weight, but may have several equivalent weights. If Fe is deposited from the Fe^{2+} state, the valence change is 2 and the equivalent weight is the atomic weight 55.85, divided by 2 or 27.93. If the reaction is the reduction of Fe^{3+} or ferric ion to the metal, the equivalent weight is 55.85/3 or 18.62; and if the reaction is merely the reduction of Fe^{3+} to Fe^{2+}, the valence change is 1 and the chemical equivalent weight is equal to the atomic weight.

The coulomb was once defined as the quantity of electricity that deposits 0.001118 g of Ag from solution. From Faraday's laws, the quantity of electricity required to deposit 1 g-eq. w of any element is the atomic weight of Ag divided by 0.001118 or 96,483 C. This quantity (usually rounded off to 96,500 C) is called the Faraday, F. The number of grams of substance reacting, G is

$$G = Iet/96,500, \qquad (17)$$

where I is the current (A), e the electrochemical equivalent weight (g), and t the time (sec).

C. Electroplating Variables

1. Current Efficiency

Although according to Faraday's laws, the total amount of chemical change produced by a given quantity of electricity can be exactly calculated, in plating and similar processes such as electrorefining and electrowinning, we are often interested only in the weight of metal deposited on the cathode or dissolved from the anode and regard any current consumed in causing other changes as "wasted." In most electroplating processes, this weight of metal will be less than predicted by Eq. (17), the side reaction (usually) being evolution of H_2 at the cathode or of O_2 at the anode. Current efficiency is defined as

$$CE = 100 \times W_a/W_t, \qquad (18)$$

where W_a is the actual amount of metal deposited or dissolved, W_t that calculated from Faraday's laws, and CE the current efficiency (%). Cathode efficiency is CE as applied to the cathode reaction, and anode efficiency is CE applied to the anode reaction.

Current efficiency in plating differs widely from one type of process to another, and also depends on such factors as bath concentration, temperature, degree of agitation, pH, and current density. In Cr plating the usual cathode CE is about 20–25%; in Ni plating it approaches 100%, and in some Cu plating processes it may actually be 100%. Other plating processes tend to fall between these extremes.

In the ideal situation, cathode and anode efficiencies would be about equal, so that as much metal dissolves from the anode as is deposited at the cathode and the bath remains in balance. Whether this condition can be attained also depends on individual factors in each case.

2. Current Density

The weight of material deposited is of less interest than the distribution of the deposit over the cathode, and its thickness. One of the principal variables in electroplating, therefore, is current density, defined simply as the total current divided by the area of the electrode. Current density is easily controlled by the operator. It may determine the character of the deposit, the current efficiency, and in some cases whether a deposit will form at all.

3. Current Distribution

The current density yields an average figure. Except for the simplest geometries, the current density over a cathode will vary from point to point. Current will tend to concentrate at edges and points, and to be low in recesses and cavities, since current will tend to flow more readily to points nearer the opposite electrode than to more distant points, unless the resistance of the solution is much lower than is met with in practice. Thus, the thickness of deposit will tend to vary over the surface of the cathode, and be thicker at edges and points than in other areas. Since specifications usually call for a minimum thickness of deposit on the thinnest portions, the excess metal required to attain this minimum is a waste of metal and time. Means to minimize the difference and to equalize the thickness over the cathode are therefore important. Design of parts is a principal factor in achieving this end [36].

Some solutions have the property of decreasing the difference between the thinnest and thickest deposits; such solutions are said to have good throwing power. Throwing power is usually measured by means of the Haring–Blum throwing power box, shown in Fig. 2. The far cathode is usually five times as far from the central anode as the near cathode, but distances of two times have been used. The throwing power is expressed by the equation:

$$T = [(P - M)/(P + M - 2)] \times 100\%, \tag{19}$$

where T is the throwing power, P the ratio distance of far cathode/distance of near cathode from the anode, and M the corresponding ratio of weight of metal deposited on the two cathodes.

Plating solutions vary considerably in throwing power when measured in this way; the exact figures have no theoretical significance, but it is

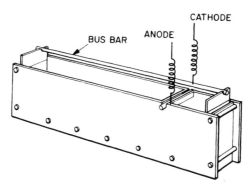

Fig. 2. The Haring–Blum throwing power box.

usually found that solutions showing good throwing power in the laboratory apparatus also evidence it in practice. Chromium plating baths measured in this way have actually negative throwing power; many acid Ni and Cu plating baths yield figures near zero, and cyanide complex and stannate Sn baths give results averaging 40–60%.

D. Electrode Potentials

Electrolytes, like all conductors, obey Ohm's law so that the current relationships are fairly straightforward. Since the resistivity of electrolytes is so much greater than that of metals, we can focus our attention upon the solution and neglect the resistances of the metallic portions of the electroplating setup such as bus bars and the electrodes themselves. (This may not be true when plating long lengths of wire or strip.)

When a metal electrode is dipped into a solution containing ions of that metal, a dynamic equilibrium is set up between the tendency of the metal to go into solution:

$$M \rightarrow M^{n+} + ne \qquad (20)$$

and the opposing tendency for the metal ions to deposit in or on the electrode:

$$M^{n+} + ne \rightarrow M. \qquad (21)$$

When no external voltage is applied, these opposing reactions cancel each other and, long before any change can be detected analytically, the system comes to equilibrium. But before this equilibrium can be attained, a charge separation will occur: if ionization is initially the faster reaction, the metal becomes negatively charged relative to the solution, and if the opposite condition holds, the metal becomes positively charged relative to the solution. The potential that results between the metal and the solution is called the electrode potential.

Such potentials are found even when no metal is involved in the reactions. If H_2 is bubbled over a platinized Pt electrode in a solution of H ions, the potential depends on the equilibrium:

$$H_2 \rightarrow 2H^+ + 2e \qquad (22)$$

with the platinum acting merely as a catalyst to allow the reaction to take place. This potential, the hydrogen electrode, is the reference point from which others are measured. If the activity of H ions is unity, the H electrode potential is considered to be zero at all temperatures.

Similar potentials may be set up between O_2 and a solution of hydroxyl ions, and between metal ions in two different oxidation states.

The magnitude of the electrode potential is given by the Nernst equation:

$$E = E^\circ + (RT/nF)\ \ln(a/a_m), \qquad (23)$$

where E° is a constant characteristic of the metal electrode, R the gas constant (8.314 J K^{-1} mole^{-1}), T the absolute temperature, n the valence change (number of electrons taking part in the reaction), a the activity of the metal ion, and a_m the activity of the metal itself. Generally a_m is taken as 1 for pure metals in their standard states, but it may be different in alloys. In general, the numerator contains the activity of the oxidized state of the reactant, and the denominator the activity of the reduced state in the logarithmic term. Often, if approximate treatment is sufficient, the concentration c may be used instead of a since activity coefficients are often not known for many solutions, and the factor appears in a logarithmic term where any errors are lessened.

If a (or c) is unity, $\ln a = 0$, so that $E = E^\circ$. E° is the standard electrode potential of an element: the potential of an electrode in contact with a solution of its ions at unit activity.

It is not possible experimentally to measure single electrode potentials directly. It is necessary to construct a cell wherein the test electrode is coupled to a reference electrode, which may be the H electrode but is more often a secondary reference electrode such as calomel or Ag–AgCl, because the H electrode is inconvenient to set up and use. The potentials of the secondary reference electrode are known versus that of the H electrode. Very often even such direct methods are not feasible, and standard electrode potentials are calculated from heats of reaction or similar thermodynamic methods.

When the standard electrode potentials for the elements are tabulated, beginning with the most negative, the result is the familiar electromotive force (emf) series (Table I). Those elements with the greatest negative potentials have the greatest tendency to ionize; they are the "base" metals; those with the greatest positive potentials are the "noble" metals. Since the convention as to signs is not universally accepted, the terms "noble" and "base" (or "active") are often preferred as being unambiguous.

There are definite limits to the practical use of the emf series. The previous history of an electrode may have an effect on its potential; large crystals have potentials differing from those of small ones; and other complications may occur. Nor is the standard electrode potential the only determining factor in whether a metal may be electrodeposited from an aqueous solution, since kinetic effects may affect the result. At first sight, it would appear that any metal more negative than H could not be deposited from water since H_2 would be preferentially evolved; but the phenomenon of "hydrogen overvoltage" (mentioned below) requires that, depending on the surface of the electrode, a considerably more negative potential can be reached before H_2 is evolved. Zinc and manganese, for example, can be deposited easily; Mn is the most negative metal depositable from aqueous solution. Moreover, from the emf series, it would appear that Ti and W could be deposited, but in fact neither metal has been electroplated from aqueous solution. Here kinetic factors such as the extreme stability of the complexes of these metals may account for the discrepancy. If T is taken as 298 K and if logarithms are taken to the base 10 instead of e, Eq. (25) reduces to

$$E = E^\circ + (0.059/n) \log a. \tag{24}$$

(Again, an approximation can be made by substituting $\log c$ for $\log a$.) Since E° and n are constant for any given electrode reaction, the potential difference between two identical electrodes in contact with solutions changes with the ion concentration of the solution. If the concentration of a univalent ion changes by a factor of 10, the potential changes by 59 mV, for a bivalent ion by 29.5 mV, etc. Therefore, a potential difference will be set up between an anode and a cathode in an electroplating cell even after the current is shut off. The anolyte will be momentarily more concentrated, and the catholyte more dilute, than the bulk of the solution. Convection or agitation will soon negate these differences and the potential will gradually drop to zero. Such a situation is known as a concentration cell.

We have discussed these effects as though they were equilibrium conditions. Actually, they are in dynamic equilibrium in nature: metal ions are being discharged and metal is being ionized, but the effects cancel each other and there is no net change. The small currents involved are known as exchange currents.

For useful reactions to occur, the system must be moved away from these equilibrium conditions by the application of an external source of electrical current, or perhaps by the contact of dissimilar metals, the effect being the same. The system is now irreversible and somewhat more complicated. The external potential needed to yield useful electrode reactions at practical rates is determined by many factors.

Most obvious is the need to overcome the resistance of the electrolyte. If we neglect the heating effect of the current, the electrolytic resistance will remain constant so long as the solution composition does. The higher the current density required, the higher is the potential needed to drive the current through the solution. In addition, more subtle changes take place at the electrode surfaces (variously termed overvoltage, polarization, and overpotential).

When the electrode reactions move away from dynamic equilibrium by the imposition of a potential or source of current, electrode kinetics must be considered in addition to thermodynamic equilibria. Thus, the total cathode reaction tells us nothing about the intermediate steps; each step may involve kinetic factors that add to the potential required to drive the reaction (or in rare cases, may reduce the potential). These additions to the resistance may be subsumed under the heading of overvoltage. When an appreciable current is passing in an electrolytic cell, the required potential to drive the current differs from the equilibrium potential E_{eq} by the overvoltage,

$$\eta = E_i - E_{eq}, \tag{25}$$

where E_i is the potential when current is flowing. The overvoltage may arise in several ways.

Concentration overpotential has already been mentioned: the catholyte has a lower concentration of metal ions than the anolyte. Overpotential may be required to overcome various kinetic barriers to the reaction. The most common is H overvoltage, but all gases evolving at electrodes exhibit some overvoltage. Certain metals, notably the Fe, Co, Ni triad, also do. These overvoltages are included in the term activation overvoltage. A minimum energy must be possessed by the reactants in order for a reaction to occur; this may be considered as an energy barrier that must be overcome, as illustrated in Fig. 3. When the electrode is cathodic, this potential barrier is easier to surmount so that more ions can cross it in a given time and so deposit. Effects at the anode are of the opposite sign but comparable.

This part of the overvoltage is called activation overvoltage η_{act}. This is partially accounted for by the fact that most metal ions are complexed, either by water or other ligands, and the coordination sphere of the ion

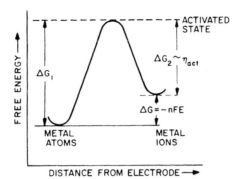

Fig. 3.   The energy barrier in electrodeposition.

must be distorted and the ligands freed of their association with the metal ion, in order for deposition to occur. If η_{act} is above about 50 mV, it can be expressed for the cathodic reaction by the Tafel equation:

$$\eta_{act} = a + b \log i, \tag{26}$$

where a and b are constants depending on the kinetics of the reaction, the activities of the reactants, and the temperature; and i is the current density. For most metals, η_{act} is fairly small and can usually be neglected unless i is very high. Overvoltage is most important in electroplating processes. It accounts for the fact that some metals, such as Zn, that are far more negative in their standard potentials, can be deposited at all from aqueous solutions. The H overvoltage on most cathodes is quite high, and the potential can reach a considerably more negative value than that predicted by simple theory.

From Eq. (26), one would expect that hydrogen, with $E° = 0$, would be discharged from an acid solution of unit activity at a cathode potential of 0 V. Actually, the potential must become considerably more negative before H_2 bubbles form:

$$E = E° + (0.059/n) \log a - \eta. \tag{27}$$

Hydrogen overvoltage depends not only on the current density and other factors already mentioned, but on the physical and chemical nature of the cathode. It is higher on smooth than on rough surfaces; and its value varies from negligible on platinum black and very low on graphite to as much as a volt or so on Hg. Among the consequences of this are that Zn is very difficult to plate on cast iron, which contains C inclusions; and that some very active metals such as the alkalis can be deposited into a Hg cathode.

Another source of overpotential results when an electrode acquires a surface film that possesses substantial resistance. The electrode may be covered with a film having a resistance different from that of the bath. This contributes to the total potential. In the anodizing of aluminum, for example, it is the major factor; and many anodes in electroplating solutions form resistant films, especially in cyanide and stannate baths.

When the anode and cathode reactions do not balance, another source of potential enters. For instance, if O_2 is being evolved at the anode, H_2O must be decomposed, and enough potential must be supplied to allow this reaction to occur. This is called the decomposition potential.

In electroplating, the compound being decomposed is usually H_2O since the potentials converting ions to metal and metal to ions cancel out. The decomposition of water requires that at the cathode the reaction $2H^+ + 2e \rightarrow H_2$ take place, and at the anode the reaction is $2H_2O \rightarrow O_2 + 4H^+ + 4e$. Since the potential of the hydrogen reaction may be regarded as zero, the decomposition potential is essentially that required for the evolution of O_2 (under the conditions that the activity of the hydroxyl ion is 10^{-14} if the H activity is one.) This potential is about 1.23 V; to which must be added, in any practical case, the H and O overvoltages, already mentioned. In theory, the decomposition potential may be calculated from its free energy of formation: $-\Delta G = nFE$. More theoretical treatments of this subject are available [32, 35].

E. Electrode Kinetics

When an electrode is dipped into a solution it attracts molecules of water and ions carrying an opposite charge. These are held near the surface of the electrode by electrostatic forces. Thus, an electric double layer is formed, having the characteristics of a capacitor. Metal ions from the double layer, during deposition, reach the surface of the metal and move to stable positions in the metal lattice, thereby releasing their ligands (water molecules or complexing agents) and neutralizing the charges on the ions and the metal. The total process comprises a spontaneous flow of cathodic current. Simultaneously, neighboring lattice ions free themselves from lattice forces and coordinate with some of the adsorbed water molecules. Finally they move to the ionic side of the double layer as hydrated ions and then on into the solution. This is an anodic or dissolution current.

When the electrode is at equilibrium these currents are equal, and there is no net reaction. These small currents are called exchange currents. These differ appreciably from one metal to another; Sn and Pb exhibit high currents, others may be lower by as much as a factor of 10^5. The magnitudes of these currents are important in the study of electrode kinetics. High values usually mean that the metal is deposited with little polarization.

During deposition the catholyte is depleted of metal ions, and if plating is to continue, these ions must be replenished. This can be accomplished in three ways.

Ionic migration is the least important. Assuming that the metal is contained in a cation, that the bath is acidic and that the metal is not complexed with an anion, the rate of deposition is controlled by the total current. However, the rate of inmigration of cations is only t times that current, where t is the transport number of the cation. For most metal ions t is considerably less than 0.5, and if the bath contains a high concentration of conducting salts, especially acids (hydrogen ion), the metal cation may carry almost no current. In alkaline and cyanide baths, where the metal is tied up in an anionic complex, it is actually migrating the wrong way.

Much more important for replenishing the cathode layer is convection, either natural, artificial, or both. The electrodes may move, the solution may move, or both. Near the cathode, as metal is deposited, the solution becomes less dense and tends to rise along the face of the cathode. The opposite effect takes place at the anode, where denser solution may tend to accumulate at the bottom of the bath unless some form of stirring is used; but the stratification also causes some convection effect.

Passage of the current through the resistance of the bath causes Joule heating. Artificial heating or cooling coils may also be used. In any case, temperature gradients are set up that oppose the stratification and aid in convective transfer of ions. Artificial agitation is often used to supplement the convective effect. In automatic plating, the cathodes are moved through the bath, and in continuous strip or wire plating, this movement may be very rapid.

Although convection is effective in bringing fresh solution to the vicinity of the electrode, it is negligible at and very near the electrode surface. The final travel from the bulk of the solution to the face of the electrode is accomplished by the forces of ion migration (if they operate in the correct sign) and diffusion. Diffusion is the result of random motion of particles (ions or molecules), and tends to produce a more uniform distribution of the dissolved species throughout the solution.

This region next to the electrode, where the concentration of a species differs from that in the bulk of the solution, is called the diffusion layer. The boundary between this layer and the solution bulk is not a sharp line. It has been defined, somewhat arbitrarily, as the region where the concentration of any species differs from that in the bulk by 1% or more.

The diffusion rate, in gram-ions or moles per square centimeter per second, is proportional to the concentration gradient at the electrode; the proportionality constant D is the diffusion constant, in square centimeters per second; thus the rate is

$$R = D(C_0 - C_E)/dN \qquad (28)$$

where C_0 is the bulk concentration, C_E the concentration at the electrode surface, and dN the effective thickness of the diffusion layer, sometimes called the Nernst thickness.

The effective thickness of the diffusion layer (dN) may be decreased by agitation, and may be very low where agitation is rapid as in high-speed strip and wire plating. Without agitation, dN is about 0.2 mm; with a great agitation it may be as little as 0.015 mm. The diffusion layer is, however, much thicker than the electric double layer, which is only about 1 nm thick.

The diffusion rate at the cathode is important in determining the limiting cathode current density that may be employed, and, consequently, the deposition rate. If the limiting current density is exceeded, either no deposit forms or it is unsatisfactory in one or more respects. The double layer and the diffusion rate also are factors in such properties as microthrowing power and leveling [37].

The individual steps, by which a metal is transformed from its ionic state in solution into an integral part of the metal lattice of the cathode, are not accounted for by the total stoichiometric reaction $M^{n+} + ne \rightarrow M$. For instance, the deposition of Cr can be formulated as

$$Cr_2O_7^{2-} + 14H^+ + 12e \rightarrow Cr + 7H_2O, \qquad (29)$$

but it is hardly likely that a reaction of such high order could take place in one step. The study of electrode kinetics attempts to investigate the individual steps by which the total reaction can be accounted for. This subject, sometimes called "electrodics," is not fully understood and cannot be discussed in full here. Only a simplified summary is offered.

The total energy for conversion of a metal ion in solution to a metal atom in the lattice comprises several components. The metal ion probably first reacts with the cathode surface at some point on the flat area of a growing crystal. It is adsorbed in a state somewhat intermediate between the ionic and the metallic. It retains part of its ionic charge and some ligands (water or complexing agents). It is now called an adion. The adion migrates over the surface by diffusion to a growth site, where it is incorporated into the growing lattice. This lateral growth produces a monolayer; but at some point the lateral growth ceases and a new layer is formed over the last one. As the process continues, a crystallite or grain is formed.

In the double layer at the cathode, the metal ion is coordinated to several ligands. These ligands may be released one by one, or the metal may temporarily increase its coordination number to accommodate extra ligands. In either case, the coordination sphere of the ion becomes distorted, and energy is expended. The coordination to the ligand in solution is replaced by coordination to the lattice. The steps may be summarized: (1) arrival at the double layer; (2) release of some ligands to form an adion; (3) deposition at a flat area of the cathode; and (4) lateral movement to a final place in the metal lattice.

As potential is increased across an electrolytic cell, the current also increases according to Ohm's law, if one subtracts electrode polarization from the total potential. When the concentration of depositing ions is fairly small and that of nondepositing ions (the supporting electrolyte such as H_2SO_4 in a $CuSO_4$ bath) is fairly large, a condition arises at which further increases in potential cause only increased cathode polarization and the current remains constant (as in polarography). This situation obtains when the depositing ions are plated out as fast as they can reach the cathode surface. At this point the current depends on the rate of diffusion of the ions to the cathode. This current, the diffusion or limiting current, is reached when the concentration of depositing ions at the double layer is almost zero. This is the limiting current density for any given electrolysis. If the potential is increased further, some new process such as H_2 evolution may begin and the current may rise again. Thus, if the limiting current density is exceeded, deposits tend to be unsatisfactory, usually "burned." Such burned deposits appear first at points or edges where the current density is higher than elsewhere.

F. Summary of Important Process Parameters

The factors influencing electroplating results may be summarized as follows.

1. pH

The hydrogen ion concentration may be vital to obtaining satisfactory results. If it is too low, only hydrogen may deposit. If it is too high, hydroxides (or hydrated metal oxides) may be included in the deposits.

2. Current Density

There is usually a range of current densities in which deposits are satisfactory. If it is too high, the limiting current density may be exceeded. If it is too low, the deposition rate may be unacceptably slow, or in some cases no deposit at all may form.

3. Temperature

Temperature influences convection, diffusion rates, and often the nature and stability of complexes, as well as the possible decomposition of additives.

4. Agitation

Agitation may be required to achieve reasonable deposition rates.

5. Bath Composition

Both the starting composition and the composition changes as a function of time are important. It is sometimes necessary to replace depleted components continuously during operation.

G. Alloy Plating Fundamentals

Two conditions must be met if an alloy is to be deposited successfully: (1) at least one of the metals must be depositable independently; (2) their deposition potentials must be fairly close together or must be capable of being brought close together by means of complexing agents. Normally, the metal requiring the least negative potential will deposit preferentially.

An electrode can have only one potential at a time, so that for two reactions to take place simultaneously at an electrode, they must take place at the same potential. Thus, for simultaneous deposition of two metals in useful form, conditions must be such that the more electronegative potential of the less noble metal can be reached without the use of excessively high current density. The potentials of the two metals must be brought close together if they are not already so.

The emf series (Table I) is a rough guide for deciding whether two metals can be codeposited from simple (noncomplexed) solutions. Metals close together in the table are usually easier to codeposit than metals that are far apart. Some easily depositable pairs, from simple acid solutions, are Pb–Sn, Cu–Bi, Ni–Co, and Ni–Fe. As a general guide, the standard potentials should be within 200 mV of each other for codeposition to take place.

To repeat the Nernst equation and relate it to alloy plating:

$$E = E° + (RT/nF) \ln a + P, \tag{30}$$

where P is the polarization or departure from equilibrium conditions,

$$E_d = E^\circ + (0.059/n) \log a + P \qquad \text{(at 25°C)}. \qquad (31)$$

For alloy plating this equation must be somewhat modified:

$$E = E^\circ + (RT/nF) \ln(a_{M^{n+}}/a_M), \qquad (32)$$

where the activity of the oxidized form (metal ions) is in the numerator of the logarithmic term and that of the reduced form (the activity of the metal in the alloy) is in the denominator. The denominator is usually neglected for single metal deposition because a_M is taken to be unity, but this may not be the case for the metal in an alloy.

That cathodic reaction will normally take place that requires the least negative potential. In a solution containing both Cu and H ions, only Cu will deposit without H_2 evolution because the deposition of Cu requires a much less negative potential than that of H. In the case of Zn, the metal can be deposited even with H ions present because the term P is large enough (hydrogen overvoltage) to allow the deposition potential of Zn to be reached.

For two different cathodic processes to occur simultaneously we need only equate the Nernst potentials to each other ($E_1 \cong E_2$):

$$E_1^\circ + (RT/nF) \ln a_1 + P_1 \cong E_2^\circ + (RT/nF) \ln a_2 + P_2. \qquad (33)$$

Therefore, for alloy deposition, either E_1° must already be about equal to E_2°, so that minor adjustments in the activities or concentrations of the two metals will permit E_1 to equal E_2; or large differences in a for the two metals must be provided; or P_1 and P_2 must be vastly different. The last case is seldom realized, and may usually be neglected.

If the standard electrode potentials of the two metals are close together, and they both can be plated satisfactorily from the same type of solution, the simplest case of alloy deposition is found. Such is the case with Ni–Co and Sn–Pb alloys. Tin and lead are only about 10 mV apart in the emf series and both can be plated from fluoroborate solutions; Ni and Co are about 27 mV apart and both can be plated from many types of solutions, especially sulfate–chloride.

These, however, are exceptional. More commonly, adjustments in a, the activities of one or both metals, must be made, usually by complexing one or both. Theoretically, it would be possible, for example, in depositing brass to provide such a low concentration of Cu ions and such a high one of Zn ions that brass could be plated; but the Cu concentration would have to be so low that it would soon be exhausted from the solution and the process would be entirely impractical. Calculations show that the

Cu concentration of the solution would have to be about 10^{-27} M, far beyond the capabilities of most analytical techniques and virtually impossible to control.

The more practical approach to equalizing the potentials is to regulate their activities by complex formation. For example, the cyanide Cu complex is so stable that although the Cu metal concentration may be appreciable, the actual Cu ion concentration is low enough so that its deposition potential becomes almost as negative as that of Zn. Although Zn is also complexed by cyanide, the complex is much weaker than the Cu cyanide complex. Thus, the potential of Zn is not affected so much, and Cu and Zn can be codeposited easily from cyanide solutions. For the Cu–Sn alloy, the two metals are complexed by different ligands: Cu as the cyanide, and Sn as the hydroxo complex or stannate.

The use of complexing agents is the most important way of equalizing the potentials of the two metals involved in alloy plating. Complexing shifts the potentials to more negative, or less noble, values. The extent of the shift depends on the concentration of the complexing agent and the strength of the complex (i.e., its dissociation constant). The most useful complexing agent is cyanide, because it forms complexes with many metals of varying stability; the shift to more negative values tends to crowd the electrode potentials closer together.

The energy released in forming a stable alloy may also aid in codeposition. An extreme case is the formation of amalgams in deposition into a mercury cathode.

The metal content of alloy plating baths must be replenished, usually by anodic solution, just as in the case of single metal deposition. If the alloy is not homogeneous, one phase may dissolve, leaving the other(s) as solid particles which slough off, thus not only failing to replenish the bath, but causing the usual difficulties presented by solid particles in the bath. In such cases, separate anodes of the two metals may be used, if they dissolve at about the same potentials. If one constitutent of the alloy is a relatively minor one, the main constituent may be used as anode and the minor one replenished chemically. Other expedients have also been used. The Sn–Ni alloy presents a special case. The compound SnNi does not appear on the equilibrium phase diagram, and it is believed that deposition takes place from a complex ion containing both Sn and Ni in a multinuclear complex.

Just as in all electroplating operations, the variables at the control of the operator are current density, temperature, agitation, pH, and the concentrations of the constituents of the bath. Usually these factors require more stringent control in alloy plating than in plating pure metals, because it is likely that a change in one of them will affect one of the constituents of the alloy more than the other and thus change the composition of the deposit. How these variables will affect the deposit must be determined empirically, because electroplating theory has not yet advanced far enough to permit generalizations; but the following may serve as a guide.

An increase in current density normally will increase the proportion of the less noble metal in the alloy deposit, although some addition agents may reverse this effect.

Agitation usually increases the proportion of the more noble metal; thus it tends to offset the effect of current density. By bringing fresh solution to the cathode face and decreasing the thickness of the cathode film, agitation tends to offset the tendency for more rapid depletion of the more noble metal in the cathode film.

An increase in temperature normally increases the content of the more noble metal in the alloy. It has about the same effect as agitation. However, it is somewhat more complicated since change in temperature may also have indirect effects, such as altering the degree of dissociation of complexes.

pH is more important in regulating the physical properties of the deposit than its composition, and the best pH is almost always determined by experiment.

The concentrations of the two metals in alloy plating baths affect the deposit composition directly, but not necessarily in the same proportion. The higher the ratio M_1/M_2 in the bath, the higher that in the deposit, but not necessarily in the same proportion.

The current efficiency in alloy plating may be higher or lower than that for each single metal. In some cases, such as the deposition of W alloys, the current efficiency of W deposition has been raised from zero to a net positive number.

Current efficiency in alloy plating is calculated using the electrochemical equivalents derived from the following equation:

$$W_a = W_1 W_2/(f_1 W_2 + f_2 W_1), \tag{34}$$

where W_a is the electrochemical equivalent of the alloy, W_1 and W_2 are the electrochemical equivalents of the two metals, and f_1 and f_2 are the fractions by mass of the respective metals in the alloy; $f_1 + f_2 = 1$.

H. Electrodeposited Coatings and Their Functions

Metals depositable from aqueous solutions occupy a compact region in the middle of the long-form periodic table, as shown in Fig. 4. Although this suggests that electronic structure and other common factors may account for this property, no such relationship has been shown. In addition

H																	He
Li	Be											B	C	N	O	F	Ne
Na	Mg											Al	Si	P	S	Cl	Ar
K	Ca	Sc	Ti	V	Cr	Mn	Fe	Co	Ni	Cu	Zn	Ga	Ge	As	Se	Br	Kr
Rb	Sr	Y	Zr	Nb	Mo	Tc	Ru	Rh	Pd	Ag	Cd	In	Sn	Sb	Te	I	Xe
Cs	Ba	La	Hf	Ta	W	Re	Os	Ir	Pt	Au	Hg	Tl	Pb	Bi	Pa	At	Rn
Fr	Ra	Ac															

Fig. 4. Long-form periodic table. The elements depositable from aqueous solution are within the heavy lines.

to these metals, several alloys are commercially plated, and many more have been the subject of research. Among those commonly plated are brass (Cu–Zn), bronze (Cu–Sn), Co–Ni, Fe–Ni, Sn–Ni, Sn–Zn, and Sn–Pb. In addition, many Au alloys are plated, usually to modify the properties of the deposit with regard to color, hardness, wear resistance, and other properties considered important for the application.

Most electrodeposited alloys have the same metallurgical constitution as those produced thermally; but there are exceptions. In particular, an intermetallic compound SnNi is not observed on the thermal diagram of the Sn–Ni system and has been produced only by electrodeposition. Alloys of W or Mo with the Fe-group metals are of interest because neither metal can be deposited in the pure condition, but alloys with Co, Fe, and Ni are fairly easily deposited. Some, especially Co–W, appear to have interesting properties.

In this section, specific directions for operating and controlling the solutions mentioned are not offered; they are available in many texts and handbooks [32, 35, 38].

Manganese is the most electronegative metal that can be deposited from aqueous solution, but it has not found commercial application except for electrowinning.

The most electronegative metal commonly plated is Zn. Zinc is the cheapest and most effective metal for protecting most basis metals from corrosion. Being more electronegative than most substrates, it protects the substrate by sacrificial, galvanic action. It corrodes in preference to the substrate and small pores or discontinuities have little deleterious effect, so long as any Zn remains in the immediate vicinity of the discontinuity.

Cadmium, although slightly less negative, protects by a similar mechanism and has some advantages. It does not produce the voluminous "white rust" characteristic of Zn which may interfere with the action of moving parts, and it is easily solderable. On the other hand, it is extremely toxic, and there is a tendency to eliminate Cd plating. Tin–zinc alloy has been used for similar purposes; its protective properties lie somewhere between those of Zn and of Cd. Lead and Pb–Sn alloys are sometimes used for protective purposes. Lead does not protect the substrate, as do Zn and Cd, by sacrificial action, but it resists sulfur-containing gases and H_2SO_4 at all concentrations up to about 200°C. Any pores or discontinuities are quickly filled with insoluble products that prevent corrosion from continuing.

Although Sn is not sacrificially protective to steel under ordinary circumstances, deposits that are sufficiently thick are protective; and inside the tin can, under anaerobic conditions, Sn and Fe reverse in potential and Sn protects the steel sacrificially.

Copper is useful in many applications. Its high conductivity, second only to that of Ag, recommends it for electrical and electronics applications. A fresh surface is easily solderable, but exposure to air usually causes tarnish films, so that it is often overplated with Sn or Sn–Pb alloys for retention of solderability. It is easily buffed to a high luster, and in plating decorative/protective deposits on steel, it is common to use Cu as an underplate, buff the Cu, and proceed. Buffing may be eliminated by use of bright leveling copper baths. Copper may also be used decoratively if protected by clear lacquers.

Most "chromium" plate is actually Cu/Ni/Cr or Ni/Cr, the Cr being only a very thin overplate intended to prevent tarnishing of the Ni. In much thicker deposits, Cr is useful for abrasion resistance on machine tools, dies, oil-well drilling equipment, etc.

So-called "black chrome" is not Cr metal but a mixture of oxides. It can be plated from several types of solution, mostly proprietary. Although originally promoted as a decorative deposit, it has lately received attention for its energy absorbent properties as a solar heat collector [18, 39].

Nickel, sometimes underplated with Cu and over-plated with Cr, has excellent corrosion resistance to many reagents. It is particularly useful in plating on brass, as a barrier layer before Sn plating to prevent diffusion of Zn from the brass into the deposit. Diffusion of Zn reduces solderability.

Many deposited coatings are used for their specific surface characteristics. The list is long and the following is illustrative, but not complete. Silver is used for reflectivity and electrical conductivity, which are somewhat reduced by its tendency to form tarnish films. Gold is used for low contact resistance, ability to withstand many makes-and-breaks in communications equipment, and its nontarnish qualities. Because of its high price, Au is often "spot-plated"; that is, the Au is deposited only on the actual functioning area; and many means have been devised to achieve this effect.

Rhodium is used for its reflectivity. Although it is not as high as that of Ag, it does not tarnish.

Palladium can potentially replace Au in some computer and allied applications. Its advantages include its lower price and density; but lack of experience with it has limited its use. The other Pt metals have, at present, few uses as electrodeposits; but all except Ir and Os can be deposited from fairly well-established processes. Osmium and iridium plating are in the experimental stage.

Indium plating is well established but little used. It has good bearing properties, and has also been used in the semiconductor industry as a doping agent [40].

The other platable metals (Bi, Sb, As, Re, etc.) have few established uses. All can be electrodeposited from published formulations if required.

Most common electroplates are pure metals, if one discounts the minor quantities of codeposited additives such as C and S from some addition agents. However, several alloy deposits are of considerable importance. Most of these are binary alloys, since the complications in control of the composition of higher-order alloys have discouraged their application. An exception is the Cu–Sn–Zn alloy, occasionally used in electronics applications; and some Sn–Pb alloys with a minor amount of a third metal, usually Cu.

Although literally hundreds of alloys have been plated on a "beaker" scale, only a few have reached commercial acceptance. To make it worthwhile to plate alloys, they must possess sufficient advantageous properties to outweigh the additional cost and control problems that alloy deposition involves. Alloys may exhibit better physical or chemical properties, better corrosion resistance, magnetic properties, heat treatability, and the ability to substitute for more expensive metals. In some cases their properties are unique. Electroplating thus differs from thermal preparation of metals, where alloys are far more common than pure metals. Cu is almost the only pure bulk metal used.

Some examples may be mentioned. The yellow color of brass is highly prized for appearance, and no single-metal deposit can duplicate it. Brass is also the preferred deposit for adhering steel to rubber, as in automative tires. Bronze can simulate the color of Au for decorative coatings at a much lower price. Tin–nickel alloy can substitute for Au in some electronic applications, or overplated with very thin Au, with a substantial saving in cost. Nickel–phosphorus and nickel–cobalt alloys have magnetic properties that render them useful as thin films in computer applications. Nickel–iron alloys (up to 35% Fe) have been promoted as a substitute for pure Ni, at a saving in cost. Alloying additions to Au can modify color and physical properties and reduce costs. Tin–lead alloys are superior to either metal alone in some corrosive atmospheres, and in solderability. Tin–cobalt alloy, with a small amount of a third metal such as Zn, has been promoted as a substitute for decorative Cr. Advantages claimed are much better throwing power and higher efficiency. Some commercial success of the process has been attained [40a].

An exhaustive study of alloy deposition is available [41]. The principal reason for the comparative lack of commercial interest in alloy plating is the difficulty in control, as discussed in Section III.G.

I. Electrodeposition from Nonaqueous Solutions

Very few practical electroplating processes using nonaqueous solutions have been developed. Several metals have been deposited from such solvents as liquid ammonia. The two principal alternatives to aqueous baths are organic solvents and fused salts, and both have had limited application.

Organic solvents pose many problems for practical application: necessity for inert atmospheres in some cases and moisture-free conditions in all; fire, explosion, and toxicity hazards; poor electrical conductivity of most organic solutions; disposal problems; and cost. Fused salts are equally unattractive. The high temperatures required and corrosion problems for the apparatus required are the principal drawbacks.

The principal reason for considering nonaqueous solvents is the presence in water solutions of depositable H so that metals too electronegative for their deposition potentials to be reached in the presence of H ions might be plated from aprotic solvents. However, although metals such as Be and Mg have been investigated from this standpoint, only Al deposition has been developed to a semicommercial stage. It must be remembered that it is not sufficient that a metal be deposited. The deposit must also be in coherent and adherent form, and satisfactory in the same way as normal aqueous-based deposits. This has been a major problem in fused salt electrolysis. Many metals, including Al, Mg, and Ca are deposited commercially from fused salt baths, but the cathodic deposits are generally either liquid (as in the case of Al and Mg) or dendritic powders (as in the case of Ca). These deposits are suitable for producing the metal, but not for electroplated coatings.

1. Organic Electrolytes

Two types of metals cannot be deposited from aqueous solutions: those whose potentials are too negative to be reached in the presence of H ion, and some, such as Mo and W, whose potentials would appear to allow deposition, but for some reason have not been deposited in pure form. Organic solvents have proved successful for Al (a member of the first group), but have not been equally adaptable to the other group.

Many organic solvents decompose under the passage of an electric current. Water does so also, of course, but the decomposition products are H_2 and O_2, gases that pass off harmlessly. On the other hand, many organic solvents produce polymers, degradation products, and other deleterious products. Brenner [42] has discussed the requirements for a satisfactory, or promising, organic solvent for electrodeposition of metals. The dielectric constant is no criterion. Diethyl ether, with a dielectric constant of only 4.5, is a much better solvent for electrodeposition than HCN, with a dielectric constant of 95 (water = 80). Conductivity of the solvent is also irrelevant. The conductivity of water is about 10^6 that of ether, but both are good solvents for electrodeposition.

The organic media that are at least potentially useful as plating solvents, are those that have a "weak coordinating center" such as ether's oxygen atom, or an aromatic hydrocarbon with double bonds in its structure. Polar solvents may form such stable complexes with the solutes that they are no more effective than water in allowing deposition.

The desirability of being able to deposit Al in useful form has long been recognized. It should be even more protective than Zn to substrates such as steel, and it retains its pleasing appearance much longer. On a volume basis, it is cheaper than Zn. Hot-dipped coatings of Al on steel are produced commercially. The early literature of electroplating contains many references and patents on the deposition of Al from aqueous solutions, but none was ever verified.

Since Al is produced by electrolysis of molten salts, attempts have been made to adapt the process to the production of Al-coated steel, and at least one such process was carried to a pilot-plant stage, but proved uneconomical, impractical, or both. Many processes for the deposition from molten organic electrolytes have been patented and some may be successful at the laboratory stage, but only one has had any commercial success, and that only by a few specialists. Aluminum is not plated in the industry on a wide scale.

The only significant commercial application of Al plating utilizes the "hydride" process developed at the National Bureau of Standards. The solvent is diethyl ether, in which are dissolved anhydrous $AlCl_3$ and $LiAlH_4$. Periodic reverse current (PR) is said to be useful in producing sounder deposits. A dry box or glove box is required for small operations; larger ones require more elaborate constructions [43].

2. Fused Salt Electrolysis

None of the so-called refractory metals can be deposited in the pure state from aqueous solutions. Of groups IVB, VB, and VIB of the periodic system, only Cr has been deposited. None of these metals have been deposited from organic solvents.

The reasons for the failure to deposit these metals are probably not all the same. Some, like Ti, Zr, and Hf, are probably too electronegative and are, therefore, in the same electrochemical class as Al. But the others, so far as their electrode potentials are known, should be depositable, and failure is attributed to kinetic factors.

Most metals can be deposited from appropriate fused salt baths; but in most cases the deposits are liquid at the operating temperature of the bath, or are dendritic and powdery. This is useful for electrowinning but of no use for electrodeposits.

A process for the deposition of a few of these metals, in the form of useful adherent plates, has been developed, and was practiced commercially for a short time. They require inert atmospheres (e.g., Ar), relatively high temperature, and careful control. From a practical standpoint, these processes are now of little importance [44].

J. Anodization

Although most electrolytic processes are concerned primarily with cathodic reactions, at least one process, anodization, relies on anodic reactions, with the cathodic processes being incidental. Anodizing is an electrolytic process in which the metal is made the anode in a suitable electrolyte. When electric current is passed, the surface of the metal is converted to a form of its oxide having decorative, protective, or other properties. The cathode is either a metal or graphite, at which the only important reaction is H_2 evolution. The metallic anode is consumed and converted to an oxide coating. This coating progresses from the solution side, inward, toward the metal, so that the last-formed oxide is adjacent to the metal. The oxygen required originates from the electrolyte used [45, 46].

Although anodizing processes are used for other metals, Al is by far the most important. Magnesium can be anodized by processes similar to those used for Al. Zinc can be "anodized," but the process is not truly comparable, depending on a high-voltage discharge that produces a fritted semifused surface. Several other metals, including Cu, Ag, Cd, Ti, and steel, can be treated anodically for decorative effects.

The mechanism of anodic oxidation has been studied extensively, but is not completely understood. Faraday's law predicts that for each faraday of electricity passed, 8.9938 g of Al would react, forming 17 g of Al_2O_3, for a "coating ratio" ($Al_2O_3/2Al$) of 1.89. This ratio is never observed; it seldom exceeds about 1.60. The ratio is lower for Al alloys than for pure metal.

Anodic oxide coatings on Al may be of two main types. One is the so-called barrier layer, which forms when the anodizing electrolyte has little capacity for dissolving the oxide. These coatings are essentially non-porous; their thickness is limited to about 1.3 nm/V applied. This thickness represents the distance through which ions can penetrate the oxide under the influence of the applied potential. Once this limiting thickness is reached, it is an effective barrier to further ionic or electron flow. The current drops to a low leakage value and oxide formation stops.

When the electrolyte has appreciable solvent action on the oxide, the barrier layer does not reach its limiting thickness; current continues to flow, resulting in a "porous" oxide structure. Porous coatings may be quite thick: up to several tens of micrometers; but a thin barrier oxide layer always remains at the metal–oxide interface.

Electron microscope studies have enabled workers to describe relatively accurately the structure of anodic coatings. They reveal the presence of close-packed cells of amorphous oxide; there may be billions of cells per square centimeter; their size is a function of the anodizing voltage.

Although many electrolytes have been suggested and used for anodizing Al, H_2SO_4 is most widely employed. Concentrations are from 12 to 25 wt %.

Ordinary or general-purpose anodic coatings from an H_2SO_4 electrolyte, applied for decorative/protective purposes, are typically 2.5–30 μm thick. Temperatures are about 21°C, current density is 130–260 A/m^2, and voltages of 12–22 V are used.

Temperature and time of anodization affect the thickness of the oxide coating. Since the oxide coating is more voluminous than the substrate metal, the thickness of the section increases unless the electrolyte has appreciable solvent action on the coating. The increase is about one-third the thickness of the coating. The weight of the oxide coating is a better measure of the solvent action of the electrolyte than is its thickness. ASTM B 137 [47] offers a method for determining this weight.

Although H_2SO_4 is the most common anodizing electrolyte, many others are in use. For special effects or attainment of special properties, chromic acid anodic coatings are opaque, limited to about 10 μm in thickness. Although they compete with conversion coatings for the purpose, they are still used as a base for paints or adhesive bonding, especially for the military. They are also useful in anodizing complex parts that are difficult to rinse since if some electrolyte is left on the part, little harm is done.

Phosphoric acid anodizing has been used as a basis for plating on Al, though it has been largely superseded by the zincate and other processes.

Oxalic acid coatings are yellow, and somewhat harder than conventional H_2SO_4 coatings. Mixtures of oxalic and sulfuric acid produce hard coatings, competing with low-temperature H_2SO_4 processes.

Sulfonated organic acids, combined with H_2SO_4, produce so-called "integrally colored" coatings on specific alloys. Various shades are used in architectural applications. These processes are proprietary.

Boric acid electrolytes, often with additions of borax, produce thin barrier oxide coatings used for electrical capacitors.

Many anodic coatings on Al can be colored, either with organic dyes or mineral pigments. Postanodic treatments, especially that known as sealing, are important in determining the final properties of the coating. Sealing is a treatment that renders the coating nonabsorptive or modifies it in other ways. It usually involves subjecting the coating to hot aqueous environments that hydrate the coating. When pure water is used at high temperatures, the water reacts with the surface of the Al_2O_3 to form boehmite.

$$Al_2O_3 + H_2O \xrightarrow{\text{heat}} 2AlOOH. \qquad (35)$$

Sealing thus dissolves some oxide and reprecipitates a voluminous hydroxide (or hydrated oxide) inside the pores. Some aqueous sealants contain metal salts whose oxides or hydrated oxides may be coprecipitated with the aluminum hydroxide.

The choice of sealant depends on the environment to which the anodized coating will be subjected and on any special requirements.

Water is the most widely used sealant. It must be of high quality: distilled or deionized, low in solids, and free of phosphates, silicates, fluorides, and chlorides. Sealing temperatures must be at least 98–100°C if sealing is to take place in a reasonable time. The pH of the sealing water should be 5.5 to 6.5. The immersion time is at least 10 min for thin (2.5 μm) coatings and up to 60 min for thicker (25 μm) coatings. Sealing with steam has the advantage of assured water purity, but the setup is more complicated and expensive.

Solutions of nickel acetate are widely used for sealing anodic coatings on Al, particularly after the anodic oxides have been dyed. It is presumed that hydrolysis products of the nickel acetate (colloidal nickel hydroxide) precipitate in the pores of the coating. The concentration of nickel acetate is 1–5 g/liter. Pure water is required. Many other metal salts have been used to seal anodic coatings. Dichromates, in particular, have an advantage when resistance to salt environments is required. Organic materials may be used when lubricity is a factor.

Anodic coatings on Al may serve one or more of the following functions: better corrosion resistance, better paint adhesion, a base for subsequent plating, electrical insulation, application of photographic and lithographic emulsions, better emissivity, better abrasion resistance, and ability to detect surface flaws.

Although other metals are anodically oxidized, Al is by far the most important. The anodic oxidation of Ti is important in the use of Pt-plated Ti anodes, because even through pores in the Pt coating, the Ti oxidizes anodically and resists the action of the electrolyte involved.

Titanium and its alloys are susceptible to siezing, galling, fretting, and wear. Some type of coating is required in applications involving any of these problems. Titanium is a difficult substance to electroplate, although it can be done. Anodizing is the preferred method of providing a coating [19]. Many electrolytes can be used to form anodic coatings on Ti. Two types of coating can be produced. Films of limited thickness (0.01 μm) form in electrolytes having little solvent action on the coating. They are called interference coatings, provide little wear resistance, and are poor dielectrics. Thicker coatings (0.6–0.8 μm) can be formed in electrolytes having some solvent action on the oxide as it forms. Coatings of this type are more protective, though more porous. Some are good electrical insulators [48].

IV. CONCLUSION

This chapter has considered several practical methods of depositing films on metals and nonmetals, all of which are in fairly wide-scale use.

Homogeneous chemical reduction from solution is used for silvering mirrors and other items; similar processes can be used for gold and some other metals, and for lead sulfide.

So-called autocatalytic chemical reduction (electroless plating) is similar except that it is applicable only to certain substrates that catalyze the reduction, hence the name. Although several metals can be so deposited, Ni alloys with P and B, and Cu, are most widely used, and constitute essential steps in electroplating on nonconductors such as plastics.

Conversion coatings, as the name implies, are compounds of the basis metal itself; chromate coatings are widely used to improve the performance of zinc and cadmium electroplates; phosphating is an excellent base for subsequent painting; zinc phosphate coatings also improve corrosion protection by the subsequent paint coat. Manganese phosphate coatings are used mainly for their lubricity, permitting drawing operations and preventing metal-to-metal contact.

Displacement deposition (also called galvanic deposition and immersion plating) has limited use, mainly for Sn coatings on Cu and its alloys. In this reaction the substrate metal goes into solution and is replaced by the metal of the solution.

Electroplating is perhaps the most commonly used of the processes discussed, and is by no means limited to thin films. The laws and theories of electroplating are discussed in some detail, and the principal uses of electroplated coatings are mentioned.

Finally, anodizing is an electrochemical reaction in which the substrate is anodic; its principal application is to Al, but several other metals are also anodized. In this process an oxide or hydrated oxide is formed on the surface of the metal. This may have several applications, including improved corrosion resistance, acceptability of dyes for coloring, electrical insulation, and others.

ACKNOWLEDGMENT

The author thanks Mr. Fred Pearlstein, Aviation Supply Office, Philadelphia, Pennsylvania, for valuable assistance in preparing this chapter.

REFERENCES

1. S. Wein, "Silver Films," Rep. No. PB 111236. Off. Tech. Serv., Washington, D.C. (1953).
2. Natl. Bur. Stand. (U.S.), Circ. No. 389 (1931).
3. W. Blum and H. Hogaboom, "Principles of Electroplating and Electroforming," 3rd ed., pp. 225–226. McGraw-Hill, New York, 1949.
4. C. Davidoff, in "Metal Finishing Guidebook Directory" (N. Hall, ed.), pp. 421–422. Met. Plast. Pub., Hackensack, New Jersey, 1977.
5. "Recommended Practices for Processing of Mandrels for Electroforming," ASTM B431-69. ASTM, Philadelphia, Pennsylvania (1977).
6. S. Wein, "Copper Films," Rep. No. PB111237. Off. Tech. Serv., Washington, D.C. (1953).
7. S. Wein, "Gold Films," Rep. No. PB111332. Off. Tech. Serv., Washington, D.C. (1953).
8. S. Wein, "Lead Sulphide Films," Rep. No. PB111331. Off. Tech. Serv., Washington, D.C. (1953).
9. N. Hall, in "Metal Finishing Guidebook Directory" (N. Hall, ed.), pp. 504–513. Met. Plast. Pub., Hackensack, New Jersey, 1977.
10. F. Pearlstein, in "Modern Electroplating" (F. A. Lowenheim, ed.), 3rd ed., pp. 710–747. Wiley, New York, 1974.
11. "Standard Definitions of Terms Relating to Electroplating," ASTM B374-75. ASTM, Philadelphia, Pennsylvania (1977).
12. R. M. Lukes, Plating 51, 969 (1964).
13. N. A. Tope, E. A. Baker, and B. C. Jackson, Plating Surf. Finish. 63(10), 30 (1976).
14. N. Feldstein, Plating 61, 146 (1974).
15. E. B. Saubestre, Plating 59, 563 (1972).
16. C. F. Coombs, ed., "Printed Circuits Handbook." McGraw-Hill, New York, 1967.
17. N. Feldstein, U.S. Patent 3,993,799 (1976).
18. Anonymous, Prod. Finish. (Cincinnati) 41(4), 68 (1977); 41(5), 52 (1977).
19. "A Handbook of Protective Finishes for Military Equipment," Chap. IV. National Association of Corrosion Engineers, Committee T-9B, Houston, Texas, 1977.

20. F. Pearlstein and R. F. Weightman, Pitman-Dunn Lab. Rep. FA-TR-76003. Frankford Arsenal, Philadelphia, Pennsylvania (1976).
21. F. Pearlstein and R. F. Weightman, Pitman-Dunn Lab. Rep. M71-22-1. Frankford Arsenal, Philadelphia, Pennsylvania (1971).
22. G. Dubpernell, in "Modern Electroplating" (F. A. Lowenheim, ed.), 3rd ed., p. 106. Wiley, New York, 1974.
23. Fukubayashi, U.S. Patent Appl. 601, 125 (1976), available from Natl. Tech. Inf. Serv., Springfield, Virginia.
24. L. W. Higley, W. M. Dressel, and E. R. Cole, U.S. Bur. Mines, Rep. Invest. RI-8111 (1976), available from Natl. Tech. Inf. Serv. Springfield, Virginia.
25. D. W. Wabner, H. P. Fritz, D. Missol, H. Rainer, and F. Hindelang, Z. Naturforsch., Teil B 31, 39 (1976).
26. V. A. Sclyapnikov, Zh. Prikl. Khim. (Leningrad) 49, 90 (1976).
27. L. M. Elina, V. M. Gitneva, and V. I. Bystrov, Elektrokhimiya 11, 1279 (1975).
28. E. H. Cook, U.S. Patents 3,940,323 and 3,943,042 (1976).
29. J. J. H. Krause, U.S. Patent 3,929,607 (1975).
30. K. J. O'Leary and T. J. Navin, in "Chlorine Bicentennial Symposium (T. C. Jeffrey, P. A. Danna, and H. S. Holden, eds.), p. 174. Electrochem. Soc., Princeton, New Jersey, 1974.
31. H. Hirata, T. Yamazaki, and S. Hiyama, Electrochem. Soc. Extend. Abstr. No. 72-2, 514 (1972).
32. F. A. Lowenheim, "Electroplating: Fundamentals of Surface Finishing." McGraw-Hill, New York, 1978.
33. F. W. Eppensteiner and M. R. Jenkins, in "Metal Finishing Guidebook Directory" (N. Hall, ed.), pp. 540–556. Met. Plast. Pub., Hackensack, New Jersey, 1977.
34. M. F. Maher and A. M. Pradel, in "Metal Finishing Guidebook Directory" (N. Hall, ed.), pp. 568–581. Met. Plast. Pub., Hackensack, New Jersey, 1977.
35. F. A. Lowenheim, ed., "Modern Electroplating," 3rd ed. Wiley, New York, 1974.
36. "Standard Recommended Practice for Design of Articles to be Plated on Racks," ASTM B507. ASTM, Philadelphia, Pennsylvania (1977).
37. O. Kardos, Plating Surf. Finish. 61, 129 (1974); 61, 229 (1974); 61, 316 (1974).
38. N. Hall, ed., "Metal Finishing Guidebook Directory," Met. Plast. Pub., Hackensack, New Jersey (annual publication).
39. J. P. Bianciaroli and P. G. Stutzman, Plating 56, 37 (1969), and references cited therein.
40. J. A. Slattery, Prod. Finish. (Cincinnati) 41(10), 74 (1977).
40a. I. Hyner, Plating Surf. Finish. 64(2), 32 (1977).
41. A. Brenner, "Electrodeposition of Alloys, Principles and Practice," Vols. 1 and 2. Academic Press, New York, 1963.
42. A. Brenner, Adv. Electrochem. Electrochem. Eng. 5, 205 (1967).
43. W. B. Harding, in "Modern Electroplating" (F. A. Lowenheim, ed.), 3rd ed., p. 63. Wiley, New York, 1974.
44. S. Senderoff, in "Modern Electroplating (F. A. Lowenheim, ed.), 3rd ed., p. 473. Wiley, New York, 1974.
45. D. J. George, in F. A. Lowenheim "Electroplating: Fundamentals of Surface Finishing," pp. 452–471. McGraw-Hill, New York, 1978.
46. S. Wernick and R. Pinner, "The Surface Treatment and Finishing of Aluminum and Its Alloys," 4th ed. Robert Draper, Teddington, England, 1972.
47. "Standard Method for Measurement of Coating on Anodically Coated Aluminum," ASTM B137-45 (1972). ASTM, Philadelphia, Pennsylvania (1977).
48. M. E. Seibert, J. Electrochem. Soc. 110, 65 (1963).

III-2

Chemical Vapor Deposition of Inorganic Thin Films

WERNER KERN AND VLADIMIR S. BAN

RCA Laboratories
Princeton, New Jersey

I. INTRODUCTION

The growth of thin films by chemical vapor deposition (CVD) has become one of the most important methods of film formation and now constitutes a corner stone for modern technologies such as solid-state electronics. The reasons for the rapidly growing importance of CVD in the
past decade lie primarily in its versatility for depositing a very large variety of elements and compounds at relatively low temperatures, in the
form of both vitreous and crystalline layers having a high degree of perfection and purity. Other unique advantages of CVD over other methods
of film formation are the relative ease for creating materials of a wide
range of accurately controllable stoichiometric composition and layer
structures that are difficult or impossible to attain by other techniques.

Several excellent surveys and reviews encompassing theoretical,
practical, general, and specialized aspects of CVD are available. The single most comprehensive work covering many aspects of CVD is the mon-
ograph edited by Powell et al. [1] published in 1966. Techniques of CVD
were described by Holzl [2], and general surveys were presented by Feist
et al. [3], Campbell [4], Haskell and Byrne [5], Hastie [6], and Bryant [7].

A large compilation of original research results, covering a wide range
of aspects of CVD, has been presented in symposia proceedings [8–13].
Several reviews especially concerned with CVD of electronic materials
are also available [14–23]. A large body of information exists on epitaxial
CVD of semiconductors, as evident from a recent bibliography [24] and
surveys [25, 26] of this specialized subject; additional references on epitaxy have been cited in Chapter I-1 of this book.

The aim of this chapter is to provide a broad outline for readers who
wish to acquaint themselves with the fundamental and practical aspects of
CVD. Selected literature references will be provided as a guide to more
detailed sources of information. Emphasis will be placed on materials and
processes for semiconductor electronics, where much of the most advanced and sophisticated thin-film CVD technology is being applied.

Fundamental aspects of CVD covering reaction chemistry, thermodynamics, kinetics, transport phenomena, low-pressure CVD, and film
growth will be discussed first. Reactor systems and process control tech-

niques will then be discussed. Finally, a brief survey of CVD applications for preparing important and representative materials will be presented. These will be grouped according to insulators, elemental and compound semiconductors, conductors, and miscellaneous types of materials.

To provide a more detailed insight into some of the complexities and problems encountered in practice, one typical CVD process has been chosen where the effects of various parameters on the deposition and properties of thin glass films will be described. Much of this presentation, which is included in the section on insulators, is a summary of recent research results obtained at our laboratories and should serve well as an interesting example of applied CVD.

II. FUNDAMENTAL ASPECTS OF CVD

With the increasingly stringent quality requirements imposed on CVD films, a thorough understanding of the underlying principles of the method becomes very important. In this section we shall briefly review these principles. Since CVD is an interdisciplinary undertaking we shall cross boundaries of several classical fields. First, we shall discuss chemical aspects of CVD (e.g., the nature, type, and extent of chemical reactions), followed by discussions of thermodynamics and kinetics of CVD processes. Next, transport phenomena (i.e., heat, mass, and momentum transport) in CVD reactors and principles of low-pressure CVD will be discussed. Finally, the film growth aspects of CVD processes (e.g., the nucleation, growth of epitaxial, polycrystalline, and amorphous deposits) will be covered briefly.

A. Chemistry of CVD

Chemical vapor deposition can be defined as a material synthesis method in which the constituents of the vapor phase react to form a solid film at some surface. Thus, the occurrence of the chemical reaction is an essential characteristic of the CVD method. In order to understand CVD processes, one must know which chemical reactions occur in the reactor and to what extent. Furthermore, the effects of process variables such as temperature, pressure, input concentrations, and flow rates on these reactions must be understood.

The nature and extent of chemical reactions can be deduced if one knows the composition of the solid and vapor phases in the CVD system. The composition of the solid can be analyzed after the experiment, but the composition of the vapor must be determined *in situ* at CVD pressures and temperatures, because many high-temperature species disappear or change upon cooling to room temperature.

1. Instrumental Analysis of the Vapor Phase

The composition of the vapor phase can be assumed on the basis of similarity with some known system or from known thermodynamic data. Since this can be incorrect, there is a growing tendency to apply some reliable, direct analytical method. We shall reference the pertinent work and discuss in some detail results of mass spectrometric analyses carried out in our laboratory.

Figure 1 shows a mass spectrometer attached to a CVD reactor which permits direct sampling of the vapor under CVD conditions for qualitative and quantitative analysis. This system was applied to CVD studies of III–V compound semiconductors [27–32], II–VI materials [33], and silicon [34–37]. Table I summarizes the results of these studies. In some cases, the results obtained are as expected, but in others they are unexpected (e.g., mixed As_xP_y species and the absence of $GaCl_3$ and $InCl_3$) [30]. The quantitative data on partial pressures are given in the corresponding references.

Other direct analytical methods applied to CVD studies are Raman spectroscopy [38, 39], absorption spectroscopy [40], and gas chromatography [41, 42]. A summary of techniques and results of *in situ* gas species detection in CVD reactions has been presented recently [42]. Additional references will be cited in Section IV on CVD of insulators.

2. Chemical Reactions in CVD Systems

In this section, we shall discuss some of the typical chemical reactions encountered in CVD. The emphasis will be placed on electronic materials, where the most detailed studies have been undertaken.

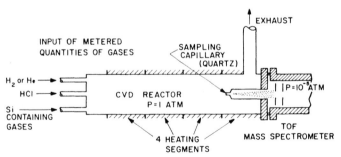

Fig. 1. Schematic representation of the system mass spectrometer—CVD reactor (from Ban and Ettenberg [34], reprinted by permission of the publisher, The Electrochemical Society, Inc.).

Table I

Summary of Results of Mass Spectrometric Studies of CVD of Semiconductors

CVD products	Chemical system	Input reactants	Vapor species found	References
GaN	Ga–Cl–H–N	HCl, NH₃, Ga, H₂	H₂, HCl, GaCl, NH₃, GaCl₃·NH₃, N₂	[28]
GaP	Ga–Cl–H–P	HCl, PH₃, Ga, H₂	H₂, HCl, GaCl, PH₃, P₂, P₄	[27]
GaAs	Ga–Cl–H–As	HCl, AsH₃, Ga, H₂	H₂, HCl, GaCl, AsH₃, As₂, As₄	[27]
GaSb	Ga–Cl–H–Sb	HCl, Sb, Ga, H₂	H₂, HCl, GaCl, Sb₂, Sb₄	[30]
InP	In–Cl–H–P	HCl, PH₃, In, H₂	H₂, HCl, InCl, PH₃, P₂, P₄	[31]
InAs	In–Cl–H–As	HCl, AsH₃, In, H₂	H₂, HCl, InCl, AsH₃, As₂As₄	[32]
GaAsP	Ga–Cl–H–P–As	HCl, PH₃, AsH₃, Ga, H₂	H₂, HCl, GaCl, PH₃, P₂, P₄, AsH₃, As₂As₄, AsP, As₂P, As₂P₂, AsP₃	[27]
GaAsSb	Ga–Cl–H–As–Sb	HCl, AsH₃, Ga, Sb, H₂	H₂, HCl, GaCl, AsH₃, As₂, As₄, Sb₂, Sb₄, AsSb, As₂Sb, As₂Sb₂, AsSb₃	[30]
InGaP	Ga–In–Cl–H–P	HCl, PH₃, Ga, In, H₂	H₂, HCl, GaCl, InCl, PH₃, P₂, P₄	[31]
InGaAs	Ga–In–Cl–H–As	HCl, AsH₃, Ga, In, H₂	H₂, HCl, GaCl, InCl, AsH₃As₂, As₄	[32]
ScN	Sc–Cl–H–N	HCl, NH₃, Sc, H₂	H₂, HCl, ScCl₂, NH₃, N₂	[30]
Si	Si–Cl–H	SiCl₂H₂, H	H₂, HCl, SiCl₂, SiCl₂H₂, SiCl₃H, SiCl₄, Si₂Cl₆	[34]
Si	Si–Cl–H	SiCl₃H, H₂	H₂, HCl, SiCl₂, SiCl₂H₂, SiCl₃H, SiCl₄, Si₂Cl₆	[35]
Si	Si–Cl–H	SiCl₄, H₂	H₂, HCl, SiCl₂, SiCl₂H₂, SiCl₃H, SiCl₄, Si₂Cl₆	[35]
Si	Si–Cl–He	SiCl₄, He	He, SiCl₂, SiCl₄	[35]
Si	Si–H	SiH₄, H₂	H₂, SiH₄, Si₂H₆	[37]
CdS	Cd–S	CdS	Cd, S₂	[33]
ZnS	Zn–S	ZnS	Zn, S₂	[33]

261

a. Pyrolysis. In many systems, CVD deposits are obtained through a pyrolytic or thermal decomposition of gaseous species at the hot susceptor surface. Organometallic compounds, hydrides, and metal hydrides are particularly suitable starting materials for this type of reaction. Typical examples are the deposition of silicon from silane and nickel from nickel carbonyl:

$$SiH_{4(g)} \rightarrow Si_{(s)} + 2H_{2(g)},$$
$$Ni(CO)_{4(g)} \rightarrow Ni_{(s)} + 4CO_{(g)}.$$

b. Reduction. Hydrogen is the most commonly used reducing agent. A well-known example of a reduction reaction is the deposition of silicon from silicon tetrachloride:

$$SiCl_{4(g)} + 2H_{2(g)} \rightarrow Si_{(s)} + 4HCl_{(g)}.$$

The HCl formed might, in turn, react with the solid Si to form other gaseous species detected in the Si–Cl–H system [34, 35] such as $SiCl_2$, $SiCl_2H_2$, and $SiCl_3H$. Thus, the situation in the reactor is considerably more complex than that suggested by the overall reaction noted above. Another example is the hydrogen reduction of tungsten hexafluoride to tungsten:

$$WF_{6(g)} + 3H_{2(g)} \rightarrow W_{(s)} + 6HF_{(g)}.$$

c. Oxidation. SiO_2 films are usually deposited by the following oxidation reaction:

$$SiH_{4(g)} + O_{2(g)} \rightarrow SiO_{2(s)} + 2H_{2(g)}$$

or, depending on conditions,

$$SiH_{4(g)} + 2O_{2(g)} \rightarrow SiO_{2(s)} + 2H_2O_{(g)}.$$

d. Hydrolysis. The deposition of polycrystalline Al_2O_3 films from aluminum chloride with CO_2 and H_2 is a good example of a hydrolysis reaction:

$$Al_2Cl_{6(g)} + 3CO_{2(g)} + H_{2(g)} \rightarrow Al_2O_{3(s)} + 3HCl_{(g)} + 3CO_{(g)}.$$

e. Nitride Formation. Silicon nitride films are formed via the nitridation or ammonolysis reaction:

$$3SiH_{4(g)} + 4NH_{3(g)} \rightarrow Si_3N_{4(s)} + 12H_{2(g)}.$$

f. Carbide Formation. The formation of titanium carbide illustrates carbidization reactions:

$$TiCl_{4(g)} + CH_{4(g)} \rightarrow TiC_{(s)} + 4HCl_{(g)}.$$

g. Disproportionation. Chemical transformation can also take place by disproportionation of high-temperature vapor species, as exemplified by the following two examples:

$$2GeI_{2(g)} \xrightarrow{\text{lower temp.}} Ge_{(s)} + GeI_{4(g)},$$

$$Si_{(s)} + SiI_{4(g)} \rightarrow 2SiI_{2(g)}.$$

h. Synthesis Reactions. Deposition of group III–V and II–VI compounds from organometallic compounds may proceed as follows:

$$(CH_3)_3Ga_{(g)} + AsH_{3(g)} \rightarrow GaAs_{(s)} + 3CH_{4(g)},$$

$$(CH_3)_2Cd_{(g)} + H_2Se_{(g)} \rightarrow CdSe_{(s)} + 2CH_{4(g)}.$$

i. Chemical Transport Reaction. The essential reaction in the deposition of GaAs by means of the chloride method is a good example of a chemical transport reaction. Transport in these processes depends upon the difference in equilibrium constants for the temperatures of the reactant source and the substrate;

$$6GaAs_{(g)} + 6HCl_{(g)} \underset{T_2}{\overset{T_1}{\rightleftharpoons}} As_{4(g)} + As_{2(g)} + GaCl_{(g)} + 3H_{2(g)},$$

where $T_1 > T_2$. Lowering of temperature in the downstream position of the reactor reverses the reaction and thus initiates the deposition of GaAs.

j. Combined Reactions. The deposition of group III–V compound semiconductors via the hydride method [43] involves a sequence of several different reactions. The first one is transport of the group III metal M:

$$M_{(l)} + HCl_{(g)} \rightarrow MCl_{(g)} + \tfrac{1}{2}H_{2(g)},$$

where M = Ga, In, or Al. Downstream from the metal transport reaction, decomposition of group V hydrides takes place:

$$MH_3 \rightarrow (1 - x) MH_3 + (1 - \alpha) \tfrac{1}{4}xM_{4(g)} + \alpha\tfrac{1}{2}xM_{2(g)} + \tfrac{3}{2}xH_{2(g)},$$

where M is As or P, x the degree of decomposition, and α the ratio of dimers (M_2) to tetramers (M_4). Finally, the deposition reaction involving the previously formed gaseous species takes place. In the case of GaAs:

$$12GaCl_{(g)} + 4AsH_{3(g)} + 2As_{2(g)} + As_{4(g)} \rightarrow 12GaAs_{(s)} + 12HCl_{(g)}.$$

From this discussion, it is obvious that various types of chemical reactions occur during CVD. In the great majority of cases, the deposition reaction is heterogeneous in character, taking place at the substrate surface. Homogeneous reactions usually affect the composition of the gaseous phase only. The creation of the solid phase via homogeneous reactions is usually undesirable because powdery or flaky deposits result instead of well-adhering films.

The reactant concentrations and temperatures necessary for CVD reactions depend on the thermodynamics and kinetics of the system in question.

B. Thermodynamics of CVD

The main functions of thermodynamics in relation to CVD are to predict the feasibility of the process under some specified conditions and to provide quantitative information about the process. Properly performed thermodynamic calculations give the theoretically obtainable amount of a deposit and partial pressures of all vapor species under specified experimental conditions, such as the temperature and pressure in the reactor and the input concentrations of reactants. Thus, thermodynamics can be used as a guideline for establishing general process parameters, but it does not provide any information on the rate of various CVD processes, thus imposing limitations on its applicability. In some systems, reactions which are thermodynamically feasible proceed at such slow rates that they are unimportant in practice. Furthermore, the application of thermodynamics implies the establishment of chemical equilibrium in the reactor; this, especially in flow reactors, is not usually the case. Nevertheless, despite these limitations, one of the first steps in considering a CVD process is to perform the necessary thermodynamic calculations to obtain the general conditions the process requires.

In order to perform the calculations, one needs reliable thermodynamic data. The most useful data are the free energy of formation of all vapor and condensed constituents of the system: ΔG_f° is the standard free energy of formation of a compound and $\Delta G_f^\circ = 0$ for all elements in their standard state. The free energy of some chemical reaction (ΔG_r°) can be calculated if the ΔG_f° values are known:

$$\Delta G_r^\circ = \Sigma \Delta G_f^\circ \text{ products} - \Sigma \Delta G_f^\circ \text{ reactants.}$$

ΔG_r° is related to the equilibrium constant k_p, which is itself related to partial pressures in the system:

$$-\Delta G_r^0 = 2.3 \, RT \log k_p; \; k_p = \frac{\prod_{i=1}^{n} P_i \text{ products}}{\prod_{i=1}^{n} P_i \text{ reactants}}.$$

In CVD, one usually deals with a multicomponent and a multiphase system. Most often there are only two phases, the vapor and the solid, although more than one condensed phase may be present. The number of components varies, e.g., from two in the simple Si–H system (i.e., the deposition of Si from SiH$_4$) to five in the In–Ga–As–H–Cl system (i.e., the deposition of In$_x$Ga$_{1-x}$ As via the hydride method).

There are several ways to calculate thermodynamic equilibrium in multicomponent systems [44]. We shall briefly discuss the optimization method and the nonlinear equation method.

1. Optimization Method

Consider a system in which a chemical reaction proceeds to some degree of completion, ϵ:

$$(1 - \epsilon)(\text{reactants}) \rightarrow \epsilon(\text{products}).$$

If one plots the free energy of this system versus ϵ, one sees that the energy curve has a minimum at some value of ϵ (Fig. 2). This value of ϵ is the equilibrium value and is characterized by the equilibrium concentrations of reactants and products. The equation describing the free energy G of the whole system consisting of m gaseous species and s solid phases is

$$G = \sum_{i=n_i}^{m} \left(n_i^g \, \Delta G_{fig}^{\circ} + RT \ln P + 2T \ln \frac{n_{ig}}{N_g} \right) + \sum_{i=1}^{s} n_i^s \, \Delta G_{fis}^{0},$$

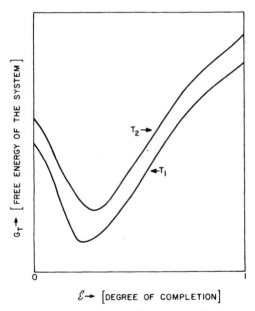

$\mathcal{E} \rightarrow$ [DEGREE OF COMPLETION]

Fig. 2. Variation of total free energy G_T of the system with degree of completion ϵ of chemical reactions in the system at two different temperatures.

where n_i^g and n_i^s are the number of moles of gaseous and solid species, respectively, N_g is the total number of moles of gaseous species, P is the total pressure, and ΔG_{fig}° and ΔG_{fis}° are the free energies of formation at CVD temperatures of gaseous and solid species, respectively. The objective of optimization calculations is to determine the set of (n_i) which minimizes G. This set of (n_i) represents the equilibrium composition of the system. The detailed method for solution of the above equation is given by Eriksson [45], who described a computer program to obtain the equilibrium compositions of gaseous and solid phases at some specified temperature, pressure, and input concentrations. Very large numbers of species can be handled by this method.

2. Nonlinear Equation Method

In this method, one first establishes a set of m (m is the number gaseous species in the system) simultaneous, generally nonlinear equations, which specify quantitative relationship between partial pressures of species present. Solution of this set of equations yields values of partial pressures of all species under the specified temperature, total pressure, and input concentrations. This method of calculation has been applied to many CVD systems, among them the Si–Cl–H system [34, 35, 46, 47] which is shown as an example. The following set of equations was used:

$$\mathrm{Si}_{(s)} + 4\mathrm{HCl}_{(g)} \rightarrow \mathrm{SiCl}_{4(g)} + 2\mathrm{H}_{2(g)}, \qquad K_{p(1)} = \frac{\mathrm{SiCl}_4 \cdot P_{\mathrm{H}_2}^2}{a\mathrm{Si} \cdot P_{\mathrm{HCl}}^4}; \qquad (1)$$

$$\mathrm{Si}_{(s)} + 3\mathrm{HCl}_{(g)} \rightarrow \mathrm{SiCl}_3\mathrm{H}_{(g)} + \mathrm{H}_{2(g)} \qquad K_{p(2)} = \frac{P_{\mathrm{SiCl}_3\mathrm{H}} \cdot P_{\mathrm{H}_2}}{a\mathrm{Si} \cdot P_{\mathrm{HCl}}^3}; \qquad (2)$$

$$\mathrm{Si}_{(s)} + 2\mathrm{HCl}_{(g)} \rightarrow \mathrm{SiCl}_2\mathrm{H}_{2(g)}, \qquad K_{p(3)} = \frac{P_{\mathrm{SiCl}_2\mathrm{H}_2}}{a\mathrm{Si} \cdot P_{\mathrm{HCl}}^2} \qquad (3)$$

$$\mathrm{Si}_{(s)} + \mathrm{HCl}_{(g)} + \mathrm{H}_{2(g)} \rightarrow \mathrm{SiClH}_{3(g)}, \qquad K_{p(4)} = \frac{P_{\mathrm{SiClH}_3}}{a\mathrm{Si} \cdot P_{\mathrm{HCl}} \cdot P_{\mathrm{H}_2}}; \qquad (4)$$

$$\mathrm{Si}_{(s)} + 2\mathrm{HCl}_{(g)} \rightarrow \mathrm{SiCl}_{2(g)} + \mathrm{H}_{2(g)}, \qquad K_{p(5)} = \frac{P_{\mathrm{SiCl}_2} \cdot P_{\mathrm{H}_2}}{a\mathrm{Si} \cdot P_{\mathrm{HCl}}^2}; \qquad (5)$$

$$\mathrm{Si}_{(s)} + 2\mathrm{H}_{2(g)} \rightarrow \mathrm{SiH}_{4(g)}, \qquad K_{p(6)} = \frac{P_{\mathrm{SiH}_4}}{a\mathrm{Si} \cdot P_{\mathrm{H}_2}^2}. \qquad (6)$$

The activity of Si is presented by a, where $a = 1$. The remaining two equations specify the Cl/H ratio and state that the total pressure in the system equals 1 atm:

$$\mathrm{Cl/H} = \frac{4P_{\mathrm{SiCl}_4} + 3P_{\mathrm{SiCl}_3\mathrm{H}} + 2P_{\mathrm{SiCl}_2\mathrm{H}_2} + 2P_{\mathrm{SiCl}_2} + P_{\mathrm{SiClH}_3} + P_{\mathrm{HCl}}}{2P_{\mathrm{H}_2} + P_{\mathrm{SiCl}_3\mathrm{H}} + 2P_{\mathrm{SiCl}_2\mathrm{H}_2} + 3P_{\mathrm{SiClH}_3} + P_{\mathrm{HCl}} + 4P_{\mathrm{SiH}_4}} \qquad (7)$$

and

$$P_{SiCl_4} + P_{SiCl_3H} + P_{SiCl_2H_2} + P_{SiClH_3} + P_{SiCl_2} + P_{SiH_4} + P_{HCl} + P_{H_2} = 1.$$
$$(8)$$

The Cl/H is the compositional variable, the value of which is determined by the composition of the chlorosilane–hydrogen input mixture. Table II shows results of calculations of equilibrium partial pressures in the Si–H–Cl system at various temperatures and Cl/H ratios. From these values, one can determine the total quantity of Si in the vapor phase and compare it with the input amount. The difference specifies the theoretically available amount of the solid Si deposit under the specified process parameters (i.e., $T, P, Cl/H$).

Since the calculations can be computerized, one can model the system through a systematic variation of pertinent parameters and find a set of optimal parameters for any CVD process in question. The accuracy and usefulness of thermodynamic calculations depend strongly on the accuracy of the thermochemical data used. We are including several references which have been found reliable and useful [48–54].

C. Kinetics of CVD

We have mentioned that the situation in CVD reactors might differ from the predictions of thermodynamical equilibrium calculations. In this section we discuss some of the factors that cause deviations from equilibrium. The deposition reaction is almost always a heterogeneous reaction. The sequence of events in the usual heterogeneous processes can be described as follows:

(1) diffusion of reactants to the surface;
(2) adsorption of reactants at the surface;
(3) surface events, such as chemical reaction, surface motion, lattice incorporation, etc.;
(4) desorption of products from the surface;
(5) diffusion of products away from the surface.

The steps are sequential and the slowest one is the rate-determining step. We shall now present some practical cases of CVD where kinetics play an important role.

1. Temperature Dependence of Si Deposition Rate

Figure 3 shows the temperature dependence of deposition rates of Si from various source gases [55]. A relatively steep temperature dependence is observed in the lower temperature range, and a milder depen-

Table II

Calculated Values of Equilibrium Partial Pressures (in atm) of Vapor Species in the Si–Cl–H System[a]

T°K	Cl/H	H_2	HCl	$SiCl_4$	$SiHCl_3$	SiH_2Cl_2	SiH_3Cl	SiH_4	$SiCl_2$
1000	10^0	6.28×10^{-1}	1.52×10^{-2}	2.74×10^{-1}	8.01×10^{-2}	2.70×10^{-3}	5.05×10^{-5}	2.55×10^{-7}	1.99×10^{-4}
1000	10^{-1}	9.39×10^{-1}	1.07×10^{-2}	3.03×10^{-2}	1.88×10^{-2}	1.34×10^{-3}	5.33×10^{-5}	5.70×10^{-7}	6.62×10^{-5}
1000	10^{-2}	9.90×10^{-1}	5.40×10^{-3}	1.74×10^{-3}	2.26×10^{-3}	3.39×10^{-4}	2.83×10^{-5}	6.34×10^{-7}	1.59×10^{-5}
1000	10^{-3}	9.98×10^{-1}	1.67×10^{-3}	1.55×10^{-5}	6.58×10^{-5}	3.22×10^{-5}	8.78×10^{-6}	6.45×10^{-7}	1.50×10^{-6}
1200	10^0	6.00×10^{-1}	5.80×10^{-2}	2.61×10^{-1}	7.36×10^{-2}	2.92×10^{-3}	5.95×10^{-5}	3.43×10^{-7}	5.01×10^{-3}
1200	10^{-1}	9.18×10^{-1}	3.97×10^{-2}	2.44×10^{-2}	1.54×10^{-2}	1.37×10^{-3}	6.23×10^{-5}	8.02×10^{-7}	1.53×10^{-3}
1200	10^{-2}	9.83×10^{-1}	1.50×10^{-2}	4.30×10^{-4}	7.72×10^{-4}	1.94×10^{-4}	2.52×10^{-5}	9.22×10^{-7}	2.03×10^{-4}
1200	10^{-3}	9.98×10^{-1}	1.98×10^{-3}	1.26×10^{-7}	1.74×10^{-6}	3.38×10^{-6}	3.37×10^{-6}	9.49×10^{-7}	3.49×10^{-6}
1400	10^0	5.37×10^{-1}	1.38×10^{-1}	2.16×10^{-1}	5.99×10^{-2}	2.63×10^{-3}	5.75×10^{-5}	3.64×10^{-7}	4.62×10^{-2}
1400	10^{-1}	8.81×10^{-1}	8.63×10^{-2}	1.22×10^{-2}	8.89×10^{-3}	1.03×10^{-3}	5.88×10^{-5}	9.77×10^{-7}	1.10×10^{-2}
1400	10^{-2}	9.81×10^{-1}	1.85×10^{-2}	2.07×10^{-5}	7.84×10^{-5}	4.71×10^{-5}	1.40×10^{-5}	1.21×10^{-6}	4.52×10^{-4}
1400	10^{-3}	9.98×10^{-1}	1.98×10^{-3}	2.66×10^{-9}	9.56×10^{-8}	5.44×10^{-7}	1.53×10^{-6}	1.25×10^{-6}	5.13×10^{-6}

[a] Total system pressure = 1 atm $\pm 1 \times 10^{-5}$.

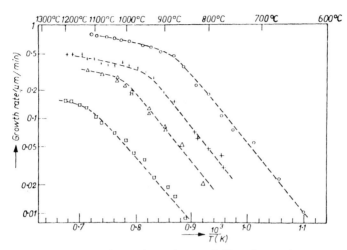

Fig. 3. Temperature dependence of growth rates of silicon from various source gases. O, SiH_4; +, SiH_2Cl_2; △, $SiHCl_3$; □, $SiCl_4$ (from Eversteijn [55], courtesy of Philips Research Reports).

dence in the upper range, indicating that the nature of the rate-controlling step changes with temperature. The observed behavior suggests that in the lower range, the rate-controlling step is some surface process, the rate of which is exponentially dependent on temperature, as seen from the Arrhenius equation

$$r = ae(-\Delta E/RT),$$

where a is the frequency factor and ΔE the activation energy—usually 25–100 kcal/mole for surface processes. The most likely rate-controlling step is the weak adsorption of reactants on the substrate surface. In the upper temperature range, the most probable rate-controlling step is the diffusion of reactants and products to and from the reacting surface, respectively. The diffusion temperature dependence is usually mild (it varies as $T^{1.5-2.0}$), and the rate of deposition follows a less steep slope. The effect of temperature on the growth rate was also studied in III–V compound semiconductors [29, 46].

2. Dependence of the Deposition Rate on Substrate Orientation

It is well known that the deposition rate can strongly depend on the crystallographic orientation of the substrate [56, 57]. Figure 4 shows the deposition rate of GaAs as a function of the substrate orientation. It can be seen that the (111) orientation causes a three to four times faster growth than the (100) orientation. There are several fundamental reasons for the observed differences, among them are variations in

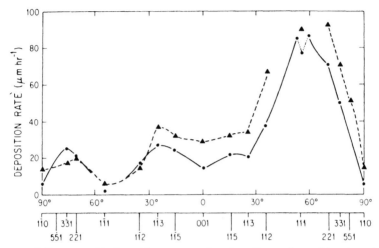

Fig. 4. Gallium arsenide deposition rate as a function of crystallographic orientation of the substrate. ▲, deposition temperature 750°C; ●, deposition temperature 755°C (from Shaw [56], courtesy of North-Holland Publishing Company).

(1)   the densities and geometric arrangements of surface sites;

(2)   the number and nature of surface bonds;

(3)   the composition of various crystallographic surfaces, [i.e., in GaAs: (111 A) terminates with Ga atoms; (111 B) with As atoms];

(4)   the number and nature of surface features, such as steps, kinks, ledges, vacancies, etc.

All of these factors can influence deposition by affecting adsorption, desorption, surface mobility, reactivity, etc. The exact nature of the various effects is frequently not known. Chiang [58] and Chang [59] have given good examples of contemporary experimental efforts in this field.

Several other kinetic effects in CVD háve been observed, e.g., the effect of dopants on the semiconductor growth rate [58, 59] and the existence of metastable phases [28]. The effect of mass transport limitations will be discussed in the next section. It is important to keep in mind that while thermodynamics specifies what ought to happen in the reactor, the kinetics determine what actually will happen.

D. Transport Phenomena in CVD

1. Introduction

In this section we shall discuss the heat, mass, and momentum transport in CVD reactors. There are two basic reasons for studying transport phenomena:

(1)   The requirement of thickness and semiconductor doping uniformity which can be fulfilled only when equal amounts of reactant gases and dopants are delivered to all substrates in the system.

(2)   The requirements of high chemical efficiency to achieve satisfactory growth rates and utilization of input chemicals; this means that sufficient amounts of reactants must be delivered to growth surfaces.

Once transport phenomena are understood, one can design a CVD reactor capable of fulfilling the above requirements.

Transport phenomena in fluids (i.e., gases and liquids) are related to the nature of the fluid flow. We shall therefore consider the nature of gas flow in CVD reactors and briefly discuss the relevant fluid dynamics. The following parameters affect the nature of gas flow in reactors: (1) velocity of flow, (2) temperature and temperature distribution in the system, (3) pressure in the system, (4) geometry of the system, (5) gas or vapor characteristics.

Chemical vapor deposition reactors will be discussed in more detail later; for the present purpose, it is sufficient to divide them into hot-wall and cold-wall reactors. Hot-wall reactors are usually tubular in design, and it is reasonable to assume that due to the low gas velocities employed, a fully developed laminar flow and fully developed thermal fields exist in the reactor. Transport phenomena under these conditions are frequently encountered in chemical engineering [60]. The existence of various flow disturbance fixtures, potential entry length problems, etc., must also be taken into account.

In cold-wall reactors the situation is somewhat more complex. The existence of rather sharp temperature gradients perpendicular to, as well as along, the susceptor affects the flow and thus the transport phenomena considerably. Furthermore, in many instances susceptors are rather short and entry effects can play a significant role. Under such complex conditions, it is difficult to choose an appropriate model for transport calculations. For this reason, a considerable effort was put into experimental studies of transport phenomena in cold-wall CVD reactors; we shall discuss some of these experiments, again presenting those conducted in our laboratories as examples.

2. Experimental Studies of Transport Phenomena in Cold-Wall CVD Reactors

a. Flow Visualization.   In these studies [55, 61, 62], TiO_2 smoke was employed to visualize flow patterns in the reactor. The main conclusions are that there exists a boundary layer adjacent to the susceptor, and that the steep temperature gradients perpendicular to the flow affect the nature of flow; furthermore, evidence of a convection-caused gas motion was found. These flow visualization experiments provide a qualitative insight into the momentum transfer in the gas phase.

b. Temperature Measurements. Measurements of the temperature and the temperature spatial variation provide information on the heat transport in the reactor. We constructed an apparatus which permitted us to measure temperature at any point in the area above the susceptor [36]. Measurements were taken under conditions of varying susceptor temperature and gas velocity and with different carrier gases (N_2, He, H_2) [62]. The main results can be summarized as follows:

(1) There is a steep temperature gradient in the first 1.5 cm above the susceptor, and a shallower gradient in the space above the 1.5 cm;

(2) for the same gas velocity and susceptor temperature, the gradients are steeper in N_2 than in He and H_2;

(3) the average temperature of the gas increases along the length of the susceptor.

Figure 5 shows isotherms in the reactor. The gas used was He; other parameters are specified in the figure.

c. Mass Transport Measurements. Information on mass transport in the reactor was obtained by measuring partial pressures and their spatial variation in the area above the susceptor using a specially designed movable mass spectrometric probe [36]. From these measurements we constructed the $SiCl_4$ isobars shown in Fig. 6. Again, in the upper area a flatter concentration gradient exists than in the lower area where isobars are bunched closer together. Other gaseous species detected were HCl and a relatively small amount of $SiCl_2$. The HCl isobars would be different from those of $SiCl_4$ since the HCl partial pressure is the highest in the vicinity of the hot susceptor where HCl is formed via the reaction shown in Section A.2.b.

d. Flow Model. The above measurements allow us to construct the flow

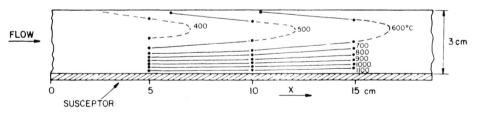

Fig. 5. Isotherms in the horizontal reactor; carrier gas with He, other conditions are specified in the figure. $T_S = 1200°C$, $V° = 50$ cm/sec (from Ban [62], reprinted by permission of the publisher, The Electrochemical Society, Inc.).

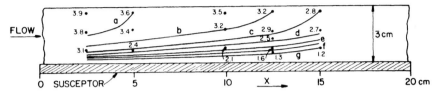

Fig. 6. SiCl$_4$ isobars in the horizontal reactor obtained by mass spectrometric measurements; experimental conditions are specified in the figure. $T_S = 1200°C$, $V° = 25$ cm/sec, $P°_{SiCl_4} = 6.1 \times 10^{-3}$ atm in 1 atm H$_2$. (from Ban [62], reprinted by permission of the publisher, The Electrochemical Society, Inc.).

Curve	a	b	c	d	e	f	g
SiCl$_4$ pressure $\times$ 10^{-3} atm	3.6	3.2	2.8	2.4	2.0	1.6	1.2

model depicted in Fig. 7 showing the directions of changes of reactant concentration, gas velocity, and temperature. The main characteristics of the model are the developing velocity and temperature boundary layers, and the existence of a well-mixed central core of gas. Reactant species diffuse from this central core to the susceptor surface, where the deposition reaction takes place. The mass transport is the rate-controlling factor under such conditions. The model can be used to develop a mathematical treatment of transport phenomena for designing CVD reactors with maxi-

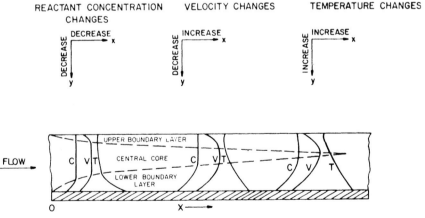

Fig. 7. Model of gas flow in the horizontal reactor showing directions of changes of reactant concentration, gas, velocity, and temperature. C, concentration profile; V, velocity profile; T, temperature profile (from Ban [62], reprinted by permission of the publisher, The Electrochemical Society, Inc.).

mum efficiency. Several such treatments have been developed in the last few years [55, 63–65].

E. Principles of Low-Pressure CVD*

The fundaments of depositing films at a low pressure are basically the same as for those deposited near atmospheric pressure. The major difference is the radical change of emphasis that is placed on the rates of mass transfer of the gaseous reactant and by-product species relative to their surface reaction rate to form the film deposit. At normal atmospheric pressure these rates are generally of the same order of magnitude. Therefore, in order to attain the objective of uniform film thicknesses and properties over extended surfaces for large production-type substrate loads, it is necessary to ensure that careful consideration be given to both types of rate-determining steps. We have seen that mass transfer rates depend mainly upon reactant concentration, diffusivity, and boundary layer thickness, which is related to reactor configuration, flow velocity, distances from edges, etc., while surface reaction rates depend mainly upon reactant concentration and temperature.

The diffusivity of a gas is inversely related to pressure. Therefore, as pressure is lowered from its atmospheric value near 760 Torr to 0.5–1 Torr, the diffusivity increases by a factor of 1000. This is only partially offset by the fact that the (laminar) boundary layer (distance across which the reactants must diffuse) increases at less than the square root of pressure.† The net effect is more than one order of magnitude increase of the gas-phase transfer of reactants to and by-products from the substrate surface, and the rate-determining step in the sequence is the surface reaction. Therefore, in such a low-pressure operation, much less attention need be paid to the mass transfer variables that are so critical at atmospheric pressure.

The practical consequences of these principles are very significant. The increased mean-free-path of the reactant gas molecules allows the substrate wafers in the low-pressure horizontal tubular hot-wall reactor to

---

* Portions of this section have been reprinted from Kern and Rosler [23], with permission of the publisher, The American Institute of Physics.

† The boundary layer (laminar) is inversely proportional to the square root of the Reynolds number (which is free stream velocity × distance × gas density ÷ gas velocity). For typical low-pressure conditions compared to atmospheric reactor conditions, the velocity is generally a factor of 10–100 higher, the density a factor of 1000 lower, the viscosity and distance basically the same. Therefore, the Reynolds number is a factor of 10–100 lower. Thus, theoretically, the boundary layer thickness should be about a factor of 3–10 thicker in low-pressure CVD than in atmospheric-pressure CVD.

be stacked on edge, instead of lying flat as in conventional atmospheric-pressure processing. No carrier gases are needed, and the resulting film deposits exhibit improved uniformity, better step coverage and conformality, and superior structural integrity with fewer pinholes. Deposits on the reactor wall are well adherent and hence do not tend to flake off, which would lead to particles with consequent substrate contamination.

The most successful applications of low-pressure CVD at present are the deposition of polycrystalline Si films from SiH_4 in the temperature range 600–650°C, for Si_3N_4 films from SiH_2Cl_2 and NH_3 at 750–920°C, and for vitreous SiO_2 films from SiH_2Cl_2 and N_2O at 910°C [66, 67]. Film deposition rates range up to 200 Å/min. Other reactants and CVD conditions can be used but generally result in lower deposition rates or greater sensitivity of substrate wafer spacing on film uniformity. Intensive research is in progress to extend low-pressure CVD for preparing doped oxide films [68, 69], for epitaxial growth of semiconductors [70], and for other applications that are beyond the capability of present systems because of fundamental or technical difficulties.

Further details of this subject, including examples, economics, and a brief historical review, have been published by Kern and Rosler [23]. The design of low-pressure CVD reactors is discussed in Section III.E.

A low-pressure pulse CVD technique operating under conditions different from those described should also be noted. In this process, the reactant gases are intermittently injected into a previously evacuated deposition chamber, producing deposits of uniform thickness over their entire extended area [70a].

F. Film Growth Aspects

Most important properties of a given composition of CVD films (e.g., electrical, magnetic, mechanical, optical, etc.) are related to their structure. Thus, understanding of factors governing structure is important. We shall briefly discuss nucleation and the formation of polycrystalline, amorphous, and epitaxial CVD films.

The older theory of nucleation is based on a thermodynamical approach. The nucleus of the new phase is formed from the supersaturated vapor. Below a certain size, the nucleus is subcritical because it has a better chance of disappearing than of growing, due to the action of surface free energy. Above a certain size, the surface becomes a relatively small portion of the total nucleus, and the bulk free energy, which is now the dominant factor, favors further growth. Details of this theory can be found in numerous classical textbooks.

The newer theory is the atomistic nucleation concept. It combines sta-

tistical mechanics with the chemical bonding characteristics of solid surfaces. Not all surface sites have equal bonding characteristics. Those with
strong bonds are particularly favorable nucleation sites. Again, after nuclei reach a certain size it becomes energetically more favorable for them
to grow than to reevaporate. The growing nuclei come into contact and
finally coalesce, thus forming a continuous film. Other mechanisms of nucleation and growth are certainly possible. A good recent review of this
subject can be found in Venables and Price [71].

Epitaxy is the regularly oriented growth of a crystalline substance on
another one. Epitaxial growth is a desirable feature in CVD films, because
epitaxial films frequently have superior charactertistics in comparison
with either polycrystalline or amorphous films. When an epitaxial film
grows on the substance of the same kind, one deals with homoepitaxy. If
the growth occurs on a different substance, one deals with heteroepitaxy.

In most instances, there are at least some differences in the lattice constant of the film and the substrate on which it grows. In homoepitaxy,
these differences can be caused by the presence of dopants or some crystalline defects. The consequences of such mismatch are depicted in Fig. 8:

(1)   The initial layers in the growing film will be strained elastically so
that its lattice constant equals that of the substrate;

(2)   after a certain thickness (h_c) has been exceeded, it is energetically
favorable for part of the strain to be accommodated by introducing so-
called misfit dislocations;

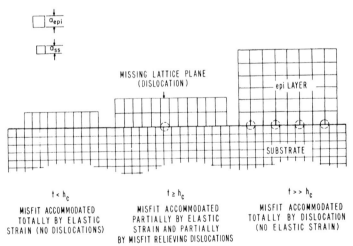

Fig. 8. Accommodation of lattice mismatch by epitaxial layers of increasing thickness
(courtesy of G. H. Olsen, RCA Laboratories).

(3)   for thicknesses much greater than h_c all of the strain is accommodated via misfit dislocations. The higher the strain, the higher is the dislocation density.

A simple relation between the strain ϵ and h_c is given by

$$\epsilon = (a_{epi} - a_{sub})/a_{sub}; \quad h_c = 2/\epsilon,$$

where a_{epi} is the lattice parameter of the epitaxial layer and a_{sub} the lattice parameter of the substrate. The larger the mismatch, the thinner is the h_c layer, and the more misfit dislocations are introduced; this generally degrades film properties. One way to minimize this problem is to grow so-called graded layers. One starts with a deposited layer matching the substrate as closely as possible and then gradually changes the composition of the growing layer until the desired thickness and composition are reached. In this way, strain from one atomic layer to the next is reduced, and thus it can be accommodated by the elastic deformation. This technique is frequently used in CVD of III–V materials (e.g., $In_xGa_{1-x}P$ on GaAs). The beneficial effect of reduced mismatch on electronic properties can be striking. At the best lattice match between the substrate and the film, the measured properties have the most desirable values [72]. In addition to the lattice parameter mismatch at the interface, one can encounter problems of chemical reactivity or diffusion at the interface. The consequence is formation of a thin transition layer, which can be compositionally different from either the substrate or the film.

Epitaxy relationships exist in many film–substrate combinations; a recent compilation has been presented by Grünbaum [24].

For many applications, polycrystalline or amorphous films are acceptable or even desirable. These two structural forms are actually more common in CVD than epitaxial films, which can be obtained only under specific conditions, such as good lattice match between the film and substrate, proper temperature range, and proper reactant concentrations. When these conditions are not satisfied, polycrystalline or amorphous modifications of the same material will be obtained. In general, lower temperature and higher gas phase concentration favor formation of polycrystalline deposits. Under these conditions, the arrival rate at the surface is high, but the surface mobility of adsorbed atoms is low. Many nuclei of different orientation are formed, which upon coalescence result in a film consisting of many differently oriented grains. Further decrease in temperature and increase in supersaturation result in even more nuclei, and consequently, in finer-grained films, eventually leading to the formation of amorphous films when crystallization is completely prevented.

Many of the protective CVD films are polycrystalline and are technically useful, where properties such as high hardness, corrosion resistance, oxidation resistance, etc., are required. The differences between the lattice parameters of the substrate and the deposit are frequently too large to allow epitaxy under any conditions, and in addition, high supersaturation and low-temperature CVD conditions are employed here. Amorphous films include, among others, oxides, nitrides, and glasses, which are of great technical importance for electronic passivation, insulation, and dielectric applications. A good discussion of structural properties of CVD deposits appears in a recent review by Bryant [7].

III. CVD REACTOR SYSTEMS

A. General Requirements

A CVD system for depositing thin film materials must provide equipment that accomplishes the following functions:

(1) Transport, meter, and time the diluent and reactant gases entering the reactor;

(2) provide heat to the site of reaction, namely, the substrate material being coated, and control this temperature by automatic feedback to the heat source;

(3) remove the by-product exhaust gases from the deposition zone and safely dispose of them.

The reactor system should be designed to fulfill these three primary functions with maximum effectiveness and simplicity of construction. It must consistently yield films of high quality, good thickness and compositional uniformity from run to run, and high purity with a minimum of structural imperfections such as pinholes, cracks, and particulate contaminants. For research applications, these requirements are usually sufficient, but for large-scale production applications the system should, in addition, perform with high throughput and economy in terms of power and chemicals, be simple and safe to operate, and be easy to maintain routinely. Labor-saving automation and computerization are attractive additional options that can be well worth the increased capital expenditure in reducing labor costs and improving the quality and reproducibility of the product.

The reactor is the basic part of any CVD setup. The geometry and construction materials of the reactor are determined by the physical and chemical characteristics of the system and by the process parameters. Reactors used for CVD of electronic materials will be described in more detail because they reach the highest degree of sophistication at this time.

We shall discuss reactors for low-temperature ($<500°C$) applications at normal pressure, for high-temperature ($>500°C$) use at normal pressure, and for low-pressure applications. Finally, control techniques and associated equipment for CVD processing will be briefly surveyed in this section. (Reactors for plasma-enhanced CVD processes are described in Chapter IV-1.)

B. Low-Temperature CVD Reactors*

Reactors of this type are used where temperature limitations of the substrate require relatively low temperatures of deposition. Primary applications are in semiconductor device processing, where CVD has become the most widely used technique for producing glassy passivating layers over aluminum-metallized semiconductor devices, particularly complex planar silicon bipolar and MOS integrated circuits for both hermetic or plastic encapsulation. The most commonly used passivating overcoat layers are SiO_2 and phosphosilicate glass (PSG), singly or in combination, deposited by oxidation of the nitrogen-diluted hydrides at a substrate temperature ranging from 325 to 450°C. (A detailed example of this process will be described in Section IV.F.) In these applications, the temperature cannot safely exceed a maximum of 500°C because solid-state dissolution of Si in Al would become a problem.

Heating is usually affected by resistance-heating techniques, and operation is at nominally atmospheric pressure. Provisions have been made in several of the commercial reactors for cooling of the reactor walls or the gas distributor to suppress homogeneous gas phase nucleation.

1. Basic Types of Reactors

Low-temperature reactors for normal-pressure operation can be classified in essentially four main categories, according to their gas flow characteristics and principle of operation:

(1) horizontal tube displacement flow reactors,
(2) rotary vertical batch-type reactors,
(3) continuous reactors employing premixed gas flow fed through an extended area slotted disperser plate, and
(4) continuous reactors employing separate nitrogen-diluted oxygen and hydride streams that are directed toward the substrate by laminar flow nozzles.

---

* Portions of this section have been reprinted from Kern [73], by permission of the publisher, Cowan Publishing Corporation.

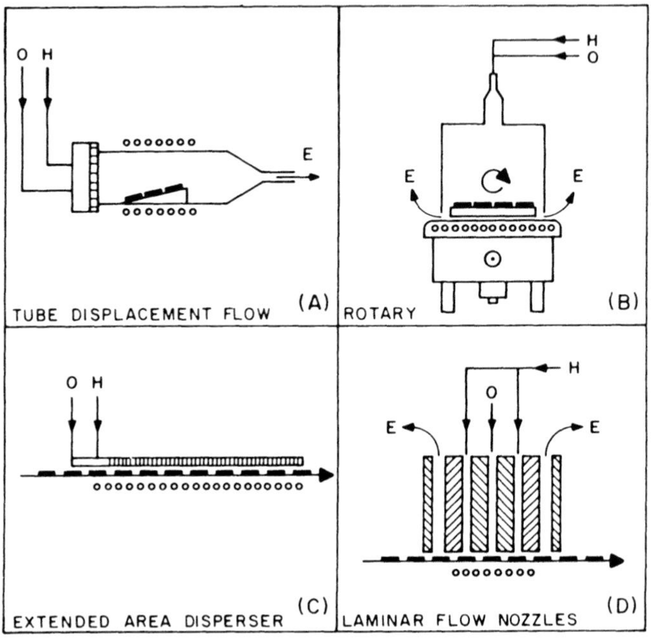

Fig. 9. Operating principles of the basic types of CVD reactors for preparing passivation overcoat layers (from Kern [73], courtesy of Cowan Publishing Corporation). (A) Horizontal tube displacement flow reactors. (B) Rotary vertical batch-type reactors. (C) Continuous reactors employing premixed gas flow fed through an extended area slotted disperser plate. (D) Continuous reactors employing separate, nitrogen-diluted oxygen and hydride streams that are directed toward the substrate by laminar flow nozzles. O = $O_2 + N_2$; H = $SiH_4 + N_2 \pm PH_3$; E, Exhaust gases; OOOO, resistance heater; ▬, substrate wafer; →, direction of gas flow; ➝, direction of travel.

The schematic diagrams presented in Fig. 9 show the basic features of these four types.

2. *Horizontal Displacement Flow Reactors*

Horizontal tube reactors consisting of an elongated quartz tube of circular or rectangular cross section, as used in high-temperature processing (Section III.C), were the first types used in CVD of oxides at low temperature. The diluent and reactive gases are continually being supplied and are usually mixed and dispersed at the point of entry into the tube. The mixed gas stream passes over the heated substrates which rest on a tilted wafer carrier. Heat may be supplied by either rf or resistance heating. Displacement-type forced laminar gas flow of this type features a low degree of gas mixing, requiring careful optimization of the gas dynamics to obtain good uniformity of film thickness and composition. Critical factors affecting film uniformity are substrate positioning, temperature profile of the deposition zone, geometry of reactor, and exhaust configuration. Several types of CVD systems based on horizontal displacement flow reactors are commercially available. A schematic of the basic system is shown in Fig. 9a.

3. Rotary Vertical Batch Reactors

In rotary vertical reactors for batch processing, the substrate wafers are placed on a plate rotating above a resistance heater (Fig. 9b). The reaction chamber consists of a cylindrical, conical, or hemispherical bell jar that may have provisions for cooling. The gases enter through single or multiple inlets from the top or side of the bell jar, pass over the substrates, and exit at the bottom of the chamber below the rotating plate. The incoming gases mix gradually with the partially reacted gas before passing over the substrates. Intentional changes in gas composition therefore proceed gradually (rather than abruptly as in the case of horizontal flow reactors), which can be advantageous if double layers with a graded interface are desired. The geometry of the reaction chamber and the configurations of the gas inlet and exit openings are critical in attaining the most effective gas flow dynamic conditions required for producing uniform deposits.

The construction and performance of a reactor of this type designed for research applications have been described [74]. The unit consists of gear-driven disks heated by conduction from a hot plate. Planetary motion of the substrates promotes maximum uniformity of the film deposits by thermal and gas dynamic averaging effects over a wide range of operating conditions. A modified single-rotation reactor we have built for research and pilot production use has also been described in detail [73]. Several commercial rotary resistance-heated vertical reactor systems for semicontinuous operation are available commercially [73].

4. Continuous Reactor Systems

Application of continuous processing concepts to CVD reactor systems makes large-scale production of oxide and glass films possible at lower operating cost than the use of batch-type reactors, especially if combined with automation that is particularly suitable for such systems. In a commercial CVD processor of this type [75], the substrate wafers are moved conveyor-fashion on Inconel trays through the reactor at constant speed. Heating is by radiation from a resistance heater. The nitrogen, oxygen, and hydride gas streams are combined before entering the manifolds to the slotted disperser plate from which the gas mixture passes by laminar flow over the wafers. The cross-flow gas dispersion and the relatively close spacing (approximately 1 cm) of the substrate to the water-cooled, extended area disperser plate minimize undesirable homogeneous gas phase nucleation. The machine has a throughput capacity of 400 5-cm diameter wafers per hour for a SiO_2 film thickness of 1 μm. A schematic is shown in Fig. 9c.

In nozzle-type reactors, streams of nitrogen-diluted hydrides and oxygen impinge on the substrate surface where they react, forming the glassy films. In one commercially available continuous reactor system [76] based on nozzle-type gas dispersion, the nitrogen-diluted oxygen and hydrides are directed to the substrate surface as separate streams flowing through laminar flow slots that form the nozzles. The water-cooled nozzle array unit is 4-cm wide and 25-cm long, extending over the width of the heater block assembly. The space between the nozzle array and the wafer surface during deposition is only about 2.5 mm and defines the reaction zone where the gases mix and react forming the oxide or glass films. Exhaust gases are removed through separate slots at the periphery of the nozzle array. The substrate wafers traverse under the nozzle unit on a resistance-heated conveyor plate at a variable speed. The plate accepts 21 wafers of 5-cm diameter. Unusually high film deposition rates are obtainable with this system. Figure 9d depicts schematically the construction of this type of nozzle reactor.

C. High-Temperature CVD Reactors

High-temperature reactors can be divided into hot-wall and cold-wall reactors. The former is predominantly used in systems where the deposition reaction is exothermic in nature, since the high wall temperature minimizes or even prevents undesirable deposition on the reactor walls. Hot-wall reactors are frequently tubular in form; heating is most often accomplished by resistance elements surrounding the reactor tube. Combinations of several heating elements make it possible to impose various temperature gradients along the tube to control CVD processes occurring in various sections of the reactor. Reactors of this type have been successfully used for the CVD of group III–V and II–VI semiconductor compounds, among others. Figure 10 shows a modern reactor employed in the CVD of III–V materials. An associated gas panel is used to supply the required amounts of chemicals, dopants, and carrier gases. Several temperature zones control the process. Time sequencers are employed to control the composition, the doping level, and the thickness of layers grown subsequently in the reactor, making it possible to grow the complex multilayer structures needed for modern microelectronic devices.

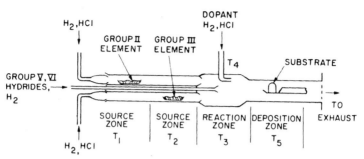

Fig. 10. CVD reactor for deposition of III–V and II–VI semiconductors via hydride method (courtesy of W. M. Yim, RCA Laboratories).

Another form of the hot-wall reactor is the high-capacity production unit shown in Fig. 11. It resembles the so-called barrel reactor employed in silicon epitaxy. Heating is accomplished by rf coupling with a graphite sleeve which surrounds the reaction chamber.

Figure 12 shows various cold-wall reactors employed in deposition of Si from the halides or the hydride. Since these are endothermic reactions,

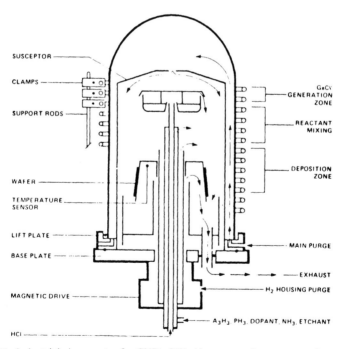

Fig. 11. Industrial-size reactor for CVD of III–V compounds (courtesy of Applied Materials, Inc.).

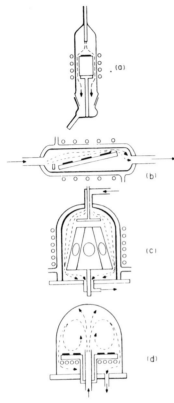

Fig. 12. Schematic representation of reactors used for CVD of silicon: (a) pedestal, (b) horizontal, (c) barrel, and (d) pancake type. Water-cooled jackets are not necessary when $SiCl_4$ or $SiCl_3H$ are source gases (from Berkman *et al.* [64], courtesy of Springer Verlag).

they will proceed most readily on the hottest surfaces of the system. The first reactor is a single-substrate reactor used for laboratory work. The next three are normally used in Si deposition: the horizontal, the barrel, and the pancake reactor. Their walls are much less hot than the susceptor surface where the substrate wafers are placed. In the case of deposition from SiH_4, the walls are water cooled to prevent the on-wall decomposition of SiH_4. When $SiCl_4$ is used, only air cooling is employed; walls reach upward of 500°C, but this is not enough to initiate deposition in the Si–Cl–H system.

The susceptor is the only intentionally heated area. Most frequently rf heating is employed, although some commercial reactors use high-intensity radiation lamps for heating. Resistance-heated graphite slabs are seldom used. Both 10 and 450 kHz frequencies serve for rf heating to temperatures of typically 1000–1300°. Susceptors are made of SiC-coated graphite; reactor tubes are almost always constructed of fused silica.

Horizontal and barrel reactors are so-called slug flow or displacement systems. The incoming gas constantly sweeps out reacted gases.

Table III

Operating Characteristics of Some Commercial Reactors for Silicon Epitaxy

Reactor type	Capacity 7.5-cm diam.	Line power (kW) at 1200°C	H₂ flow (standard liter/min)	Vol % SiCl₄ in H₂	Growth rate (μm/min)
Horizontal	24	150	200	1	1
Barrel	18	100	100	1	1
Barrel	30	130	100	1	0.6
Pancake	9	75	50	1	1

The flow dynamics and transport phenomena in displacement reactors have been discussed in some detail already. Reactant species are supplied to the growth interface via diffusion through the boundary layer adjacent to the susceptor. The thickness of this layer is a function of the gas velocity and the reactor geometry. One of the problems of the displacement reactors is the depletion of the reactant in the downstream direction which results in thinner downstream deposits. One can correct this situation by slightly tilting the susceptor, thus diminishing the reactor cross section in the downstream direction. As a consequence, the velocity increases at downstream positions, which leads to a thinner boundary layer there and thus to higher mass transport. In this way one compensates for the downstream depletion of the reactant. The pancake reactor is operating in a mixed flow mode, where the incoming gas mixes partially with the reacted gases. Mathematical treatment of this case has been worked out by Sugawara [77].

Table III summarizes important characteristics of the current Si deposition reactors. Despite the considerable advance in reactor design and construction, the presently used CVD reactors leave much to be desired. As one can see from Table III, large amounts of chemicals and power are needed. In the present geometries with flat exposed susceptor faces, the radiation heat losses are very large, and high gas velocities are needed to ensure downstream uniformity. In fact, some 60% of the incoming reactant is exhausted without reaching the deposition interface. Chemical vapor deposition of epitaxial Si, at present, is therefore an expensive process where significant improvement could still be realized.

D. Low-Pressure CVD Reactors*

One of the most significant recent innovations in CVD processing is the introduction of commercial reactor systems for low-pressure CVD in

* Portions of this section have been reprinted from Kern and Rosler [23] by permission of the publisher, The American Institute of Physics.

the semiconductor industry. The design of these reactors is based on the principles of low-pressure CVD processes discussed in Section II.E. With the restraint of reactor configuration largely removed in the case of low-pressure operation, the design of a reactor mainly involves optimization for good temperature uniformity and high substrate wafer throughput. The uniform temperatures inherent in a tubular resistance-heated furnace, combined with the very high packing density possible with stand-up closely packed wafer loading, make this a very attractive approach.

A simplified cross-sectional view of a typical reactor is shown in Fig. 13. A round quartz tube, heated by standard wire-wound elements, forms the reaction chamber. Silicon wafers, typically at 3–5-mm spacing, stand vertically facing the ends of the tube. It is preferred to load/unload wafers at the gas inlet end of the tube to avoid contamination of them by hot unreacted gases and to avoid any inlet pressure change due to a different tube pressure drop caused by varying load size or wafer diameter. Inlet gas flow rates are measured as in atmospheric-pressure systems. For any pressure-sensitive flow measurement device, such as a rotameter, the control valve must be placed downstream of the sensor. For upstream pressures of greater than 1500 Torr (30 psia), the flow through the control valve will be sonic. Thus, the upstream sensor and the flow rate will not be affected by a downstream pressure, whether it is 760 Torr, or less.

There are several commercially available mechanical vacuum pumps in the 15–50 cfm capacity range that can readily handle the 50–500 standard cm^3/min flows at 0.5 Torr pressure that many low-pressure CVD processes require. However, some processes (450°C SiO_2 and 900°C Si_3N_4, for example) require higher gas flows and/or lower pressures. In that case,

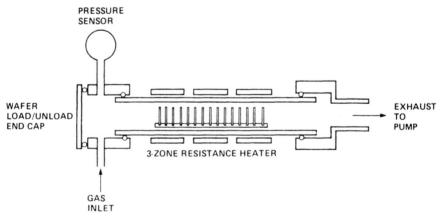

Fig. 13. Schematic of a low-pressure CVD reactor (from Kern and Rosler [23], courtesy of the American Institute of Physics).

rather than use a greater than 50 cfm mechanical pump, it is more efficient to use a blower as a booster pump between the reactor and the mechanical pump to increase the capacity and low-pressure capability of the system. The use of a variable speed motor on the blower allows independent control of reactor pressure (a most important variable) by either manual adjustment of the motor speed or the use of an automatic feedback loop from the pressure sensor output. Several problems with blowers should be noted: (1) blowers do not pump H_2 very well; (2) mechanical pumps may cause contamination from the oil; (3) special precaution must be taken to avoid pump corrosion and to eliminate the possibility of pump explosion by insufficiently diluted reactive vapors or gases.

We have already noted that a significant difference between atmospheric-pressure and low-pressure CVD concerns reactor wall deposits, which can lead to particulates on wafers. Atmospheric-pressure reactors are usually built to avoid the problem by cooling the walls sufficiently and using very high flows of main stream carrier gas so that little deposit occurs on them. The low-pressure reactor has hot walls so that films deposit on the wall just as they do on the wafers. Provided the film deposit is kept thinner than about 20 μm, this has not been a problem, most likely due to the following combination of favorable circumstances:

(1)   The temperature is held constant;
(2)   the deposit on the wall is dense and adherent; and
(3)   any particles present have a low chance of falling on a vertically oriented substrate wafer.

E. Control of CVD Processes

The stringent quality requirements (e.g., thickness and compositional uniformity, crystallinity, electrical properties) imposed on CVD deposits imply the need for stringent control of the CVD processes. In particular the following must be controlled.

(1)   temperature in the reactor (one or more zones),
(2)   the quantities and compositions of all gases or vapors entering the reactor,
(3)   the time sequencing of variables under (a) and (b), and
(4)   the pressure in the case of low-pressure CVD.

The temperature is measured either pyrometrically or by means of a thermocouple. Pyrometer measurements are particularly convenient in cold-wall reactors where the hot areas are visible through a quartz reactor tube. Thermocouples are frequently used with hot-wall reactors; they are usually placed between the reactor tube and the heating element at vari-

ous places along the tube. The outputs of the sensing element are connected to temperature controllers, which maintain the temperature within
the prescribed range (typically ± 5°C).

Input gas flow rates are measured and controlled by either rotameter-
type flow meters or by electronic mass controllers. The rotameters are
frequently coupled to precision valves which are used to preset flows.
Various diameter tubes and float materials accommodate a wide range of
flow rates. The accuracy of rotameters is typically ± 5% of the full scale.
Electronic flow meters operate by measuring changes in temperature of
the flowing gas subjected to heat flux. The temperature change is related
to the amount of gas and the heat capacity of the gas, which is in turn
related to the composition of the gas. The accuracy of electronic flow
meters is typically ± 2% of the full scale. They can be calibrated in either
volume or mass units (e.g., liter/min or g/min). A schematic of a typical
gas flow control system used in conjunction with low-temperature CVD of
SiO_2 and PSG films (Section III.B) is shown in Fig. 14 as an example.

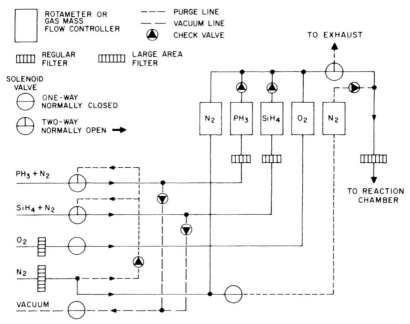

Fig. 14. Schematic of an automated gas flow and metering system for CVD of SiO_2 and
PSG films (from Kern [73], courtesy of Cowan Publishing Corporation). (Less expensive
mass flow meters or inexpensive rotameters with high-accuracy flow control needle valves
can be substituted for the electronic mass flow controllers, especially in the case of the N_2
and O_2 lines. The vacuum lines can be omitted if two-stage gas regulators with provisions for
purging are used at the hydride source cylinders.)

A significant advantage of electronic controllers is the ease of automation and programming for CVD processes. A computer or a microprocessor can be connected to the controllers to determine, with separate digital control, the amount of each input gas, the temperature, as well as the duration and time of the input in the deposition cycle. In this way, complex deposit structures can be synthesized with good control, since the amount and the duration of gaseous reactants determine the composition and the thickness of the deposited layer.

Figure 15 shows the process sequence employed in a deposition of a hypothetical multilayer silicon device structure (in this case substrate/n layer/p layer). Several firms supply components as well as whole systems capable of producing very complex CVD structures through automation and step programming. There have been some new developments for *in situ* measurements of thickness [78]. Optical methods have been employed using infrared or laser radiation, but at present only thin layers (<5 μm) of suitable materials can be measured.

Instrumental *in situ* gas analysis methods, mentioned already in Section II.A.1, are powerful techniques for compositional control. Two reviews on this subject have been presented recently [37, 42].

Finally, mechanization and automation in CVD processing [78a], particularly in semiconductor device wafer fabrication, promises to lead to improved process control coupled with greater cost effectiveness.

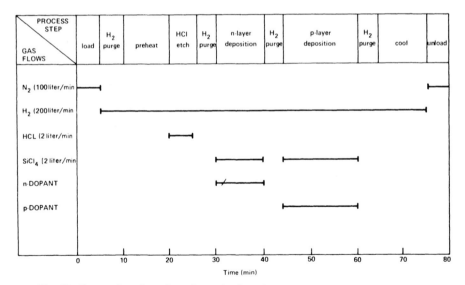

Fig. 15. Sequencing of gas flows in production of a two-layer (n-layer/p-layer) epitaxial semiconductor structure grown by CVD techniques in a barrel reactor.

IV. CVD OF INSULATORS

A. Introduction

Thin films of insulating materials prepared by CVD processes have found numerous important technical applications, particularly in modern microelectronics technology and as protective coatings for high-temperature materials. The reason for the popularity of CVD dielectric films is the relative ease and economics of their formation, the high structural quality attainable, the wide compositional ranges that can be realized, and the flexibility of deposition temperatures that can be selected for specific requirements by suitable choice of a CVD reaction.

In this section we shall briefly survey processes for depositing films of single and mixed oxides, silicate glasses, nitrides, and oxynitrides, emphasizing basic and recent work. In Section IV.E we shall discuss in detail the low-temperature deposition of SiO_2 and PSG films from the hydrides as an example of a technically important and typical CVD process.

B. Single Oxides

1. Silicon Dioxide

One of the earliest CVD processes for SiO_2 film formation is based on the pyrolysis of silicon alkoxides in an inert carrier gas (Ar or N_2) at normal pressure in the temperature range of typically 700–765°C [79–87]. Particularly suitable and widely used source materials are ethyltriethoxysilane, $C_2H_5 Si(OC_2H_5)_3$ [79, 81, 82, 87], and tetraethoxysilane, $Si(OC_2H_5)_4$ [80, 83, 86]. Pyrolysis of the latter compound is catalyzed by triethylstibine, $(C_2H_5)_3Sb$, allowing SiO_2 film deposition in O_2–Ar at temperatures of only 250–500°C [88]. Film deposition from ethyltriethoxysilane can take place at temperatures below 200°C, but the films are organic polymers of silicon rather than SiO_2 [82, 87, 89].

The application of low-pressure techniques for pyrolytic deposition of SiO_2 films at 700–800°C from silicon alkoxides [84, 90–93] has become important recently because of the economic advantages inherent to low-pressure CVD [23]. Pulsing techniques in an evacuated system, as described in Section III.E [70a], have also been used [93].

Oxidation of tripropylsilane, $(C_3H_7)_3SiH$, with O_2 at 750° under atmospheric-pressure conditions has been claimed to yield SiO_2 films of excellent quality [94].

Oxidation or hydrolysis of silicon halides is a second type of reaction used for forming SiO_2 films. Nitric oxide can serve as the oxidant for $SiCl_4$

at 1150°C, or for $SiBr_4$ at 850°C, in the presence of H_2 [95]:

$$SiX_4 + 2H_2 + 2NO \rightarrow SiO_2 + N_2 + 4HX,$$

where X is Br or Cl.

A lower deposition temperature (800°C) is possible by reacting $SiBr_4$ with CO_2 [96], expressed by the simplified overall reaction equation

$$SiBr_4 + 2CO_2 + 2H_2 \rightarrow SiO_2 + 2CO + 4HBr.$$

The analogous reactions with $SiCl_4$ proceed at temperatures at and above 1100°C [97, 98]. Forming the water *in situ* from the relatively stable mixture of H_2–CO_2, rather than introducing water vapor directly, minimizes homogeneous gas phase nucleation. The second step in the reaction is the hydrolysis of the silicon halide with the generated water vapor. If O_2 is used as the oxidant for $SiCl_4$, a substrate temperature of only 800°C is adequate [99]:

$$SiCl_4 + O_2 + 2H_2 \rightarrow SiO_2 + 4HCl.$$

Dichlorosilane, SiH_2Cl_2, can also be used as the silicon source material with either O_2 or N_2O as the oxidant at a temperature of typically 900°C. The deposition rate of these reactions is less temperature sensitive than in other reactions for SiO_2 film formation [100]. The SiH_2Cl_2–N_2O system is particularly successful if operated by the recently introduced low-pressure CVD technique at 900°C [23, 66].

A process that is widely used at present for depositing SiO_2 films is based on the oxidation of diluted silane, SiH_4, with O_2 at 250–500°C, to be discussed in detail in Section IV.E. Higher temperatures [101] and different oxidants [102] have also been investigated for silane oxidation, as well as low-pressure CVD techniques [23, 66]. If higher substrate temperatures are permissible, excellent-quality SiO_2 films can be deposited from SiH_4–CO_2–H_2 at 800–1050°C [103–105], preferably in the presence of HCl vapor as an impurity getter [106].

2. Aluminum Oxide

Chemical vapor deposition of Al_2O_3 films can be based on pyrolysis or oxidation of organometallic reactants, or on hydrolysis of aluminum halides at generally higher temperature.

Pyrolytic CVD of amorphous Al_2O_3 films is particularly suitable with aluminum isopropoxide, $Al(OC_3H_7)_3$, as the starting material at a reaction temperature of 420°C in N_2 or O_2 [107–109]. Much higher deposition rates result by increasing the temperature to 700–800°C and reducing the system pressure to 10 Torr in dry N_2 [110]. Pyrolysis of aluminum triethoxide at 300–500°C [111] and at 400–750°C [112, 113], as well as of aluminum acetylacetonate at about 480°C [114, 115], have also been used.

Oxidative CVD reactions based on tri-isobutylaluminum and O_2 at 250 –500°C [116], trimethylaluminum and nitrous oxide at 650°C [117, 118], and trimethylaluminum with O_2 at 275–475°C [108, 119], all yield Al_2O_3. The films derived from trimethylaluminum are of much better quality than those from the tri-isobutyl derivative. In general, Al_2O_3 films deposited at low temperature are amorphous but convert to crystalline modifications when heat treated at or above 800°C, with substantial consequent changes of their properties.

Hydrolytic deposition of Al_2O_3 films by the $AlCl_3$–CO_2–H_2 reaction at 850–1200°C has been used extensively for preparing dense, polycrystalline deposits of high quality [98, 120–126]. The generally accepted overall equation for this two step reaction is

$$2AlCl_3 + 3H_2 + 3CO_2 \rightarrow Al_2O_3 + 6HCl + 3CO.$$

Carbon dioxide and hydrogen react above 750°C forming H_2O, which then reacts with $AlCl_3$, yielding Al_2O_3 films. As in the case of the analogous reaction for SiO_2 previously noted, the formation of H_2O *in situ* prevents premature hydrolysis of the metal halides that would lead to particulate contamination. Epitaxial Al_2O_3 films have been deposited on sapphire substrates at 1500°C [127].

Mass spectrometric vapor analyses during CVD of Al_2O_3 by hydrolysis–oxidation of $AlCl_3$ identified the major vapor molecule as $AlCl$ for the temperature and pressure conditions of the system studied [128]. Al_2O_3 films have also been prepared by low-pressure CVD (7 Torr, 900°C), using the system $AlCl_3$–NO–H_2 [92].

3. Other Metal Oxides

Amorphous, polycrystalline, and in a few cases, epitaxial films of other insulating and dielectric oxides with a single metal component have been prepared by CVD. These include primarily TiO_2, ZrO_2, HfO_2, Ta_2O_5, and Nb_2O_5. Some CVD work has been reported for miscellaneous oxide films of Fe, Zn, Pb, Ni, Co, Mn, V, Cr, Ge, B, and P, which will also be noted.

In general, processes analogous to those described for CVD of SiO_2 and Al_2O_3 can be applied for preparing these oxides. The most frequently used reactions are pyrolysis and pyrolytic oxidation of the metal alkoxides at relatively low temperatures, and hydrolysis or oxidation of the metal halides at generally higher temperatures. Several other types of reactions have also been employed, as will be noted.

Alkoxy pyrolysis, often in the presence of O_2, is conducted in the temperature range of about 350–750°C. Alkoxy compounds are generally preferred over other organometallic reactants because of their relatively high vapor pressures; however, they are sensitive to hydrolysis. The following oxides have been prepared by this type of reaction: TiO_2 [129–133], ZrO_2 [134], HfO_2 [132, 134], Ta_2O_5 [119, 135, 136], Nb_2O_5 [134], and ZnO [113, 137].

Metal halide hydrolysis and oxidation are generally effected in the temperature range of about 400–1000°C. Oxide films prepared by these reactions include the following: TiO_2 [119, 136, 138, 139], ZrO_2 [140], Ta_2O_5 [141–142a], NiO [143], CoO [144], and MnO_2 [145]. Epitaxial growth of TiO_2 has also been reported [139].

Hydrolysis, at very low temperatures (< 300°C), of metal halides has been used to prepare TiO_2 films [136, 146, 147]. Low-temperature hydrolysis, typically at 150°C, has been used for forming amorphous oxide films from the metal alkoxy compounds, leading to TiO_2 [119, 146, 148–150], and to ZrO_2, Nb_2O_5, Sb_2O_3, VO_x, and BO_x [148]. Postdeposition heat treatments improve the quality of these films substantially.

Oxidation and pyrolysis reactions of organometallic compounds have been used extensively to prepare metal oxide films: TiO_2 [151], ZnO [152, 153], and PbO [153a, 153b] from the metal alkyls; ZrO_2 [133], HfO_2 [133], Cr_2O_3 [137], and V_2O_3 with NH_3 as carrier gas [115] from the metal acetylacetonates at 450–700°C; HfO_2 [154] from hafnium diketonates at 450–550°C; and Ta_2O_5 [155] from tantalum dichlorodiethoxy acetylacetonate at 300–500°C.

Pyrolytic decomposition of metal acetylacetonates at 400–500°C has yielded epitaxial films of Fe_2O_3 and NiO on single crystals of Al_2O_3 and MgO, respectively [156], and polycrystalline films of Fe_2O_3 at a temperature of about 400°C [157]. The use of tris(trifluoroacetylacetonate)iron(III), $Fe(F_3CCOCHCOCH_3)_3$, at a deposition temperature of 300°C, in the presence of O_2, has yielded films of β-Fe_2O_3 [158]. Poly(vinyl ferrocene) at 380°C has also been used to deposit Fe_2O_3 films [159]. Amorphous Fe_2O_3, for use as hard semitransparent coatings on photolithographic glass masks, has been prepared from iron pentacarbonyl, $Fe(CO)_5$, in Ar or N_2 at deposition temperatures of only 90–160°C, in the presence of CO_2, with or without O_2 and H_2O [160–163].

Niobium pentoxide films have been prepared, in addition to the reactions described, by an interesting halogen transport process at 700–1100°C based on the reaction:

$$4NbOCl_3 + 3O_2 \rightarrow 2Nb_2O_5 + 6Cl_2.$$

The Nb_2O_5 films were deposited on Pt and Si substrates [164].

Finally, low-temperature oxidation of hydrides can form oxide films: GeO_2 from $GeH_4-O_2-N_2$ at 400–450°C [153a], B_2O_3 from $B_2H_6-O_2-N_2$ at 400–450°C [165], and P_2O_5 from PH_3-O_2-Ar (but not N_2) at a deposition temperature of 300°C [166].

For additional information on CVD of oxides, the reader is referred to the comprehensive review by Powell [167] which covers the older literature to 1966 that is not referenced in detail in our survey.

C. Mixed Oxides and Silicate Glasses

Combination of two or more oxide source reactants in a given type of CVD reaction often (but not always) leads to films of mixed oxides or silicate glasses. The CVD reactions employed are those we discussed for single component oxides. However, the resulting ratio of oxides is frequently different from that predicted, due to catalytic or inhibiting effects of added reactants.

One important use of these films is in semiconductor technology where they serve as excellent solid-to-solid diffusion sources [168]. SiO_2 is usually the matrix for codeposited dopant oxides. These films are known as "doped oxides", but they are actually silicate glasses that are chemically formed in the vapor phase, rather than being mere physical mixtures of the oxides, as evidenced by their chemical and physical properties.

1. Phosphosilicates

Films of phosphosilicate glasses (PSG) can be generated at 900°C by hydrolysis of $SiCl_4$ and $POCl_3$ (or PCl_3) vapors in O_2 saturated with H_2O [169]. Vapors of these liquid reactants are carried with Ar or N_2 into the reactor where they are mixed with O_2-H_2O and react according to the equations:

$$SiCl_4 + 2H_2O \rightarrow SiO_2 + 4HCl,$$

and

$$2POCl_3 + 3H_2O \xrightarrow{O_2} P_2O_5 + 6HCl.$$

The second reaction proceeds only in O_2. The SiO_2 deposition rate is not affected by the $POCl_3$ cohydrolysis. The second hydrolysis reaction can also be combined with thermal oxidation of Si to produce PSG [169]. Alternatively, $POCl_3$ can be reacted separately with O_2 to form P_2O_5 vapor which is then reacted with preformed SiO_2 layers at temperatures of 800–

1200°C, converting some of the SiO_2 into PSG [170, 171]. Precisely controllable thin layers of PSG have been deposited from $SiBr_4-PH_3-O_2-H_2-N_2$ at 600–1000°C [172].

Organic reactants have also been used for PSG formation. These systems are based on oxidative copyrolysis of tetrethoxysilane or tetrapropoxysilane with trimethylphosphate, $PO(OCH_3)_3$, at 700–800°C. Both atmospheric-pressure CVD [153b, 173, 174] and low-pressure CVD techniques [68, 92, 153b] are applicable.

A different type of reaction uses diluted silane as the source for the SiO_2, which allows deposition at low temperatures. $POCl_3$ can be used with O_2 as the source for the phosphorus oxide [175]. However, PSG films of better quality can be attained more easily by the widely used $SiH_4-PH_3-O_2$ system at 325–450°C [168]; this system is discussed in detail in Section IV.E. Low-pressure techniques for this system have also been described recently [23, 66, 69].

2. Borosilicates

Films of borosilicate glasses (BSG) have been deposited by oxidative copyrolysis of organosilicate and organoborate compounds, normally in the presence of O_2 and with Ar or N_2 as the carrier and diluent gas. The following systems have been described: (1) tetraethoxysilane and trimethylborate, $B(OCH_3)_3$, at 700°C under normal pressure conditions [153b, 173, 176]; (2) tetraethoxysilane and tri-n-propylborate, $B(OC_3H_7)_3$, at 688°C under normal pressure [177]; (3) tetrapropoxysilane and trimethylborate under low-pressure conditions (0.45–0.55 Torr) at 750–775°C [92].

Deposition of BSG from the hydrides at lower temperature [325–450°C] is used much more widely. The system $SiH_4-B_2H_6$ (diborane)–O_2 with N_2 or Ar as the diluent is capable of producing BSG films over the entire composition range of $(SiO_2)_{1-x}(B_2O_3)_x$, and has been thoroughly investigated [152, 153, 165, 168, 178–186]. The reaction chemistry for this system is very similar to that for PSG from the hydride oxidation system. Some of the water formed as a by-product becomes incorporated in the films, but can be readily removed by a brief heat treatment above the temperature of deposition [153].

Borosilicate glass layers can also be formed by reacting BBr_3 vapor with O_2 and passing the boron oxide vapor formed over SiO_2 at 800–1200°C [170]. Part of the SiO_2 is transformed into BSG, analogous to the reaction described for PSG [170, 171].

Infrared spectroscopic studies have clearly shown that BSG films deposited from the hydrides at temperatures of 325–475°C are indeed chemically formed silicates rather than physical mixtures of B_2O_3 and SiO_2 ; the same holds for PSG and AsSG [92, 153, 165, 185–187]. There is evidence that, most probably, other "doped SiO_2" or "mixed SiO_2–oxide" systems are silicate compounds as well, regardless of the CVD reactions employed for their synthesis [153a].

3. Arsenosilicates

Films of arsenosilicate glasses (AsSG) have been prepared by a variety of CVD reactions which are all based on $AsCl_3$ or AsH_3 (arsene) as the source of the As, and organic or inorganic silicon reactants as the source of Si, all at deposition temperatures of about 500°C. The following systems have been described: (1) $Si(OC_2H_5)_4$–$AsCl_3$–O_2–N_2 or Ar [188–192]; (2) $Si(OC_2H_5)_4$–$AsCl_3$–CO_2–H_2 [189]; (3) $Si(OC_2H_5)_4$–$AsCl_3$–N_2 [190]; (4) $SiCl_4$–$AsCl_3$–CO_2–H_2 [189]; (5) SiH_4–$AsCl_3$–O_2–Ar [193, 194]; and (6) SiH_4–AsH_3–O_2–Ar or N_2 [192, 195, 196]. The hydride reaction in the last system is generally preferred since it proceeds, at a deposition temperature of 500°C, at convenient deposition rates of 500–1500 Å/min for concentrations of up to 15 mole % As_2O_3 in the glass [192]. This reaction has also been conducted in the presence of GeH_4 or $GeCl_4$ to effect Ge doping of the AsSG for the purpose of enhancing subsequent As diffusion [195, 196a]. Doubly and triply dopes oxides have also been reported which were prepared, for strain-free Si diffusion, by simultaneous reactions from SiH_4–PH_3–O_2–Ar–AsH_3 and/or SbH_3 at 500°C [197].

4. Aluminosilicates

Films of aluminosilicate glasses (AlSG) have been deposited by reaction of tetraethoxysilane and tri-isobutylaluminum vapors in an O_2 atmosphere at temperatures in the range of 250–500°C [116, 119, 198]. Pyrolytic decomposition of aluminum tri-isopropoxide and tetraethoxysilane in a N_2 atmosphere at 450°C has also been used [199]. In both systems, the decomposition of the tetraethoxysilane is greatly accelerated by the presence of the organoaluminum compounds, allowing much lower temperatures of deposition than would be expected [116, 198]. Pyrolysis of trimethylsiloxyaluminum isopropoxide has also been used at deposition temperatures ranging from 300 to 800°C, yielding AlSG films from a single source compound [199, 200]. Aluminosilicate glass films can be prepared (with some difficulties) from SiH_4–$Al(CH_3)_3$–O_2–N_2 at 450°C [152, 153]. The hydrolysis reaction CO_2–H_2–$AlCl_3$–$SiCl_4$ at 850–1100°C yields AlSG films of excellent quality over a wide range of composition [122].

5. Other Mixed Oxides and Silicates

Antimony-doped oxide CVD diffusion sources for Si have been prepared from $(CH_3)_3Sb$ (trimethylstibine), SiH_4, and O_2 in N_2 at 315°C [201]. Stibine, SbH_3, can be used but has an inhibitory effect on SiH_4 oxidation, and hence is best deposited as a pure Sb_2O_5 layer capped with SiO_2 [202].

Tin-doped oxide diffusion sources for GaAs have been deposited pyrolytically from tetraethoxysilane and tetraethylstannane [202] or tetramethylstannane at 500°C in N_2-H_2 carrier gas [203].

Thin-film Zr_2SiO_4 : Mn phosphors have been prepared by halide hydrolysis at 1100–1200°C using $SiCl_4$, $ZnCl_2$, $MnCl_2$, H_2, and CO_2 [204]. Even higher temperatures (1800–2100°C) were employed reacting combinations of $SiCl_4$, $GeCl_4$, $POCl_4$, and $SbCl_5$ vapors with O_2 for depositing some glasses for optical waveguides [205].

Tantalum-doped Al_2O_3 films have been deposited by pyrolysis of tantalum pentaethoxide, $Ta(OC_2H_5)_5$, and $Al(C_2H_5)_3$ [119] or $AlCl_3$ [206], as well as by hydrolysis at 800–950°C of $AlCl_3$ and $TaCl_5$ with CO_2 and H_2 [126]. Codeposition of Al_2O_3 and P_2O_5 from vapors of aluminium isopropoxide and $POCl_3$ at 500°C in N_2-O_2 has been described [207]. Mixtures of Al_2O_3 and Cr_2O_3 have been made by thermal decomposition in O_2 at 650–750°C, of vapors from previously codeposited Al and Cr acetylacetonates [127].

Luminescent films of rare-earth doped YVO_4, Y_2O_3, and Y_2O_2S phosphors have been prepared by pyrohydrolysis of the chelates at 500°C (with H_2S in the case of Y_2O_2S) [208].

Finally, we have synthesized in our laboratory a number of interesting glasses worth mentioning: lead borosilicate films of excellent quality from $Si(OC_2H_5)_4-B(OCH_3)_3-Pb(C_2H_5)_4-O_2-Ar$ at 730°C, both at normal pressure and by low-pressure CVD techniques [23, 152, 153b]; lead aluminosilicates and aluminophosphosilicates were deposited similarly, using $Al(OC_3H_7)_3$ vapor as additional reactant. Hydride-based systems with organometallics have been utilized to deposit films at much lower temperatures, typically 450°C. For example, zinc silicate and zinc borosilicate have been prepared from $Zn(C_2H_5)_2-SiH_4-O_2-N_2$ without and with B_2H_6, respectively; and aluminoborosilicates have been deposited by this system using $Al(CH_3)_3$ as the metal reactant [152, 153, 178]. Unusual glasses that we have prepared from the hydrides in the temperature range of 325–450°C include phosphoborosilicates, deposited from $SiH_4-PH_3-B_2H_6-O_2-N_2$ [152, 153]; and borophosphates, deposited from $B_2H_6-PH_3-O_2-N_2$ or Ar [153a]. We have prepared, similarly, germanoborosilicates from the $GeH_4-B_2H_6-SiH_4-O_2-N_2$ system at 450° [153a], using rotary hot-plate CVD reactors [73, 74].

D. Nitrides and Oxynitrides

1. Silicon Nitride

Of the nitride films used for insulating and dielectric applications, silicon nitride (Si_3N_4) is by far the most important one. Extensive use has been made of Si_3N_4 films in semiconductor device technology, mainly because of their high density (compared to SiO_2) with associated excellent barrier properties against Na ions and other deleterious contaminants, and the very high dielectric strength these films exhibit. Extensive research has been carried out over the years on the preparation, properties, and applications of Si_3N_4. Most of the published results have been reviewed in several articles between 1969 and 1977 by the following authors: Feist *et al.* [3], Gregor [209], Chu [14], Duffy and Kern [210], Milek [211], Balk [212], Niihara and Hirai [213], and Kern and Rosler for low-pressure CVD [23]. The most comprehensive of these are the two monographs by Milek [211] covering most of the literature to 1972. Relatively few additional papers have appeared on CVD Si_3N_4 since then. We will therefore note briefly the main CVD methods used, referencing representative papers and recent contributions.

The two most commonly used CVD processes for forming Si_3N_4 films are (1) the silane–ammonia reaction and (2) the silicon tetrachloride–ammonia reaction under normal-pressure conditions according to the following overall equations:

$$3SiH_4 + 4NH_3 \rightarrow Si_3N_4 + 12H_2$$

and

$$3SiCl_4 + 4NH_3 \rightarrow Si_3N_4 + 12HCl.$$

The halide reaction actually takes place by polymerization of $Si(NH)_2$, followed by pyrolysis of the polymers formed at different temperatures [213].

The hydride reaction proceeds at 700–1150°C with a large stoichiometric excess of anhydrous NH_3, and with H_2 as the diluent [214–219]. Molar ratios of NH_3/SiH_4 greater than 20 or 30 should be used to avoid formation of Si-rich deposits; these are likely to form, especially at ratios < 10, as shown by IR and UV spectroscopy [220]. Ammonia/silane ratios of up to 10000:1 have been investigated and found to yield films with superior properties [210]. At temperatures of 900°C and above, a large excess of H_2 tends to lead to undesirable crystallites [214, 215]. Argon, helium, and nitrogen at large ratios have also been used as diluent gases [216], with resulting changes in the electronic properties of the Si_3N_4 that were beneficial, particularly in the case of N_2 [210]. Deposition of thick, amorphous coatings of Si_3N_4 from $SiH_4-NH_3-N_2-H_2$ at 800–1200°C has been described for ceramics applications [221]. Low-pressure CVD (3 Torr, 800°C) of Si_3N_4 films from $SiH_4-NH_3-N_2(H_2)$ has also been noted [92].

A detailed analysis of the thermodynamics of both the $SiH_4-NH_3-H_2$ and $SiCl_4-NH_3-H_2$ systems has been made [222]. Gas phase analysis techniques *in situ* have been described for studying CVD parameters of the SiH_4-NH_3 system [223] and of the SiF_4-NH_3 system [224]. Recently reported results of Auger electron spectroscopy for Si_3N_4 films, deposited at 750°C by the NH_3-SiH_4-Ar system, showed oxygen concentrations ranging from 0.4 to 7 at. % in various samples [225]. Multiple internal reflection spectroscopic measurements showed the dependence of chemically bound H on the NH_3/SiH_4 ratio and on annealing of films deposited by the same system at 700°C [226].

The reaction utilizing $SiCl_4$ and NH_3 can be conducted in the temperature range 550–1250°C (typically 850°C), using either H_2 [219, 227] or N_2 [228–230] as the carrier gas. One advantage of this lower-cost process lies in its capability to yield Si_3N_4 films with lower electrical conductivity and polarization, but it has the disadvantage of requiring a higher deposition temperature than the SiH_4 reaction [219]. Another important advantage is the built-in gettering capability (due to the chloride ions) for removing ionic contaminants [230]. A tube-furnace CVD system based on the $SiCl_4-NH_3-N_2$ reaction at about 900°C has been described for large-scale production applications [231]. High-temperature amorphous and crystalline α-Si_3N_4 coatings for ceramic applications have been deposited at 1100–1350°C from $SiH_4-NH_3-N_2(Ar)$ at normal pressure [221], from $SiCl_4-NH_3-H_2$ at 1100–1500°C under reduced pressure [213], and from both SiF_4-NH_3 [232] and $SiCl_4-NH_3$ [233] at low pressure and between 1100–1550°C.

Silicon nitride films of excellent uniformity have been prepared recently on a production scale for semiconductor applications from $SiH_2Cl_2-NH_3$ at temperatures in the range 700–800°C by low-pressure CVD techniques [23, 66].

2. Silicon Oxynitride

Films of silicon oxynitride, $Si_xO_yN_z$, combine some of the useful electrical, physical, and chemical properties of SiO_2 and Si_3N_4. Many different CVD reaction systems have been used to prepare widely ranging compositions of this material, as has been discussed in several reviews [14, 208, 209, 211, 212]. The chemical systems most frequently used are based on SiH_4 as the Si-bearing source compound, NH_3 as the nitriding agent, and various oxidants as the source for O. Oxidizing agents used include O_2 [234, 235], NO [236, 237], N_2O [238], and CO_2 [239]. The NH_3/oxidant ratios required for typical films must be in the range of about 50–5000. Both H_2 and N_2 have served as diluent gases. Ammonia/silane ratios of up to 40000 : 1 have been used to minimize the inclusion of excess and chemically bonded H [236]. Temperatures of deposition have ranged from about 850 to 1000°C under normal pressure conditions.

Significantly better control over the process and the film properties result by using CO_2 as the oxidant at about 1000°C, rather than the other oxidizing agents noted [239]. The reason for this lies in the similar activation energies for the $SiH_4–CO_2$ and the $SiH_4–NH_3$ reactions, resulting in similar rates of oxidation and nitriding. Another advantage of the first-order $SiH_4–CO_2–NH_3–H_2$ reaction system is its ability to readily deposit $Si_xO_yN_z$ films with a refractive index greater than 1.73 [239]. Hydrogen content and annealing effects of $Si_xO_yN_z$ films have been determined recently by using multiple internal reflection spectroscopy [240].

3. Other Nitrides and Oxynitrides

Nitrides of a number of metals are of technical importance, particularly those of Ti, Zr, Hf, Nb, Ta, and Be (and to a lesser extend B, Al, and V), which form hard and highly stable refractory nitrides with very high melting points. Widely used applications of several of these coatings (especially TiN) are for improving the wear resistance of cemented carbide tools, for achieving decorative effects, and as protective coatings [240–242]. Brief mention will be made only since the primary use of these coatings is for thick-film applications.

The chemical systems for producing CVD refractory nitrides are based on reacting volatile metal halides with N_2, without or with H_2, at high temperatures, i.e., 2500–2700°C for the $ZrCl_4–N_2$ or $HfCl_4–N_2$ systems, and 1100–1700°C for the $TiCl_4–N_2–H_2$ system [167]. However, lower temperatures (900–1100°C) for depositing TiN films by the latter reaction have been reported recently [243, 244]. Films of hafnium nitride, HfN, from the reaction of chlorides of Hf (generated *in situ*) with $N_2–H_2$, have been deposited on W, also at relatively low temperatures in the range of 900–1300°C; thermodynamic and kinetic studies showed that the process is mass transport limited at low H_2 concentrations [245].

Boron nitride, BN, prepared under optimal CVD conditions, is an excellent insulator and *p*-type dopant source for junction formation in Si device fabrication. Several CVD methods have been investigated for depositing thin BN films of both microcrystalline and amorphous structure [208, 246, 247]. Clear vitreous films have been deposited on a variety of substrates at 600–1000°C by reacting B_2H_6 and NH_3 in H_2 or inert carrier gases, according to the over-all reaction

$$B_2H_6 + 2NH_3 \rightarrow 2BN + 6H_2$$

via formation of several intermediate polymeric compounds [246]. Thorough characterization of these films has been conducted which showed, for example, that BN deposited at 800°C on Si is electrically similar to Si_3N_4 but is not as good a Na^+ barrier and is not as stable chemically [246].

Another refractory nitride of interest is aluminum nitride, AlN, which has potential as a good dielectric for active and passive components in semiconductor devices because of its large energy gap and high thermal and chemical stability [248]. Aluminum nitride films have been deposited by reacting NH_3 with $AlCl_3$ [248–250] or $Al(CH_3)_3$ [251]. A particularly attractive technique is pyrolysis of the complex $AlCl_3 \cdot 3NH_3$ at 800–1200°C with H_2 and anhydrous NH_3 as the carrier gases, yielding polycrystalline films [248, 249]. Mixed compositions of AlN and Si_3N_4 prepared from $AlCl_3$, SiH_4, and NH_3 at 600–1100°C have been characterized [252].

Amorphous and polycrystalline dielectric films of aluminum oxynitride, $Al_xO_yN_z$, have been deposited recently on Si by using the system $AlCl_3-CO_2-NH_3$ in N_2 as the carrier gas at 770 and 900°C [253]. The film composition and structure could be varied over the entire range of the pseudobinary $AlN-Al_2O_3$ system by controlling the NH_3/CO_2 ratio and deposition temperature [254].

Films of gallium nitride, GaN, have been grown most commonly by the ammonolysis of gallium monochloride, GaCl [255]. Pyrolysis of the $GaBr_3 \cdot NH_3$ complex [256] and ammonolysis of trimethyl gallium, $Ga(CH_3)_3$ [257], have also been employed. Amorphous films of germanium nitride, Ge_3N_4, have been prepared by ammonolysis of $GeCl_4$ at 400–600° [258] and have been evaluated for electronic applications [259].

Low-temperature deposition of films of nitrides of Ti, Zr, and Nb by pyrolysis of the dialkylamides has been reported for temperatures in the range 250–800°C [260].

E. Practical Example of a CVD Process

In this section we shall discuss effects of deposition and substrate variables on the growth rate and properties of two typical CVD films as a practical example of a CVD process. We have chosen the low-temperature deposition at normal pressure of SiO_2 and PSG from the hydrides as a representative example. A more detailed treatment has been presented elsewhere [23, 261].

1. Basic Reactions

The formation on heated substrates of vitreous SiO_2 films by reacting diluted SiH_4 with excess O_2 proceeds by a complex heterogeneous free-radical branching-chain mechanism that involves surface adsorption and catalysis leading progressively to SiO_2 as the solid end product. High partial pressures of SiH_4 and high temperatures (600–1000°C) tend to lead to H_2O as the other reaction product [101, 262], whereas low SiH_4 partial pressures and low temperatures (< 500°C) tend to form predominatly H_2 as the gaseous reaction product [102, 180, 263]. Mass spectrographic analysis of the exhaust gases from the reaction conducted under the usual conditions for SiO_2 and PSG film deposition in a rotary hotplate reactor [73] at 450°C showed that both H_2 and H_2O are formed, the quantities of each varying with the total gas flow rate [23]. The reaction is sensitive to substrate surface conditions and deposition-related factors [186, 263–269].

The desired glassy films of SiO_2 are formed by heterogeneous surface reaction. However, homogeneous gas phase nucleation can also occur and result in a smog of colloidal particles which tend to form a white powdery coating on the reactor surfaces of lower temperature [23, 261, 263–269, 270]. These particles, which are a mixture of SiO_2 and an H-containing polymer [263], are a source of contamination and reduce the yield of the desirable glassy product. Homogeneous gas phase nucleation in normal-pressure CVD can be suppressed by using high carrier gas flow rates, minimum temperature, and cold-wall operation.

Cooxidation of diluted PH_3 with SiH_4 plus excess O_2 at 300–500°C leads to phosphosilicate glasses, $(SiO_2)_{1-x}(P_2O_5)_x$ [152, 153, 178, 180, 186, 266, 268, 270–273], rather than a physical mixture of SiO_2 and P_2O_5 [153, 185–187].

*2. Effects of CVD Parameters on Deposition Rate and
    Film Properties*

The following factors in the hydride oxidation-based CVD of SiO_2 and PSG films can affect the rate of growth and the uniformity, composition, and intrinsic stress of the films: (1) substrate temperature, (2) O_2/hydride ratio, (3) SiH_4/PH_3 ratio, (4) hydride flow rate, (5) diluent gas type and flow rate, (6) reactor geometry, (7) reactor wall temperature, (8) impurities in CVD system, (9) gas additives, (10) substrate surface conditions, and (11) differences in reactor operating principles.

a. Substrate Temperatures and O_2/Hydride Ratio. These two factors are closely interrelated and are of fundamental importance in controlling the deposition rate and the film properties. SiH_4 diluted with N_2 begins to form SiO_2 films at a substrate temperature of about 240°C if the O_2/SiH_4 mole ratio is in the range of 3:1 [264]. The rate of film growth increases rapidly as the temperature is increased to 310°C, and very gradually as it is further increased to 450°C. To attain an increase in the maximum deposition rate with temperature, the O_2/SiH_4 ratio must be progressively increased with the temperature. For example, at 475°C, an O_2/SiH_4 ratio of at least 14:1 is required. Larger ratios of up to 33:1 cause no change, but ratios beyond this limit inhibit the reaction, decreasing the rate of film growth. Thus, a plateau region forms that is insensitive to the O_2/SiH_4 ratio and widens as the temperature is increased, independently of the reactor type used [75, 102, 186, 263, 265–270, 274]. The reduction in SiO_2 deposition rate at high O_2/SiH_4 ratios has been explained on the basis that O_2 may retard the reaction by being adsorbed on the substrate surface [268], or that a relatively stable intermediate product forms in the gas phase, retarding the chain process [275]. Up to the point of O_2 inhibition the SiH_4/O_2 reaction kinetics is first-order overall [268].

The corresponding parameter effects for PSG formation by cooxidation of diluted $SiH_4 + PH_3$ with excess O_2 are very similar to those discussed for SiO_2 [268, 276]. Figure 16 shows comparative plots of the effect of O_2/hydride mole ratio on the deposition rate of PSG films at several temperatures. As the temperature is increased, the maximum deposition rate increases if progressively larger O_2/hydride ratios are used. The ratio range of the maximum region widens with increasing temperature, as in the case of SiO_2 deposition.

Figure 17 illustrates the effects of O_2/hydride ratio and temperature on the composition of PSG films. The P concentration in the films generally increases with increasing O_2/hydride ratio and with decreasing temperature. The composition of films deposited at 450°C is less critically dependent on the ratio than at lower temperature, making the higher temperature more desirable for practical applications.

Stress in PSG films on Si substrates increases strongly with increasing O_2/hydride ratios; relatively small differences result for SiO_2 films [261].

b. SiH_4/PH_3 Ratio. The relationship of the mole ratio of SiH_4/PH_3 in the gas and of SiO_2/P_2O_5 in the resulting glass is nearly linear up to about 10 mole % P_2O_5 [166, 272, 273, 277]. Comparative plots of data [166, 261, 277] are presented in Fig. 18; they show satisfactory agreement, despite wide differences in experimental conditions. Lower temperatures of deposition at constant SiH_4/PH_3 ratio increase the P content in the PSG if the O_2/hydride ratio is adjusted for the plateau maximum deposition rate for each particular temperature (Figs. 16 and 18). The PSG contains more P than expected from stoichiometry because the conversion efficiency of

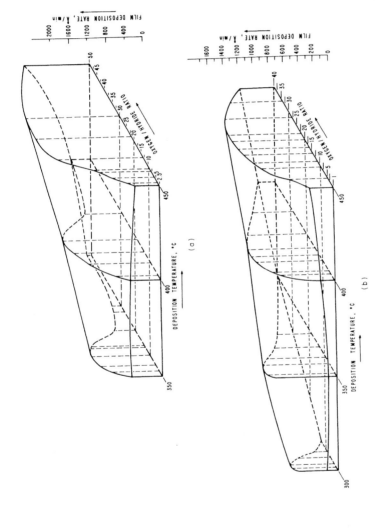

Fig. 16. Deposition rate of PSG films as a function of $O_2/(SiH_4 + PH_3)$ ratio and substrate deposition temperature. Upper graph was constructed from data by Kern *et al.* [261] $SiH_4/PH_3 = 20:1$, diluent N_2, total gas flow 11 liter/min; lower graph from data by Shibata and Sugawara [276] $SiH_4/PH_3 = 14.1$, diluent N_2, total gas flow 20 liter/min. General curvature in both plots is remarkably similar despite the indicated difference in CVD parameters and in geometry of the CVD reactors used (from Kern and Rosler [23], courtesy of the American Institute of Physics).

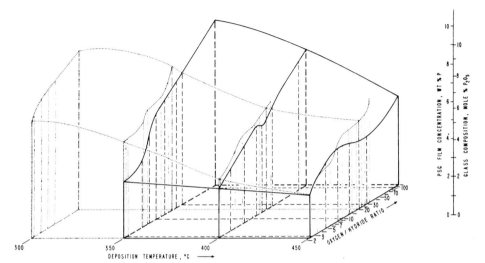

Fig. 17. Composition of PSG films in wt % P and in mole % P_2O_5 as a function of $O_2/(SiH_4 + PH_3)$ ratio and substrate deposition temperature for the same sets of samples used in Fig. 16. The two graphs were constructed from data by Kern *et al.* [261] (solid and dashed lines) $SiH_4/PH_3 = 20:1$, diluent N_2, total gas flow 11 liter/min, and from data by Shibata and Sugawara [276] (dotted lines) $SiH_4/PH_3 = 14:1$, diluent N_2, and total gas flow 20 liter/min. Again, the general tendency for both sets of data is similar, despite difference in CVD parameters (from Kern and Rosler [23], courtesy of the American Institute of Physics).

SiH_4 to SiO_2 during cooxidation with PH_3 is apparently lowered. This effect increases with decreasing temperature of deposition [186, 261, 277].

Too high a P concentration (i.e., >5 wt % P in a passivation glass) may result in a hygroscopic glass, which can cause current leakage across a device surface and metal corrosion. Too low a P content (i.e., <2 wt % P) may result in cracking of the glass on Si because of excessive stress.

c. Hydride Flow Rate. At a constant substrate temperature and O_2/hydride ratio, the hydride flow rate affects the rate of film growth linearly up to some saturation level which is dictated by the capacity of the reaction chamber. Film thicknesses of many micrometers will eventually cause a decrease in the thermal conductance, with consequent decreases of the deposition rate, and in the case of PSG, an increase of the P content.

PSG deposition studies in the continuous nozzle-type CVD reactor described in Section III.B.4 indicated that the cooxidation of SiH_4 plus PH_3 can be influenced by this unusual reactor configuration [261]. Slowing the travel speed of the Si substrates beneath the nozzle array under otherwise constant conditions caused an increase in the P concentration in the films and, in contrast to SiO_2, the PSG deposition rate increased substantially.

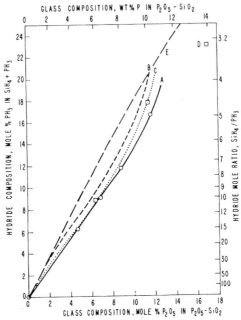

Fig. 18. Composition of PSG films as a function of hydride composition of the substrate deposition temperatures and $O_2/(SiH_4 + PH_3)$ ratios (R) indicated. Data A by Kern *et al.* [261] 445°C, $R = 20$; B,C by Shibata *et al.* [277] 450°C, $R \approx 18$; 400°C, $R \approx 12$; D by Wong and Ghezzo [166] 400°C, $R \approx 10$; E based on stoichiometric reaction. All compositions are richer in P than calculated for stoichiometric conversion (from Kern and Rosler [23], courtesy of the American Instit. of Physics).

The relative conversion efficiencies for SiH_4 and PH_3 during cooxidation in this restricted reactor configuration apparently differ from those normally observed.

d. Diluent Gas. The diluent or carrier gas—usually N_2—serves to reduce the concentration of each reactive gas so that their streams can be safely mixed without spontaneous reaction. It provides sufficient volume to the mixture to force it over the heated substrate, and allows optimization of gas convection and turbulence conditions for establishing good film uniformity. Too low or too high a flow rate lowers the film deposition rate and causes poor film uniformity. Excessive flow rates decrease the substrate temperature which, in the case of PSG, leads to an increase of the P content. The highest acceptable diluent gas flow rate should be used to depress homogeneous gas phase nucleation. The deposition rate of very high-P PSG is inhibited by N_2, but not with Ar [166]. The rate of SiO_2 deposition on chrome-steel substrates is slower when CO_2 is used as diluent instead of N_2, whereas with He it is faster by approximately the square root of the molecular weight of the gases [263].

e. Geometry and Wall Temperature of Reactor. The exact shape and dimensions of the reaction chamber and the gas inlet/outlet affect the gas flow dynamics and hence the deposition rate and the thickness uniformity of the films. A cylindrical bell jar reactor with a single top inlet and concentric bottom outlets [73] was used in much of the work described. Cooling the reactor parts that are not intended for heating the substrate suppresses homogeneous gas phase nucleation, resulting in cleaner film deposits.

f. Impurities and Gas Additives. The surface-catalyzed reaction underlying low-temperature CVD of oxide and glass films is sensitive to particulate contaminants in the gas stream. These can cause microbubbles, pinholes, and other localized structural defects in the films. Submicron filters in the gas lines alleviate this condition.

Stress in SiO_2 and PSG films can be lowered by adding water vapor to the gas stream entering the reactor. Stress at 23°C in the SiO_2 films deposited on Si wafers at 450°C by this technique was reduced from 3.4×10^9 dyn/cm^2 (tensile) to values of as low as 2.7×10^9 dyn/cm^2. Aluminum etching to demarcate cracks [278] in SiO_2 coating deposited on patterned Al films on oxidized Si showed that oxide prepared with wet N_2 had no cracks, while numerous cracks occurred on the samples prepared with dry N_2 [261].

g. Substrate Surface Conditions. The composition of the substrate and adsorbed contaminants can deleteriously affect the growth rate and also the structural quality of CVD films, as noted in the preceding paragraph. Chemical impurities on the substrate surface can act as growth catalysts or inhibitors. In general, defects such as pinholes, pits, thin spots, blisters, microbubbles, localized devitrification, cloudiness, and local variations in film thickness can result.

The topography of the substrate also affects the film quality due to localized disturbances in the gas flow pattern. Unless special care is exercised, the step coverage in low-temperature, normal-pressure CVD can be poor, characterized by a dogbone-type structure on the upper edge of the step and reentrant angles at the base of the step. The quality of step coverage can be improved by reducing the total gas flow rate, but tapering of the structure prior to overcoating is more effective [279]. The chemical composition of an overcoating material can have a marked influence on the step coverage. For example, PSG films containing 2–4 wt % P afford a substantially better conformal edge coverage over Al patterns than does SiO_2 deposited under comparable conditions [273].

h. Different CVD Reactor Systems. Comparative experiments with several types of low-temperature normal-pressure CVD reactor systems, both laboratory and industrial types, showed good agreement in the deposition parameters and the resulting film properties, despite substantial differences in geometry and system operation. One exception in the case of an extreme reactor geometry (close-spaced nozzle array) has been noted.

3. Summary

The parameters underlying SiO_2 and PSG film formation by gas-phase oxidation of the diluted hydrides with O_2 at temperatures between 300 and 475°C under normal-pressure conditions have been examined in some detail as practical examples of a CVD process. The most critical factors determining film deposition rates, intrinsic stress, composition, and film quality are substrate temperature of deposition, O_2/hydride ratio, and SiH_4/PH_3 ratio. Several other factors have secondary effects on the properties of the film deposits. The effects of CVD parameters are essentially independent of the reactor systems used, even though the operating principles can differ greatly for various types of systems. Intrinsic stress in CVD SiO_2 and PSG films can be lowered by depositing the films in the presence of excess water vapor. Thicker layers can be deposited crack free by this technique than is otherwise possible.

The essential effects of interrelated CVD parameters on PSG film dep-

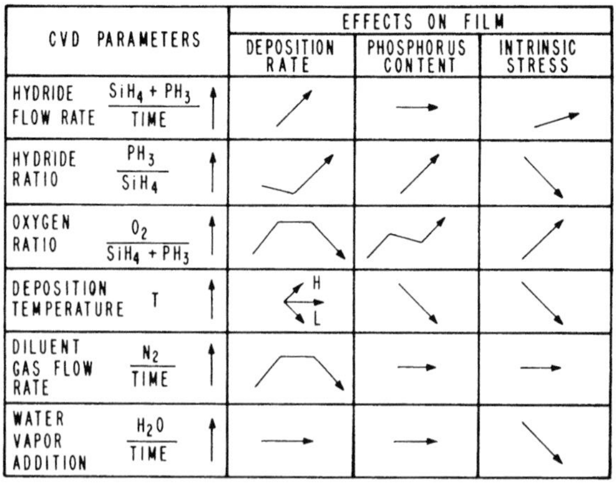

CVD PARAMETERS	EFFECTS ON FILM		
	DEPOSITION RATE	PHOSPHORUS CONTENT	INTRINSIC STRESS
HYDRIDE FLOW RATE $\dfrac{SiH_4 + PH_3}{TIME}$ ↑	↗	→	↗
HYDRIDE RATIO $\dfrac{PH_3}{SiH_4}$ ↑	↗	↗	↘
OXYGEN RATIO $\dfrac{O_2}{SiH_4 + PH_3}$ ↑	⌒↘	↗	↗
DEPOSITION TEMPERATURE T ↑	⇄ H L	↘	↘
DILUENT GAS FLOW RATE $\dfrac{N_2}{TIME}$ ↑	⌒↘	→	→
WATER VAPOR ADDITION $\dfrac{H_2O}{TIME}$ ↑	→	→	↘

Fig. 19. Graphic representation of essential effects of low-temperature CVD key parameters on PSG deposition rate, composition, and stress. Direction of arrows indicates relative increase or decrease. ↗↘; strong;⇄, slight; →, none. H, high oxygen ratio; L, low oxygen ratio (from Kern and Rosler [23], courtesy of the American Institute of Physics).

osition rate, composition, and intrinsic film stress are summarized schematically in Fig. 19.

V. CVD OF SEMICONDUCTORS

A. Introduction

Chemical vapor deposition films of semiconductors are crucial parts of many modern electronic devices. Hundreds of scientific articles have been written on various aspects of CVD of semiconductors. Clearly, it is impractical to review even a significant fraction of this literature. We thus decided to present essentials of important CVD processes in a tabular form, including only a few pertinent references for each particular process. Each of these tables will be devoted to one important group of semiconductors, e.g., III–V and II–VI compounds. Processes and compounds of practical importance will be emphasized. Important reviews and surveys of semiconductor CVD have been noted in the introductory Section I, and many specific examples have been discussed already in Section II on fundamental aspects, allowing us to keep this treatment very brief.

B. Group IV Semiconductors

Table IV lists CVD processes for semiconductors containing elements from group IV, i.e., Si, Ge, and C [66, 280–294]. Since Si is the most important semiconductor today, it will be treated in more detail. Silicon sources of practical importance are silane (SiH_4) [66, 280, 280a, 283–285] and various chlorosilanes, i.e., $SiCl_2H_2$ [280, 280a], $SiCl_3H$ [280–281, 286], and $SiCl_4$ [280–281]. The chief advantages of chlorosilanes in comparison with silane are the lower price, the applicability at higher temperature where higher growth rates are possible, and the absence of need for water cooling of reactor walls. Silane is advantageous when low deposition temperatures are required and when the presence of chlorine is harmful, as, e.g., in the case of silicon deposition on sapphire substrates.

The thickness and complexity of Si films can vary greatly. A typical epitaxial structure for power devices can contain three to four differently doped layers with a total thickness of ~ 100 μm. The epitaxial structures of integrated circuits contain one or two layers with a total thickness of ~ 10 μm. Polycrystalline Si films in integrated circuit devices are typically 0.5–10 μm thick.

Chemical vapor deposition films of Ge are primarily of historical in-

Table IV

CVD Film of Group IV Semiconductors

Semicond.	Input materials	Deposition temp. (°C)	Crystallinity	Substrate	References
Si	$SiCl_4$, H_2	~1200	Epitaxial	Single-crystal Si wafer	[280, 281]
Si	$SiCl_3H$, H_2	~1150	Epitaxial	Single-crystal Si wafer	[280, 281]
Si	$SiCl_2H_2$, H_2	1100	Epitaxial	Single-crystal Si wafer	[282]
Si	SiH_4, H_2	1050	Epitaxial	Single-crystal Si wafer	[283]
Si	SiH_4, He	900	Epitaxial	Single-crystal Si wafer	[284]
Si	SiH_4, H_2	960	Epitaxial	Sapphire	[285]
Si	SiH_4, H_2	960	Epitaxial	Spinel	[285]
Si	SiH_4, N_2 or H_2	700–900	Polycrystalline	SiO_2	[280]
Si	$SiCl_3H$ in H_2	900–1000	Polycrystalline	Graphite, metal	[286]
Si	SiH_4	600–700 (low pressure)	Polycrystalline	IC structures	[66]
Ge	$GeCl_4$, H_2	600–900	Epitaxial	Single-crystal Ge wafer	[287]
Ge	GeH_4, H_2	600	Epitaxial	Sapphire, spinel	[285]
Ge	$(CH_3)_4Ge$, H_2	700–1000	Epitaxial	Sapphire, spinel	[288]
SiGe	SiH_4, GeH_4, H_2	800–850	Epitaxial	Sapphire, spinel	[289]
SiC	$SiCl_4$, toluene, H_2	~1100	Polycrystalline	Single-crystal Si wafer	[290]
SiC	$SiCl_4$, hexane, H_2	1850	Epitaxial	Single-crystal Si wafer	[291]
SiC	CH_3SiCl_3, H_2	1150–1600 (65–500 Torr)	Polycrystalline	Single-crystal Si wafer	[292]
C(diamond)	Cl_4 or CH_4	~1000	Epitaxial	Diamond powder	[293, 294]

terest. Other group IV semiconductor films, e.g., SiC and C (diamond), are at this point mainly of research importance.

C. Group III–V Compound Semiconductors

Table V lists CVD processes for III–V semiconductors [295–314]. These compounds are useful for various applications: electrooptic de-

Table V *CVD Films of Group III–V Compound Semiconductors*

Compound	Input materials	Deposition temp. (°C)	Crystallinity	Substrate	References
AlN	$AlCl_3$, NH_3, H_2	1000	Epitaxial	Sapphire	[295]
AlN	$(CH_3)_3Al$, NH_3, H_2	700	Epitaxial	Sapphire	[296]
AlP	AL, HCl, PH_3, H_2	1200	Epitaxial	Si	[297]
AlAs	Al, HCl, AsH_3, H_2	1000	Epitaxial	GaAs	[298]
AlAs	$(CH_3)_3Al$, AsH_3, H_2	700	Epitaxial	GaAs	[299]
GaN	Ga, HCl, NH_3, H_2	850	Epitaxial	Sapphire	[300]
GaN	$(CH_3)_3Ga$, NH_3, H_2	950	Epitaxial	Sapphire	[296]
GaP	Ga, HCl, PH_3, H_2	800	Epitaxial	GaAs	[302]
GaP	$(C_2H_5)_3Ga$, $(C_2H_5)_3P$, H_2	500	Epitaxial	Si	[301]
GaAs	Ga, $AsCl_3$, H_2	760	Epitaxial	GaAs	[303]
GaAs	Ga, HCl, AsH_3, H_2	725	Epitaxial	GaAs, Ge	[304]
GaAs	$(CH_3)_3Ga$, AsH_3, H_2	700	Epitaxial	Spinel	[305]
GaSb	Ga, HCl, SbH_3, H_2	650	Epitaxial	GaAs	[306]
InP	In, PCl_3, H_2	650	Epitaxial	InP	[307]
InAs	$(C_2H_5)_3In$, AsH_3, H_2	675	Epitaxial	Sapphire	[308]
InAs	On, HCl, AsH_3, H_2	700	Epitaxial	InAs	[309]
AlGaAs	$(CH_3)_3Al$, $(CH_3)_3Ga$, AsH_3, H_2	700	Epitaxial	GaAs, Sapphire	[299]
GaAsP	Ga, HCl, AsH_3, PH_3, H_2	750	Epitaxial	Ge, GaAs	[304]
GaAsP	Ga, $AsCl_3$, PCl_3, H_2	800	Epitaxial	GaAs	[310]
GaAsSb	Ga, HCl, AsH_3, SbH_2, H_2	700	Epitaxial	GaAs	[306]
GaInP	Ga, In, HCl, PH_3, H_2	700	Epitaxial	GaAs	[311]
GaInAs	Ga, In, HCl, AsH_3, H_2	725	Epitaxial	GaAs	[312]
GaInAs	$(CH_3)_3Ga$, $(C_2H_5)_3In$, AsH_3, H_2	600	Epitaxial	GaAs	[313]
InAsP	In, HCl, AsH_3, PH_3, H_2	700	Epitaxial	InAs	[309]
GaInAsP	Ga, In, HCl, PH_3, AsH_3, H_2	650	Epitaxial	InP	[314]

311

vices, solar cells, microwave devices, electron emitters, etc. Particularly interesting is the possibility of tailoring the bandgap to a desired value by producing ternary alloys such as, e.g. $Ga_xIn_{1-x}P$ or $GaAs_{1-x}P_x$. The value of the bandgap is determined by the compositional variable x. Lately, this method has been extended to quaternary alloys, such as $Ga_xIn_{1-x}As_yP_{1-y}$. In this case it is not only possible to choose a bandgap, but also to select a desired lattice constant value, because there are several compositions with different lattice constants, which possess the chosen bandgap value. A proper choice of lattice constants minimizes the interlayer strain in multilayer structures, and thus results in improved device performance.

Practically all of the listed processes belong to one of the following three methods:

(1)   the hydride method where group V elements are introduced as hydrides, and group III elements as monochlorides, which result from the HCl interaction with the appropriate metals;

(2)   the chloride method where group V trichlorides interact with the appropriate III–V compound or group III metal;

(3)   the organometallic method where group III elements are introduced as trimethyl or triethyl compounds, while the group V elements are usually introduced as hydrides.

Examples of these methods were discussed in Section II.A on chemical reactions in CVD.

D. Group II–VI Compound Semiconductors

Table VI lists CVD films of several II–VI compounds [315–320]. Three different methods have been employed in preparation of these films:

(1)   reaction of metal vapors with group VI hydrides;

(2)   open tube transport of II–VI compounds by HCl or HBr; and

(3)   organometallic method where diethyl or dimethyl compounds of metals interact with group VI hydrides.

The II–VI semiconductors are not as widely used as the III–V compounds because of difficulties in achieving p-type doping in these compounds. Among the better known applications are electroluminescent displays and photovoltaic devices.

E. Miscellaneous Compound Semiconductors

Table VII lists CVD films of semiconductors not discussed so far [137, 321–331]. The versatility of the method is demonstrated by the wide vari-

Table VI

CVD Films of Group II–VI Compound Semiconductors

Compound	Input materials	Deposition temp. (°C)	Crystallinity	Substrate	References
ZnS	Zn, H_2S, H_2	825	Epitaxial	GaAs, GaP	[315]
ZnS	$(C_2H_5)_2Zn$, H_2S, H_2	750	Epitaxial	Sapphire	[316]
ZnS	ZnS, HCl, H_2	800	Epitaxial	GaAs	[317]
ZnSe	Zn, H_2Se, H_2	890	Epitaxial	GaAs, Sapphire	[315]
ZnSe	ZnSe, HBr, H_2	550	Epitaxial	GaAs, Ga	[318]
ZnSe	$(C_2H_5)_2Zn$, H_2Se, H_2	750	Epitaxial	Sapphire, BeO	[316]
ZnTe	$(C_2H_5)_2Zn$, $(CH_3)_2Te$, H_2	500	Epitaxial	Sapphire	[316]
CdS	Cd, H_2S, H_2	690	Epitaxial	GaAs, Sapphire	[315]
CdS	$(CH_3)_2Cd$, H_2S, H_2	475	Epitaxial	Sapphire	[316]
CdSe	Cd, H_2Se, H_2	700	Epitaxial	CdS, Sapphire	[319]
CdSe	$(CH_3)_2Cd$, H_2Se, H_2	600	Epitaxial	Sapphire	[316]
CdTe	$(CH_3)_2Cd$, $(CH_3)_2Te$, H_2	500	Epitaxial	Sapphire, Spinel, BeO	[316]
CdTe	Cd, Te, He	700–960	Polycrystalline	Carbon	[320]

313

Table VII *CVD Films of Miscellaneous Semiconductors*

Compound	Input materials	Deposition temp. (°C)	Crystallinity	Substrate	References
ScN	Sc, HCl, NH_3, H_2	900	Epitaxial	Sapphire	[321]
ScP	Sc, HCl, PH_3, H_2	850	Epitaxial	Si	[322]
ScAs	Sc, HCl, AsH_3, H_2	850	Epitaxial	Si	[322]
ScAsP	Sc, HCl, AsH_3, PH_3, H_2	850	Epitaxial	Si	[322]
ZnS–GaP	Zn, Ga, HCl, H_2S, PH_3, H_2	810	Epitaxial	GaAs	[323]
ZnSe–GaP	Zn, Ga, HCl, H_2Se, PH_3, H_2	890	Epitaxial	GaAs	[323]
ZnSe–GaAs	Zn, Ga, HCl, H_2Se, AsH_3, H_2	890	Epitaxial	GaAs	[323]
CdS–InP	Cd. In, HCl, H_2S, PH_3, H_2	690	Epitaxial	GaAs	[323]
$ZnSiP_2$	Zn, $SiCl_4$, PH_3, H_2	850	Polycrystalline	GaP, Si	[324]
YN	Y, HCl, NH_3, H_2	950	Epitaxial	Sapphire	[325]
DYN	Dy, HCl, NH_3, H_2	950	Epitaxial	Sapphire	[325]
ErN	Er, HCl, NH_3, H_2	950	Epitaxial	Sapphire	[325]
YbN	Yb, HCl, NH_3, H_2	950	Epitaxial	Sapphire	[325]
LnN	Ln, HCl, NH_3, H_2	950	Epitaxial	Sapphire	[325]
$CdCr_2S_4$	Cd, $CrCl_3$, S, He	740	Epitaxial	Sapphire, spinel	[326]
$ZnCr_2S_4$	Zn, $CrCl_3$, S, He	740	Polycrystalline	Sapphire, spinel	[326]
SnTe	$(C_2H_5)_4Sn$, $(CH_3)_2Te$, H_2	625	Polycrystalline	BaF_2, PbTe	[327]
PbSnTe	$(CH_3)_4Pb$, $(C_2H_5)_4Sn$, $(CH_3)_2Te$, H_2	600	Epitaxial	PbTe, BaF_2	[327]
PbS	$(CH_3)_4Pb$, H_2S, H_2	550	Epitaxial	BaF_2, sapphire	[327]
PbSe	$(CH_3)_4Pb$, H_2Se, H_2	550	Epitaxial	BaF_2, sapphire	[327]
PbTe	$(C_2H_5)_2Pb$, $(CH_3)_2Te$, H_2	500	Epitaxial	MgO, NaCl, BaF_2	[327]
SnO_2	$(C_4H_9)_2Sn(OOCH_3)_2$, O_2, H_2O, N_2	420	Amorphous	Si, quartz, glass	[328]
SnO_2:Sb	$(C_4H_9)_2Sn(OOCH_3)_2$, $SbCl_5$, O_2, H_2O, N_2	450	Amorphous	Si, quartz, glass, sapphire	[329]
In_2O_3:Sn	In- chelate, $(C_4H_9)_2Sn(OOCH_3)_2$, H_2O, O_2, N_2	500	Amorphous	Si, quartz, glass	[330]
V_2O_3, VO_2, V_2O_5	$(C_5H_7O_2)_4V$, N_2, O_2	400	Polycrystalline	Glass, sapphire, quartz,	[331]
ZnO	$(C_3H_5O_2)_2Zn$	500–600	Amorphous	Metals	[137]
		650	Single crystal	Metals	[137]
In_2O_3	$(C_5H_7O_2)_3In$	450–550	Amorphous	Metals	[137]
		630	Single crystal	Metals	[137]

314

ety of materials produced, ranging from sulfospinels ($CdCrS_4$) to rare-earth nitrides (YbN, LuN, etc.). None of these materials is in wide use, although some might be useful for specific applications. This list is by no means complete or closed; undoubtedly additional semiconductor compounds will be synthesized by CVD methods in the future.

VI. CVD OF CONDUCTORS

Electrically conductive materials that can be prepared by CVD comprise elemental metals, metal alloys, superconductive compounds, and films of optically transparent conductors.

A. General Reactions For Metals and Metal Alloys

Nearly all metals and many alloys have been deposited by chemical vapor reaction; exceptions are the alkali metals and alkaline earth metals. The major processes employed for CVD of metals and alloys are as follows:

(1)   thermal decomposition or pyrolysis of organometallic compounds, generally at low temperatures;

(2)   hydrogen reduction of metal halides, oxyhalides, carbonyl halides, and other oxygen-containing compounds; and, to a lesser extent,

(3)   reduction with vapors of thermodynamically appropriate metals (instead of H_2) of metal hydrides, and

(4)   chemical transport reactions based on temperature differentials in heterogeneous chemical reaction systems.

These processes have been exemplified in Section II.A.2 dealing with chemical reactions used in CVD.

B. Tungsten Films

Chemical vapor deposition of W films has been investigated more thoroughly than CVD of any other metal, and is widely employed in industry for many important applications. The most frequently used CVD process is the hydrogen reduction of the halides, WF_6 and WCl_6. Numerous papers have been published on chemical reaction parameters, the kinetics, thermodynamics, pressure and temperature factors, substrate effects, deposit orientation, and industrial applications [1–7, 322–335].

Applications in electronics are for manufacturing electrodes for thermionic energy conversion devices [336, 337], for preparing Si Schottky-barrier diodes [338, 339], and for depositing refractory metallization for Si solid-state devices [340–346]. The latter may serve as a specific example. Tungsten films have been deposited at 700°C on SiO_2 and sapphire surfaces using a pulsing technique [342]. The chemical reaction is based on the usual H_2 reduction of WF_6 vapor to W metal and HF as the by-product. On substrates with exposed Si areas, the direct contact reduction of WF_6 at temperatures above 400°C can be used according to the following self-limiting surface reaction [338, 342, 343]:

$$2WF_6 + Si \rightarrow 2W + 3SiF_4 .$$

Combination of the two processes allows deposition of W for electric contacting of Si devices and as refractory interconnect metallization over the insulator surfaces [342].

C. Films of Other Metals and Alloys

The comprehensive surveys of CVD processes for metals published in 1966 by Powell [347] and by Oxley [348] serve as excellent basic source references. Representative papers that have appeared since then (with a few exceptions) are categorized in Table VIII.

D. Superconductors

The most widely used superconductive compound is niobium stannide, Nb_3Sn. Films can be prepared in single-crystalline or polycrystalline form by the hydrogen reduction of $NbCl_4$–$SnCl_4$ mixtures at 900–1200°C [376, 377]. Superconductive nitrides of Nb (NbN, Nb_4N_5) have been deposited by reacting NH_3 with $NbCl_5$ [378].

More recently, Nb_3Ge has received a great deal of attention because of its high transition temperature. This compound can be prepared by H_2-reduction of the chlorides, typically at 900°C [379, 380].

E. Transparent Conductors

Thin films of optically transparent and electrically conductive materials are used in electrooptic device technology. Transparent conductors prepared by CVD are, besides polycrystalline Si, certain metal oxides, such as SnO_2 [328], In_2O_3 [137], VO_2, V_2O_3, and V_2O_5 [331]. Doped metal oxide films, such as SnO_2:Sb [329] and In_2O_3:Sn [330], can exhibit considerably higher conductivity. All of these films are prepared either by hydrolysis of the metal chlorides or by pyrolysis of metal organic compounds, as indicated in Table VII (they have been included in the section

Table VIII

CVD of Metals and Metal Alloys[a]

CVD metal	Metal halide reductions	Organometallic reactions	Inorganic pyrolysis and other reactions
Ag	[349]		
Al		[119, 350, 351]	
Al₃Ta		[352]	
Au			[119]
Be		[353]	
Cr	[354, 355]	[119, 356, 357]	
Cu	[119]	[119, 358]	
Ir	[359]	[352, 359]	[359]
Mo	[142a, 337, 339, 340, 344, 360–364]	[119, 343, 352]	
Nb	[365]		
Ni		[119, 358]	
Pt		[366]	[366]
Re	[336, 368]	[368]	[336, 367, 368]
Rh		[358]	
Ta	[119, 142a, 340, 365, 369]	[119]	
Ta–W	[369, 370]		
W	[70a, 119, 332–342, 344, 346, 364, 369]	[119, 343, 352]	
W–Mo–Re	[371]		
W–Re	[372–374]		
V	[340, 375]		

[a] For work published prior to 1966 of these and additional metals, see Powell [347] and Oxley [348].

on semiconductors, since these materials are usually classified as such). A comprehensive review of materials and processes used to fabricate transparent conductive films is available [381].

VII. CVD OF MISCELLANEOUS MATERIALS

In this last section we shall briefly note various types of technically important CVD materials which have not been included in our discussions. These include Al_2O_3–Al cermets, elemental boron, borides, boron phosphide, carbon of various modifications, carbides, oxycarbides, carbonitrides, silicides, chalcogenide glasses of the type Si–As–Te, rare-earth phosphors, Si-rich Si_3N_4 and SiO_2, and oxygen-containing Si for semiconductor electronics applications. A survey of theoretical CVD studies and of representative materials and CVD processes in these categories, with literature references [382–421] since 1966, is presented in

Table IX. Several of these processes were developed for thick-film coating or bulk deposition but could be adapted for thin-film applications. A comprehensive review covering the prior literature and reaction chemistry of several of these groups was published by Powell [167]. The industrially most important types of the materials noted in Table IX are the borides, carbon, and carbides.

The most frequently used process for depositing films of B and metal borides is H_2-reduction of the halide compounds at high temperature. The deposition of TiB_2 from $TiCl_4-BCl_3-H_2$, typically at 1225°C and 0.26 atm pressure, is a good example of an extensively studied system [407]. Borides of metals (i.e., Mo, Nb, Ta, W) that are not readily deposited by this process can be prepared by boriding. In this case a film of elemental B is chemically vapor deposited and diffused in to the metal substrate, forming a boride layer [167].

Carbon in its various modifications is produced generally by pyrolysis of hydrocarbon compounds, such as simple aliphatics (i.e., CH_4, C_2H_6, C_2H_2) and aromatics (i.e., C_6H_6). The detailed reaction mechanisms of this industrially important processes are quire involved.

Numerous metal carbides have been prepared by CVD from a gaseous mixture of the volatile metal halide and a carbon compound (i.e., CO, CCl_4, hydrocarbons, or by pyrolysis of C-to-metal bonded organometallic compounds.

Table IX

CVD of Miscellaneous Materials[a]

CVD material	Reference	CVD material	Reference
Al_2O_3-Al	[382]	$Si-(O)$	[401]
B	[383, 383a]	TaB_2	[393]
BC_4	[384-387]	TaC; Ta_2C	[402]
BP	[388]	TiB_2	[403-407]
C	[293, 294, 385, 389-391]	TiC	[243, 383, 384, 408]
HfC	[245, 392]	Ti(CN)	[409-411]
NbB_2	[393]	$Ti-C-O$	[412, 413]
NbC	[394]	$Ti-Ge-C$	[414, 415]
Nb-, Mo-, Ta-Si	[395]	$Ti-Si$	[416]
Si-As-Te	[395a]	$Ti-Si-C$	[414-416]
Si-B	[396]	WC; W_2C	[417, 418]
SiC	[291, 292, 384, 397-399]	WSi_2; W_5Si_3	[419]
$Si_3N_4-(Si)$; $SiO_2-(Si)$	[400]	ZrC	[394, 420, 421]

[a] For work published prior to 1966 of these and additional materials, see Powell [167].

An important group of compounds not included is Table IX are magnetic materials prepared by CVD. These materials are used for magnetic bubble memory or information storage devices in computers. Both ferrites and the more versatile garnets have been synthesized by CVD techniques. A fairly recent example of a ferrite film is single-crystal nickel ferrite grown on MgO substrates by vapor oxidation of $FeCl_2-NiCl_2$ [422]. Single-crystal epitaxial films of rare-earth garnets are usually deposited on a single-crystal nonmagnetic garnet substrate (i.e., $Gd_3Ga_5O_{12}$) by oxidation or hydrolysis of volatile metal halides at temperatures above 1100°C. Examples of magnetic bubble materials grown by these techniques are gadolinium iron garent [423], gallium-substituted yttrium iron garnet [424], and terbium erbium iron garnet ($Tb_{3-x}Er_xFe_5O_{12}$) [425]. Organometallics have also been used as starting reactants, allowing lower deposition temperatures [426]. Several reviews and bibliographies of CVD work of these materials are available [3, 19, 24, 427–429]. The last reference is a bibliography with 211 report abstracts of federally funded CVD research (1964–1975) dealing with a variety of miscellaneous materials.

VIII. CONCLUSIONS

In this chapter we have presented a broad survey of CVD, with particular emphasis on electronic materials, where some of the most sophisticated thin-film deposition processes are required. A significant portion has been devoted to the underlying principles of CVD processes. One of the chief purposes of this approach was to illustrate the interdisciplinary nature of the method which is basic for understanding CVD, because no single classical scientific or engineering discipline adequately covers all of the pertinent areas. Such interdisciplinary approach is not frequently encountered in the literature; rather a disproportionately large number of CVD articles emphasize the immediate application of CVD without due reference to the underlying principles.

Relatively few CVD systems have been studied by means of modern analytical techniques despite the fact that such studies could in many cases result in an improved and more economical CVD product. We hope that our discussion of fundamentals have indicated some of the areas where basic studies would be fruitful.

Modern thin-film technology is inconceivable without many of the CVD layers which we have discussed. Most likely CVD will continue this important role in the future. We can expect an evolutionary development in the control of dimensions, purity, crystalline quality, etc., of CVD products; this development will be based on the improvement of CVD equipment and starting materials, as well as on the refinement and control in technology. Adequate fundamental research in CVD will result in CVD processes significantly different from those in current use and will reduce the cost of the present CVD products, as well as extend the applicability of CVD processes to new areas. These developments will further enhance contributions of this versatile method for modern material science and technology.

ACKNOWLEDGMENTS

We wish to thank Dr. G. L. Schnable, Dr. D. Richman, and Dr. G. H. Olsen for critically reading the manuscript.

REFERENCES

1. C. F. Powell, J. H. Oxley, and J. M. Blocher, Jr., eds., "Vapor Deposition." Wiley, New York, 1966.
2. R. A. Holzl, in "Techniques of Metals Research" (R. F. Bunshah, ed.), Vol. 1, p. 1377. Wiley Interscience, New York, 1968.
3. W. F. Feist, S. R. Steele, and D. W. Ready, *Phys. Thin Films* 5, 237 (1969).
4. D. S. Campbell, in "Handbook of Thin Film Technology" (L. I. Maissel and R. Glang, eds.), Ch. 5. McGraw-Hill, New York, 1970.
5. R. W. Haskell and J. G. Byrne, in "Treatise on Materials Science and Technology" (H. Herman, ed.), Vol. 1, p. 293. Academic Press, New York, 1972.
6. J. W. Hastie, "High Temperature Vapors: Science and Technology," Ch. 3. Academic Press, New York, 1975.
7. W. A. Bryant, *J. Mater. Sci.* 12, 1285 (1977).
8. "Chemical Vapor Deposition of Refractory Metals, Alloys and Compounds," Proceedings of the Conference (A. W. Shaffhauser, ed.). Am. Nucl. Soc., Hinsdale, Illionis, 1967.
9. "Chemical Vapor Deposition—Second International Conference" (J. M. Blocher, Jr. and J. C. Withers, eds.). Electrochem. Soc., New York, 1970.
10. "Chemical Vapor Deposition—Third International Conference" (F. A. Glaski, ed.). Am. Nucl. Soc., Hinsdale, Illinois, 1972.
11. "Chemical Vapor Deposition—Fourth International Conference" (G. F. Wakefield and J. M. Blocher, Jr., eds.). Electrochem. Soc., Princeton, New Jersey, 1973.
12. "Chemical Vapor Deposition—Fifth International Conference" (J. M. Blocher, Jr., H. E. Hintermann, and L. H. Hall, eds.). Electrochem. Soc., Princeton, New Jersey, 1975.
13. "Chemical Vapor Deposition—Sixth International Conference" (L. F. Donaghey, P. Rai-Choudhury, and R. N. Tauber, eds.). Electrochem. Soc., Princeton, New Jersey, 1977.
14. T. L. Chu, *J. Vac. Sci. Technol.* 6, 25 (1969).
15. J. A. Amick and W. Kern, ref. 9, p. 551.
16. "Chemical Vapor Phase Deposition of Electronic Materials," Special Issue. *RCA Rev.* 31, No. 4 (1970).
16a. P. Wang and R. C. Bracken, ref. 10, p. 755.
17. E. L. MacKenna, *Proc. Semicond./IC Proc. Prod. Conf. Ind. Sci. Conf. Manage.*, Chicago, p. 71 (1971).

18. T. L. Chu and R. K. Schmeltzer, *J. Vac. Sci. Technol.* **10**, 1 (1973).
19. J. J. Tietjen, *Annu. Rev. Mater. Sci.* **3**, 317 (1973).
20. B. F. Watts, *Thin Solid Films* **18**, 1 (1973).
21. J. Bloem *et al.*, Deposition and Growth of Silicon and Insulating Films, *in* "Semiconductor Silicon 1973" (H. R. Huffand and R. R. Burgess, eds.), Ch. 2. Electrochem. Soc., Princeton, New Jersey, 1973.
22. B. O. Seraphin, *Thin Solid Films* **39**, 87 (1976).
23. W. Kern and R. S. Rosler, *J. Vac. Sci. Technol.* **14**, 1082 (1977).
24. E. Grünbaum, *in* "Epitaxial Growth" (J. W. Matthews, ed.), Part B, p. 611. Academic Press, New York, 1975.
25. D. W. Shaw, *in* "Epitaxial Growth" (J. W. Matthews, ed.), Part A, p. 89. Academic Press, New York, 1975.
26. B. E. Barry, *Thin Solid Films* **39**, 35 (1976).
27. V. S. Ban, *J. Electrochem. Soc.* **118**, 1473 (1971).
28. V. S. Ban, *J. Electrochem. Soc.* **119**, 761 (1972).
29. V. S. Ban, H. F. Gossenberger, and J. J. Tietjen, *J. Appl. Phys.* **43**, 2471 (1972).
30. V. S. Ban, *J. Cryst. Growth* **17**, 19 (1972).
31. V. S. Ban and M. Ettenberg, *J. Phys. Chem. Solids* **34**, 1119 (1973).
32. V. S. Ban and M. Ettenberg, ref. 11, p. 31.
33. V. S. Ban and E. A. D. White, *J. Cryst. Growth* **31**, 284 (1975).
34. V. S. Ban and S. L. Gilbert, *J. Electrochem. Soc.* **122**, 1382 (1975).
35. V. S. Bank *J. Electrochem. Soc.* **122**, 1389 (1975).
36. V. S. Ban and S. L. Gilbert, *J. Cryst. Growth* **31**, 284 (1975).
37. V. S. Ban, ref. 13, p. 66.
38. T. O. Sedgwick, J. E. Smith, Jr., R. Ghez, and M. E. Cowher, *J. Cryst. Growth* **31**, 264 (1975).
39. T. O. Sedgwick and J. E. Smith, Jr., *J. Electrochem. Soc.* **123**, 254 (1976).
40. D. Richman, *RCA Rev.* **24**, 596 (1963).
41. M. L. Lieberman and G. T. Noles, ref. 11, p. 19.
42. T. Sedgewick. ref. 13, p. 59.
43. J. J. Tietjen and J. A. Amick, *J. Electrochem. Soc.* **113**, 724 (1966).
44. F. Van Zeggeren and S. H. Storey, "The Computation of Chemical Equilibria," Cambridge Univ. Press, London and New York, 1970.
45. G. Eriksson, *Acta Chem. Scand.* **25**, 2651 (1971).
46. F. R. Lever, *IBM J. Res. Dev.* **8**, 460 (1964).
47. L. P. Hunt and E. Sirtl, *J. Electrochem. Soc.* **119**, 1741 (1972).
48. "JANAF Table of Thermochemical Data" (D. R. Stull, ed.). Dow Chem. Co., Midland, Michigan (1965); and Quarterly Suppl. PB-168370-1, -2, and -3, Clearinghouse Fed. Sci. Tech. Inf., Springfield, Virginia.
49. D. D. Wagman, W. H. Evans, V. B. Parker, I. Halow, S. M. Bailey, and R. H. Schumm, "Selected Values of Chemical Thermodynamic Properties." *Natl. Bur. Stand. (U.S.), Tech. Note* No. 270-3 (1968).
50. "The Chacterization of High Temperature Vapors" (J. Margrave, ed.). Wiley, New York, 1967; see especially R. Honig, Appendix A and M. S. Chandrasekharaiah, Appendix B.
51. O. Kubaschewski, E. L. Evans, and C. B. Alcock, "Metallurgical Thermochemistry," 4th Ed. Pergamon, Oxford, 1967.
52. C. E. Wicks and F. E. Block, "Thermodynamic Properties of 65 Elements—Their Oxides, Halides, Carbides and Nitrides." *U.S. Bur. Mines, Bull.* No. 605 (1963).
53. T. B. Reed, "Free Energy of Formation of Binary Compounds," MIT Press, Cambridge, Massachusetts, 1971.

54. D. R. Stull and G. C. Sinke, "Thermodynamic Properties of the Elements." *Adv. Chem. Ser.* No. 18. Am. Chem. Soc., Washington, D.C., 1956.
55. F. C. Eversteijn, *Philips Res. Rep.* **29**, 45 (1974).
56. D. W. Shaw, *J. Cryst. Growth* **31**, 130 (1975).
57. M. E. Jones, *in* "Reactivity of Solids" (R. W. Roberts and R. C. Devries, eds.), p. 443. Wiley, New York, 1965.
58. Y. S. Chiang, *in* "Semiconductor Silicon 1973" (H. R. Huff and R. R. Burgess, eds.), p. 285. Electrochem. Soc., Princeton, New Jersey, 1973.
59. C.-A. Chang, *J. Electrochem. Soc.* **123**, 1245 (1976).
60. R. B. Bird, W. E. Stewart, and E. N. Lightfoot, "Transport Phenomena." Wiley, New York, 1960.
61. R. Takahashi, Y. Koga, and K. Sugawara, *J. Electrochem. Soc.* **119**, 1406 (1972).
62. V. S. Ban, *J. Electrochem. Soc.* **125**, 317 (1978).
63. S. E. Bradshaw, *Int. J. Electron.* **23**, 381 (1967).
64. S. Berkman, V. S. Ban, and N. Goldsmith, *in* "Heteroepitaxial Semiconductors for Electronic Devices" (G. W. Cullen and C. C. Wang, eds.), p. 264, Springer-Verlag, Berlin and New York, (1978).
65. C. W. Manke and L. F. Donaghey, *J. Electrochem. Soc.* **124**, 561 (1977).
66. R. S. Rosler, *Solid State Technol.* **20**(4), 63 (1977).
67. R. J. Gieske, J. J. McMullen, and L. F. Donaghey, ref. 13, p. 183.
68. K. Sugawara, T. Yoshimi, and H. Sakai, ref. 12, p. 407.
69. R. E. Logar, M. T. Wauk, and R. S. Rosler, ref. 13, p. 195.
70. J. L. Deines and A. Spiro, *Electrochem. Soc. Extend. Abstr.* **74-1**, p. 161 (1974).
70a. W. A. Bryant, *J. Cryst. Growth* **35**, 257 (1976).
71. J. A. Venables and G. L. Price, *in* "Epitaxial Growth" (J. W. Matthews, ed.), Part B, p. 382. Academic Press, New York, 1975.
72. G. H. Olsen, *J. Cryst. Growth* **31**, 223 (1975).
73. W. Kern, *Solid State Technol.* **18**(12), 25 (1975).
74. W. Kern, *RCA Rev.* **39**, 525 (1968).
75. W. C. Benzing, R. S. Rosler, and R. W. East, *Solid State Technol.* **16**(11), 37 (1973).
76. AMS-1000 Silox Reactor System. Applied Materials, Santa Clara, California.
77. K. Sugawara, *J. Electrochem. Soc.* **119**, 1749 (1972).
78. K. Sugawara, Y. Nakazawa, and T. Yoshimi, *J. Electrochem. Soc.* **123**, 759 (1976).
78a. W. C. Benzing and R. Fisk, *Solid State Technol.* **18**(1), 39 (1975; R. S. Rosler and W. C. Benzing, *Solid State Technol.* **20**(7), 27 (1977).
79. E. L. Jordan, *J. Electrochem. Soc.* **108**, 478 (1961).
80. P. L. Jorgensen, *J. Chem. Phys.* **37**, 874 (1962).
81. H. Egagawa, Y. Morita, and S. Maekawa, *Jpn. J. Appl. Phys.* **2**, 765 (1963).
82. J. Klerer, *J. Electrochem. Soc.* **112**, 503 (1965).
83. G. W. Heunisch, *Anal. Chim. Acta* **48**, 405 (1969).
84. W. Kern and J. P. White, *RCA Rev.* **31**, 771 (1970).
85. P. M. Dunbar and J. R. Hauser, *J. Electrochem. Soc.* **117**, 674 (1970).
86. V. D. Wohlheiter and R. A. Whitner, *Electrochem. Soc. Extend. Abstr.* **75-1**, p. 424 (1975).
87. J. M. Albella, A. Criado, and E. M. Merino, *Thin Solid Films* **36**, 479 (1976).
88. Y. Nakai, *Electrochem. Soc. Extend. Abstr.* No. 84, Spring Meeting, p. 215 (1968).
89. J. Klerer, *J. Electrochem. Soc.* **108**, 1070 (1961).
90. J. Sandor, *Electrochem. Soc. Extend. Abstr.* No. 96, Spring Meeting, p. 228 (1962).
91. E. Tanikawa, T. Okabe, and K. Maeda, *Denki Kagaku* **41**, 491 (1973).
92. E. Tanikawa, O. Takayama, and K. Maeda, ref. 11, p. 261.
93. D. Burt, R. Taraci, and T. Zavion, U.S. Patent 3,934,060 (1976).

94. Y. Avigal, I. Beinglass, and M. Schieber, *J. Electrochem. Soc.* **121**, 1103 (1974).
95. M. J. Rand, *J. Electrochem. Soc.* **114**, 274 (1967).
96. M. J. Rand and J. L. Ashworth, *J. Electrochem. Soc.* **113**, 48 (1966).
97. W. Steinmaier and J. Bloem, *J. Electrochem. Soc.* **111**, 206 (1964).
98. S. K. Tung and R. E. Caffrey, *Trans. Metall. Soc. AIME* **233**, 572 (1965).
99. V. Y. Doo and D. R. Kerr, "Investigation of Refractory Dielectrics for Integrated Circuits," Contract NAS 12-105, Final Repts., IBM, Cambridge, Massachusetts (1965, 1968).
100. M.-J. Lim, *Electrochem. Soc. Extend. Abstr.* **76-2**, p. 631 (1976).
101. T. L. Chu, J. R. Szedon, and G. A. Gruber, *Trans. Metall. Soc. AIME* **242**, 532 (1968).
102. K. Strater, *RCA Rev.* **29**, 618 (1968).
103. R. C. G. Swann and A. E. Pyne, *J. Electrochem. Soc.* **116**, 1014 (1969).
104. W. J. Kroll, R. L. Titus, and J. B. Wagner, Jr., *J. Electrochem. Soc.* **122**, 573 (1975).
105. A. K. Gaind, G. K. Ackermann, V. J. Lucarini, and R. L. Bratter, *J. Electrochem. Soc.* **123**, 111 (1976).
106. A. K. Gaind, G. K. Ackermann, A. Nagarajan, and R. L. Bratter, *J. Electrochem. Soc.* **123**, 238 (1976).
107. J. A. Aboaf, *J. Electrochem. Soc.* **114**, 948 (1967).
108. M. T. Duffy and W. Kern, *RCA Rev.* **31**, 754 (1970).
109. M. Mutoh, Y. Mizokami, H. Matsui, S. Hagiwara, and N. Ino, *J. Electrochem. Soc.* **122**, 987 (1975).
110. R. L. Hough, ref. 10, p. 232.
111. M. Matsushita and Y. Yoga, *Electrochem. Soc. Extend. Abstr. No. 90*, Spring Meeting, p. 230 (1968).
112. N. N. Tvorogov, *Zh. Prikl. Khim. (Leningrad)* **34**, 2203 (1961).
113. L. A. Ryabova and Y. S. Savitskaya, *J. Vac. Sci. Technol.* **6**, 934 (1969).
114. V. F. Korzo, N. S. Ibraimov, and B. D. Halkin, *J. Appl. Chem.* **42**, 989 (1969).
115. A. Politycki and K. Hieber, in "Science and Technology of Surface Coating" (B. N. Chapman and J. C. Anderson, eds.), p. 159. Academic Press, New York, 1974.
116. F. C. Eversteijn, *Philips Res. Rep.* **21**, 379 (1966).
117. L. Hall and B. Robinette, ref. 9, p. 637.
118. L. Hall and B. Robinette, *J. Electrochem. Soc.* **118**, 1624 (1971).
119. D. Peterson, "Non-Vacuum Deposition Techniques For Use in Fabricating Thin Film Circuits," No. NObsr 91336. Final Rep. (1967).
120. S. K. Tung and R. E. Caffrey, *J. Electrochem. Soc.* **114**, 275C (1967).
121. V. Y. Doo and P. J. Tsang, *Electrochem. Soc. Extend. Abstr. No. 16*, Spring Meeting, p. 33 (1969).
122. S. K. Tung and R. E. Caffrey, *J. Electrochem. Soc.* **117**, 91 (1970).
123. T. Tsujide, S. Nakanuma, and Y. Ikushima, *J. Electrochem. Soc.* **117**, 703 (1970).
124. P. Wong and McD. Robinson, *J. Am. Ceram. Soc.* **53**, 617 (1970).
125. K. M. Schlesier, J. M. Shaw, and C. W. Benyon, Jr., *RCA Rev.* **37**, 358 (1976).
126. P. J. Tsang, R. M. Anderson, and S. Crikevich, *J. Electrochem. Soc.* **123**, 57 (1976).
127. D. R. Messier and P. Wong, *J. Electrochem. Soc.* **118**, 772 (1971); see also ref. 9, p. 803.
128. S. S. Lin, *J. Electrochem. Soc.* **122**, 1405 (1975).
129. M. Yokozowa, H. Iwasa, and I. Teramoto, *Jpn. J. Appl. Phys.* **7**, 96 (1968).
130. D. R. Harbison and H. L. Taylor, in "Thin Film Dielectrics" (F. Vratny, ed.), p. 254. Electrochem. Soc., New York, 1969.
131. C. C. Wang and K. H. Zaininger, *Proc. Electron. Components Conf.* p. 345 (1969).
132. S. Pakswer and P. Skoug, ref. 9, p. 619.
133. M. Balog, M. Schieber, S. Patai, and M. Michman, *J. Cryst. Growth* **17**, 298 (1972).
134. J. R. Szedon and R. M. Handy, *J. Vac. Sci. Technol.* **6**, 1 (1969).

324   WERNER KERN AND VLADIMIR S. BAN

135. M. T. Duffy, C. C. Wang, A. Waxman, and K. H. Zaininger, *J. Electrochem. Soc.* **116**, 234 (1969).
136. C. C. Wang, K. H. Zaininger, and M. T. Duffy, *RCA Rev.* **31**, 728 (1970).
137. L. A. Ryabova and Y. S. Savitskaya, *Thin Solid Films* **2**, 141 (1968).
138. R. N. Goshtagore, *J. Electrochem. Soc.* **117**, 529 (1970).
139. R. N. Goshtagore and A. J. Noreika, *J. Electrochem. Soc.* **117**, 1310 (1970).
140. R. N. Tauber, A. C. Diumbri, and R. E. Caffrey, *J. Electrochem. Soc.* **118**, 747 (1971).
141. T. Takahashi and H. Itoh, *J. Less-Common Met.* **38**, 211 (1974).
142. W. H. Knausenberger and R. N. Tauber, *J. Electrochem. Soc.* **120**, 927 (1973).
142a. K. Hieber and M. Stolz, ref. 12, p. 426.
143. M. W. Vernon and F. J. Spooner, *J. Mater. Sci.* **4**, 112 (1969).
144. F. J. Spooner and M. W. Vernon, *J. Mater. Sci.* **5**, 734 (1969).
145. F. J. Spooner and M. W. Vernon, *J. Mater. Sci.* **5**, 731 (1970).
146. A. E. Feuersanger, *Proc. IEEE* **52**, 1463 (1964).
147. Y. W. Hsueh and H. C. Lin, *Annu. Rep., Conf. Electr. Insul. Dielectr. Phenom.* p. 515 (1974).
148. K. J. Sladek and W. W. Gibert, ref. 10, p. 215.
149. E. T. Fitzgibbons, K. J. Sladek, and W. H. Hartwig, *J. Electrochem. Soc.* **119**, 735 (1972).
150. K. L. Hardee and A. J. Bard, *J. Electrochem. Soc.* **122**, 739 (1975).
151. I. Weitzel and K. Kempter, *Electrochem. Soc. Extend. Abstr.* **76-2**, p. 642 (1976).
152. W. Kern and R. C. Heim, *J. Electrochem. Soc.* **117**, 562 (1970).
153. W. Kern and R. C. Heim, *J. Electrochem. Soc.* **117**, 568 (1970).
153a. W. Kern, unpublished observations.
153b. A. Mayer, N. Goldsmith, and W. Kern, "Surface Passivation Techniques for Compound Solid State Devices," Tech. Rep. AFAL-TR-65-213. Air Force Syst. Command, Wright-Patterson Air Force Base, Ohio (1965).
154. M. Balog, M. Schieber, M. Michman, and S. Patai, *Thin Solid Films* **41**, 247 (1977).
155. E. Kaplan, M. Balog, and D. Frohman-Bentchowsky, *J. Electrochem. Soc.* **123**, 1570 (1976).
156. L. Ben-Dor, R. Druilhe, and P. Gibart, *J. Cryst. Growth* **24/25**, 1972 (1974).
157. K. L. Hardee and A. J. Bard, *J. Electrochem. Soc.* **123**, 1024 (1976).
158. L. Ben-Dor, E. Fischbein, I. Felner, and Z. Kalman, *J. Electrochem. Soc.* **124**, 451 (1977).
159. L. F. Thompson, *J. Electrochem. Soc.* **122**, 108 (1975).
160. J. B. MacChesney, P. B. O'Connor, and M. V. Sullivan, *J. Electrochem. Soc.* **118**, 776 (1971).
161. M. Sullivan, *J. Electrochem. Soc.* **120**, 545 (1973).
162. P. K. Gallagher, W. R. Sinclair, R. A. Fastnacht, and J. P. Luongo, *Thermochim. Acta* **8**, 141 (1974).
163. D. R. Mason, *J. Electrochem. Soc.* **123**, 519 (1976).
164. H. R. Brunner, F. P. Emmenegger, M. L. A. Robinson, and H. Rötschi, *J. Electrochem. Soc.* **115**, 1287 (1968).
165. W. Kern, *RCA Rev.* **32**, 429 (1971).
166. J. Wong and M. Ghezzo, *J. Electrochem. Soc.* **122**, 1268 (1975).
167. C. F. Powell, ref. 1, Ch. 11.
168. A. W. Fisher, J. A. Amick, H. Hyman, and J. H. Scott, Jr., *RCA Rev.* **29**, 533 (1968).
169. Y. Miura, S. Tanaka, Y. Matukura, and H. Osafune, *J. Electrochem. Soc.* **113**, 399 (1966).
170. J. M. Eldridge and P. Balk, *Trans. Metall. Soc. AIME* **242**, 539 (1968).
171. L. H. Kaplan and M. E. Lowe, *J. Electrochem. Soc.* **118**, 1649 (1971).

172. R. A. Gdula and P. C. Li, *Electrochem. Soc. Extend. Abstr.* **76-2**, p. 634 (1976).
173. J. Scott and J. Olmstead, *RCA Rev.* **26**, 357 (1965).
174. D. Flatley, N. Goldsmith, and J. Scott, *Electrochem. Soc. Extend. Abstr. No. 69*, Spring Meeting, p. 170 (1964).
175. M. Ghezzo, *J. Electrochem. Soc.* **119**, 1428 (1972).
176. D. M. Brown, W. E. Engeler, M. Garfinkel, and P. V. Gray, *Solid-State Electron.* **11**, 1105 (1968); *J. Electrochem. Soc.* **115**, 874 (1968).
177. K. M. Whittle and G. L. Vick, *J. Electrochem. Soc.* **116**, 645 (1969).
178. W. Kern and R. C. Heim, *Electrochem. Soc. Extend. Abstr. No. 5*, Spring Meeting, p. 234 (1968).
179. A. W. Fisher and J. A. Amick, *RCA Rev.* **29**, 549 (1968).
180. K. Strater and A. Mayer, *in* "Semiconductor Silicon" (R. R. Haberecht and E. L. Kern, eds.), p. 469. Electrochem. Soc., New York, 1969.
181. D. M. Brown and P. R. Kennicott, *J. Electrochem. Soc.* **118**, 293 (1971).
182. A. S. Tenney, *J. Electrochem. Soc.* **118**, 1658 (1971).
183. E. A. Taft, *J. Electrochem. Soc.* **118**, 1985 (1971).
184. A. S. Tenney and J. Wong, *J. Chem. Phys.* **56**, 5516 (1972).
185. D. M. Brown, M. Garfinkel, M. Ghezzo, E. A. Taft, A. Tenney, and J. Wong, *J. Cryst. Growth* **17**, 276 (1972).
186. G. Wahl, ref. 12, p. 391.
187. J. Wong, *J. Electron. Mater.* **5**, 113 (1976).
188. D. B. Lee, *Solid-State Electron.* **10**, 623 (1967).
189. H. Teshima, Y. Tarui, and O. Takeda, *Denki Shikenjo Iho* **33**, 631 (1969).
190. A. Cuccia, G. Shrank, and G. Queirolo, *in* "Semiconductor Silicon" (R. R. Haberecht and E. L. Kern, eds.), p. 506. Electrochem. Soc., New York, 1969.
191. E. Arai and Y. Terunuma, *Jpn. J. Appl. Phys.* **9**, 691 (1970).
192. P. C. Parekh, D. R. Goldstein, and T. C. Chan, *Solid-State Electron.* **14**, 281 (1971).
193. T. Wong and M. Ghezzo, *J. Electrochem. Soc.* **118**, 1540 (1971); **119**, 1413 (1972).
194. J. Wong, *J. Electrochem. Soc.* **119**, 1071, 1080 (1972); **120**, 122 (1973).
195. R. B. Fair, *J. Electrochem. Soc.* **119**, 1389 (1972).
196. M. Ghezzo and D. M. Brown, *J. Electrochem. Soc.* **120**, 110 (1973).
196a. T. Abe, K. Sato, M. Konaka, and A. Miyazaki, *Jpn. J. Appl. Phys., Suppl.* **39**, 88 (1970).
197. T. Kato, M. Nakamura, T. Yonezawa, and W. Watanabe, *Electrochem. Soc. Extend. Abstr.* **70-1**, p. 306 (1970).
198. D. Peterson, *IEEE Trans. Component Parts* **CP-10**, 119 (1963).
199. Y. Koga, M. Matsushita, M. Kobayashi, Y. Nakaido, and S. Toyoshima, *in* "Thin Film Dielectrics" (F. Vratny, ed.), p. 355. Electrochem. Soc., New York, 1969.
200. Y. Nakaido and S. Toyoshima, *J. Electrochem. Soc.* **115**, 1094 (1968).
201. F. L. Gittler and R. A. Porter, *J. Electrochem. Soc.* **117**, 1551 (1970).
202. W. von Muench, *Solid-State Electron.* **9**, 619 (1966).
203. C. F. Gibbon and D. R. Ketchow, *J. Electrochem. Soc.* **118**, 975 (1971).
204. G. M. di Giacomo, *J. Electrochem. Soc.* **116**, 313 (1969).
205. G. Gliemeroth, ref. 13, p. 477.
206. T. Matsuo, *Jpn. J. Appl. Phys.* **12**, 1862 (1973).
207. N. Hashimoto, Y. Koga, and E. Yamada, *in* "Thin Film Dielectrics" (F. Vratny, ed.), p. 327. Electrochem. Soc., New York, 1969.
208. J. P. Dismukes, J. Kane, B. Binggeli, and H. P. Schweizer, ref. 11, p. 275.
209. L. V. Gregor, *in* "Thin Film Dielectrics" (F. Vratny, ed.), p. 447. Electrochem. Soc., New York, 1969.
210. M. T. Duffy and W. Kern, *RCA Rev.* **31**, 742 (1970).

211. J. T. Milek, "Silicon Nitride for Microelectronic Applications," Parts 1 and 2, "Handbook of Electronic Materials," Vols. 3 and 6. IFI/Plenum, New York, 1971-1972).
212. P. Balk, *Solid State Devices, Eur. Solid State Device Res. Conf., 3rd Conf. Ser.* No. 19, pp. 51-82. Inst. Phys., London, (1973).
213. K. Niihara and T. Hirai, *J. Mater. Sci.* **11,** 593, 604 (1976).
214. V. Y. Doo, D. R. Nichols, and G. A. Silvey, *J. Electrochem. Soc.* **113,** 1279 (1966).
215. K. E. Bean, P. S. Gleim, R. L. Yeakley, and W. R. Runyan, *J. Electrochem. Soc.* **114,** 733 (1967).
216. V. Y. Doo, D. R. Kerr, and D. R. Nichols, *J. Electrochem. Soc.* **115,** 61 (1968).
217. J. R. Yeargan and H. L. Taylor, *J. Electrochem. Soc.* **115,** 273 (1968).
218. P. S. Schaffer and B. Swaroop, *Am. Ceram. Soc., Bull.* **49,** 536 (1970).
219. W. A. Kohler, *Trans. Metall. AIME* **1,** 735 (1970).
220. E. A. Taft, *J. Electrochem. Soc.* **118,** 1341 (1971).
221. A. C. Airey, S. Clark, and P. Popper, *Proc. Br. Ceram. Soc.* **22,** 305 (1973).
222. T. Arizumi, T. Nishinaga, and H. Ogawa, *Jpn. J. Appl. Phys.* **7,** 1021 (1968).
223. G. Cochet, H. Mellottée, and R. Delbourgo, ref. 12, p. 43.
224. S. S. Lin, *J. Electrochem. Soc.* **124,** 1945 (1977).
225. P. H. Holloway and H. J. Stein, *J. Electrochem. Soc.* **123,** 723 (1976).
226. H. J. Stein and H. A. R. Wegener, *J. Electrochem. Soc.* **124,** 908 (1977).
227. T. L. Chu, C. H. Lee, and G. A. Gruber, *J. Electrochem. Soc.* **114,** 717 (1967).
228. J. V. Dalton and J. Drobek, *J. Electrochem. Soc.* **115,** 865 (1968).
229. M. J. Grieco, F. L. Worthing, and B. Schwartz, *J. Electrochem. Soc.* **115,** 525 (1968).
230. E. MacKenna and P. Kodama, *J. Electrochem. Soc.* **119,** 1094 (1972).
231. V. D. Wohlheiter and R. A. Whitner, *J. Electrochem. Soc.* **119,** 945 (1972).
232. F. Galasso, U. Kuntz, and W. J. Croft, *J. Am. Ceram. Soc.* **55,** 431 (1972).
233. J. J. Gebhardth, R. A. Tanzilli, and T. A. Harris, ref. 12, p. 786; *J. Electrochem. Soc.* **123,** 1578 (1976).
234. T. L. Chu, J. R. Szedon, and C. H. Lee, *J. Electrochem. Soc.* **115,** 318 (1968).
235. A. Bhattacharyya, C. T. Kroll, and P. C. Velasquez, *Electrochem. Soc. Extend. Abstr.* **76-2,** p. 800 (1976).
236. D. M. Brown, P. V. Gray, F. K. Heumann, H. R. Phillipp, and A. E. Taft, *J. Electrochem. Soc.* **115,** 311 (1968).
237. M. J. Rand and J. F. Roberts, *J. Electrochem. Soc.* **120,** 446 (1973).
238. N. C. Tombs, F. A. Sewell, Jr., and J. J. Comer, *J. Electrochem. Soc.* **116,** 862 (1969).
239. A. K. Gaind, G. K. Ackermann, V. J. Lucarini, and R. L. Bratter, *J. Electrochem. Soc.* **124,** 599 (1977); ref. 12, p. 30; see also A. K. Gaind and E. W. Hern, *J. Electrochem. Soc.* **125,** 139 (1978).
240. R. Kieffer, D. Fister, H. Schoof, and K. Mauer, *Powder Metall. Int.* **5**(4), 1 (1973).
241. R. Kieffer and P. Ettmayer, *High Temp.—High Pressures* **6,** 253 (1974).
242. W. Schintlmeister and O. Pacher, *Metall* **28,** 690 (1974).
243. W. Schintlmeister, O. Pacher, and K. Pfaffinger, ref. 13, p. 523.
244. T. Takahashi and H. Itoh, *J. Electrochem. Soc.* **124,** 797 (1977).
245. M. J. Hakim, ref. 12, p. 634.
246. M. J. Rand and J. F. Roberts, *J. Electrochem. Soc.* **115,** 423 (1968).
247. M. Hirayama and K. Shohno, *J. Electrochem. Soc.* **122,** 1671 (1975).
248. T. L. Chu and R. W. Kelm, Jr., *J. Electrochem. Soc.* **122,** 995 (1975).
249. D. W. Lewis, *J. Electrochem. Soc.* **117,** 978 (1970).
250. W. M. Kim, E. J. Stofko, P. J. Zanzucchi, J. I. Pankove, N. Ettenberg, and S. L. Gilbert, *J. Appl. Phys.* **44,** 292 (1973).
251. H. M. Manasevit, F. M. Erdmann, and W. I. Simpson, *J. Electrochem. Soc.* **118,** 1864 (1971).

252. S. Zirinsky and E. A. Irene, *Electrochem. Soc. Extend. Abstr.* **76-2**, p. 639 (1976).
253. V. J. Silvestri, E. A. Irene, S. Zirinsky, and J. D. Kuptsis, *J. Electron. Mater.* **4**, 429 (1975).
254. E. A. Irene, V. J. Silvestri, and G. R. Woolhouse, *J. Electron. Mater.* **4**, 409 (1975).
255. N. Ilegems, *J. Cryst. Growth* **13/14**, 360 (1972).
256. T. L. Chu, *J. Electrochem. Soc.* **118**, 1200 (1971).
257. H. M. Manasevit, F. N. Erdman, and W. I. Simpson, *J. Electrochem. Soc.* **118**, 1865 (1971).
258. H. Nagai and T. Niimi, *J. Electrochem. Soc.* **115**, 671 (1968).
259. T. Yashiro, *J. Electrochem. Soc.* **119**, 780 (1972).
260. K. Sugiyama, S. Pac, Y. Tahashi, and S. Motojiama, *J. Electrochem. Soc.* **122**, 1545 (1975); ref. 12, p. 147.
261. W. Kern, G. L. Schnable, and A. W. Fisher, *RCA Rev.* **37**, 3 (1976).
262. H. J. Emeleus and K. Stewart, *J. Chem. Soc.* Part I, p. 1182 (1935); Part II, p. 677 (1936).
263. J. Graham, *High Temp.—High Pressures* **6**, 577 (1974).
264. N. Goldsmith and W. Kern, *RCA Rev.* **28**, 153 (1967).
265. M. L. Hammond and G. M. Bowers, *Trans. Metall. Soc. AIME* **242**, 546 (1968).
266. W. Kern and A. W. Fisher, *RCA Rev.* **21**, 715 (1970).
267. M. L. Barry, ref. 9, p. 595.
268. B. J. Baliga and S. K. Ghandhi, *J. Appl. Phys.* **44**, 990 (1973).
269. J. Middelhoek and A. J. Klinkhamer, ref. 12, p. 30.
270. R. Möller, L. Fabian, H. Weise, and C. Weissmantel, *Thin Solid Films* **29**, 349 (1975).
271. T. Tokuyama, T. Miyazaki, and M. Horiuchi *in* "Thin Film Dielectrics" (F. Vratny, ed.), p. 297. Electrochem. Soc., New York, 1969.
272. G. L. Schnable, *IEEE Int. Conv. Dig.* pp. 586–587 (1971).
273. M. M. Schlacter, E. S. Schlegel, R. S. Keen, R. A. Lathlaen, and G. L. Schnable, *IEEE Trans. Electron Devices* **ED-17**, 1077 (1970).
274. U. Feltl-Pohl, J. Herbst, and H. Splittgerber, *Electrochem. Soc. Extend. Abstr.* **74-2**, p. 498 (1974).
275. L. L. Vasil'eva, V. N. Drozdov, S. M. Repinskii, and K. K. Svitashev, *Mikroelektronika (Akad. Nauk SSR)* **5**, 448 (1976).
276. M. Shibata and K. Sugawara, *J. Electrochem. Soc.* **122**, 155 (1975).
277. M. Shibata, T. Yoshimi, and K. Sugawara, *J. Electrochem. Soc.* **122**, 157 (1975).
278. W. Kern and R. B. Comizzoli, *J. Vac. Sci. Technol.* **14**, 32 (1977).
279. W. Kern, J. L. Vossen, and G. L. Schnable, *Annu. Proc. Reliab. Phys., 11th* p. 214 (1973).
280. Articles on CVD of Silicon in "Semiconductor Silicon 1969, 1973, and 1977." Electrochemical Soc., Princeton, New Jersey, 1969, 1973, 1977.
280a. J. P. Duchemin, *Rev. Tech. Thomson–CFS* **9**(1), 33 (1977).
281. D. C. Gupta, *Solid State Technol.* **14**(10), 33 (1971).
282. N. Goldsmith and P. H. Robinson, *RCA Rev.* **34**, 358 (1973).
283. D. Richman, Y. S. Chiang, and P. H. Robinson, *RCA Rev.* **31**, 613 (1970).
284. Y. S. Chiang and D. Richman, *Metall. Trans.* **2**, 743 (1971).
285. D. J. Dumin, P. H. Robinson, G. W. Cullen, and G. E. Gottlieb, *RCA Rev.* **31**, 620 (1977).
286. T. L. Chu, *J. Vac. Sci. Technol.* **10**, 912 (1975).
287. E. F. Cave and B. R. Czorny, *RCA Rev.* **24**, 523 (1963).
288. Y. Avigal, D. Itzhak, and M. Schieber, *J. Electrochem. Soc.* **122**, 1226 (1975).
289. G. M. Oleszek and R. L. Anderson, *J. Electrochem. Soc.* **120**, 554 (1973).
290. K. E. Bean and P. S. Gleim, *J. Electrochem. Soc.* **114**, 1150 (1967).

291. W. V. Muench and I. Pfaffeneder, *Thin Solid Films* **31**, 39 (1976).
292. J. Chin, P. K. Gantzel, and R. G. Hudson, *Thin Solid Films* **40**, 57 (1977).
293. B. V. Derjaguin and B. V. Spitsyn, USSR Patent 339134 (1956).
294. W. G. Eversole, U.S. Patents 3030187 and 3030188 (1962).
295. W. M. Yim, E. J. Stofko, P. J. Zanzucchi, J. I. Pankove, M. Ettenberg, and S. L. Gilbert, *J. Appl. Phys.* **44**, 292 (1973).
296. H. M. Manasevit, F. M. Erdman, and W. I. Simpson, *J. Electrochem. Soc.* **118**, 1864 (1971).
297. D. Richman, *J. Electrochem. Soc.* **115**, 945 (1968).
298. M. Ettenberg, A. G. Sigai, A. Dreeben, and S. L. Gilbert, *J. Electrochem. Soc.* **118**, 1355 (1971).
299. H. M. Manasevit, *J. Electrochem. Soc.* **118**, 647 (1971).
300. H. P. Maruska and J. J. Tietjen, *Appl. Phys. Lett.* **15**, 327 (1969).
301. R. W. Thomas, *J. Electrochem. Soc.* **116**, 1450 (1969).
302. D. Richman and J. J. Tietjen, *Trans. Metall. Soc. AIME* **239**, 418 (1967).
303. J. R. Knight, D. Effer, and P. R. Evans, *Solid-State Electron.* **8**, 178 (1965).
304. J. J. Tietjen, R. E. Enstrom, V. S. Ban, and D. Richman, *Solid State Technol.* **15**(10), 42 (1972).
305. C. C. Wang, F. C. Dougherty, P. J. Zanzucchi, and S. H. McFarlane, III, *J. Electrochem. Soc.* **121**, 571 (1974).
306. R. B. Clough and J. J. Tietjen, *Trans. Metall. Soc. AIME* **245**, 583 (1969).
307. B. D. Joyce and E. W. Williams, *Int. Conf. GaAs Relat. Compounds, 3rd* p. 57. Inst. Phys., London (1972).
308. H. M. Manasevit and W. I. Simpson, *J. Electrochem. Soc.* **120**, 135 (1973).
309. J. J. Tietjen, H. P. Maruska, and R. B. Clough, *J. Electrochem. Soc.* **116**, 492 (1969).
310. W. F. Finch and E. W. Mehal, *J. Electrochem. Soc.* **111**, 814 (1964).
311. R. E. Enstrom, C. J. Nuese, V. S. Ban, and J. R. Appert, *Int. Conf. GaAs Relat. Compounds, 4th* p. 37. Inst. Phys., London (1972).
312. R. E. Enstrom, D. Richman, M. S. Abrahams, J. R. Appert, D. G. Fisher, A. H. Sommer, and B. F. Williams, *Int. Symp. GaAs Relat. Mater., 3rd* p. 30. Inst. Phys., London (1970).
313. B. J. Baliga and S. K. Ghandhi, *J. Electrochem. Soc.* **122**, 638 (1975).
314. G. H. Olsen, RCA Lab., personal communication (1977).
315. W. M. Yim and E. J. Stofko, *J. Electrochem. Soc.* **119**, 381 (1972).
316. H. M. Manasevit and W. I. Simpson, *J. Electrochem. Soc.* **118**, 644 (1971).
317. P. M. Kay and P. Lilley, *J. Cryst. Growth* **31**, 339 (1975).
318. S. G. Parker, J. E. Pinnell, and L. N. Swink, *J. Phys. Chem. Solids* **32**, 139 (1971).
319. W. M. Yim and E. J. Stofko, *J. Electrochem. Soc.* **121**, 965 (1974).
320. H. L. Tuller, K. Uematsu, and H. K. Bowen, *J. Cryst. Growth* **42**, 150 (1977).
321. J. P. Dismukes, W. M. Yim, and V. S. Ban, *J. Cryst. Growth* **13/14**, 365 (1972).
322. W. M. Yim, E. J. Stofko, and R. T. Smith, *J. Appl. Phys.* **43**, 254 (1972).
323. W. M. Yim, J. P. Dismukes, and H. Kressel, *RCA Rev.* **31**, 662 (1970).
324. B. J. Curtis, F. P. Emmeneger, and R. N. Nitsche, *RCA Rev.* **31**, 647 (1970).
325. J. P. Dismukes, W. M. Yim, J. J. Tietjen, and R. E. Novak, *RCA Rev.* **31**, 680 (1970).
326. H. L. Pinch and L. Ekstrom, *RCA Rev.* **31**, 692 (1970).
327. H. M. Manasevit and W. I. Simpson, *J. Electrochem. Soc.* **122**, 444 (1975).
328. J. Kane, W. Kern, and H. P. Schweizer, *J. Electrochem. Soc.* **122**, 1144 (1975).
329. J. Kane, W. Kern, and H. P. Schweizer, *J. Electrochem. Soc.* **123**, 270 (1976).
330. J. Kane, W. Kern, and H. P. Schweizer, *Thin Solid Films* **29**, 155 (1975).
331. L. A. Ryabova, I. A. Serbinov, and A. S. Darevsky, *J. Electrochem. Soc.* **119**, 427 (1972).

332. Numerous articles on CVD of tungsten appearing in refs. 8–12.
333. J. F. Berkeley, A. Brenner, and W. E. Reid, Jr., *J. Electrochem. Soc.* **114**, 561 (1967).
334. R. W. Haskell and J. G. Byrne, in "Treatise on Materials Science and Technology" (H. Hermann, ed.), Vol. 1, p. 293. Academic Press, New York, 1972.
335. W. A. Bryant and G. H. Meier, *J. Electrochem. Soc.* **120**, 559 (1973).
336. P. Rouklove and F. A. Glaski, ref. 10, p. 647.
337. R. Faron, M. Barques, J. P. Durand, and J. Gillardeau, ref. 11, p. 375.
338. C. Crowell, J. Sarace, and S. Sze, *Trans. Metall. Soc. AIME* **233**, 478 (1965).
339. K. Sato, M. Yoshida, and T. Matsui, *Toshiba Rev. (Int. Ed.)* p. 37 (1967).
340. N. Hashimoto and Y. Koga, *Electrochem. Soc. Extend. Abstr. No. 100,* Spring Meeting, p. 252 (1968).
341. A. F. Mayadas, J. J. Cuomo, and R. Rosenberg, *J. Electrochem. Soc.* **116**, 1742 (1969).
342. J. M. Shaw and J. A. Amick, *RCA Rev.* **30**, 306 (1970).
343. L. H. Kaplan and F. M. D'Heurle, *J. Electrochem. Soc.* **117**, 693 (1970).
344. G. G. Pinneo, ref. 10, p. 462.
345. J.-S. Lo, R. W. Haskell, J. G. Byrne, and A. Sosin, ref. 11, p. 74.
346. C. M. Melliar-Smith, A. C. Adams, R. H. Kaiser, and R. A. Kushner, ref. 11, p. 245; also *J. Electrochem. Soc.* **121**, 298 (1974).
347. C. F. Powell, ref. 1, pp. 277–342.
348. J. H. Oxley, ref. 1, pp. 452–483.
349. R. J. H. Voorhoeve and J. W. Merewether, *J. Electrochem. Soc.* **119**, 364 (1972).
350. G. J. Kostas, *Proc. Int. Symp. Decomposition Organometallics.* Air Force Material Laboratory, Dayton, Ohio (1967).
351. H. O. Pierson, *Thin Solid Films* **45**, 257 (1977).
352. J. A. Papke and R. D. Stevenson, ref. 8, p. 193.
353. J. M. Wood and F. W. Frey, ref. 8, p. 205.
354. E. F. Wakefield, *J. Electrochem. Soc.* **116**, 5 (1969).
355. W. Hänni and H. E. Hintermann, *Thin Solid Films* **40**, 107 (1977).
356. N. G. Ananthan, V. Y. Doo, and D. K. Seto, *J. Electrochem. Soc.* **118**, 163 (1971); also ref. 9, p. 649.
357. T. J. Truex, R. B. Saillant, and F. M. Monroe, *J. Electrochem. Soc.* **122**, 1396 (1975).
358. R. L. VanHemert, L. B. Spendlove, and R. E. Sievers, *J. Electrochem. Soc.* **112**, 1123 (1965).
359. B. A. Macklin and J. C. Withers, ref. 8, p. 161.
360. D. K. Seto, V. Y. Doo, and S. Dash, ref. 9, p. 569.
361. J. J. Casey, R. R. Verderber, and R. R. Granache, *J. Electrochem. Soc.* **114**, 201 (1967).
362. T. Sugano, H.-K. Chou, M. Yoshida, and T. Nishi, *Jpn. J. Appl. Phys.* **7**, 1028 (1968).
363. A. H. El-Hoshy, *J. Electrochem. Soc.* **118**, 2028 (1971).
364. R. Faron, M. Barques, J. Gillardeau, R. Hasson, G. Dejachy, and J. P. Durand, ref. 10, p. 439.
365. E. Fitzer and D. Kehr, ref. 11, p. 144.
366. M. J. Rand, *J. Electrochem. Soc.* **120**, 686 (1973).
367. L. Yang, R. G. Hudson, and J. J. Ward, ref. 10, p. 253.
368. H. J. Anderson and A. Brenner, ref. 9, p. 355.
369. W. A. Bryant and G. H. Meier, ref. 12, p. 161.
370. F. A. Glaski and A. Crowson, ref. 13, p. 542.
371. C. I. Fairchild, ref. 8, p. 149.
372. C. E. Hamrin, Jr. and E. M. Foster, ref. 8, p. 243.
373. J. I. Federer and J. E. Spruiell, ref. 8, p. 443.
374. J. I. Federer and A. C. Schaffhauser, ref. 10, p. 242.

375. K. J. Miller, M. J. Grieco, and S. M. Sze, *J. Electrochem. Soc.* **113**, 902 (1966).
376. R. E. Enstrom, J. J. Hanak, and G. W. Cullen, *RCA Rev.* **31**, 702 (1970).
377. R. E. Enstrom, J. J. Hanak, J. R. Appert, and K. Strater, *J. Electrochem. Soc.* **119**, 743 (1972).
378. G. Oya and Y. Onodera, *Jpn. J. Appl. Phys.* **10**, 1485 (1971).
379. L. R. Newkirk, F. A. Valencia, and T. C. Wallace, *J. Electrochem. Soc.* **123**, 425 (1976); see also ref. 12, p. 704.
380. C. F. Wan and K. E. Spear, ref. 13, p. 47.
381. J. L. Vossen, *Phys. Thin Films* **9**, 1 (1977).
382. H. Gurev, ref. 11, p. 321.
383. Numerous articles on CVD of boron and titanium carbide appearing in refs. 9–13.
383a. L. Vandenbulcke and G. Vuillard, *J. Electrochem. Soc.* **123**, 278 (1976); **124**, 1931, 1937 (1977).
384. L. C. McCandless and J. C. Withers, ref. 9, p. 423.
385. E. Fitzer and M. Rohm, ref. 11, p. 133.
386. H. E. Hintermann, R. Bonetti, and H. Breiter, ref. 11, p. 536.
387. M. Ducarroir and C. Bernard, ref. 12, p. 72.
388. J. Lindstrom, K. Fundell, and A. Lind, ref. 11, p. 546.
389. M. L. Lieberman, ref. 10, p. 95.
390. R. P. Gower and J. Hill, ref. 12, p. 114.
391. J. Guilleray, R. L. R. Lefevre, M. S. T. Price, and J. P. Thomas, ref. 12, p. 727.
392. R. B. Kaplan, ref. 10, p. 176.
393. B. Armas, C. Combescure, and F. Trombe, *J. Electrochem. Soc.* **123**, 308 (1976); see also ref. 12, p. 695.
394. A. J. Caputo, *Thin Solid Films* **40**, 49 (1977).
395. D. E. R. Kehr, ref. 13, p. 511.
395a. P. A. Tick, N. W. Wallace, and M. P. Teter, *J. Vac. Sci. Technol.* **11**, 709 (1974).
396. B. Armas and C. Combescure, ref. 13, p. 181.
397. J. M. Harris, H. C. Gatos, and A. F. Witt, ref. 9, pp. 795, 799.
398. J. E. Spruiell, ref. 9, p. 279.
399. J. R. Weiss and R. J. Diefendorf, ref. 11, p. 488.
400. D. Dong, E. A. Irene, and D. R. Young, ref. 13, p. 483.
401. H. Mochizuki, T. Aoki, H. Yamoto, M. Okayama, M. Abe, and T. Ando, *Jpn. J. Appl. Phys.* **15**(Suppl.), 41 (1976).
402. W. J. Heffernan, I. Ahmad, and R. W. Haskell, ref. 11, p. 498.
403. T. M. Besman and K. E. Spear, *J. Cryst. Growth* **31**, 60 (1975).
404. H. O. Pierson and E. Randich, ref. 13, p. 304.
405. G. Blandenet, Y. Lagarde, J. P. Morlevat, and G. Uny, ref. 13, p. 330.
406. J. Cueilleron, G. Lahet, F. Thevenot, and R. A. Pâris, *J. Less-Common Met.* **24**, 317 (1971).
407. T. M. Besmann and K. E. Spear, *J. Electrochem. Soc.* **124**, 786, 790 (1977).
408. H. Gass and H. E. Hintermann, *Thin Solid Films* **40**, 81 (1977).
409. G. F. Wakefield, C. L. Yaws, and J. A. Bloom, ref. 11, p. 173.
410. C. L. Yaws and G. F. Wakefield, ref. 11, p. 57.
411. H. O. Pierson, *Thin Solid Films* **40**, 41 (1977).
412. G. F. Wakefield and J. A. Bloom, ref. 10, p. 397.
413. N. Kikuchi, H. Doi, and T. Onishi, ref. 13, p. 403.
414. J. J. Nicki, K. K. Schweitzer, P. Luxenberg, and A. Weiss, ref. 10, p. 4.
415. K. K. Schweitzer and J. J. Nicki, ref. 13, p. 417.
416. J. J. Nicki and K. K. Schweitzer, ref. 9, p. 297.
417. H. Mantle, H. Gass, and H. E. Hintermann, ref. 12, p. 540.

418. N. J. Archer, ref. 12, p. 556.
419. J. S. Lo, R. W. Haskell, J. G. Byrne, and A. Sosin, ref. 11, p. 74.
420. Articles appearing in ref. 11: T. C. Wallace, p. 91; A. R. Driesner, E. K. Storms, P. Wagner, and T. C. Wallace, p. 473; J. R. Weiss and R. J. Diefendorf, p. 488; H. Gass and H. E. Hintermann, p. 563.
421. C. M. Hollabaugh, R. D. Reiswig, P. Wagner, L. A. Wahman, and R. W. White, ref. 13, p. 419.
422. A. G. Fitzgerald and R. Engin, *Thin Solid Films* **20**, 317 (1974).
423. B. F. Stein, *J. Appl. Phys.* **42**, 2336 (1971).
424. D. M. Heinz, P. J. Besser, J. M. Owens, J. E. Mee, and G. R. Pulliam, *J. Appl. Phys.* **42**, 1243 (1971).
425. M. Robinson, A. H. Bobeck, and J. W. Nielson, *IEEE Trans. Magn.* **7**, 464 (1971).
426. M. E. Cowher, T. O. Sedgwick, and J. Landerman, *J. Electron. Mater.* **3**, 621 (1974).
427. J. E. Mee, G. R. Pulliam, J. L. Archer, and P. J. Besser, *IEEE Trans. Magn.* **5**, 717 (1969).
428. P. Chaudhari, J. J. Cuomo, R. J. Gambino, and E. A. Giess, *Phys. Thin Films* **9**, 263 (1977).
429. E. J. Lehmann, "Chemical Vapor Deposition," NTIS/PS-75/477 (1977), Natl. Tech. Inf. Serv., U.S. Dep. Commer., Springfield, Virginia.

Part IV

PHYSICAL-CHEMICAL METHODS OF FILM DEPOSITION

IV-1

Plasma Deposition of Inorganic Thin Films

J. R. HOLLAHAN AND R. S. ROSLER*

Applied Materials Incorporated
Santa Clara, California

I. INTRODUCTION

The technology of thin films produced by plasma deposition processes is fast outrunning the scientific understanding of their detailed properties. The physics of the deposition process, the chemical knowledge of the kinetics of formation, and the nature of film microstructural detail are just now emerging. While not a new field, plasma deposition has yet to reach the state of refinement attained in sputtering and in conventional chemical vapor deposition (CVD) processes. The production of thin film materials by electric discharge (plasma) processes has long been a laboratory study

* Present Address: Motorola, Inc. Pheonix, Arizona 85002.

of interest, with published works appearing before the turn of the century. The majority of these studies relate to the formation of thin organic polymeric films, largely because of the ready availability of the organic reactant gases. Early papers on inorganic thin films produced in a plasma appeared in the 1930s. Studies and publications on both inorganic and organic films accelerated as these films became technologically important in such diverse areas as optical coatings, packaging, and in films in microelectronics.

It is the purpose here to summarize what is known about the different film materials that can be produced, to describe the probable pathways for their formation, and to note some of their general properties. A complete list of the properties found cannot be described here. Rather, reference to the original papers must be made since emphasis will be placed on the methods for film deposition and the attendant plasma conditions for the production of various types of films.

The uniqueness of the plasma to generate chemically reactive species at low temperatures is due to the nonequilibrium nature of the plasma state. By nonequilibrium, we mean a gas plasma typically sustained at 0.1 to several Torr, which exhibits temperatures of the free electrons of tens of thousands of degrees Kelvin, while the temperature of the translational and rotational modes of free atoms, radicals, or molecules will be only hundreds of degrees Kelvin. A good and concise discussion of the principles of generating low temperature plasma species, their ion and neutral balance, and their continuous loss by diffusion, attachment, and recombination may be found in Kaufman [1]. No attempt will be made here to review the general fundamentals of plasma chemistry since works are available which accomplish this in detail [2, 3].

One of the prime motivating factors in utilizing plasma deposition processes is that the substrate temperature can be kept relatively low, typically 300°C or lower. Conventional CVD processes usually require temperatures substantially higher that may be inappropriate for certain substrate materials or device structures. The films that are deposited by plasma reactions are usually amorphous in nature, with very little short-range structural ordering. The stoichiometry of the films can be made to vary in a controlled fashion by variation of dominant plasma parameters, such as reactant gas flow ratios. Because there is a range in film stoichiometry, it is to be expected, and is indeed observed, that electrical, mechanical, and chemical properties can also vary. It has been suggested by some authors that some of these materials may be referred to as "polymeric" in nature, since their structures are in most cases very likely three-dimensional networks of randomly bonded atoms and, possibly, molecular pendants. Hence, without crystalline order, structural detail can only be inferred from optical, spectroscopic, and electrical measurements.

II. EXPERIMENTAL REQUIREMENTS AND TECHNIQUES

A. Dominant Plasma Parameters

Anyone becoming familiar with plasma thin film deposition processes will realize that there are as many deposition reactor designs and experimental conditions to consider as there are published studies describing plasma deposited thin film materials. This makes the comparison of data concerning properties and measurements of the film materials difficult, if not impossible. With the exception of production type reactors now becoming widely used, plasma reactors in which thin films have been produced are small laboratory systems with a limited usable substrate area. Obviously, it is not possible to describe or catalog here in detail all of the designs and experimental approaches that have been taken; rather, general principles and some laboratory examples will be given to exemplify the requirements and techniques of plasma thin film deposition.

The results of thin film deposition (i.e., film properties) will be subject to many variables which may be interdependent: reactor geometry, electrode configuration and separation, power level and frequency, gas composition and flow rate, gas flow direction and flow pattern, pressure, substrate temperature, and any added gas diluents. Table I categorizes the major parameters controlling plasma deposition. Of the three major groups affecting deposition, the electrical element of plasma systems is the least well understood. For example, although the literature described plasma deposition occurring under a wide variety of source frequencies, practically no published studies exist which describe film properties resulting from a systematic variation of plasma frequency for a given experimental setup and process. There are indications that the level of frequency can affect some materials properties.

B. Deposition System Requirements

1. General Requirements

Every glow discharge system for thin film production is comprised of the following basic components: (1) the reactor or deposition chamber; (2) electronics, i.e., power generation source, usually at some fixed frequency; (3) a possible impedance matching network to transfer the power more efficiently from the generator to the gas load; (4) gas flow regulators

Table I

Parameters Controlling Plasma Deposition[a]

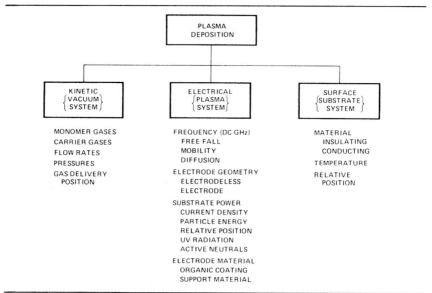

[a] After M. Hudis [3a].

(automatic or manual, such as rotameters) and gas control section; (5) pressure measurement plus other panel instrumentation. Commercial units exist for plasma dry etching and stripping or cleaning. However, those systems have, at present, no capability for inorganic thin film deposition, usually because of the barrel design approach; but they do have the above mentioned components conveniently packaged into one system.

2. Electrode Designs

Electrode design approaches for achieving the power to the plasma at the voltage level suitable for gas breakdown are shown in Fig. 1. Many variations are possible on these basic configurations. The electrode shown in Fig. 1a has been found suitable for larger scale production-type systems for deposition (see Section III.A.2). In the latter configuration, the substrates usually lie flat (for heating purposes), while in Figs. 1b and c, substrates normally are in the stand-up position.

While the tube design is satisfactory for photoresist stripping and some etching applications [4, 5] where the substrates typically are in the stand-up position, the requirements for thin film deposition are best met

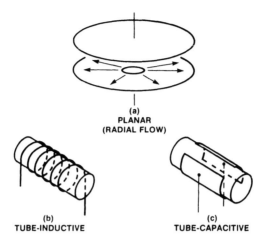

**(b)
TUBE-INDUCTIVE**   **(c)
TUBE-CAPACITIVE**

Fig. 1. Various basic types of electrode configurations. In (a) the electrodes are normally inside the reactor directly interfacing with the plasma, while in (b) and (c) the electrodes are external to the plasma process (electrodeless excitation).

by the planar reactor approach. There are various reasons for this: (1) the substrates usually lie flat for ease of heating the substrate uniformly during deposition; (2) the substrates themselves and substrate-to-substrate relative positions experience a high degree of electric field uniformity, and hence power density (watts per square centimeter) distribution. As a result, the deposition proceeds quite uniformly. In tube-type reactors, the substrates are generally difficult to heat uniformly to temperatures of 300–350°C required for proper film densities. Additionally, the tube design is usually provided with either external inductive or capacitive electrode geometries for plasma generation, and this can lead to an asymmetric electric field contour across the wafer and from front to back of the vertically standing wafers, resulting in significant radial and longitudinal thickness variations. This nonuniformity has been observed with etching processes.

3. Power Supplies

Power supplies, either at high or low frequency, are commercially available in a large range of power level outputs. The basic decision that has to be made is the choice of frequency, if information should exist on an optimum frequency for the particular process in question. Deposition rate does appear to be affected by the pulsing of the high frequency power instead of operating at continuous wave output. This is certainly true for organic compounds in plasma polymerization [6], and very likely holds true for inorganic systems. In the case of organic compounds, if the power were pulsed, the polymerization rate could be either increased or decreased by as much as 100%, depending on the structure of the starting compound.

4. Vacuum Requirements

Most plasma systems do not require critical high vacuum or pumping components. Usually, process pressures are in the range 0.1–1.0 Torr at total gas flows of a few to several hundred cubic centimeters per minute (STP), i.e., "soft" vacuum. Thus, diffusion pumps are not necessarily required, although these have been incorporated in some experiments to establish an initial high vacuum ($10^{-5}–10^{-6}$ Torr) prior to the start of the experiment. Often a rotary mechanical pump is all that is used. To prevent backstreaming of the high-vapor-pressure fluids used in these pumps, it is necessary to insert a trap in the vacuum line and/or gas ballast the pump so that its inlet pressure never exceeds about 150 mTorr. In addition, safety precautions are required when certain gases are used. Oxidation resistant pump fluids must be used with oxidizing gases (e.g., O_2, F_2). If the plasma reactions are expected to liberate large quantities of explosive gases (e.g., H_2), the exhaust of the pump may have to be diluted (e.g., with N_2).

In general, the choice of pumping system must be considered carefully based on considerations of pumping rate for all gases introduced and generated in the process, inlet flow rates, background contamination from residual gases outgassing and backstreaming, and safety.

5. System Examples

Typical systems which may be described are depicted in the following figures. In Fig. 2 [7] a straightforward vertical tube reactor is shown that is powered inductively by an rf source. In this case, silicon nitride films are produced on a small heated sample support pedestal. These smaller test reactors are usually made of quartz; they are convenient and of low cost. In Fig. 3 a vertical tube reactor is shown, again suitable for silicon nitride formation [8]. In this case, a microwave power supply is utilized with the power being delivered to the cavity out of the direct substrate region. The coil around the reactor is for the inductive heating of the substrate. The silicon-containing starting compound SiI_4, manufactured in situ at point C by reaction of silicon with iodine at point B, enters the reactor encountering a downward flowing nitrogen plasma produced in the cavity.

Another example is that of Fig. 4 [9], in which epitaxial silicon is being

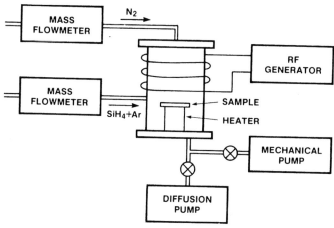

Fig. 2. Simple laboratory coil-excited vertical tube reactor with a sample support pedestal which can be heated (reprinted from Helix *et al.* by permission of the publisher, The Electrochemical Society, Inc.).

grown at several hundred degrees centigrade with substrates on a heated susceptor. At the same time, high frequency breakdown of the gas is accomplished by circular capactive band electrodes on either side of the inductive heating coil. This method couples both thermal and plasma assisted reactions at the same time to produce films of good quality.

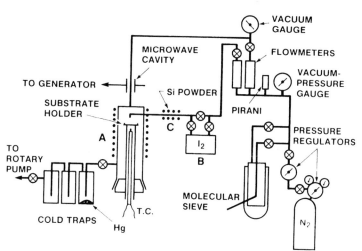

Fig. 3. Vertical tube reactor with microwave excitation of the plasma indirectly from the substrate location (Shiloh *et al.* [8] originally presented at the 152nd Fall Meeting of the Electrochemical Society, Inc., in Atlanta, Georgia).

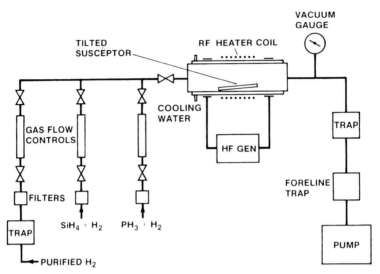

Fig. 4. Horizontal tube reactor around which an rf heater coil is placed. At either end of the coil are located electrode bands to provide the plasma throughout the reactor zone (after Townsend and Uddin [9], by permission of the publisher, Pergammon Press, Ltd.).

6. *Process Control*

The control of plasma-enhanced CVD processes is assuming increasing importance, especially with the advent of production-type reactors. Although very little has been published in this area specifically for plasma processes, process control should be achievable by the use of film thickness monitors, mass spectroscopy, and optical spectroscopy for detecting gas phase intermediate species and for monitoring their concentration. Measurement of film composition is another monitoring approach. For example, in the case of plasma "silicon nitride" films, correlation of optical absorption edges with composition in terms of Si/N ratios can be used [10]. A rapid optical inspection of the film produced after each deposition run provided better assurance that film qualities are within specification.

III. DEPOSITION OF THIN INORGANIC FILMS

A. Silicon Nitride

Plasma deposition of "silicon nitride" films from SiH_4, N_2, and/or NH_3 has been extensively studied and is now at the stage of production applications in semiconductor device manufacture, mainly for device final passivation. Films on the order of 3000–10,000 Å thickness are excellent diffusion masks for alkali ion contaminants and other impurities. The recognition of the potential of low-temperature deposited nitrides to·semiconductor applications is not new. The delay in the widespread use for this material has resulted, in part, not from an emphasis on competing technologies but, more particularly, from the lack of equipment adequate to the film uniformity and production requirements of the industry. The chemistry of plasma silicon nitride deposition is extremely complex and the detailed reaction kinetics are not understood, largely because such deposition reactions are difficult to probe diagnostically as opposed to processes taking place exclusively in the gas phase.

"Silicon nitride" produced via plasma is quite unlike silicon nitride produced by conventional chemical or physical vapor deposition techniques. In the latter case, the stoichiometry is virtually assured to be Si_3N_4, while in the former process the stoichiometry (Si/N ratio) can be controllably varied depending on ratios of entrant reactant gas flows, power level, substrate temperature during deposition, and reactor pressure.

1. Film Deposition and Properties

Amorphous silicon nitridelike films can be produced by reaction of nitrogen and/or ammonia with silane or other substituted silane or derivative reactants. Nitrogen atoms derived from ammonia may be more advantageous in reaction systems because of the lower ionization energy of NH_3 relative to N_2. The relevant appearance potentials (AP) are given in Table II [11].

If we assume a predominately silicon nitride phase in the solid material and propose kinetic pathways for the formation of Si—N, Si=N, or Si≡N bonds, we immediately arrive at an overwhelming number of reaction possibilities. This is because the gas phase dissociation of silane produces SiH_3, SiH_2, and SiH, and dissociation of nitrogen or ammonia produces N or NH_2 and NH.

Table II

Appearance Potentials for Nitrogen Containing Ions from NH_3 and N_2 [a]

Parent molecule	NH_3			N_2	
Ion	NH_3^+	NH_2^+	NH^+	N_2^+	N^+
AP (eV)	10.5	15.8	19.5	15.6	24.3

[a] After Reinberg [11].

Such speculative intermediate reactions involving N or NH_2 species as

$$SiH_4 + N \quad \rightarrow SiN + 2H_2$$
$$SiH_4 + NH_2 \rightarrow SiN + 3H_2$$
$$SiH_3 + N_2 \quad \rightarrow SiN + NH_3$$
$$SiH_3 + NH_2 \rightarrow SiN + \tfrac{5}{2}H_2$$

may be important, but undoubtedly are oversimplifications of the true reaction processes. This is evidenced by the fact that plasma "silicon nitride" films can vary in stoichiometry, contain large amounts of hydrogen, and may incorporate impurity constituents, such as a few percent oxygen from the background gases or water and possible carbon content from background hydrocarbons (pump oil, for example). It is possible then that the "silicon nitride" film has a compositional nature of $Si_aN_bO_x$-H_yC_z, where indeed x, y, and z may be very small numbers and a and b are not necessarily 3 and 4, respectively. A similar argument for departure from ideal stoichiometry likely holds for plasma produced films in general.

In early studies in small laboratory reactors, usually of the tube type, reports on silicon nitride deposition described the macroscopic properties of the film material. Only recently have more detailed analyses of the films evolved along with an attempt to infer the chemistry of film formation from the gas phase. The number of published reports describing silicon nitride production is now steadily increasing.

The initial nitride deposition process is generally attributed to Sterling and Swann [12], who in 1965 reported on films of "silicon nitride" they had produced in a quartz tube by decomposition of SiH_4 and NH_3 gas mixtures in a 1-MHz rf discharge at 500 W and 0.1 Torr. Deposition rates of 500 Å/min were measured.

This initial work briefly presented macroscopic property measurements of plasma nitride as well as of plasma silicon oxide and was followed by an expanded report [13]. Additional detailed reports appeared [14–19] which reported laboratory studies in unique reactor designs.

It is not possible to compare all data on an absolute basis in these references from one reactor study to another because of differences in experimental design. The comparison of deposition data or plasma parameters becomes more meaningful with the advent of parallel plate reactors which have become reasonably standardized in production of films on a larger scale basis (see Section III.A.2). Although diverse reactors, plasma source frequencies, electrode configurations, power, pressure, and flow ranges have been investigated, some unmistakably consistent trends are evident in plasma nitride film properties. The refractive index can often serve as one criterion of film quality, and correlate with electrical properties (dielectric constant, breakdown strength, etc.), mechanical, and chemical properties, such as etch rate.

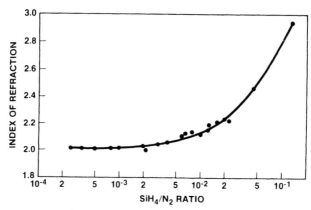

Fig. 5. Variation of the refractive index with the SiH_4/N_2 ratio (after Gereth and Scherber [16], by permission of the publisher, The Electrochemical Society, Inc.).

As representative examples [16], Fig. 5 shows the variation of refractive index with the SiH_4/N_2 ratios of the entrant gases. At higher SiH_4 flows, the films are becoming quite rich in silicon. Therefore, the SiH_4/N_2 ratio must be kept relatively low for the usually required refractive index range (2.0–2.1) to be achieved. The data of Fig. 5 were obtained from a quartz reactor 90-cm long and 7 cm in diameter. The plasma of 1–5% SiH_4 in N_2 was established at 500 kHz. Growth rates at a substrate temperature of 500°C up to 400 Å/min were measured.

Figure 6 shows the refractive index and deposition rate dependence on total pressure [20]. These data were obtained from a large 66-cm parallel plate reactor (to be described in detail in Section III.A.2). The refractive index is shown not to be very sensitive to total pressure. Figure 7 shows the index dependence on substrate (silicon wafer) temperature [20]. Above a temperature of 300°C, the index remains essentially constant, and a 50°C change in temperature causes only a 3% rate change.

Plasma silicon nitride invariably contains large amounts of hydrogen, which can be removed to some extent by annealing. The lower the deposition temperature, the greater will be the amount of entrained or bonded hydrogen, which has been correlated with a lowering of the index of refraction and an increase in etch rate [21].

The hydrogen appears to be largely bonded to Si entities in SiH_x groups where $x = 1, 2, 3$. In a detailed infrared spectroscopic study, evidence was presented that under preparative conditions where the films are Si rich, N–H infrared absorption bands appear [21]. Plasma nitride film density, composition, and refractive index have been correlated through the Lorentz–Lorenz relationship, as shown in Fig. 8 [22]. The information in Fig. 8 is helpful for correlating refractive index with density

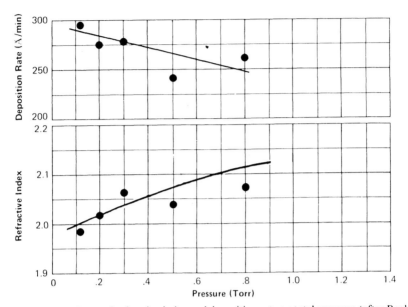

Fig. 6. Dependence of refractive index and deposition rate on total pressure (after Rosler *et al.* [20]).

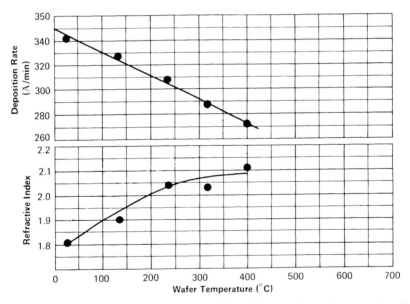

Fig. 7. Dependence of refractive index and deposition rate dependence on deposition temperature (after Rosler *et al.* [20]).

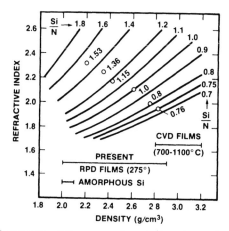

Fig. 8. Lorentz–Lorenz correlation curves for reactive plasma deposited (RPD) films of silicon nitride (after Sinha [22]).

for various Si/N ratios found in the film, which relates directly to the Si/N ratio of reactor entrant silicon and nitrogen containing gases.

Optical property studies of silicon nitride have shown [23] that different refractive indices and optical gaps can be obtained for the same film stoichiometry, depending upon the film densification. The influence of the reacting gas concentration and flow rate on the variation of the optical gap (3.5–4.9 eV) and the refractive index (1.8–2.1) is interpreted [24] in a detailed study as a variation of the stoichiometries of the deposited films. Therefore, it is necessary to separate out the substrate temperature effect from other plasma parameters (flow, pressure, and power) when considering optical properties of silicon nitride, and very likely plasma-produced thin films in general.

Electron spectroscopy for chemical analysis (ESCA) or Auger characterization [25] of substrate surfaces and plasma deposited films permits one to maintain better control of film growth which are affected by interfacial phenomena, background gas contaminants, and other sources of impurity incorporation. Since the density and optical properties can vary through the film, these surface limited analytical techniques become quite valuable and provide a true profile of film composition by repeated ion etching and ESCA or Auger analysis, provided that the results are interpreted to account for artifacts introduced in these measurements by the ion etching.

Silicon nitride films have also been prepared by reaction or active nitrogen and SiI_4 according to the overall representative reaction [8]

$$4N + 3SiI_4 \rightarrow Si_3N_4 + 6I_2.$$

which was found to be exothermic. The SiI_4 was prepared *in situ* outside the plasma region by passing I_2 vapor over heated silicon powder. The product then entered the reactor which contained a microwave excited nitrogen plasma. No provision was made to heat the substrate to elevated temperatures, and except for infrared spectra, no physical or chemical evaluation has yet been published on nitride prepared via this plasma process. Infrared spectra showed basically the same dominant Si–N absorption based centering at ~ 870 cm^{-1} that is found for CVD films produced from SiH_4–NH_3 and reactive sputtering of Si in N_2. Rather high rates of nitride film formation were observed (> 3300 Å/min) to produce films of 1.4 to 2.0 index of refraction.

The deposition of silicon nitride from SiH_4 has been described [26] using a microwave excited plasma produced at some distance from the substrate, thus minimizing the direct field of excitation around the substrate. At 680 W, a substrate temperature of 350°C, and a SiH_4 concentration ranging from 1 to 20% in N_2, deposition rates approaching 400 Å/min were obtained. A maximum of 1.95 was found for the refractive index. The significance of these types of experimental arrangements is that the substrate is out of the field of direct plasma generation where this may be a problem for very radiation sensitive substrates.

A phenomenon of a weak visible luminescence appearing over the surface of silicon wafers was attributed to emissions from silicon and nitrogen atoms by correlation of the emission band frequencies with known values. If this is true, the following might be proposed:

$$SiH_4 \rightarrow Si + 2H_2,$$
$$2N \rightarrow 2N \quad ,$$
$$Si + N \rightarrow SiN \quad ,$$
$$SiH_4 + N \rightarrow SiN + 2H_2, \quad etc.$$

2. Production Reactors

The deposition of "silicon nitride" films has now been optimized for production reactors designed for multisubstrate throughput (e.g., silicon device wafers). Reinberg first announced the radial flow reactor design in 1973 [27]. A variation on this concept was later commercialized [20]. In the latter case, the reactor is a parallel plate system, as shown in cross-sectional form in Fig. 9.

The reactor, constructed largely of aluminum, contains two parallel 66-cm diameter electrodes separated by a distance of 5 cm. The high-voltage upper electrode is connected through a matching network to a 5 kW, 50 kHz power supply. The lower electrode, which serves as the substrate support, is grounded. A magnetic drive assembly permits rotation of this

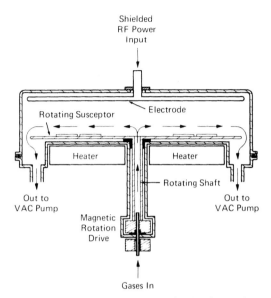

Fig. 9. Planar, radial flow reactor (after Rosler *et al.* [20]).

lower electrode to permit randomization of substrate position for maximum substrate-to-substrate uniformity. This electrode is typically heated
to 300–325°C by an externally mounted three-zone heater.

The reactant gas mixture ($SiH_4/NH_3/N_2$) enters upward through the
rotating central shaft and flows radially outward between the two electrodes. Reaction by-products and unreacted gases exhaust into a circum-
ferential plenum below the outer edge of the grounded electrode and then
through four uniformly spaced exhaust ports to the pumping system.

Depending on gas flow and composition, the deposition rate can be
strongly dependent or independent of power level above a certain point.
At higher SiH_4 flow rates, this relationship is linear within the range of the
reported experimental conditions. At lower SiH_4 flows, particularly when
combined with reduced total gas flow, the rate of film deposition tends
toward an asymptotic value at higher power input levels. Reactor characterization for deposition of materials other than silicon nitride would have
to take these possibilities into consideration. Under flow conditions that
provide uniform deposition, the rate of film deposition can be controlled
by variation in the input power without materially affecting the radial uniformity of deposition. The conditions of 600 cm³/min N_2, 300 cm³/min
NH_3, and 150 cm³/min SiH_4 at 500 W provides a ± 1.6% film thickness
uniformity in the deposition zone. Deposition rates of 300–350 Å/min can

Table IIIa

Physical Properties of Silicon Nitride Films from $SiH_4 + NH_3 + N_2$[a]

Property	Si_3N_4 HT–CVD–NP 900°C	$Si_xN_yH_z$ PE–CVD–LP 300°C
Density	2.8–3.1 g/cm³	2.5–2.8 g/cm³
Refractive index	2.0–2.1	2.0–2.1
Dielectric constant	6–7	6–9
Dielectric strength	1×10^7 V/cm	6×10^6 V/cm
Bulk resistivity	10^{15}–10^{17} Ω-cm	10^{15} Ω-cm
Surface resistivity	$>10^{13}$ Ω/sq	1×10^{13} Ω/sq
Stress at 23°C on Si	1.2–1.8×10^{10} dyn/cm²	1–8×10^9 dyn/cm²
	Tensile	Compressive
Thermal expansion	4×10^{-6}/°C	$>4 < 7 \times 10^{-6}$/°C
Color, transmitted	None	Yellow
Step coverage	Fair	Conformal
H_2O permeability	Zero	Low–none
Thermal stability	Excellent	Variable >400°C

Table IIIb

Chemical Properties of Silicon Nitride Films from $SiH_4 + NH_3 + N_2$[a]

Property	HT–CVD–NP 900°C	PE–CVD–LP 300°C
Composition	Si_3N_4	$Si_xN_yH_z$
Si/N ratio	0.75	0.8–1.0
Solution etch rate		
HFB 20–25°C	10–15 Å/min	200–300 Å/min
49% HF 23°C	80 Å/min	1500–3000 Å/min
85% H_3PO_4 155°C	15 Å/min	100–200 Å/min
85% H_3PO_4 180°C	120 Å/min	600–1000 Å/min
Plasma etch rate		
70% CF_4/30% O_2, 150 W, 100°C	200 Å/min	500 Å/min
Na^+ penetration	<100 Å	<100 Å
Na^+ retained in top 100 Å	>99%	>99%
IR absorption		
Si–N max	~870 cm⁻¹	~830 cm⁻¹
Si–H minor	—	2180 cm⁻¹

[a] After Kern and Rosler [28].

be attained. A ± 40% change in power level causes only a modest increase in the radial uniformity profile.

As the pressure in the reactor increases, the deposition rate decreases steadily over the range studied of 0.1–0.8 Torr. This general behavior is not necessarily common in plasma deposition. However, it can result from depletion and gas phase recombination at higher pressures, reducing the availability of reactive film-producing species. Typical chemical and physical properties of plasma deposited nitride produced in the planar reactor are listed in Tables IIIa and IIIb [28].

B. Silicon Oxide and Oxynitride

1. Silicon Oxide

Unless great care is taken with the vacuum and wall conditions during plasma deposition, most silicon oxides will contain a small, but measurable, amount of nitrogen derived from background sources. The reactive competition of oxygen for silane, however, is much greater than that of nitrogen. Although several publications describe the plasma deposition of ''SiO_2,'' in many instances the true stoichiometry was never determined. As with the plasma nitride, an amount of hydrogen is certain to exist, with possible carbon content as well.

Silicon oxides may be prepared by:

(1) Coreaction of silane or silane derivatives with oxygen, N_2O, CO, or other sources of oxygen.

(2) Decomposition of an oxygenate, such as $Si(OR)_4$, where R is an organic alkyl group, such as $—C_2H_5$ or $—CH_3$.

(3) *In situ* film growth on the substrate, i.e., ''plasma anodization.''

The basic laboratory or production reactor design that has been described for the deposition of the plasma nitride in most cases is suitable for oxide deposition. Other than the gas composition, major plasma parameters (power, pressure, and total gas flows) may not vary appreciably from nitride deposition requirements for a given reactor configuration.

The reaction of nitrous oxide with silane in the overall reaction

$$SiH_4 + 2N_2O \rightarrow SiO_2 + 2N_2 + 2H_2$$

is among the earliest studied plasma reaction system to produce SiO_2 [12]. The possible mechanism for the reaction might be explained as follows, consistent with bond energy considerations. Nitrous oxide has two resonant electronic structures, $^-N{=}N^+{=}O$ and $N{\equiv}N^+{—}O^-$. The nitrogen–nitrogen bonds have a higher bond energy than nitrogen–oxygen or silicon–hydrogen bonds. If SiH_4 is predominantly dissociated in the plasma, SiH_2 or SiH radicals could form intermediate complexes of the types, as shown below, which decompose to SiO_2 and N_2 + H_2.

$$N\equiv N-O\cdots\underset{\displaystyle H}{\overset{\displaystyle H}{\underset{|}{\overset{|}{Si}}}}\cdot \quad \text{and} \quad N\equiv N-O-\underset{\displaystyle H}{\overset{\displaystyle H}{\underset{|}{\overset{|}{Si}}}}-O-N\equiv N, \quad \text{etc.}$$

Nitrogen–oxygen single bond energies are low by a factor of 2 or 3 compared to Si–O bond energies so that Si–O bonds may form. What part the substrate surface versus the gas phase contributes to the process is unknown, at least in the published literature.

Radio frequency discharge of N_2O + SiH_4 [12] produced silicon oxide films readily up to 1.0 μm in thickness. Again, as the relative gas flow ratios are varied, the electrical and structural properties also vary. Dielectric constant values varied from 3.8 to 10 in the above referenced study for N_2O/SiH_4 ratios ranging from 1:1 to 10:1. Over a similar range of gas ratios, the refractive index varied from 1.46 to 1.9. Infrared studies [21] showed predominantly Si—O bonds at 9.3, 12.4, and 22 μm.

Silicon tetrachloride reacting with an oxygen plasma, which was sustained by a microwave discharge, produced glassy SiO_2 films up to several millimeters in thickness in small diameter quartz tubes at 950–1000°C deposition temperature [29] by the overall reaction

$$SiCl_4 + O_2 \rightarrow SiO_2 + 2Cl_2.$$

In an oxygen plasma, the primary reactive species is $O(^3P)$, and, in addition, a rather long lived excited molecular oxygen state exists [$O_2(^1\Delta_g)$, ~0.96 eV] in appreciable concentration [3]. This species has been termed "singlet" oxygen and undoubtedly also participates in the formation of oxide films, if oxygen plasmas are utilized.

If TEOS [tetraethoxy silane, $Si(OC_2H_5)_4$] is decomposed in the presence of an oxygen plasma amorphous silicon oxide films are formed [30–32]. Tetraethoxy silane is a liquid and must be transported by a carrier gas (O_2) into the reaction zone.

If TEOS is decomposed with argon as the carrier gas, organosilicon films are formed. A detailed infrared study [30] of films prepared at 1 MHz and 0.2 Torr total pressure in a vertical tube reactor showed evidence of CH_2, CH_3, Si–OH, Si–O–C_2H_5, and many other hydrocarbon moieties. The primary mechanism for deposition then appears to be based on an oxygen plasma combustion of the hydrocarbon part of the molecule to CO_2 and H_2O, with the remaining SiO_x fragments probably forming types of cluster complexes with adsorption and growth on the substrate surface. Excess TEOS vapor (that is, more than can be consumed by the oxidizing plasma) will lead to a hydrocarbon content in the film depending on the substrate surface temperature. If the temperature is $\sim 500°C$, then simultaneous plasma decomposition and pyrolysis of the excess TEOS will occur, leading to a more nearly SiO_2 stoichiometry.

In a microwave discharge (2450 MHz) study [33], vapors of either TEOS or tetraethyl silane were decomposed in an oxygen plasma to produce films on substrates at 200°C. At 100 W and 0.24 Torr the deposition rates were rather low, about $10-80$ Å/min. A bell jar apparatus was employed permitting the use of a quartz microbalance to measure film growth.

Oxide films prepared from $SiH_4 + N_2O$ under plasma conditions are reported [34] to be better suited for use in semiconductor fabrication than those deposited from TEOS. Hydrogen and hydrocarbon content is lower at a deposition temperature of 450°C, and the films are reportedly cleaner. By whatever preparative technique employed, SiO_x films do not exhibit the excellent alkali ion diffusion barrier properties displayed by the silicon nitride films.

2. Silicon Oxynitride

Nitrogen atoms may be included into an SiO_x network by such plasma processes as coreaction of $SiH_4 + NO + NH_3$, or reactively evaporating SiO with an ionized plasma of nitrogen [35]. This latter reaction produces a SiO_xN_y structure that has described [35] as a glassy mixture of elemental Si, N, and O, rather than a mixed phase of SiO_2 and Si_3N_4. As the amount of N is increased, infrared absorptions of Si–N appear while Si–O intensities decrease. Multiple internal infrared reflection measurements have shown [36] as much as 7% hydrogen concentrations. Hydrogen bound to silicon can be annealed out after annealing 1 hr at 900°C, with a 20% reduction in the N–H bond centers [36]. Thus, a very complex structure is being produced in the formation of oxynitrides.

The same general techniques to produce silicon nitride and oxide are appropriate for the oxynitride. In spite of the complexity of the resulting material, basic properties of the film are reproducible, if plasma conditions are carefully reproduced. The refractive index is claimed, for example, to be reproducible to ± 0.01 over the range $1.5-2.0$ [34]. In principle, by adjustment of gas composition and deposition temperature at given plasma conditions, one can tune in desirable electrical and chemical properties intermediate between all-oxide and all-nitride compositions.

C. Silicon Carbide

Amorphous SiC films can be readily prepared from silane and a volatile carbon containing coreactant, such as C_2H_4 [37], CH_4 [38], or CF_4

[39]. In these examples, the overall reactions may be written

$$2SiH_4 + C_2H_2 \rightarrow 2SiC + 5H_2$$

and

$$SiH_4 + CH_4 \text{ (or } CF_4) \rightarrow SiC + 2H_2 \text{ (or 4HF)}.$$

The variations of the SiC optical gap $\mathscr{E}_g$ and the refractive index were correlated with pressure, gas flow rates, power, and substrate temperature. In a detailed reactor performance and materials study [38], $\mathscr{E}_g$ decreased from 2.5 to 2.1 eV and the refractive index increased from 1.9 to 2.1 as the SiH_4/CH_4 ratio increased from 0.24 to 0.80. Typical optimum deposition conditions in the 7-cm diameter $\times$ 40-cm length cylindrical reactor were 0.2 Torr pressure, 100 W power (5 MHz), 40 cm^3/min (STP) total flow, and a 200°C substrate temperature. It is interesting that if the pressure is above 1.0 Torr, a fine brownish powder is deposited along with the film. Rapid gas phase polymerization with nucleation of particles is occurring, similar to the case found in the plasma deposition with ethylene, where different plasma regimes would produce films, particles, or oils [40].

The optical properties of plasma SiC have been considered in some detail [37]. A general conclusion is that amorphous plasma produced thin films of SiC have an appreciably lower overall density of gap states compared to sputtered or pyrolytically deposited material. The optical gap was found to reach a maximum at a film composition of $Si_{0.32}C_{0.68}$ at deposition temperatures of 225 and 525°C.

D. Silicon and Germanium

Amorphous silicon or germanium are the simplest materials that can be produced in a plasma of their respective hydrides. Mass spectrometric fragmentation patterns reveal that the primary product ion is SiH_2^+; thus, the term "polysilane" might be appropriate for films containing large amounts of hydrogen where SiH, SiH_2, and terminal SiH_3 are certain to exist [40]. Films of low hydrogen content can be viewed as a random three-dimensional matrix of tetrahedrally bonded silicon atoms with an occasional hydrogen atom or atoms at a silicon bond site. Large amounts of hydrogen will be retained at lower deposition temperature higher-pressure regimes, while at higher temperatures and lower pressures, the amount retained is much less. For example, the atomic fraction of hydrogen in films prepared at 250°C and 0.1 Torr is 0.14 (0.7 $\times$ 10^{22} atoms/cm^3), while at 25°C and 1.0 Torr the atomic fraction is 0.35 (1.7 $\times$ 10^{22} atoms/cm^3).

The amount of hydrogen contained in amorphous silicon has been analytically followed by infrared spectroscopy, and also by the nuclear resonance reaction $^{15}N + {}^1H + {}^{13}C = {}^4HC + \gamma$, microprobe, and mass spectrographic methods of analysis [41]. Doping of amorphous silicon with phosphine or diborane was achieved [42, 43] by the plasma decomposition of silane with these dopants (5×10^{-6} PH_3 and 10^{-2} B_2H_6 per unit volume SiH_4). The substrate temperature was held at 250°C and the total plasma pressure at 0.8 Torr, which produced a relatively low 50 Å/min deposition rate. An amorphous silicon p-n junction was finally fabricated whose rectification and photovoltaic responses were qualitatively similar to those of a crystalline p-n junction.

Epitaxial silicon has been grown from SiH_4 in H_2 in the presence of an electric discharge (350 W at ~ 27 MHz) at 0.2–0.6 Torr in the apparatus discussed in Fig. 4 [9]. Epitaxial silicon can be produced, of course, in such an apparatus without an electric discharge at 1050–1200°C at 1 atm. However, when the deposition occurred in the presence of the plasma, temperatures could be reduced to as low as 800°C, while still maintaining good epitaxial growth for both doped and undoped N-type layers. That good films can be grown at 800°C was augmented by the marked reduction in stacking fault density ($\sim 20\%$ of the process without glow discharge), and improvement in surface quality of the epitaxial silicon layers was observed. The main advantage of the discharge was thought to be the continuous cleanup of the substrate provided during deposition. In addition, however, there may be energy transfer processes available not possible at 1 atm that produce films of better integrity at low pressure.

E. Other Oxides

1. Aluminum Oxide

Amorphous films of Al_2O_3 have been formed by vaporizing $AlCl_3$ into an oxygen plasma [44]. The pressure of the discharge and the $AlCl_3$ vaporization rate greatly affected the rate of film formation, which varied from 70 to 500 Å/min, and was linearly dependent on rf power over the range studied. Adherent films up to thicknesses of several microns were prepared at the optimum substrate temperature of 480°C. The Al–Al_2O_3–Si structures were characterized as having large positive flat band voltage shifts of up to 40 V, under negative bias at elevated temperature. This implies these films will show high resistance to Na^+ ion diffusion, unlike SiO_2. Other film properties such as resistivity and dielectric constant and strength were acceptable or compared well with oxide prepared from thermal decomposition of aluminum triethoxide.

2. Oxides of Ge, B, Ti, and Sn

Clear, crystalline, and glassy films of several oxides were prepared by microwave discharge decomposition of the corresponding alkyls or alkoxides [32]. Films were produced typically at 0.24 Torr on NaCl substrates (for infrared spectroscopic studies) at 200°C. Growth rates were very low, about 20 Å/min. Table IV summarizes the M–O band infrared absorption and index of refraction measurements. Thus, these oxides were formed with indices (and presumably densities) very close to their thermal counterparts.

3. Plasma Oxides by Anodization

Plasma anodization is the formation of thin oxide films by placing the metal or semiconductor substrate under bias in the field of an oxygen plasma. Thus, plasma anodization is not a deposition process but rather an *in situ* film growth process. A positive bias of a few to several hundred volts is placed on the sample to affect film growth. Ions in the discharge are now more actively participating in the film growth, in contrast to much less participation by ions on floating or grounded substrates. Reviews of plasma anodization have summarized a good deal of this work [45, 46] to which the reader is referred.

Table IV

Comparison of the Optical Properties of Plasma-Deposited and Glassy or Crystalline Oxide Films[a]

	Material	Principal infrared frequency (cm^{-1})	Index of refraction
GeO_2	Plasma-formed film	850	1.582 ± 0.002
	Fusion-formed glass	850	1.534–1.607
B_2O_3	Plasma-formed film	1350	1.470 ± 0.002
	Fusion-formed glass	1350	1.464(3)
Ti_xO_y	Plasma-formed film	950–700	1.7
TiO_2	Anatase	1200–500	1.7
Sn_xO_y	Plasma-formed film	1425	1.536 ± 0.002
SnO_2	Cassiterite	850–500	1.7
SiO_2	Plasma-formed film	1045; 800	1.458 ± 0.002
	Fusion-formed glass	1080; 800	1.458 ± 0.002

[a] After Secrist and Mac Kenzie [32].

F. Miscellaneous Films

1. Boron Nitride

Boron nitride thin films have been prepared by reacting B_2H_6 and NH_3 in a plasma at 1000°C on various substrates lying flat on a graphite susceptor in a horizontal tube reactor [47]. Only 4 W rf power at 13.56 MHz were required. Pressures ranged from 0.3 to 1.0 Torr. Transmission and reflection electron diffraction revealed that some crystallites exist within the film. The films were smooth and transparent, and exhibited better crystalline quality than BN films obtained from the same reagents by high-temperature CVD. At a $NH_3 : B_2H_6$ ratio of 8.1 : 1 nearly stoichiometric but slightly boron-rich BN was found by electron microprobe analysis. At a projected gas ratio of 7.1 : 1, stoichiometric BN should be obtained.

2. Phosphorus Nitride

By directly vaporizing elemental phosphorus in a nitrogen plasma, transparent glasslike films of P_3N_5 ranging from 0.5 to 10 μm in thickness were produced on a variety of metal and glass substrates [48]. Conditions of film formation were as follows: horizontal tube; pressure, 1.0 Torr; power, to 500 W; frequency, 25 MHz; deposition temperatures, 265°C. Electrical properties were evaluated, and permittivity at room temperature was found to be 4.4, independent of frequency. Dielectric strength values between 0.3 and 0.5 × 10^7 V/cm were found for samples which had been cycled through temperature extremes of − 196 to 230°C.

IV. CONCLUSIONS, APPLICATIONS, AND PROSPECTS

The foregoing examples indicate the diversity of thin film processes obtainable under low temperature plasma conditions. By adjusting gas flows, their ratios, power, and pressure of the plasma, one can, in principle, alter and optimize film properties at will. A major limitation today is the transition from laboratory R&D efforts to production-type or standardized reactors. While this has been accomplished for silicon nitride, many other materials which have been qualified as useful structural or functional thin films require the scaled up processor to be able to produce them in practice economically.

The major reason for the acceleration of interest in plasma processing and production-type reactors has been for the use of plasma silicon nitride films as a final device passivation overcoat. Because these nitride films are a significant improvement over doped or undoped SiO_2 films, their use is now rapidly becoming the standard within the semiconductor industry. Description of the superior passivating properties of plasma nitride with actual devices shows that its use as an overcoating of trimetal devices has eliminated electroplating between closely spaced oppositely biased conductors [49], and that it has improved the lifetime by a factor of 3 as an overcoat for plastic encapsulated nichrome link PROMS [50]. The use of plasma nitride as an excellent interlevel dielectric has been described [51, 52]. This application, if high speed (and thus a smaller dielectric constant) is desired, may be better filled by plasma silicon dioxide, provided its step coverage, pinhole density, and adhesion are comparable to plasma nitride.

Plasma nitride films have potential uses as an antireflection coating over solar cells, LED's or LCD's, or optics lenses, as an implantation annealing barrier, as III–V structure diffusion masks, and as a photomask coating for scratch protection and sticking reduction.

Plasma deposited amorphous silicon has potential as a material for low cost solar cells [53] and for the thermal growth of thin epitaxial layers of Si on Si from the amorphous phase.

There are many film materials that may yet be synthesized via plasma processes. Virtually all the refractory compounds of carbides, phosphides, silicides, borides, and sulfides can be considered via various reaction pathways.

Even more complex is the prospect of mixed or ternary element films. Silicon oxynitride has already been discussed. There are possible films of oxycarbides, sulfides, silicides, borides, etc., or other mixed, nonoxy, ternary systems. Additionally, the inclusion by vaporization of metal atoms or dopant atoms into the films certainly exists, leading to semiconducting films or films with other unusual electrical properties [54]. The number and types of films that can be produced in a low temperature plasma with their as yet unevaluated mechanical, electrical, and optical properties offer the materials scientist new dimensions of research and development.

REFERENCES

1. F. Kaufman, *Adv. Chem. Ser.* No. 80, p. 29 (1969).
2. F. K. McTaggart, "Plasma Chemistry in Electrical Discharges." Elsevier, Amsterdam, 1967.
3. "Techniques and Applications of Plasma Chemistry" (J. R. Hollahan and A. T. Bell, eds.). Wiley (Interscience), New York, 1974.
3a. M. Hudis, unpublished summary, NASA–Ames Research Center, Moffett Field, California.

4. E. O. Degenkolb, C. J. Mogab, M. R. Goldrick, and J. E. Griffiths, *Appl. Spectrosc.* **30**, 520 (1976).

5. R. Kumar, C. Ladas, and G. Hudson, *Solid State Technol.* **19**(10), 54 (1976).

6. H. Yasuda and T. Hsu, *J. Polym. Sci., Polym. Chem. Ed.* **15**, 81 (1977).

7. M. J. Helix, K. V. Vaidyanathan, and B. G. Streetman, *Electrochem. Soc. Extend. Abstr.* No. 77-2 (1977).

8. M. Shiloh, B. Gayer, and F. E. Brinckman, *J. Electrochem. Soc.* **124**, 295 (1977).

9. W. G. Townsend and M. E. Uddin, *Solid-State Electron.* **16**, 39 (1973).

10. M. J. Rand and D. R. Wonsidler, *Electrochem. Soc. Extend. Abstr.* No. 77-2 (1977).

11. A. R. Reinberg, *Int. Round Table Surf. Treat. Plasma Polymer., IUPAC, Limoges, Fr.,* (1977).

12. H. F. Sterling and R. C. G. Swann, *Solid-State Electron.* **8**, 653 (1965).

13. R. C. G. Swann, R. R. Mehta, and T. P. Cauge, *J. Electrochem. Soc.* **114**, 713 (1967).

14. Y. Kuwano, *Jpn. J. Appl. Phys.* **8**, 876 (1969).

15. Y. Kuwano, *Jpn. J. Appl. Phys.* **7**, 88 (1968).

16. R. Gereth and W. Scherber, *J. Electrochem. Soc.* **119**, 1248 (1972).

17. R. Kirk and I. I. Gurev, *Symp. Appl. Electr. Discharge Chem., Am. Inst. Chem. Eng., Annu. Meet. 64th, 1971.*

18. C. R. Barnes and C. R. Geesner, *J. Electrochem. Soc.* **107**, 98 (1960).

19. A. W. Horsley, *Electronics* January 20, p. 3 (1969).

20. R. S. Rosler, W. C. Benzing, and J. Baldo, *Solid State Technol.* **19**(6), 45 (1976).

21. E. A. Taft, *J. Electrochem. Soc.* **118**, 1341 (1971).

22. A. K. Sinha, *Electrochem. Soc. Extend Abstr.* No. 76-2, p. 625 (1976).

23. A. K. Sinha, *J. Electrochem. Soc.* **123**, 262 (1976); *J. Electron. Mater.* **5**, 441 (1976).

24. Y. Catherine and G. Turban, *Int. Round Table Surf. Treat. Plasma Polymer., 3rd, IUPAC, Limoges, Fr., 1977.*

25. P. H. Holloway and H. J. Stein, *Electrochem. Soc. Extend. Abstr.* No. 75-2, p. 218 (1975).

26. M. Shibagaki, Y. Horuke, T. Yamazaki, and M. Kashiwagi, *Electrochem. Soc. Extend. Abstr.* No. 77-2, Abstr. 152 (1977).

27. A. R. Reinberg, *Electrochem. Soc. Extend. Abstr.* No. 74-1, p. 4 (1974); U.S. Patent 3,757,733 (1973).

28. W. Kern and R. S. Rosler, *J. Vac. Sci. Technol.* **14**, 1082 (1977).

29. D. Kuppers, J. Koenings, and H. Wilson, *J. Electrochem. Soc.* **123**, 1079 (1976).

30. S. P. Mukherjee and P. E. Evans, *Thin Solid Films* **14**, 105 (1972).

31 S. W. Ing and W. Davern, *J. Electrochem. Soc.* **111**, 120 (1964).

32. D. R. Secrist and J. D. MacKenzie, *J. Electrochem. Soc.* **113**, 914 (1966).

33. D. R. Secrist, *Adv. Chem. Ser.* No. 80, p. 242 (1969).

34. R. Kirk, *in* "Techniques and Applications of Plasma Chemistry" (J. R. Hollahan and A. T. Bell, eds.), Ch. 9, Wiley (Interscience), New York, 1974.

35. M. Mashita and K. Matsushima, *Jpn. J. Appl. Phys., Suppl.* **2**, Part 1, p. 761 (1974).

36. H. J. Stein, *J. Electron. Mater.* **5**, 161 (1976).

37. D. A. Anderson and W. E. Spear, *Philos. Mag.* **35** 1A (1976).

38. Y. Catherine and G. Turban, *Int. Round Table, 3rd Symp. Plasma Chem., IUPAC, Limoges, Fr.* Pap. RT2 (1977).

39. J. R. Hollahan, unpublished results (1977).

40. H. Kobayashi, A. T. Bell, and M. Shen, *J. Appl. Polym. Sci.* **17**, 885 (1973).

41. M. H. Brodsky, M. A. Frisch, and J. F. Ziegler, *Appl. Phys. Lett.* **30**, 561 (1977).

42. W. E. Spear, P. G. LeComber, S. Kinmond, and M. H. Brodsky, *Appl. Phys. Lett.* **28**, 105 (1976).

43. W. E. Spear and P. G. LeComber, *Philos. Mag.* **33**, 935 (1976).

44. H. Katto and Y. Koga, *J. Electrochem. Soc.* **118**, 1619 (1971).
45. S. B. Hyder and T. O. Yep, *J. Electrochem. Soc.* **123**, 1721 (1976).
46. J. F. O'Hanlon, *J. Vac. Sci. Technol.* **7**, 330 (1970).
47. C. J. Dell'Oca, D. L. Pulfrey, and L. Young, in "Physics of Thin Films" (M. H. Francombe and R. W. Hoffman, eds.), Vol. 6, p. 1. Academic Press, New York, 1971.
48. S. Veprek and J. Roos, *J. Phys. Chem. Solids* **37**, 554 (1976).
49. H. Khajezadeh and A. S. Rose, *Annu. Proc. Reliab. Phys.* [*Symp.*], *15th* p. 244 (1977).
50. J. A. Ferro, *Annu. Proc. Reliab. Phys.* [*Symp.*] *15th* p. 125 (1977).
51. F. W. Hewlett, Jr. and W. D. Ryden, *Int. Electron. Devices Meet., Washington, D.C.* pp. 304–307 (1976).
52. W. D. Ryden, E. F. Sabuda, and W. van Gelder, *Int. Electron. Devices Meet., Washington, D.C.* pp. 597–600 (1976).
53. D. E. Carlson and C. R. Wronski, *Appl. Phys. Lett.* **28**, 671 (1976).
54. V. M. Kolotyrkin, A. B. Gilman, and A. K. Tsapuk, *Russ. Chem. Rev.* **36**, 579 (1967).

IV-2

Glow Discharge Polymerization

H. YASUDA

Department of Chemical Engineering
University of Missouri–Rolla, Rolla, Missouri

I. INTRODUCTION

When an organic vapor is injected into a glow discharge of an inert gas such as argon, or when a glow discharge of a pure organic vapor is created, the deposition of polymeric films onto an exposed surface is often observed. Polymer formation that occurs in such a process is generally referred to as plasma polymerization or glow discharge polymerization.

The recognition of thin film formation by glow discharge polymerization can be traced back to 1874 [1, 2]. However, in most cases the polymers were considered as by-products of an electric discharge [3-9] and, consequently, little attention was paid either to the properties of those polymers (undesirable by-products) or to the process as a means of forming polymers.

Only in relatively recent years (about the 1950s) has glow discharge been utilized in a practical way to make a special coating on metals. Once some of the advantageous features of plasma coating (e.g., flawless thin coatings, good adhesion to the substrate, chemical inertness, and low dielectric constant) were recognized, much applied research on the use of the process was done. The literature cited here [10-63] represents only some of the early investigations.

In this chapter, the significance of glow discharge polymerization as a process for forming thin films is discussed. The emphasis is placed on the process aspect, but not on reactions or mechanisms of polymerization. For details of general chemical reactions in plasma and the fundamental aspects of plasma or glow discharge, the reader is advised to consult the general references cited in Baddour and Timmins [64], McTaggert [65], Gould [66], Venugopalan [67], Hollahan and Bell [68], and Shen [69].

In order to distinguish the term *plasma polymerization,* which is used to describe a special kind of polymer formation mechanism in glow discharge, the term *glow discharge polymerization* is used in this chapter to refer to plasma polymerization in the wider meaning. Therefore, in the context used in this chapter, *plasma polymerization* refers to polymerization mechanisms that constitute a portion of the glow discharge polymerization.

II. CHARACTERISTIC ASPECTS OF GLOW DISCHARGE POLYMERIZATION

A. Glow Discharge Polymerization

Although the phenomenon of polymer formation in a glow discharge is referred to as *glow discharge polymerization,* the terminology of *polymerization* may not represent the actual process of forming a polymer, or the word "polymerization" may even be misleading. The conventional meaning of *polymerization* is that the molecular units (monomers) are linked together by the *polymerization* process. Therefore, the resultant polymer is conventionally named by "poly + (the monomer)". For instance, the polymer formed by the polymerization of styrene is named *polystyrene*. In this conventional context, polymerization refers to molecular polymerization—i.e., the process of linking molecules of a monomer.

In a strict sense, polymerization in the conventional context does not represent the process of polymer formation that occurs in a glow discharge—although such polymerization may play a role, depending on the chemical structure of a monomer and also on the conditions of the glow discharge.

In contrast to conventional polymerization—i.e., molecular polymerization—polymer formation in glow discharge may be characterized as elemental or atomic polymerization. That is, in glow discharge polymerization, the molecular structure of a monomer is not retained, and the original monomer molecules serve as the source of elements which will be used in the construction of large molecules. Therefore, glow discharge polymerized styrene is not polystyrene. Also, glow discharge polymerized benzene is not polybenzene, but glow discharge polymers of styrene and of benzene are very much alike.

Because of this characteristic nature of the polymer formation process, the starting compound cannot be considered to be the monomer in the conventional context used in relation to the corresponding polymer. A compound used in glow discharge polymerization is merely a starting material in the process.

A polymer formed by glow discharge polymerization cannot be identified by the starting material since the molecular structure of the starting material is not retained in the polymer structure. This leads to another important point—the glow discharge polymerization or polymers formed by glow discharge of a starting material are highly dependent on the system or conditions under which the polymer is formed.

In other words, glow discharge polymerized styrene is not polystyrene, and there is no material that can be fully identified as *glow discharge polymerized styrene*. Not enough emphasis has been placed on the latter aspect, perhaps due to the somewhat misleading use of the word *polymerization,* and also due to an a priori concept of *polymerization* and lack of the distinction mentioned above.

It should be noted here that a polymer similar to polystyrene can be formed by using an electric discharge process if conditions are chosen to favor conventional polymerization of styrene. Even in such a case, the polymer that is formed is generally not quite equivalent to the conventional polystyrene because polymer formation that is characterized by atomic polymerization usually occurs simultaneously. The balance between different polymer formation mechanisms is indeed an important factor which contributes to the system-dependent nature of glow discharge polymerization.

B. Overall Mechanism of Polymer Formation in a Glow Discharge

The individual steps or reactions that are involved in the process of polymer formation in a glow discharge are extremely complex; however, several important types of phenomena can be identified in order to construct a general picture of glow discharge polymerization. The process involved can be represented schematically, as in Fig. 1 [70]. Glow discharge polymerization can be considered to consist of two major types of polymerization mechanisms. The direct route is *plasma-induced polymerization* and another is *plasma polymerization*. Plasma-induced polymerization is essentially the conventional (molecular) polymerization triggered by a reactive species that is created in an electric discharge. In order to form polymers by plasma-induced polymerization, the starting material must contain polymerizable structures such as olefinic double bonds, triple bonds, or cyclic structures.

Plasma (atomic) polymerization is a unique process which occurs only in a plasma state. This polymerization can be represented by:

initiation or reinitiation
$$M_i \rightarrow M_i^*$$
$$M_k \rightarrow M_k^*$$

propagation and termination
$$M_i^* + M_k^* \rightarrow M_i - M_k$$
$$M_i^* + M_k \rightarrow M_i - M_k,$$

where i and k are the numbers of repeating units (i.e., $i = k = 1$ for the starting material) and M* represents reactive species which can be an ion of either charge, an excited molecule, or a free radical, produced from M but not necessarily retaining the molecular structure of the starting material (i.e., M can be a fragment, or even an atom detached from the original starting material).

In *plasma polymerization,* the polymer is formed by the repeated stepwise reaction described above. It should be noted that *plasma-induced polymerization* does not produce a gas phase by-product, since the polymerization proceeds via utilization of polymerizable structure. Plasma-induced polymerization may be schematically represented by a chain propagation mechanism as follows:

$$M^* + M \rightarrow MM^*$$
$$M_i^* + M \rightarrow M_i^* + 1 \qquad \text{propagation}$$
$$M_i^* + M_k^* \rightarrow M_i - M_k \qquad \text{termination.}$$

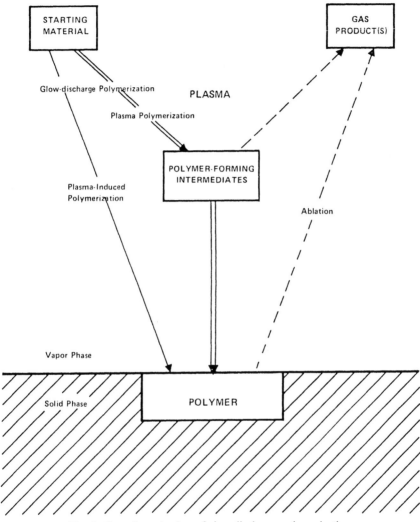

Fig. 1. Overall mechanism of glow discharge polymerization.

It should be emphasized that general polymerization in a glow discharge consists of both *plasma-induced polymerization* and *plasma polymerization*. Which one of these two polymerization mechanisms plays the predominant role in the polymer formation in a glow discharge is dependent not only on the chemical structure of the starting materials, but also on the conditions of the discharge.

Nonpolymer-forming gas products are produced in the process of forming reactive (polymer-forming) intermediates and also in the process of decomposition (etching) of the polymer deposit or of the substrate material. Since polymer-forming species do not stay in the gas (plasma) phase long enough, the major portion of the gas phase of a polymer-forming glow discharge consists of the product gas when a high conversion ratio of a starting material to a polymer is obtained. This is an extremely important factor; however, it has been dealt with lightly or has been completely neglected in most work appearing in the literature.

The characteristics of the product-gas plasma play the predominant factor in determining the extent of the ablation process, which is shown in Fig. 1. Since the majority of work on glow discharge polymerization appearing in the literature is with hydrocarbons—which produce hydrogen as the product gas—the effect of ablation happened not to be great. Consequently, the complete neglect of ablation did not make a significant difference in the overall picture of glow discharge polymerization. However, when a fluorine- or oxygen-containing compound is used as the starting material, the image of glow discharge polymerization that is built around that of hydrocarbons is completely shattered. With such a compound as the starting material, the extent of ablation becomes the predominant factor, and the extent of polymer formation is entirely dependent on the extent of the product-gas formation.

Perhaps, the most dramatic demonstration of the ablation effect has been recently shown by Kay [71] for glow discharge polymerization of CF_4. It had been thought that CF_4 was one of the very few organic compounds that does not polymerize in glow discharge. On the other hand, CF_4 has been used as one of the most effective gases in plasma etching. Kay has observed that no polymer deposition occurs under normal conditions in spite of the fact that C—F bonds are broken in the glow discharge, which is confirmed by mass-spectroscopic analysis of the gas phase. However, when a small amount of hydrogen is introduced into the discharge, deposition of polymer is observed. When the hydrogen flow is stopped, ablation of the polymer deposit occurs.

The situation observed in the above example may be visualized by the comparison of bond energies shown in Table I. It should be noted that the energy level involved in a glow discharge is high enough to break any

Table I

Comparison of Bond Energies

Bond	Energy, kcal/mole	Bond	Energy, kcal/mole
C—C	80	C—F	102
C=C	142	H—H	104
C≡C	186	F—F	37
C—H	99	H—F	135

bond [72, 73] (i.e., C—F is also broken although C—F is stronger than C—H). The important point is the stability of the product gas. The bond energy for F—F is only 37 kcal/mole, whereas H—F is 135 kcal/mole, which is higher than 102 kcal/mole for the C—F bond. The introduction of H_2 into the monomer flow evidently produces HF and removes F from the discharge system, thus reducing the etching effect by F_2 plasma. Although the term F_2 plasma is used to describe the effect of detached F in plasma, F_2 is not detected in the plasma state, perhaps due to its extremely high reactivity [74].

It is interesting to note that F and O are two elements which reduce the rate of polymer formation from compounds that contain one of these elements. These two elements are the two most electronegative elements among all elements. Of course, the bond energy itself is not a measure of the etching effect of plasma. For instance, the N—N bond is only 32 kcal/mole. However, N_2 plasma does not etch polymer surfaces; instead, the incorporation of N into the surface predominates [75]. Nevertheless, the importance of the ablation process shown in Fig. 1 seems to be well demonstrated by the poor polymer formation in glow discharge polymerization of CF_4 and oxygen-containing compounds [76].

The polymer formation and properties of polymers formed by glow discharge polymerization are controlled by the balance among *plasma-induced polymerization, plasma polymerization,* and ablation; i.e., polymer formation is a part of the competitive ablation and polymerization (CAP) scheme shown in Fig. 1. The kind of conditions that affect these balances will be explained in the following section. Because of this CAP scheme of glow discharge polymerization, gas evolved from substrate materials also plays an important role, particularly at the early stage of coating.

III. PROCESSING FACTORS OF GLOW DISCHARGE POLYMERIZATION

It is extremely important to recognize the difference between polymer-forming and nonpolymer-forming plasmas in order to understand the true meaning of the processing factors of glow discharge polymerization. Not all glow discharges yield polymer deposition. For instance, plasmas of Ar, Ne, O_2, N_2, and air are typical nonpolymer-forming plasma. The significance of polymer-forming plasmas, such as glow discharges of acetylene, ethylene, styrene, benzene, etc., is that a considerable portion or the majority of molecules of starting material leave the gas (plasma) phase and deposit as a solid polymer.

In contrast to polymer-forming plasmas, the total number of gas phase molecules in nonpolymer-forming plasmas do not change. Only a portion of gas molecules are repeating the process of being ionized, excited, and quenched. However, the total number of gas molecules remains constant. This situation can be visualized by the pressure change that occurs before, during, and after the glow discharge. In the case of a nonpolymer-forming plasma, no pressure change is observed unless a material which reacts with excited species of plasma is placed in the discharge system [77, 78]. The system pressure of a polymer-forming plasma changes as soon as discharge is initiated. The pressure change is dependent on the characteristic nature of the starting material, which is related to the product-gas formation described in the previous section. With starting materials that yield very little product gas (e.g., acetylene, benzene, styrene), the system pressure drops to nearly zero when a high polymerization yield is obtained. In other words, an efficient plasma polymerization is an excellent vacuum pump, whereas a nonpolymer-forming plasma has no characteristic of this nature.

Unfortunately, most fundamental work on the plasma state was done with nonpolymer-forming plasmas, and the concept of the operational parameters used in such studies cannot be applied directly to polymer-forming plasmas.

Characteristic polymer deposition by glow discharge polymerization occurs onto surfaces exposed to (directly contacting) the glow. Some deposition of polymer occurs on surfaces in nonglow regions (but the deposition rate is orders of magnitude smaller). The surface on which a polymer deposits could be an electrode surface, a wall surface, or a substrate surface suspended in the glow region. Another important factor that must be considered in dealing with operational factors of glow discharge polymerization is that glow discharge polymerization is system dependent. Consequently, polymer deposition rates are dependent on the ratio of surface to volume of glow. Therefore, other operational parameters such as flow rate, system pressure, and discharge power are insufficient parameters for the complete description of glow discharge polymerization. Such parameters serve as empirical means of describing operational conditions of glow discharge polymerization in a particular system, but they should not be taken beyond this limitation.

The following operational factors are important; however, all factors influence glow discharge polymerization in an interrelated manner. Therefore, any single factor cannot be taken as an independent variable of the process.

A. Modes of Electric Discharge

Electric power sources with frequencies in the 0 (dc) to gigahertz (microwave) range can be used for glow discharge polymerization. The use of a low frequency electric power source (up to about the audio frequency range) requires internal electrodes. With higher frequencies, external electrodes or a coil also can be used. Typical combinations of discharge modes and reactor design are shown schematically in Fig. 2.

The use of internal electrodes has the advantage that any frequency can be used. The glow discharge is more or less restricted to the space between electrodes. The best glow discharge is obtained with internal electrodes at a relatively high pressure (>0.1 Torr). At lower pressure, the glow discharge expands beyond the space between electrodes. At low pressure (<0.02 Torr), the glow occurs mainly in the space outside of the gap between the electrodes, and the system becomes inefficient for glow discharge polymerization. In order to restrict the glow to the space between the electrodes in the low pressure range, it is necessary to employ magnetic enhancement. Under typical conditions, polymer deposition occurs mainly onto the electrode surface. With a high frequency (rf range)

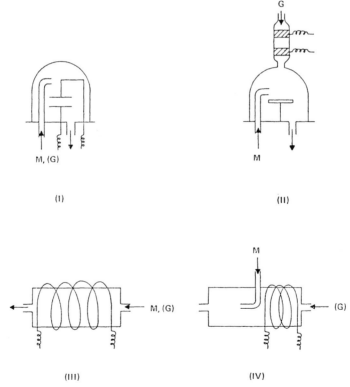

Fig. 2. Schematic representation of some typical arrangements of electric discharge, flow of starting material M, (and carrier gas G), and the location of polymer deposition.

power source, the glow tends to stray away from the space between the electrodes; however, because of this tendency, polymer deposition onto a substrate surface placed in between the electrodes increases [79]. The systems that employ external electrodes or a coil are suited for large volume glow discharges. They are particularly suited for the utilization of the tail-flame portion of the glow discharge. The tailflame refers to the glow discharge away from the energy input region (under external electrodes or coil).

Whether a substrate is placed in the energy input region or placed in the tail flame—or in the case of an internal electrodes system, whether a substrate is placed directly onto the electrode surface or placed in between electrodes—plays an important role in the properties of the polymer formed [70, 80–82]. The relative location of the energy input and the polymer deposition is an important factor to be considered in view of the CAP scheme of glow discharge polymerization (Fig. 1) in which the substrate material also plays an important role in glow discharge polymerization.

B. Flow Rate

The flow rate in most cases of glow discharge polymerization simply refers to the feeding-in rate of the starting materials into the total vacuum system, and it does not necessarily mean the rate at which the starting material is fed to the region of the system where polymerization occurs.

It should be pointed out also that flow rates of a gas in a vacuum system merely represent the total flux of gas but do not represent the velocity of molecules as visualized in the flow of a liquid. The parameter F/p (where F is flow in cubic centimeters (STP) per minute and p the system pressure in atmospheres) is proportional to the velocity of gas molecules in a given flow rate F at pressure p.

C. System Pressure

The system pressure is perhaps the most misunderstood and ill-treated parameter of glow discharge polymerization. This misunderstanding or mistreatment largely stems from the lack of distinction between nonpolymer-forming and polymer-forming plasmas. As mentioned earlier, efficient glow discharge polymerization is an excellent pump. Consequently, the polymerization itself changes the system pressure. Another factor contributing to the misunderstanding is the failure to recognize the effect of product gas. In many cases, the system pressure observed before glow discharge p_0 is cited as though it represents the system pressure during glow discharge polymerization p_g. Some authors claim that p_g is adjusted to p_0 by controlling the pumping rate. Since p_g is dependent on the production rate of product gas, such an operation is not always possible. Furthermore, in view of the ablation process, which is highly dependent on the amount of product gas, such an operation does not seem to have any advantages or significance in controlling the process.

The following points may clarify the meaning of system pressure in glow discharge polymerization:

(1) The system pressure before glow discharge p_0 at a given flow rate is entirely dependent on the pumping rate [83]; the higher the pumping rate, the lower is the value of p_0.

(2) The pumping rate of a system is dependent on the nature of the gas and is particularly important when a liquid nitrogen trap or a turbomolecular pump is employed in a vacuum system, as shown in Fig. 3. These are excellent pumps for most organic vapors (starting material of glow dis-

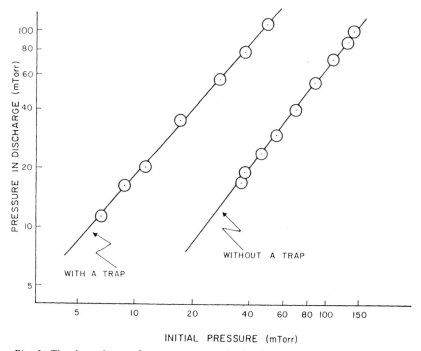

Fig. 3. The dependence of system pressure in the discharge p_g on the initial system pressure p_0, for glow discharge polymerization of ethylene. The initial pressure at a given flow rate is dependent on the pumping rate of the system as shown by the two lines representing systems with and without liquid nitrogen traps.

charge polymerization) and some gases; however, they offer virtually no pumping action for H_2, which is the main product gas when hydrogen-containing compounds are used as the starting material.

(3) As far as the gas phase is concerned, glow discharge polymerization acts as an additional pump.

(4) Glow discharge polymerization changes the gas phase from the starting material to the product gas.

(5) Consequently, the system pressure with the glow discharge on, p_g, is largely determined by the pumping efficiency of the product gas, the efficiency of the polymerization, and the production rate of gas.

(6) Therefore, there is no unique relationship between p_0 and p_g.

In a system where the polymerization yield is maintained at nearly 100%, p_g is determined by the flow rate but not by the value of p_0, as shown in Fig. 4.

Since the velocity of gas molecules is dependent on pressure, the value of p_g (but not p_0) is important in controlling the distribution of polymer deposition and the properties of polymers formed in glow discharge

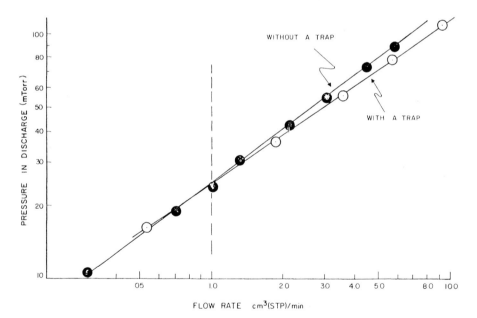

Fig. 4. The dependence of system pressure during the discharge p_g on the flow rate of starting material (ethylene). Despite the difference in p_0's for systems with or without a liquid nitrogen trap, p_g is mainly dependent on the flow rate, indicating that the production rate of the product gas (H_2) and the pumping rate of a system for the product gas determine the value of p_g.

polymerization; however, p_g cannot be considered as a manipulatable processing factor. The value of p_g can be manipulated to a certain extent, but it is largely determined by the nature of the starting material (i.e., gas production rate).

D. Discharge Power

The significance of discharge power in glow discharge polymerization is quite different from that for nonpolymer-forming plasmas. In essence, (the absolute value of) discharge power itself cannot be considered as an independent variable of the operation, since a certain level of discharge power (e.g., 60 W) in a given set of discharge conditions for one starting material (e.g., ethylene) could not even initiate a glow discharge with another starting material (e.g., n-hexane) under otherwise identical conditions. In other words, a relative level of discharge power which varies according to the characteristics of starting materials is needed to describe the discharge power for glow discharge polymerization.

In order to understand the importance of the discharge power parameter for glow discharge polymerization, it is very important to recognize the following characteristics of glow discharge polymerization: (1) the starting material is in the gas phase, but the main product is in the solid phase; (2) glow discharge polymerization occurs mainly in the glow region of a reactor; and (3) the glow region of the gas phase is not a simple plasma of the starting material but contains significant amounts of nonpolymer-forming gas product(s). Therefore, in order to describe the discharge power of glow discharge polymerization, it is necessary to express the characteristic power density in the glow volume of a flow system. Consequently, the discharge power level to describe glow discharge polymerization is a system-dependent parameter, not simply the power input into the system.

For instance, the discharge power necessary for glow discharge polymerization (based on the maximum change which occurs in gas phase [84]) of various hydrocarbons is shown in Figs. 5 and 6 as a function of flow rate of the starting material. As seen in these figures, the discharge power necessary for glow discharge polymerization depends on both the molecular weight and chemical structure of the compounds.

The best first-order approach to dealing with this situation is to use the parameter given by W/FM, where W is the power input, F the flow rate given in cubic centimeters (STP) per minute, and M the molecular weight of the starting material [84]. The parameter W/FM represents the power input per unit mass of the starting material. The parameter W/FM does not contain terms that describe the geometric factor of and flow pattern

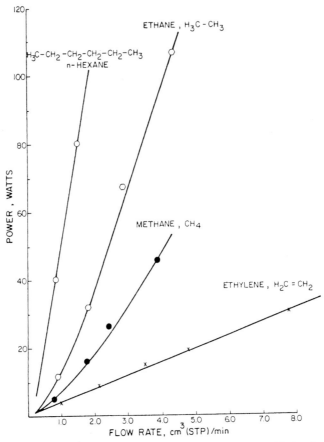

Fig. 5. The dependence of discharge power to obtain a comparable level of glow discharge polymerization on the flow rates of starting materials. The discharge power is greatly dependent on the molecular weights of the starting materials.

within a reactor, and consequently, the absolute value cannot be used in general cases. However, it is a useful parameter to describe glow discharge polymerization of different starting materials in a polymerization reactor.

The wide spread of discharge power shown in Figs. 5 and 6 for various compounds becomes roughly comparable values when $(W/FM)_c$ is plotted against F, as shown in Figs. 7 and 8. The parameter $(W/FM)_c$ represents the values of W/FM given by lines shown in Figs. 5 and 6. The values $[(W/FM)_c]_{F \to 0}$ for various hydrocarbons are nearly constant, dependent only on structures of starting materials. It is worth noting here that the slope observed in the plots of $(W/FM)_c$ versus F is proportional to the

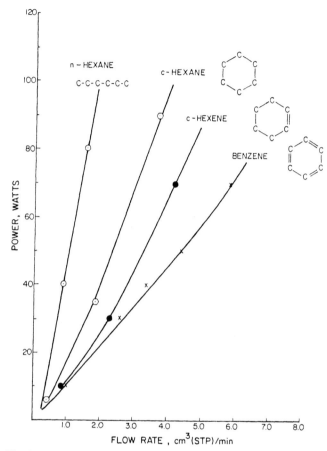

Fig. 6. The dependence of discharge power to obtain a comparable level of glow dis-charge polymerization on the flow rates of the starting materials for hydrocarbons containing six carbons. The discharge power is also dependent on the structures of starting materials.

hydrogen yield of compounds, as shown in Fig. 9. In order to obtain com-parable glow discharge polymerization, the discharge power must be selected according to F and M.

E. Geometrical Factor of Reactor

1. Bypass Ratio of Flow

Not all starting materials fed into a glow discharge polymerization reactor are utilized in the polymer formation. The bypass ratio represents the portion of flow which does not contribute to glow discharge polymer-

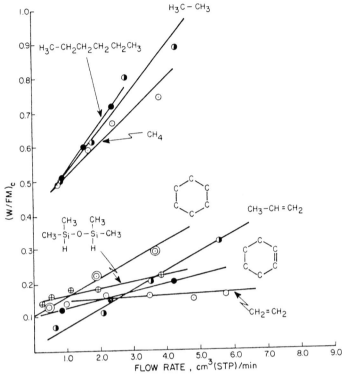

Fig. 7. Plots of $(W/FM)_c$ against the flow rate for various compounds, where W is the discharge power for glow discharge polymerization, F the flow rate, and M the molecular weight of the starting material, and $(W/FM)_c$ represents the values of W/FM given by lines shown in Figs. 5 and 6.

ization. Consequently, the higher the bypass ratio of a reactor is, the lower the conversion of the starting material to the polymer. Clearly, this ratio depends on the ratio of the volume occupied by discharge to the total volume.

2. Relative Position of Energy-Input and Polymer Deposition

In glow discharge polymerization which utilizes internal electrodes, either the substrate is placed directly on an electrode surface or in the space between the electrodes.

With external electrodes or a coil, the location of the substrate can be chosen in a variety of ways. Since the polymer properties and the deposition rate are dependent on the location within a reactor, this is an extremely important factor in practical applications. The relative position is further complicated by the factor described below.

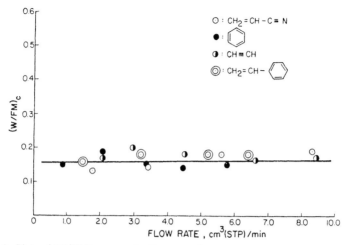

Fig. 8. Plots of $(W/FM)_c$ versus the flow rate of compounds which contain triple bonds and/or aromatic structures. $(W/FM)_c$ is nearly independent of the flow rate for these compounds (see Fig. 7 caption for letter definitions).

3. Relative Location of the Feed-In of the Starting Material and Flow Pattern

The location where the starting material is introduced is very important for polymer deposition. The importance of flow pattern with respect to the location of energy input and of polymer deposition can be visualized in an example of glow discharge polymerization in a straight tube reactor with an external coil placed in the middle portion of the tube. In such a system, the volume of glow discharge is generally much larger than the volume of the portion of tube which is directly under the coil. Consequently, polymer deposition occurs even at the upstream side of the coil. The flow can best be established by avoiding all starting materials passing through the energy-input region, as seen in the examples shown in I, II, and IV of Fig. 2. This factor is less obvious in a system with internal electrodes (e.g., in a bell jar).

IV. ORGANIC COMPOUNDS FOR GLOW DISCHARGE POLYMERIZATION

As mentioned in the introduction, nearly all organic compounds can be polymerized by glow discharge polymerization; however, the starting material should not be considered as the monomer (the starting material of polymerization in the conventional concept) of a polymerization process.

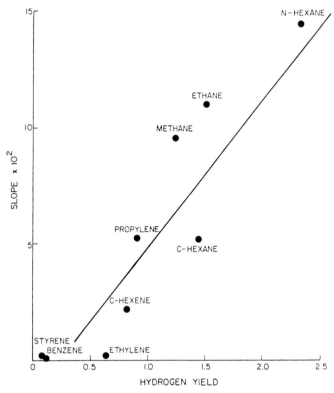

Fig. 9. The dependence of the slopes of $(W/FM)_c$ versus F plots on the hydrogen yield of compounds. The hydrogen yield is the number of hydrogen molecules evolved when a molecule of starting material is incorporated into the glow discharge polymer.

Since glow discharge polymerization can be characterized best as elemental or atomic polymerization, organic compounds can be classified based on elements contained in organic compounds.

A. Hydrocarbons

Hydrocarbons produce H and H_2 as the major nonpolymer-forming gas products. Since H and H_2 plasmas have little etching effect on polymers formed by glow discharge polymerization, the process of forming polymer is least affected by ablation.

Hydrocarbons can be grouped, according to their behavior in glow discharge polymerization, into the following three major groups [85, 86]:

Group I.   Triple-bond-containing and aromatic compounds,
Group II.   Double-bond-containing and cyclic compounds, and
Group III.   Compounds without the above-mentioned structures.

Group I compounds form polymers by utilizing the opening of triple bonds or aromatic structures with the least evolution of hydrogen.

Group II compounds form polymers via both the opening of double bonds or cyclic structures and hydrogen abstraction. Production of hydrogen is considerably higher than in group I compounds.

Group III compounds polymerize primarily by *plasma polymerization* based on hydrogen abstraction. Consequently, hydrogen production is much higher than in those for group II compounds.

Hydrogen production per mole of starting material for typical hydrocarbons is shown in Fig. 10. The discharge power necessary for glow discharge polymerization of hydrocarbons is also dependent on the types of compound. The same groupings mentioned above apply [84]. Groups I and II compounds require approximately the same energy input. However, the dependence on flow rate is nearly zero for group I compounds, but an appreciable increase in the required energy input is observed for group II compounds for increasing flow. Group III compounds require the highest energy input and their dependence on flow rate is much greater than that for group II compounds (see Figs. 7 and 8).

B. Nitrogen-Containing Compounds

Results obtained with various amines and nitriles indicate that N remains in the polymer (nitrogen does not evolve as the nonpolymer-forming product gas) [86]. This tendency is in accordance with another trend that N_2 gas used in a glow discharge is easily incorporated into either glow discharge polymers or polymers used as substrates [75].

C. Fluorine-Containing Compounds

Glow discharge polymerization of fluorine-containing compounds, particularly of perfluoro compounds, is very sensitive to the conditions of polymerization. The use of a relatively low discharge power level is an extremely important factor for obtaining polymers from fluorine-containing compounds. High discharge power causes the detachment of F and enhances ablation. Consequently, it is often observed that no polymer is formed when glow discharge polymerization is carried out at a high discharge power, while the same compound yields polymers at a low discharge power [88]. In order to obtain a polymer by glow discharge poly-

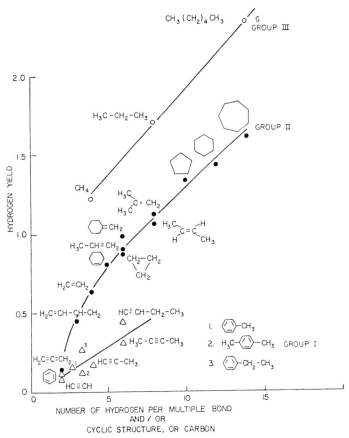

Fig. 10. Number of hydrogen molecules evolved per molecule of starting material when hydrocarbons polymerize (hydrogen yield) as a function of chemical structure.

merization of fluorine-containing compounds, it is advantageous to use compounds which belong to groups I and II mentioned above for hydrocarbons [89], and/or to employ techniques that suppress the etching effect of the detached fluorine (plasma), such as the addition of a small amount of H$_2$ [71] or hydrogen-producing compounds [90], and use of pulsed or intermittent discharges [89].

D. Oxygen-Containing Compounds

Oxygen is another of two elements (i.e., F and O) that tend to be evolved from either the starting material or substrate material, causing significant ablation of organic materials. Consequently, oxygen-containing compounds are generally poor starting materials for glow discharge polymerization. The deposition rate of polymers is generally much smaller than that of nonoxygen-containing compounds of similar molecular weight [76]. When oxygen is incorporated in chemical structures mentioned in groups I and II for hydrocarbons (i.e., easily polymerizable structures), glow discharge polymerization proceeds with ease [59, 60].

E. Si-Containing Compounds

Silicon is one of the elements that has a high tendency to stay in the solid phase; therefore, glow discharge polymerization of Si-containing compounds, such as silanes and siloxanes, proceeds extremely well [49, 50, 60]. This, together with the additional factors of relatively high molecular weight and relatively high vapor pressure of silicon-containing compounds, leads to the deposition rate obtainable from silicon-containing compounds being perhaps the highest among a variety of starting materials [60].

F. Compounds Containing Other Elements

Although not enough data are available to judge behavior of compounds that contain other elements in glow discharge polymerization, a simple rule of thumb may be drawn from the CAP scheme of glow discharge polymerization. Elements which are reactive and exist in the gas phase in the normal temperature range favor the ablation process and do not contribute to polymer formation. However, it is possible that many unusual elements which are not incorporated in conventional polymers could be incorporated into thin films formed by glow discharge polymerization.

V. DEPENDENCE OF GLOW DISCHARGE POLYMERIZATION ON PROCESSING FACTORS

In glow discharge polymerization as a means of thin film formation, the following aspects and their dependence upon the operational or processing factors seem to be of utmost practical importance. As mentioned earlier, however, many operational factors affect glow discharge polymerization in a complexly interrelated manner, and none of them can be singled out as being an independent variable or the most important factor of the process. It should also be kept in mind that glow discharge polymerization is system dependent, and consequently, the trends or conclusion based on data obtained in a particular system may not be extended to another system.

A. Rate of Polymer Deposition

In the practical sense, the rate at which a polymer deposits is an extremely important aspect of the process. The rate of polymer deposition can be increased by increasing the characteristic rate of polymer formation (often called polymerization rate), or by increasing the yield of polymer formation (reducing the amount of starting material that leaves the system without being polymerized). Unfortunately, the terms "polymer deposition rate" and "polymerization rate" are often used synonymously.

1. Discharge Power

The polymer deposition rate generally increases with discharge power in more or less linear fashion in a certian range of discharge power, and reaches a plateau, as shown in Fig. 11. Further increase of discharge power often decreases the polymer deposition rate.

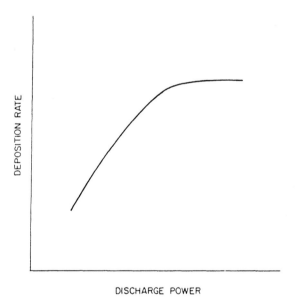

DISCHARGE POWER

Fig. 11. Schematic representation of the dependence of polymer deposition rate on discharge power when a constant flow rate is employed.

2. Flow Rate

The polymer deposition rate increases linearly with the flow rate of a starting material under ideal conditions where the conversion ratio of starting material to polymer is high or remains at a constant level. However, the change of flow rate is often associated with changes in flow pattern (affecting the yield of polymer formation or bypass ratio of the flow) and/or the efficiency of discharge power input. Therefore, the apparent dependence of the polymer deposition rate on flow rate is often characterized by a decrease of the polymer deposition rate after passing a maximum or a narrow plateau, as shown in Fig. 12.

3. W/FM Parameter

As mentioned earlier, the effect of W or F cannot be determined independently since glow discharge polymerization is dependent on the combined parameter of W/FM. As long as the W/FM value remains above a critical level $(W/FM)_c$ where energy input is sufficient for polymerization, the major effect of increasing the flow rate is to increase the feed-in rate,

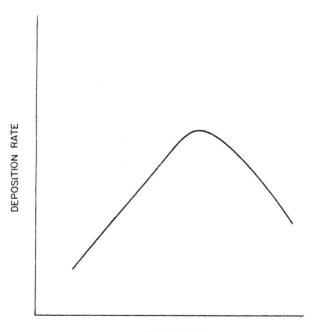

FLOW RATE

Fig. 12. Schematic representation of the dependence of polymer deposition rate on flow rate of a starting material when a constant discharge power is employed.

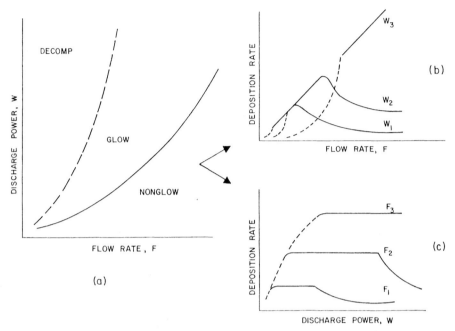

Fig. 13. Schematic representation of the interrelationship of polymer deposition rate with flow rate and discharge power. (a) Indicates the power-flow domains of decomposition (poor polymer deposition due to predominanting ablation process), normal glow where glow discharge polymerization occurs, and nonglow region. (b) and (c) At a fixed level of W or F, change of F or W crosses the domain shown in (a), and consequently, the apparent dependence of polymer deposition rate on either F or W is determined by where the change of domain occurs.

which increases the polymer deposition rate. However, if the W/FM level drops to a certain level as F increases at a constant W, where the discharge power is not sufficient to polymerize all starting materials coming into the reaction system, the polymerization mechanism itself changes. Consequently, the polymer deposition rate decreases despite the fact that more starting materials are supplied to the reaction system. The general situation is shown in the schematic diagrams given in Fig. 13.

According to the W/FM parameter, the discharge power W must be increased as the flow rate of starting materials increases, and/or as the molecular weight of the starting materials increases.

B. Distribution of Polymer Deposition

The distribution of polymer deposition is directly related to the uniformity of the thin film formed by glow discharge polymerization. Distribution of polymer deposition is dependent on (1) the geometrical arrangement of inlet of starting material, outlet of the system, and region of energy input; (2) the operating pressure of the discharge (not the initial pressure); and (3) the reactivity of a starting material to form polymers. The effects of these factors on the distribution of polymer deposition may be visualized from the data shown in Figs. 14–19 obtained from an rf (inductively coupled) discharge [91–93]. The general trends are as follows:

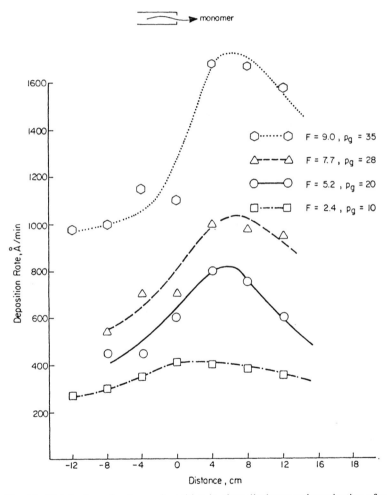

Fig. 14. Distribution of polymer deposition in glow discharge polymerization of acetylene at various flow rates. The letter F denotes flow rate in $cm^3(STP)/min$, and p_g the system pressure in the glow discharge given in mTorr. The distance is taken from the point of the starting material inlet in the direction of flow (see Yasuda and Hirotsu [91] for details of the reactor).

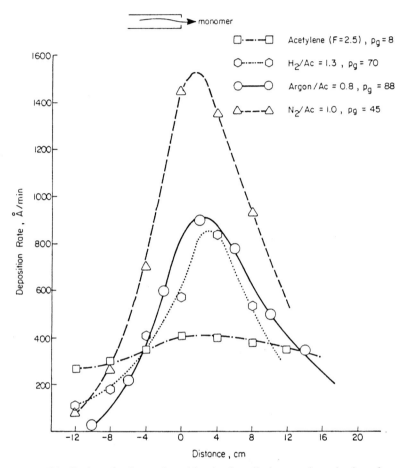

Fig. 15. Distribution of polymer deposition in glow discharge polymerization of acetylene with the addition of a carrier gas. H_2/Ac, Argon/Ac, and N_2/Ac denote the mole ratios of carrier gas to acetylene. The flow rate of acetylene is maintained constant in all cases. Other notation and units are the same as those in Fig. 14.

(1) The lower the discharge pressure, the wider is the distribution of polymer deposition. The lower the pressure, the larger is the mean free path of gas molecules and the diffusional displacement becomes more efficient. Therefore, the polymer formation is not localized at either the region of excitation or site of introduction of the starting material.

(2) The higher the reactivity of the starting material (to form polymer), the narrower is the distribution curve of polymer deposition, which has the maximum in the vicinity of the starting material inlet.

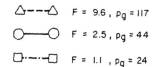

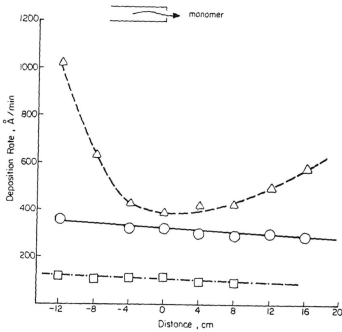

Fig. 16. Distribution of polymer deposition in the glow discharge polymerization of ethylene. All notation and units are the same as those in Fig. 14. Ethylene is a less reactive material than acetylene, as far as glow discharge polymerization is concerned. The increase of flow rate yields a maximum of deposition in the downstream side of the inlet, consequently, the apparent minimum is observed near the inlet.

(3) With starting materials that have low reactivity, the maximum peak is shifted towards the downstream side of the inlet. Consequently, the minimum (rather than the maximum) in the distribution curve is often observed at the vicinity of the inlet.

(4) Addition of nonpolymer-forming gas (e.g., Ar) tends to narrow the distribution curve.

The distribution of polymer deposition onto the surface of internal electrodes is generally very smooth, unless the starting material inlet is placed too close to the electrodes or too small an electrode gap (in relation

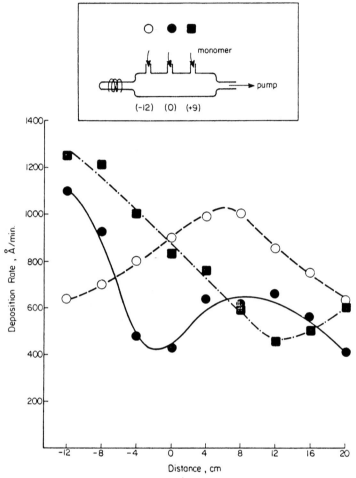

Fig. 17. The effect of location of the starting material (ethylene) inlet on the distribution of polymer deposition. The location of the inlet is shown in the insert. The flow rate of ethylene is maintained constant ($F = 9.8$ cm³(STP)/min) in all cases.

to the mean free path of gases) is employed. The effect of inlet–outlet locations in a bell-jar-type reactor is shown in Fig. 20 [94].

Regardless of the mode of electric discharge or type of reactor, the region where glow discharge polymerization occurs is located in the direct or tortuous pathway of the starting material from the inlet to the outlet. On this pathway, starting material is consumed to form a polymer, and simultaneously the gas phase changes from the starting material to the gas

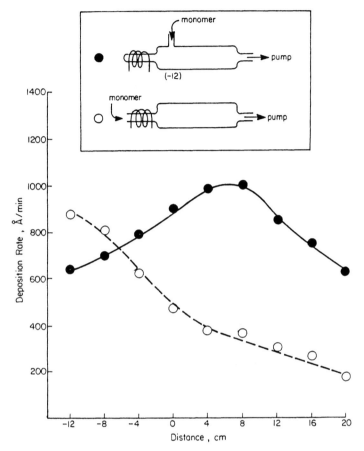

Fig. 18. The effect of flow passing through the rf coil (energy input region) on the distribution of polymer deposition in glow discharge polymerization of ethylene. $F = 9.8$ cm³(STP)/min for both cases.

product as polymerization proceeds. Therefore, an uneven distribution always exists if the polymer is collected on a stationary substrate surface. A moving substrate will average out this inherent uneven distribution of polymer deposition, and provide a practical means of yielding a uniform coating.

The distribution of polymer deposition should be taken into consideration when the polymer formation is monitored at a fixed location. The shift of the distribution curve due to changes in operational factors could be misinterpreted as a change in the polymer deposition rate itself.

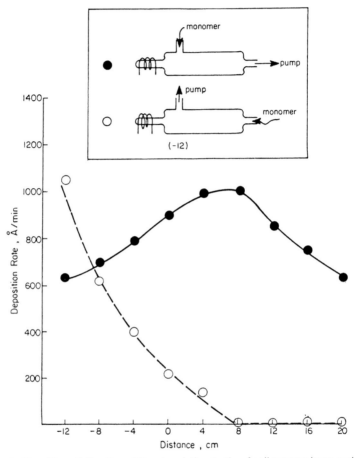

Fig. 19. The effect of direction of flow in relation to the rf coil (energy input region) on the distribution of polymer deposition in glow discharge polymerization of ethylene. All other conditions are maintained constant for both cases.

C. Properties of Polymers

Since glow discharge polymerization is system dependent, the properties of polymers formed by glow discharge polymerization are also dependent on the conditions of the process. The properties of polymers are dependent not only on the kind of reactor used but also on the location within a reactor where polymer deposition occurs.

The diagrams presented in Fig. 21 [95] show what kinds of polymers are formed from a given starting material, depending upon the apparent operational factors described. Because the strict meaning of parameters,

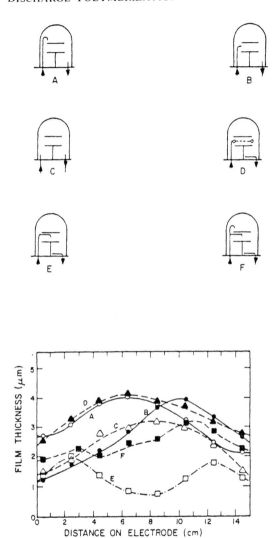

Fig. 20. Effect of relative location of starting material inlet, outlet, and electrodes on the distribution of polymer deposition onto electrode surfaces. p, 2 Torr; P, 100W; F, 80cm^3/min (see Kobayashi *et al.* [97] for details of conditions).

such as flow rate and pressure, depends on the geometrical factors of a reactor and the type of starting material, generalization of trends should not be made from such a diagram. However, it clearly shows the important fact that the properties of polymers formed by glow discharge polymerization are entirely dependent on how the polymerization is carried out.

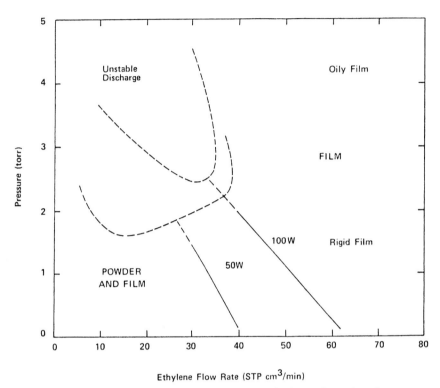

Fig. 21. An example of the dependence of the type of polymer formed on the apparent operational factors of power, pressure, and flow rate.

Analysis of polymers collected in different sections of a system, and of polymers formed by different electric discharges, also shows considerable differences in their properties [96, 97].

A study of the properties of polymers formed from tetrafluoroethylene by glow discharge polymerization and investigated by electron spectroscopy for chemical analysis (ESCA) [98] provides further evidence of the importance of processing factors. Tetrafluoroethylene is an ideal starting material to illustrate the CAP scheme of glow discharge polymerization. Therefore, some results are shown in Figs. 22–24.

The ESCA C1s spectrum of conventionally prepared polytetrafluoroethylene shows a single intense peak at 292 eV corresponding to the —CF_2— carbon bond. The peaks at binding energy levels of less than 291 eV represent the presence of cross-links (>CF—, >CC<) and carbons bonded to other substituents, including nitrogen- and oxygen-containing groups.

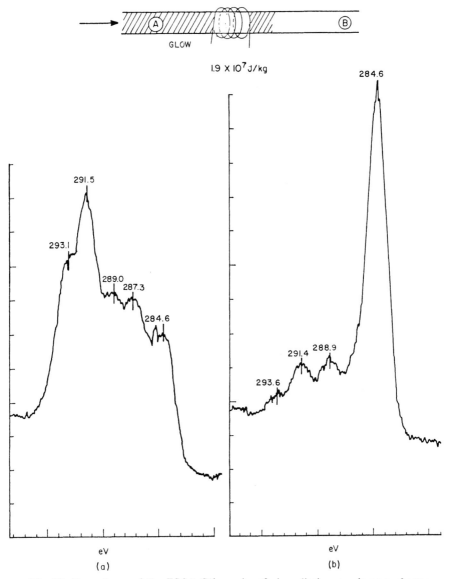

Fig. 22. Dependence of the ESCA C1's peaks of glow discharge polymers of tetra-fluoroethylene on discharge conditions and the location of polymer deposition. Polymer deposit occurred at two locations (a) before the rf coil and (b) after rf coil. Discharge power level is 1.9×10^7 J/kg.

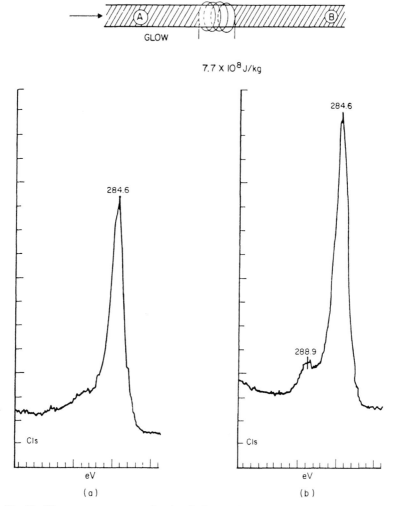

Fig. 23. Electron spectroscopy for chemical analysis C1's peaks of glow discharge poly-mers of tetrafluoroethylene in the same reactor shown in Fig. 22, but at the higher discharge power level of 7.7×10^8 J/kg.

Characteristic shapes of the C1s peaks shown in Fig. 22 indicate that polymers that are formed at locations on the upstream and downstream sides of the rf coil are quite different when a relatively low discharge power is used. The polymer formed in the upstream side contains con-siderable amounts of CF_3 and CF besides the expected CF_2. This is un-doubtedly due to the elemental or atomic nature of glow discharge poly-merization rather than conventional molecular polymerization.

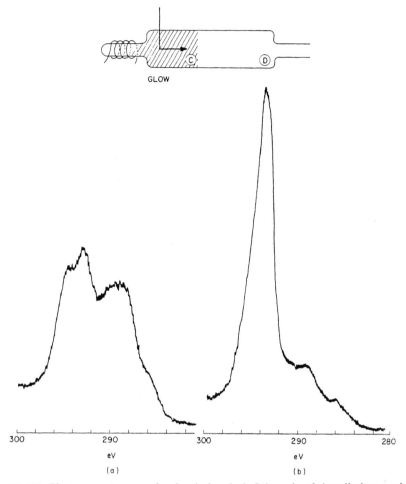

Fig. 24. Electron spectroscopy for chemical analysis C1's peaks of glow discharge polymers of tetrafluoroethylene prepared in a reactor shown in the insert: (a) at the end of the glow region and (b) at the end of the tube in the nonglow region.

The polymer formed in the downstream side of the rf coil at this low discharge power contains much less F (i.e., much smaller peaks for higher binding energy), and the peak at 284.6 eV becomes the major peak. This is a dramatic display of the effect of the energy input zone (i.e., tube directly under rf coil) on the properties of glow discharge polymers. As the discharge power is increased, this severe effect of the energy input zone expands eventually to the entire length of the tube and at a high discharge power, the polymer formed in the upstream side of the rf coil becomes similar to the polymer formed in the downstream side, as seen in Fig. 23.

When a system in which the flow does not pass through the energy input zone is used and glow discharge polymerization is carried out in the tail-flame portion of the glow discharge, the polymer formed at the downstream end of a reactor is not necessarily the same as that formed at the downstream end of a straight tube. Results given in Fig. 24 show that the polymer formed in the nonglow region, although it is located at the downstream end of a reactor, is nearly identical to the conventional polytetrafluoroethylene. This means that polymers formed under such conditions are formed mainly by *plasma-induced polymerization*.

As mentioned earlier, tetrafluoroethylene is a special starting material which reflects the effects of operationl factors in a very sensitive manner. Therefore, some effects (e.g., the increase of ablation by the increase of discharge power and by the location within a reactor) might be much smaller with other starting materials. However, the important aspects of (1) elemental or atomic polymerization and (2) system dependent polymerization, would be undoubtedly applicable to many other starting materials. These effects should be taken into consideration in designing the reactor, during processing, and in interpretation of results.

REFERENCES

1. P. de Wilde, *Ber. Dtsch. Chem. Ges.* 7, 4658 (1874).
2. A. Thenard, *C. R. Acad. Sci.* 78, 219 (1874).
3. C. S. Schoepfle and L. H. Connell, *Ind. Eng. Chem.* 21, 529 (1929).
4. J. B. Austin and I. A. Black, *J. Am. Chem. Soc.* 52, 4552 (1930).
5. E. G. Linder and A. P. Davis, *J. Phys. Chem.* 35, 3649 (1931).
6. W. D. Harkins and J. M. Jackson, *J. Chem. Phys.* 1, 37 (1933).
7. H. Koenig and G. Helwig, *Z. Phys.* 129, 491 (1951).
8. K. Otazai, S. Kume, S. Nagai, T. Yamamoto, and S. Fukushima, *Bull. Chem. Soc. Jpn.* 27, 476 (1954).
9. P. B. Weisz, *J. Phys. Chem.* 59, 464 (1955).
10. J. Goodman, *J. Polym. Sci.* 44, 551 (1960).
11. G. J. Argnette, U.S. Patent 3,061,458 (1962).
12. J. H. Coleman, U.S. Patent 3,068,510 (1962).
13. M. Stuart, *Nature (London* 199, 59 (1963).
14. A. Bradley and J. P. Hammes, *J. Electrochem. Soc.* 110, 15 (1963).
15. P. L. Kronick and K. F. Jesch, *J. Polym. Sci., Part A* 1, 767 (1963).
16. F. J. Vastola and J. P. Wightman, *J. Appl. Chem.* 14, 69 (1964).
17. E. M. DaSilva and R. E. Miller, *Electrochem. Technol.* 2, 147 (1964).
18. N. M. Bashara and C. T. Doyty, *J. Appl. Phys.* 35, 3498 (1964).
19. R. A. Connell and L. V. Gregor, *J. Electrochem. Soc.* 112, 1198 (1965).
20. T. Williams and M. W. Hayes, *Nature (London)* 209, 769 (1966).
21. R. M. Brick and J. R. Knox, *Mod. Packag.* January, p. 123 (1965).
22. K. Jesch, J. E. Bloor, and P. L. Kronick, *J. Polym. Sci., Part A-1* 4, 1487 (1966).
23. T. Willams and M. W. Hayes, *Nature (London)* 216, 614 (1967).
24. S. Tsuda, *Kobunshi* 16, 937 (1967).

25. T. Hirai and O. Nakada, *Jpn. J. Appl. Phys.* **7**, 112 (1968).
26. I. Sakurada, *Macromolecules* **1**, 265 (1968).
27. P. L. Kronick, *J. Appl. Phys.* **39**, 5806 (1968).
28. A. T. Denaro, P. A. Owens, and A. Crawshaw, *Eur. Polym. J.* **4**, 93 (1968).
29. A. R. Denaro, P. A. Owens, and A. Crawshaw, *Eur. Polym. J.* **5**, 471 (1969).
30. P. L. Kronick, K. Jesch, and J. E. Bloor, *J. Polym. Sci. Part A-1* **7**, 767 (1969).
31. J. R. Hollahan and R. P. McKeever, *Adv. Chem. Ser.* No. 80, p. 272 (1969).
32. M. W. Ranney and W. F. O'Connor, *Adv. Chem. Ser.* No. 80, p. 297 (1969).
33. D. D. Neiswender, *Adv. Chem. Ser.* No. 80, p. 338 (1969).
34. P. M. Hay, *Adv. Chem. Ser.* No. 80, p. 350 (1969).
35. C. Simionescu, N. Asandei, F. Dénes, M. Sandulovici, and G. Popa, *Eur. Polym. J.* **5**, 427 (1969).
36. A. R. Denaro, P. A. Owens, and A. Crawshaw, *Eur. Polym. J.* **6**, 487 (1970).
37. F. Dénes, C. Ungurenasu, and I. Haidue, *Eur. Polym. J.* **6**, 1155 (1970).
38. K. R. Buck and V. K. Davor, *Br. Polym. J.* **2**, 238 (1970).
39. J. R. Hollahan and C. F. Emanuel, *Biochim. Biophy. Acta* **208**, 317 (1970).
40. A. R. Westwood, *Eur. Polym. J.* **7**, 363 (1971).
41. A. R. Westwood, *Eur. Polym. J.* **7**, 377 (1971).
42. R. Liepins and J. Kearney, *J. Appl. Polym. Sci.* **15**, 1307 (1971).
43. H. Yasuda and C. E. Lamaze, *J. Appl. Polym. Sci.* **15**, 2277 (1971).
44. S. Morita, T. Mizntani, and M. Leda, *Jpn. J. Appl. Phys.* **10**, 1275 (1971).
45. R. Liepins and H. Yasuda, *J. Appl. Polym. Sci.* **15**, 2957 (1971).
46. J. R. Hollahan, *Makromol. Chem.* **154**, 303 (1972).
47. L. F. Thompson and G. Smolinsky, *J. Appl. Polym. Sci.* **16**, 1179 (1972).
48. A. F. Stancell and A. T. Spencer, *J. Appl. Polym. Sci.* **16**, 1505 (1972).
49. M. J. Vasile and G. Smolinsky, *J. Electrochem. Soc.* **119**, 451 (1972).
50. L. F. Thompson and K. G. Mayhan, *J. Appl. Polym. Sci.* **16**, 2291 (1972).
51. L. F. Thompson and K. G. Mayhan, *J. Appl. Polym. Sci.* **16**, 2317 (1972).
52. C. T. Wendel and M. H. Wiley, *J. Polym. Sci., Part A-1* **10**, 1069 (1972).
53. P. K. Tien, G. Smolinsky, and R. J. Martin, *Appl. Opt.* **11**, 637 (1972).
54. R. Liepins and K. Sakaoku, *J. Appl. Polym Sci.* **16**, 2633 (1972).
55. H. Yasuda, C. E. Lamaze, and K. Sakaoku, *J. Appl. Polym. Sci.* **17**, 137 (1973).
56. H. Yasuda and C. E. Lamaze, *J. Appl. Polym. Sci.* **17**, 201 (1973).
57. M. Duval and A. Theoret, *J. Appl. Polym. Sci.* **17**, 527 (1973).
58. H. Kobayashi, A. T. Bell, and M. Shen, *J. Appl. Polym. Sci.* **17**, 885 (1973).
59. H. Yasuda and C. E. Lamaze, *J. Appl. Polym. Sci.* **17**, 1519 (1973).
60. H. Yasuda and C. E. Lamaze, *J. Appl. Polym. Sci.* **17**, 1533 (1973).
61. M. M. Millard, J. J. Windle, and A. E. Pavlath, *J. Appl. Polym. Sci.* **17**, 2501 (1973).
62. H. Yasuda, *Appl. Polym. Symp.* No. 22, p. 241 (1973).
63. J. R. Hollahan and T. Wydeven, *Science* **179**, 500 (1973).
64. R. F. Baddour and R. S. Timmins, eds., "The Application of Plasmas to Chemical Processing." MIT Press, Cambridge, Massachusetts, 1967.
65. F. K. McTaggart, "Plasma Chemistry in Electrical Discharges." Elsevier, Amsterdam, 1967.
66. R. F. Gould, ed., *Adv. Chem. Ser.* No. 80 (1969).
67. M. Venugopalan, ed., "Reactions Under Plasma Conditions," Wiley (Interscience), New York, 1971.
68. J. H. Hollahan and A. T. Bell, eds., "Techniques and Application of Plasma Chemistry." Wiley, New York, 1974.
69. M. Shen, ed., "Plasma Chemistry of Polymers." Dekker, New York, 1976.
70. H. Yasuda and T. Hsu, *Surf. Sci.* **76**, 232 (1978).

71. E. Kay, *Int. Round Table Plasma Polymer. Treat., IUPAC Symp. Plasma Chem. Limoges, Fr., 1977*.

72. D. T. Clark and A. Dilks, "Characterization of Metal and Polymer Surfaces," Vol. 2, "Polymer Surfaces." Academic Press, New York, 1977.

73. G. K. Wehner and G. S. Anderson, *in* "Handbook of Thin Film Technology" (L. I. Maissel and R. Glang, eds.), McGraw-Hill, New York, 1970.

74. G. Smolinsky, Bell Telephone Lab., personal communication (August 1977).

75. H. Yasuda, H. C. Marsh, E. S. Brandt, and C. N. Reilley, *J. Polym. Sci., Polym. Chem. Ed.* **15**, 991 (1977).

76. H. Yasuda, *Int. Round Table Plasma Polymer. Treat., IUPAC Symp. Plasma Chem., Limoges, Fr., 1977*.

77. H. Yasuda, H. C. Marsh, M. O. Bumgarner, and N. Morosoff, *J. Appl. Polym. Sci.* **19**, 2845 (1975).

78. H. Yasuda, H. C. Marsh, E. S. Brandt, and C. N. Reilley, *J. Appl. Polym. Sci.* **20**, 543 (1976).

79. H. Yasuda and N. Morosoff, *J. Appl. Polym. Sci.* to be published (1978).

80. D. F. O'Kane and D. W. Rice, *J. Marcomol. Sci., Chem.* **10**, 567 (1976).

81. H. Yasuda and T. Hirotsu, *Radiat. Phys. Chem.* in press (1978).

82. H. Yasuda and N. Morosoff, *J. Appl. Polym. Sci.* to be published (1978).

83. H. Yasuda and T. Hirotsu, *J. Appl. Polym. Sci.* **22**, 1195.

84. H. Yasuda and T. Hirotsu, *J. Polym. Sci., Polym. Chem. Ed.* **16**, 743 (1978).

85. H. Yasuda, *J. Macromol. Sci., Chem.* **10**, 383 (1976).

86. H. Yasuda, M. O. Bumgarner, and J. J. Hillman, *J. Appl. Poly. Sci.* **19**, 1531 (1975).

87. H. Yasuda, M. O. Bumgarner, and J. J. Hillman, *J. Appl. Polym. Sci.* **19**, 1403 (1975).

88. H. Yasuda and T. Hsu, *J. Polym. Sci., Polym. Chem. Ed.* **16**, 415 (1978).

89. H. Yasuda and T. Hsu, *J. Polym. Sci., Polym. Chem. Ed.* **15**, 2411 (1977).

90. K. Nakajima, S. Bourguard, A. T. Bell, and M. Shen, *Int. Round Table Plasma Polymer. Treat., IUPAC Symp. Plasma Chem., Limoges, Fr., 1977*.

91. H. Yasuda and T. Hirotsu, *J. Polym. Sci., Polym. Chem. Ed.* Part I, **16**, 229 (1978).

92. H. Yasuda and T. Hirotsu, *J. Polym. Sci., Polym. Chem. Ed.* Part II, **16**, 313 (1978).

93. H. Yasuda and T. Hirotsu, *J. Polym. Sci., Polym. Chem. Ed.* Part III, in press (1978).

94. H. Kobayashi, A. T. Bell, and M. Shen, *J. Macromol. Sci., Chem.* **10**, 491 (1976).

95. H. Kobayashi, M. Shen, and A. T. Bell, *J. Macromol. Sci., Chem.* **8**, 373 (1974).

96. M. Duval and A. Theoret, *J. Appl. Polym. Sci.* **17**, 527 (1973).

97. M. Duval and A. Theoret, *J. Electrochem. Soc.* **122**, 581 (1975).

98. H. Yasuda and N. Morosoff, *J. Appl. Polym. Sci.* to be published (1978).

Part V

ETCHING PROCESSES

V-1

Chemical Etching

WERNER KERN AND CHERYL A. DECKERT

RCA Laboratories
Princeton, New Jersey

I. INTRODUCTION

Chemical etching in thin-film technology plays a prominent role in both the preparation and the utilization of thin films. Regardless of the method of film deposition or formation, the substrate must first be suitably prepared, either by removal of work damaged surface layers or by creating a relief structure of specific geometry. In the first case, chemical polish etching is usually the method of choice; in the second case, structural etching is required. Once a thin film has been deposited, chemical etching is often used again, this time to create patterns in the appropriately masked films.

The aim of this review is to provide a broad outline of the subject of chemical etching and to present tables, with references, of etchants and etching conditions for inorganic materials.

Numerous excellent books, treatises, and reviews are available on theoretical and practical aspects of chemical etching, covering the chemistry [1–28] and electrochemistry [29–42] of etching processes. A few partial bibliographies have been published on some aspects of etching [43, 43a]. However, most information on specific etchants for different materials, with the possible exception of semiconductors, is widely scattered throughout the scientific literature and is often difficult to retrieve because etching is most frequently a means to an end and is usually not the primary subject matter of an investigation. An attempt has been made to bring together essential information that should prove useful to the scientist or engineer who must select an etching process for a specific material. It is obviously impossible to list all etchants for all materials. Instead, a selection has been attempted which is based, in the authors' opinion, on the practical usefulness of an etchant and a solid material in thin-film technology. The most recent and advanced information is generally given preference. Special emphasis is placed on materials and processes used in semiconductor microelectronics because a substantial part of thin-film technology is applied in this area with which we are particularly familiar from practical experience.

One important application of chemical etching is in the structural characterization of materials, especially the detection of lattice defects in semiconductors, the study of distribution of localized impurities, the delineation of layer structures and $p-n$ junctions, and the determination of composition. This specialized field of analytical etching is outside the scope of the present review. Physical-chemical "dry" etching processes such as sputter etching, plasma etching, and ion milling, are covered in Chapter V-2. What *will* be covered is chemical and electrolytic etching of insulators, semiconductors, and conductors in solution and in the gas phase.

Chemical formulas noted for reagents refer to the chemicals in the usual concentrated form, as defined in Section IV.A; parts are by volume. The crystallographic notifications used are those quoted by the author(s) of the reference cited.

II. PRINCIPLES AND TECHNIQUES OF ETCHING

A. Chemistry of Etching

Chemical etching may occur by any of several different processes [1, 2, 8]. The simplest mode of etching involves dissolution of the material in a liquid solvent without any change in the chemical nature of the dissolved species. Relatively few industrially important materials are etched in this manner. Although this is the only etching process for which the word "dissolution" is properly used, the term has come into common use for describing any etching procedure carried out in liquid media.

Most etching processes involve one or more chemical reactions. In order to be truly an etching reaction, the product formed must be soluble in the etchant medium, or must at least be carried away from the surface by the medium. Various types of reactions which may be involved are oxidation–reduction, of which electrochemical etching is a special case, complexation, and gas phase etching.

An oxidation-reduction, or redox, etching process involves conversion of the material being etched to a soluble higher oxidation state:

$$M \rightarrow M^{n+} + ne^-.$$

Redox etching may occur either in a completely chemical system, by the use of certain chemical oxidizing agents, such as Ce^{4+}, or in an electrochemical cell [29], by making the material to be etched the anode, and by applying a suitable external electromotive force. A typical plot of current density versus cell voltage is given in Fig. 1; the various stages of attack are noted. Electrochemical effects can lead to certain problems in etching. Corrosion [24, 25, 42] is a special case of electrochemical etching which occurs when surface variations produce local anodes and cathodes. When films of two dissimilar metals are in contact, the resultant galvanic action can cause undercutting at the interface during etching [44].

Fig. 1. Current density versus cell voltage. Region A–B etching; B–C stable plateau with polishing; C–D slow gas evolution with pitting; D–E polishing with rapid gas evolution.

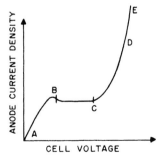

Complex formation is frequently involved in etching processes, often in conjunction with a redox reaction. The ligand groups surround and bond chemically to the etched species, forming a complex ion or molecule that is readily soluble in the etchant medium.

Gas phase etching may involve vaporization of the material being etched in a vacuum or inert atmosphere or may involve reaction of gaseous etchants with the surface to produce volatile products. Elevated temperatures are usually required.

B. Factors Affecting Etching Reactions

Etching reactions typically occur by a process involving several sequential steps [3, 8]. The observed dissolution kinetics depend upon the nature of the rate-limiting step of the process. If the rate of this step is determined by the chemical reactivity of the species involved, the process is said to be activation limited. On the other hand, if the rate is determined by the speed at which fresh reactant can be supplied to the surface, the process is said to be diffusion limited.

If a series of materials are all etched in the same solution by a diffusion-controlled process, then the same etch rate is observed for all [8]. Some etching processes are diffusion limited at low concentrations, but are activation limited at higher concentrations [3]. An increase in etching temperature may cause a change in the etching kinetics [45]. The presence of catalytic species in the etchant can also affect the etch rates markedly. Agitation of the solution may increase etch rate if the reaction is diffusion limited; it may decrease etch rate if, for example, localized solution heating occurs; or it may have no effect if activation control is involved. In pattern etching, the slope of the pattern edges depends on the type of kinetics involved [46].

Adsorption and desorption processes can affect the etching kinetics profoundly. Adsorption of reactant from the etchant solution onto the substrate may produce surface complexes which will facilitate the etching process; however, in many cases adsorption of nonreactive species or formation of passivating surface films can slow down or stop further etching [3]. Oxide films on metals are a good example of this phenomenon. Certain types of impurities in the etchant solution, even though present at low concentrations, may be adsorbed onto the substrate and hinder etching [8]. Desorption of gaseous reaction products sometimes limits the rates of etching processes [3].

The kinematic aspects of etching [8, 47] should also be mentioned briefly. This refers mainly to the tendency of various crystallographic planes to etch at different rates. Various orientations of single-crystal substrates may thus etch very differently in a given etchant, and substrates of varying roughness may also exhibit large differences in etch rates.

Several additional specific factors affecting etching reactions in various types of materials will be noted in the discussions of insulators and semiconductors (Sections III.A and III.B, respectively).

C. Etching Techniques and Processes

The choice of the etching technique to be used for a given situation depends upon the material to be etched, the requirements of pattern generation, the necessary etching reagents, the etching processes involved, and other factors such as economic considerations.

1. Immersion Etching

The simplest technique is liquid chemical immersion or dip etching where the masked or unmasked object is submerged in the etch solution. Mechanical agitation is usually desirable as it improves the uniformity and control of the etching process by enhancing the exchange and mixing of spent etching solution at the solid surface with fresh solution. This also avoids local overheating in the case of exothermic reactions, thereby maintaining a uniform and controllable etching rate. Bubbles of gas (usually H_2) that may form as a reaction product can cling to the solid surface and inhibit uniform etching. The addition of a surface-active agent to the etch solution can prevent bubble accumulation. A sufficiently large ratio of etchant to material being etched should be employed to minimize reactant depletion and to maintain the reaction temperature and the rate of attack.

2. Spray Etching

Spray etching is useful for generating patterns in relatively thick films or substrates, especially if steep pattern walls are desired, since the impinging etch solution imparts a variable degree of directionality to the process. The etching rate is increased over that of immersion etching, and can be regulated by the amount of pressure applied and the size of the droplets. Good process control and uniformity can be attained because fresh etchant is rapidly and constantly supplied to the reaction site, while the reaction products are continuously removed. Spray etching lends itself to automation, and commercial etching machines are available for many specific applications.

3. Electrolytic Etching

Electrically conductive or semiconductive materials are frequently etched by application of external emf potentials. Electropolishing of metals and semiconductors is a good example of this technique. The rate and selectivity of etching can be controlled by the potential and/or the current density applied. Electrolytic etching is considerably more complicated than other techniques, but can yield results not otherwise attainable. Specific conditions will be described for various materials in the text and in the etching tables.

4. Gas-Phase Etching

High-temperature etching in the gas or vapor phase is generally used for chemically inert materials that cannot be etched readily in liquid reagents. A different application is for in situ etching of semiconductor substrates immediately prior to epitaxial film growth in the same reactor to avoid surface contamination that would result by other techniques.

5. Mechanical–Chemical Polishing

This technique is used in semiconductor wafer preparation when a relatively defect-free surface is required. The combination of slow liquid chemical surface etching with gentle mechanical abrasion to continuously remove products from the etching reaction can result in a high-quality surface polish if carefully optimized conditions are observed, as will be described in Section III.B.

6. Isotropic versus Anisotropic Processes

Isotropic or nonpreferential etching proceeds at an equal rate in all directions. Amorphous materials of uniform composition etch isotropically, whereas many crystalline materials etch both isotropically and anisotropically. Anisotropic or preferential etching depends on the crystallographic orientation of the material and on the etching reagent used. If polishing action is desired, isotropic etching conditions must be selected to achieve a structureless surface. If structural shaping is the objective, as in the formation of deep depressions having side walls of a specific taper angle, anisotropic conditions are required. Both liquid and gas-phase etching can be used for these two types of etching processes.

7. Selective Etching Processes

Selectivity refers to the differences in etch rate between different materials, or between compositional or structural variations of the same material. It is one of the most important factors in applied etching. Most technological etching processes must be controllably selective because the material to be etched is usually part of a structure that consists of several material components. Selectivity in etching is achieved by proper choice of etching technique and etchant composition within the constraints of the systems.

Various degrees of etching selectivity are desirable for particular purposes. For example, pattern etching of Si_3N_4 or Al_2O_3 films in hot H_3PO_4 using an etch-resistant deposited SiO_2 film as the etch mask illustrates a high degree of etching selectivity. On the other hand, controlled partial etching selectivity of dielectric layer composites is important in taper etching, where a desired edge contour can be attained on the basis of etch rate differences of the component layers. In this case, a faster etching dielectric "taper-control" layer is formed over the dielectric to be beveled [46]. Numerous other important applications of selective etching have been described [26, 27, 48, 49].

8. Fusion Techniques and Other Processes

Certain highly etch-resistant materials can be etched by treatment with molten salts (often caustics or borax) at high temperature. Several examples will be noted in Sections IV.B and IV.D.

Surface oxidation by thermal or anodic treatments, followed by chemical stripping of the oxide films formed, can also be considered an etching process. However, only the second step, the etching of the oxides, will be discussed here (Section III.A).

D. Pattern Delineation Etching for Thin Films

In many instances, etching processes are used to produce certain patterns in thin films. Selected portions of the film are masked by another thin film coating material which is unaffected by the etchant to be used for patterning. Etching is then carried out so as to remove all the film material in the unprotected regions. The protective coating film is then usually stripped, leaving the desired pattern in the underlying thin film.

Pattern etching is obviously a much more complex process than simple overall surface etching. In addition to selecting the etchant, choosing a masking material is of prime importance; good adhesion of this coating to the substrate, coating integrity, adequate resolution, and resistance to the etchant are the main considerations. Ease in patterning the mask coating is important; otherwise this procedure becomes an etching process itself, requiring yet another mask.

1. Masking Materials

The most often used masking materials for high resolution thin film patterning are photoresists [50], organic polymers whose solubilities in certain solvents change drastically as a result of exposure to uv radiation. Usually, exposure is carried out by placing a glass plate bearing the desired pattern in an opaque material (such as photographic emulsion or chromium) over the photoresist-coated substrate and irradiating through the glass plate. Negative photoresists become less soluble in the developing solution in areas that were irradiated, thus producing a negative image of the pattern on the glass plate. Positive photoresists become more soluble in exposed areas and thus produce a positive image of the original pattern. Excellent photoresists are available commercially from a number of sources. Negative photoresists are generally tougher than positive resists and can usually withstand more rigorous etching processes. Positive resists are noted for their superior resolving power, and patterns as fine as 1 μm have been resolved using positive photoresist. Electron beam and x-ray resists can produce very fine resolution but they have not yet come into widespread use because of high processing costs.

When the etching process to be used in patterning the substrate involves extremes such as elevated temperatures or strong acids, photoresist masks may not provide adequate protection. In these cases, metal or dielectric masks, which can withstand the etching process more effectively, are often used. In such cases, the mask is first patterned using a photoresist process. For example, chemically vapor-deposited (CVD) SiO_2 is used as a masking material for CVD Si_3N_4 films, which are typically etched at 180°C in H_3PO_4, conditions which would quickly degrade photoresist films. The SiO_2 itself is readily patterned using a room-temperature etching process with a photoresist mask.

Sometimes, a high temperature or extremely degrading chemical etching process can be replaced by an electrochemical procedure which utilizes a much milder solution, thus allowing a photoresist mask to be employed [51].

In cases where high resolution is not a requirement, very simple masking procedures are possible. Ordinary cellophane tape is used to mask against a variety of etchants. Other masking films such as positive photoresist or silver paste can be applied in the areas to be protected using an artist's paint brush. Certain waxes which melt at temperatures of 100–250°C can be painted onto a hot substrate and will resist many etchants.

2. Adhesion and Interface Problems

Good adhesion to the substrate film during etching is the prime requirement of the mask material. Loss of adhesion usually occurs in one of two ways [52]: (a) edge attack at the interface by the etchant (undercutting) or (b) failure over a large area (lifting, peeling, crazing).

a. Edge Attack. If mask-to-film adhesion remains perfect throughout etching, and if the etching process is isotropic, a delineated pattern like that in Fig. 2a will result. If the etchant attacks the interface between mask and substrate film, however, the top edge of the patterned film can become sloped quite gradually (Fig. 2b). This phenomenon is called undercutting. Undercutting is a common occurrence because most mask/substrate film combinations involve no chemical bonding, relying solely on van der Waals forces for interfacial adhesion. These forces are strong enough to give good bond strengths under ordinary conditions (i.e., no etching), but species in the etching solutions also tend to form van der Waals bonds to the mask and substrate film surfaces, and in some instances these interactions can be stronger than the mask/substrate film bond, thus causing adhesion failure at the edges of the pattern. Adhesion promoters such as hexamethyldisilazane, which render SiO_2 surfaces essentially nonpolar, lead to better etch resistance [53] with photoresist masks simply because the polar etchants are less attracted to the interface, even though the actual van der Waal's forces between the modified SiO_2 and the photoresist may be smaller than without adhesion promoter.

Quite often it is desirable that a small amount of controlled undercut-

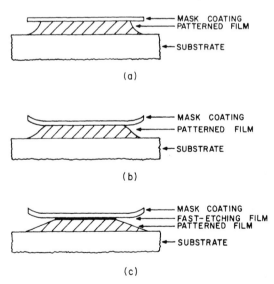

(a)

(b)

(c)

Fig. 2. Different edge profiles produced from various degrees of undercutting. (a) Good mask-to-film adhesion produces this type of edge; (b) undercutting has occurred at mask-film interface; (c) use of a fast-etching film to achieve controlled undercutting.

ting take place, since a sloped substrate edge is easier to coat uniformly than a sharp edge, if additional layers are to be deposited subsequently. In such cases, a very thin layer of material, which dissolves in the etchant more rapidly than the substrate film, may be deposited prior to masking in order to achieve controlled undercutting. This is depicted in Fig. 2c. A recent example of this method is the beveling of permalloy films using a Ti overcoat [54].

b. Large-Area Failure. Sometimes mask/substrate film adhesion failure occurs over a large area of the interface. This failure can show up in several ways. A portion of mask coating may be lifted completely from the surface, it may peel up either from the edges only or else craze and peel over the whole surface, or it may blister or bubble across the surface. These failures are usually due to differential stress buildup in the substrate film and mask layers. Thermal or chemical treatments can cause the masking film to go into tensile stress relative to the substrate layers, in which case peeling, crazing, or lifting can occur. On the other hand, if stresses in the mask become highly compressive compared with the thin film/substrate composite, blisters and bubbles will appear in the mask layer. These problems can be minimized by using a mask material either with a similar coefficient of thermal expansion to that of the substrate or with sufficient elasticity to conform more easily to the substrate.

3. Factors Affecting Image Resolution

The most obvious factor influencing image resolution is, of course, the resolving capability of the masking material. As already mentioned, negative photoresists have considerably poorer resolving power (~ 3 μm line widths or spacings) than positive resists (~ 1 μm). Metal and dielectric mask coatings are capable of generally finer resolution, down to the order of the grain size. Electron beam and x-ray resists can also be imaged to very fine dimensions (~ 80 Å resolution has been reported [55]).

The thicknesses of both the masking material and substrate film limit their resolution capability when chemical developing and etching procedures are used. Since isotropic chemical dissolution produces sloped edges, a good rule of thumb is that the thickness of the layer to be patterned should be no more than one-third of the resolution to be achieved. Dry etching processes, such as plasma and sputter etching [56], can produce very steep pattern edges, and thus finer resolution can be attained with a given film thickness.

Etching processes which involve gas evolution can lead to poor image resolution because of gas bubbles clinging to the substrate, particularly along the edges. This problem can usually be alleviated by the use of a suitable surfactant in the etching solution or, less satisfactorily, by heating or good agitation throughout the etching procedure.

In etching processes where the substrate film becomes oxidized (e.g., metals, silicon), masks of a nobler metal can lead to accelerated etching, thus producing pronounced undercutting and loss of resolution [44]. By the use of certain carefully chosen chemical etchants, the undercutting can be markedly reduced [57, 58].

E. Surface Contamination and Cleaning Techniques

The important subject of contamination and cleaning of surfaces before and after etching is closely associated with practical etch processing. A detailed discussion of this separate topic is outside the scope of this chapter; however, some general guidelines and a brief literature survey are presented.

Surface contamination as related to etching can be considered from two aspects: (1) initially present contaminants and their removal prior to etching, and (2) residual contaminants arising from etching treatments and their removal as a postetch step if traces of impurities cannot be tolerated on the etched surface.

Contamination on surfaces prior to etching may consist of particulate materials, organic residues, or inorganic surface films different from the material to be etched. These impurities should be removed since they may interfere with the etching by masking or undesirable reaction with the etchant. Particulate removal can be accomplished by ultrasonic treatments in cleaning solutions, use of compressed gas jets, application of liquid sprays or jets, or simple mechanical means such as scrubbing or brushing. Organic residues are removable down to monolayer levels by dissolution in suitable organic solvents, or by vapor refluxing in organic solvents or azeotropic solvent mixtures. Complete removal is generally possible only by plasma ashing, glow discharge sputtering, or chemical reaction leading to dissolution. Inorganic surface films must be attacked by specific chemical reagents designed to produce soluble reaction products that can be flushed away.

Whereas the treatments noted above are generally simple and relatively noncritical, the effective removal of residual trace contaminants resulting from etching processes is far more difficult to accomplish. Surface contaminants are very critical in technologies, such as solid-state device processing, where the nucleation, growth, adhesion, structure, and perfection of a deposited film can be critically affected by impurities on the substrate surface. High-temperature processing may cause residual surface impurities to penetrate into the substrate and give rise to undesirable effects, such as electrical instability of semiconductor devices.

The deposition of impurities, especially heavy metals, from liquid etchants onto semiconductor surfaces is well known since the early experimental work by Holmes *et al.* [5, 59], and the reviews by Gatos and Lavine [8] and later by Faust [60]. Kane and Larrabee [17] reviewed the literature up to 1969 on the deposition of chemical impurities from solution onto semiconductors. More recently, Kern reported results of comprehensive radioactive tracer adsorption studies of anionic and cationic etch components [61] and trace contaminants [62, 63] on Si, Ge, GaAs, and SiO_2 surfaces [64]. In addition, a decontamination method based on sequential oxidative desorption and complexing with $H_2O_2-NH_4OH-H_2O$ followed by $H_2O_2-HCl-H_2O$ was devised [64, 65]. Its remarkable effectiveness was verified specifically [66–69a] and indirectly [70–72] by several authors.

Various additional aspects of cleaning Si surfaces have been reported [66, 71–77]. Surface contamination of GaAs has been reviewed by Stirland and Straughan [23]. Meek [78] used Rutherford ion backscattering of high-energy ions as a sensitive surface analysis tool to determine the impurities left on clean Si surfaces from various etch components and organic solvents. Neutron activation analysis of Si slices that had been exposed to buffered HF etchant was employed to identify problematic trace contaminants in the NH_4F component as As and Cu [79]; purification of the etchant by treatment with Si chips [80] effectively removed the impurities.

A series of 18 symposium papers on the preparation and characterization of clean surfaces includes theoretical and practical aspects related to surface contamination on a variety of materials [81]. Ryan *et al.* [82] and De Forest [83] have described the preparation of clean surfaces prior to photoresist coating. Holland [84] discussed cleaning treatments for glass surfaces. Brown [85], and more recently Mattox *et al.* [86, 87] have reviewed those for thin film substrates of many types. Short-wave uv radiation has been found effective for removing hydrocarbons from glass surfaces [87, 88] and for removing photoresist residues [89]. Ozonization is an alternate method that offers several advantages [90]. Selection, specifications, and other aspects of surface preparation processes for numerous materials have been compiled by Snogren [91].

Surface cleaning by glow discharge sputtering techniques can also be very effective [85]. Most organic surface contaminants are removable by chemical sputtering in O_2 [92–94]. Sputter etching in Ar removes residual oxide layers on metals, as noted in Chapter 1, Section V.C. However, surface recontamination due to backscattering [95–97] or ion migration

[97] can occur during rf sputtering treatments, unless carefully optimized processing conditions are employed [98]. Ultrahigh vacuum heating after sputter cleaning is effective for desorbing gases that may become incorporated into the substrate surface during this operation [99–101].

Finally, glow discharge plasma cleaning should be noted as one of the most effective methods for surface decontamination. Kirk [102] reviewed its application to semiconductor device processing, and Kominiak and Mattox described reactive plasma cleaning of metals [103]. Many aspects associated with plasma reactions at solid surface have been discussed at a recent symposium [104].

Two additional very important aspects of surface cleaning are terminal treatment and storage of the cleaned material. The final rinsing in wet cleaning and etching processes is usually done with water. The purity of the water is therefore extremely critical. Deionized and distilled water should be used to avoid recontamination of the surface. High-purity electronic-grade isopropyl alcohol is a good alternative final rinse after water washing. The removal of residual water or alcohol is best effected by gentle centrifugation rather than by baking. Finally, storage of cleaned materials should be minimized or preferably avoided altogether by carrying out the cleaning treatment immediately before the next processing operation. If storage is necessary, chemically cleaned closed glass containers (such as Petri dishes) should be used and kept in a contamination-free clean-room atmosphere. The frequently used plastic containers are inadequate for this purpose, as they usually cause recontamination of clean surfaces due to the emission of organic vapors [79].

III. CHEMICAL ETCHING OF SPECIFIC MATERIALS

A. Insulators and Dielectrics

1. General Considerations

Important insulating and dielectric materials include grown and deposited vitreous and crystalline oxides, chemically vapor-deposited (CVD) binary silicates, fused multicomponent silicate glasses, CVD or sputter-deposited nitrides and oxynitrides, and several other compounds in thin-film or bulk form. The chemical etching properties of each of these groups of compounds will be discussed in this section.

In general, insulating and dielectric materials are relatively inert chemically and hence require highly reactive media for etching. Etchants of technical importance are ammonium fluoride–buffered hydrofluoric acid used for patterning by photolithographic techniques, strong aqueous hydrofluoric acid at room temperature, hot 85% phosphoric acid for pattern etching with oxide or metal masks, and miscellaneous other etchants, usually strong mineral acids or bases. Vapor or gas phase etching is used only in the preparation of insulator substrates.

The majority of insulator and dielectric compounds, being amorphous or extremely microcrystalline, are classified as glasses. Therefore, etching in these cases proceeds isotropically, and variations in the etch rate of a specific material in a given etchant are functions of chemical composition, film density, residual stress, defect density, and microstructure. The etch rate generally decreases as the density or crystallinity of a material increases.

As in all etching processes, selectivity is one of the most important etchant parameters in practical applications. A survey of the uses of selective etching of dielectrics in semiconductor device processing and in analytical applications for compositional and structural characterization has been published recently [27].

A qualitative summary of etchants for important insulators and dielectrics is presented in Table II of Section IV. A more concise compilation would be of questionable value because the etch rates depend very strongly on the exact conditions of film formation. Furthermore, materials consisting of more than one single component, such as silicate glasses, vary continuously in their etch rate according to composition, so that a graphical etch rate presentation is more instructive. Emphasis in this section is therefore placed on the discussion of general trends and a survey of specific results and references from the literature.

2. Single Oxides

a. SiO_2. Etchants for SiO_2 are based almost exclusively on aqueous fluoride solutions, usually HF with or without the addition of NH_4F. The exact chemical mechanism of dissolution is quite complex; it depends strongly on the ionic strength, the solution pH, and the etchant composition which determine the available quantities of solution species including HF_2^-, HF, F^-, H^+, and various fluoride polymers. Raman spectroscopy has indicated the presence of numerous reaction product species (such as hexafluorosilicate ions) in etch solution [28]. Detailed studies of the reaction mechanism underlying etching of SiO_2 have been reported by several investigators [28, 105–111].

Addition of NH_4F to HF to control the pH yields so-called buffered HF (BHF); it is important in pattern etching of SiO_2 films using photoresist masks [14] where attack of the photoresist masking layer and the polymer/dielectric interface must be minimized. Ammonium fluoride addition also prevents depletion of the fluoride ions, thus maintaining stable etching characteristics. The actual role of NH_4F may be one of an $(NH_4)_2SiF_6$ precipitating or complexing agent rather than that of a true buffer [106].

Selectivity in pattern etching of SiO_2 layers on Al device metallization can be improved over BHF by addition of a dihydroxyalcohol [112] or of glycerol [113] to the BHF to inhibit attack of the metal.

Pattern etching of SiO_2 films in vapors from aqueous HF, although rarely used, is an interesting alternative to liquid etching and can yield comparable results at reasonable rates [114]. It proceeds by formation of fluosilicic acid, which dissociates in the vapor phase to SiF_4 and HF.

As with insulator films in general, the liquid etch rate depends not only on etchant composition, agitation, and temperature, but also on the density, porosity, residual stress, microstructure, defect density, exact stoichiometry, and purity of SiO_2. A substantial increase in the etch rate of SiO_2 films has been observed as a result of defect generation by electron beam irradiation [115], Ar ion implantation [116], and ion bombardment [117]. Heat treatments of SiO_2 films deposited or grown at low temperature decrease the etch rate due to densification of the structure [118–120]; the same holds for other dielectric films.

Silicon dioxide films formed by different processes vary widely in their etch rate, mainly because of differences in stoichiometry, microstructure, and/or film density. For example, the etch rate of SiO_2 at 25°C in P etch (2 vol HNO_3 70%, 3 vol HF 49%, 60 vol H_2O) [121] for thermally grown (1000°C) or densified films is 2.0 Å/sec, for rf sputtered films 4–12 Å/sec, for organopyrolytic (undensified) films 6–20 Å/sec, for electron-gun evaporated films 20–70 Å/sec, and for anodized films 18–228 Å/sec [118].

Additional etch rate values for variously prepared SiO_2 films in HF and in BHF etchants have been reported for SiO_2 grown by thermal oxidation [65, 105, 106, 108, 110, 111, 122–124], deposited by sputtering [125–127], evaporation [16], halide reaction [128], pyrolysis or vapor oxidation of organics [16, 129–131], oxidation of SiH_4 at low (<500°C) [119, 120, 132–138] and high temperature [139–142], anodization [16, 143], by spin-on reagent solutions [144], and by plasma reactions [145–147]. Normalized etch rates for various types of SiO_2 films have been published [147a]. The effects of HF concentration and temperature [105, 107–111, 148–150] and the effects of agitation [109, 123, 148] on etching of SiO_2 films have been examined.

In addition to HF-containing etchants, SiO_2 is slightly soluble in hot H_3PO_4 and hot caustic solutions. For example, thermally grown SiO_2 is etched by 5M KOH at 85°C at ~50 Å/min [151], and 0.1M NaOH at room temperature at ≤ 0.2 Å/min [152]. The etch rate in 10 wt % NaOH at 23°C is 0.1 Å/min, at 55°C 5 Å/min, and at 90°C 500 Å/min [153]. The etch rate of oxygen-deficient SiO_2 in HF solutions decreases, and SiO requires the addition of HNO_3 to attain etchability. Alternatively, hot solutions of concentrated NH_4F mixed with NH_4OH or alkali hydroxides can be used for etching SiO films [14].

b. TiO_2, Ta_2O_5, and ZrO_2. As a general rule, dielectric films deposited at low temperature exhibit high etch rates (often due to their low density and amorphous structure), whereas films of the same compound that are annealed or deposited at high temperature exhibit consistently lower etch rates. For example, low-temperature (150–300°C) CVD TiO_2 [154–156] is readily etchable in 0.5% HF or in warm 98% H_2SO_4, whereas films annealed at 1000°C etch only slowly in 48% HF or in hot H_2SO_4 or H_3PO_4 [154, 154a, 155].

Pyrolytic Ta_2O_5 films deposited at 500°C are soluble in dilute HF [157]. Films of amorphous Ta_2O_5 (but not high-temperature crystalline Ta_2O_5 films [158]) formed by anodization of deposited Ta films can be etched in $HF–NH_4F$ solutions [159, 160]. Electron irradiation of Ta_2O_5 (and Al_2O_3) films decreases their etch rates [161], in contrast to SiO_2 films. Tantalum pentoxide films can be patterned with 9 vol NaOH or KOH (30%) plus 1 vol H_2O_2 (30%) at 90°C using a Au mask; the etch rate ranges from 1000 to 2000 Å/min [162, 163].

Monoclinic ZrO_2 films prepared by CVD from $ZrCl_4$ at 800–1000°C are slowly etchable only in hot H_3PO_4 [164].

c. Al_2O_3. Films of Al_2O_3 prepared by CVD below 500°C [165–169], grown by plasma oxidation [147, 170] formed anodically [161, 171, 172], or deposited at low temperature by evaporation [161, 173], obtained on Al by boiling in H_2O [174], or deposited by sputtering [175–177], are etchable in HF, BHF, warm H_3PO_4, and etchants based on H_3PO_4. Thermal densification at 700–800°C tends to form crystalline modifications that exhibit much lower etch rates [165, 168].

Aluminum oxide films deposited by the $AlCl_3$ hydrolysis process at 900–1000°C are nearly unetchable even in concentrated HF solution and require boiling 85% H_3PO_4 [135, 178–180]. The etch rate in 85% H_3PO_4 at 180°C is typically 100 Å/min; etch masks of CVD SiO_2 are useful for patterning these films [181].

Selective etching of anodic Al_2O_3 on Al in multilevel integrated circuits can be accomplished, without attacking the Al, by use of a solution containing H_3PO_4 and CrO_3 [182].

d. Bulk Oxides. Sapphire (α-Al_2O_3), spinel ($MgAl_2O_4$), and beryllia (BeO) used as substrates for heteroepitaxial CVD of silicon layers are slowly etchable in boiling concentrated $H_3PO_4-H_2SO_4$ mixtures [183, 184]. Gas-phase etching at high temperatures has also been used successfully for polishing sapphire [185] and spinel [186]. Dissolution of surface irregular-ities from crystalline Al_2O_3 has been accomplished by treatments with molten V_2O_5 above 800°C [187]; melts of $K_2S_2O_7$, PbO—PbF_2, and V_2O_5 have been used for dissolving sapphire [188]. BeO can be etched in hot (120°C) HCl [184].

e. Other Oxides. Films of Nb_2O_5 [189] and CVD HfO_2 [155, 189a] are etchable in HF. The etchability of GeO_2 depends on its crystallographic modification, as noted in Table I of Section IV. Several other oxides of importance are classified under compound semiconductors and are dis-cussed in Section III.B.4.

3. Binary and Ternary Silicate Glasses and Oxides

a. Phosphosilicates. The incorporation of P_2O_5 in the SiO_2 network yields technically very important phosphosilicate glasses (PSG). The etch rate in both HF and BHF increases with increasing P, as shown in Fig. 3 for a variety of film and etchant compositions. The low-temperature CVD films were prepared by chemical vapor reaction from SiH_4, PH_3, and O_2 in N_2 at 450°C [190–192]. Densification treatments were done in N_2 at 1000°C for 60 min, as indicated [138]. The thermal PSG layers were ob-tained by reacting vapors of $POCl_3$ [193] or P_2O_5 [194] with SiO_2 at 1000°C. The graphs illustrate at a glance the pronounced effects of composition and density for a given phosphorus concentration, and the difference in selectivity of a typical BHF composition and a variety of unbuffered HF mixtures. Note that the etch rate of PSG in BHF is much less affected by the film composition [137, 195] than in the unbuffered mixtures. However, the logarithm of the etch rate varies in all cases linearly with the P concen-tration. More complete graphs relating etch rate with film composition and CVD conditions have been published elsewhere [137, 195].

The etch rates of PSG films deposited from SiH_4, PH_3, and O_2 in Ar at 350°C, and containing 0–9 mole % P_2O_5, exhibit a maximum in 50%-diluted buffer etch (48% HF:40% NH_4F:H_2O, 1:10:11 by volume). The increase in etch rate with increasing P content in the glass suggests that the SiO_2 component of the PSG is dissolved by the buffered HF and the P_2O_5 component principally by water [196].

Additional etch rate studies have been reported for PSG films pre-pared by low-temperature (300–500°C) oxidation of the hydrides [136, 197–201], by high temperature (1000–1100°C) reaction of SiO_2 with $POCl_3$ or P_2O_5 [150, 193, 194, 197, 202, 203], and by pyrolysis of organometallics (700–800°C) [204].

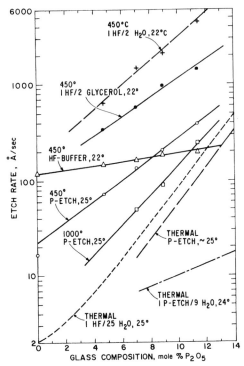

Fig. 3. Etch rates of phosphosilicate glass films versus mol % P_2O_5 in the glass. 450°C: CVD films from SiH_4–PH_3–O_2–N_2; 1000°C: CVD film densified at 1000°C in N_2 for 1 hr. Thermal, HF/H_2O, P/H_2O: SiO_2 + $POCl_3(O_2)$ at 1000°C [193]. Thermal, P-etch: SiO_2 + $P_2O_5(N_2)$ at 1020°C [194] (from Kern [27]. This figure was originally presented at the 149th Spring Meeting of The Electrochemical Society, Inc. in Washington, D.C.).

b. Borosilicates. Binary borosilicate glasses (BSG) are important as dopant sources and in silicon passivation [77, 205]. In early literature, it was reported that the HF etch rate of CVD, BSG, typically deposited at 450°C from SiH_4, B_2H_6, and O_2 in N_2, increases strongly with increasing B content, whereas in BHF it decreases sharply with increasing B content to a minimum and then increases with further B increase [190, 199]. These results have since been confirmed [126, 196, 206–209]. A series of graphs exemplifying the etching behavior of BSG films is shown in Fig. 4 [196]. Boron oxide at moderate concentrations in BSG, existing as a borosilicate rather than as B_2O_3, appears to protect the SiO_2 from attack by buffered HF; at high concentrations B_2O_3 bonding may be impaired and solubility in water becomes a more important factor, resulting in an increased etch rate [196]. Several other explanations have been proposed [28, 199, 206, 207].

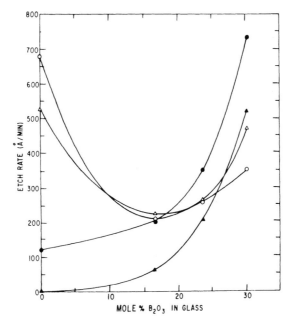

Fig. 4. Etch rates at 26° ± 1°C of heat-treated borosilicate glass films in BHF of various compositions versus mole percent B_2O_3 in the glass. Etching solutions were prepared by adding distilled H_2O in the proportions indicated to BHF (10 vol NH_4F 40%: 1 vol HF 48%). After 15 min at 1000°C in Ar: ○, 100% buffered HF; △, 50% buffered HF; ●, 10% buffered HF; ▲, 1% buffered HF (from Tenney and Ghezzo [196], reprinted by permission of the publisher, The Electrochemical Society, Inc.).

Etchant compositions for BSG have been reported that are particularly selective with respect to the B content [118, 196, 199, 208, 210, 211]; these are useful in analytical and processing applications. Densification by thermal treatments lowers the etch rate of BSG in all etchants [196, 198, 199, 212].

c. Arsenosilicates. Chemically vapor-deposited arsenosilicate glasses (AsSG), used as diffusion sources in silicon device technology, can be pattern etched readily in BHF [200, 213]. The etch rates of CVD AsSG films densified at 1100°C in Ar increase logarithmically and monotonically with increasing As_2O_3 content from 0 to 8 mole % As_2O_3, as shown in Fig. 5, with a nearly twofold increase in the etch rate over this concentration range [196, 214]. The incorporation of GeO_2 in the AsSG structure during CVD, used to improve the arsenic diffusion characteristics, enhances the etch rate of the film [215].

d. Aluminosilicates. Chemically vapor-deposited aluminosilicate glass (AlSG) films are also usually etchable in HF and in BHF [167, 199, 216,

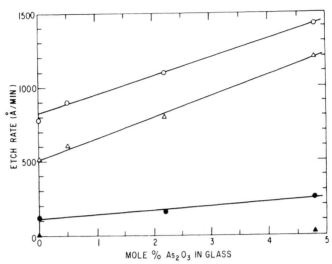

Fig. 5. Etch rates at 26° ± 1°C of heat-treated arsenosilicate glass films in BHF versus mole percent As_2O_3 in the glass. Etching solutions were prepared as noted in Fig. 4. After 5 hr at 1100°C in Ar: ○, 100% buffered HF; △, 50% buffered HF; ●, 10% buffered HF; ▲, 1% buffered HF (from Tenney and Ghezzo [196], reprinted by permission of the publisher, The Electrochemical Society, Inc.).

217]. The incorporation of Al_2O_3 into the glass structure tends to decrease the P-etch rate of the AlSG, whereas the addition of PbO increases it in comparison with SiO_2 [16]. High-temperature CVD films containing more than 50% Al_2O_3 are resistant to HF but are etchable in hot H_3PO_4 similar to Al_2O_3.

e. Other Silicates and Oxides. Additional binary and ternary silicate glass films synthesized by CVD below 500°C are all etchable in aqueous HF solutions, and include zinc silicates, zinc borosilicates, alumino-borosilicates, aluminophosphosilicates, lead silicates, and lead borosilicates [190, 191, 199]. Chemically vapor-deposited Al_2O_3 containing several percent Ta_2O_5 becomes amorphous and etchable with HF or BHF [218]. Germanosilicates are etchable in BHF.

Anodically grown native oxide films on GaAs are readily etched in dilute HCl solutions. Films heat treated at 600°C become unetchable in HCl, HNO_3, NH_4OH, or NaOH solutions, but they can be etched in hot concentrated H_3PO_4 [219] or concentrated HF.

Plasma-grown oxide films on GaAs are practically insoluble in acids and alkalis except boiling HCl (50–80 Å/min) [220]. Plasma-grown oxides of complex composition on $GaAs_{0.6}P_{0.4}$ and GaP, on the other hand, are easily soluble in acids and alkalis [220].

4. Multicomponent Silicate Glasses

Literature references on etching of multicomponent silicate glasses are very scant and are mostly concerned with chemical durability and corrosion effects [221, 222]. The essential component in all silicate glass etchants is HF. In general, addition of B_2O_3, ZnO, and PbO to the SiO_2 network increases the etch rate, whereas incorporating Al_2O_3 has a decreasing effect and leads to improved chemical resistance of the glass [222].

Certain glass compositions require the addition of complexing agents to minimize precipitation of insoluble metal fluoride reactants that mask and hinder smooth etching. Agitation, important in all etching processes, may therefore prove particularly important in attaining good uniformity. Ultrasonic treatment during etching can also be useful. Acid-stable surfactants functioning at low pH can be added to the etchant to enhance wetting characteristics and, in the case of pattern etching, further improve pattern resolution. Addition of HCl and H_2SO_4 to aqueous HF etchant can convert insoluble fluorides into soluble salts [223].

Some silicate glasses with very high Pb content are soluble in aqueous HNO_3 [224]. A reagent containing catechol and ethylenediamine tetraacetic acid etches soda-lime glass at a uniform rate of 10-15 interference fringes per hour [225].

Etch rates in P-etch of fused BSG and Pb–BSG films have been reported [118, 226]. Some lead borosilicates have etch rates in P-etch up to 500 Å/sec, as compared to Corning 7740 Pyrex glass with 8 Å/sec, and thermal SiO_2 with 2 Å/sec [226]. Etch rates have also been reported for rf sputtered Corning code 1720 AlSG [227] and General Electric GSC-1 BSG [228].

Most silicate glasses are slowly attacked by hot 85% H_3PO_4; Pyrex, for example, etches at a rate of 0.05 μm/min at 150°C [229]. Pyrex glass is also etched slowly by hot 98% H_2SO_4–30% H_2O_2 (1 : 1), and by hot concentrated solutions of NaOH or KOH [79].

Etch rates of several types of glasses in the form of plates were obtained in our laboratory [79]; a summary of the results is presented in Table II (Section IV.A).

5. Nitrides and Oxynitrides

Silicon nitride in the form of thin films is of great practical importance in semiconductor electronics because of its effectiveness as an alkali diffusion barrier. Chemically vapor-deposited Si_3N_4 films are etchable at room temperature in concentrated HF or BHF, in H_3PO_4 at 140–200°C, in 49% HF–70% HNO_3 (3 : 10) at 70°C, and in molten NaOH at 450°C [14,

230]. The etch rate is strongly affected by the presence of any oxygen linkages in the films; in HF and BHF it increases with increasing oxygen content, while in H_3PO_4 it decreases.

The dissolution process for CVD Si_3N_4 films in acidic fluoride media follows the same rate law as does thermal SiO_2 [231].

$$R = A[HF] + B[HF_2^-] + C,$$

where R is in angstroms per minute and the concentrations are molar. The rate constants for the dissolution processes are summarized in the accompanying tabulation.

Film	Temp. °C	A	B	C
Si_3N_4	25	0.16	0.31	<0.0001
	60	1.9	3.7	-0.02
SiO_2	25	2.50	9.66	-0.14
	60	10.4	48.6	-1.02

Pattern etching of Si_3N_4 films is usually carried out by reflux boiling of 85% H_3PO_4 at 180°C, with CVD SiO_2 as an etch mask. The etch rate under these conditions is typically 100 Å/min for CVD Si_3N_4, but only 0–25 Å/min for the CVD SiO_2 etch mask and 3 Å/min for any exposed single-crystal Si [232]. It should be noted, in this connection, that boiling H_3PO_4 must be treated with special care because of its complex chemistry [232, 233]. Etch rate plots as a function of temperature and H_3PO_4 concentration are presented in Fig. 6 [232].

Equal or similar etch rates for SiO_2 and Si_3N_4 are required in applications where the etched composite structure must have patterned walls with uniform taper angle. Hydrogen fluoride–water mixtures of optimal ratio at elevated temperature can etch composite layers of Si_3N_4 and SiO_2 at an equal rate [110, 111]. A 0.25 wt % HF solution at 90°C etches thermally grown SiO_2 and typical CVD Si_3N_4 films at an equal rate of 70 Å/min [234]. A 0.20 wt% HF solution at 90°C etches SiO_2 at 45 Å/min and Si_3N_4 at 60 Å/min; a near-unity etch rate ratio is generally preferable for film patterning of this type [234]. Comparison of activation energies for Si_3N_4 and SiO_2 dissolution processes suggests [231] that achievement of equal etch rates is also facilitated by use of low pH etchants, which maximize [HF] and minimize [HF_2^-].

Films of "silicon nitride" deposited by plasma-enhanced CVD at low temperature [235] have very much higher etch rates than high-temperature CVD Si_3N_4 [137, 146, 147, 236–240]. These rates depend strongly upon the film composition, which may be expressed as $Si_xN_yH_z$ [235]. It has been suggested that these plasma silicon nitride films should be re-

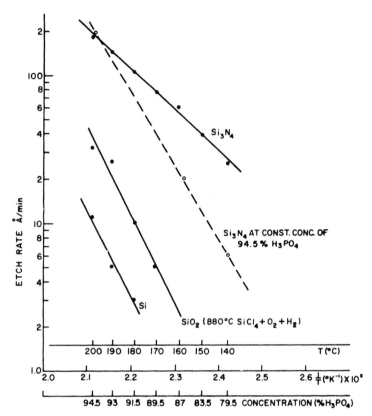

Fig. 6. Solid lines: etch rate of Si_3N_4, SiO_2, and Si in refluxed boiling phosphoric acid at atmospheric pressure as a function of boiling temperature and acid concentration. Dashed line: etch rate of Si_3N_4 at a constant concentration of 94.5% H_3PO_4 as a function of temperature only (from Van Gelder and Hauser [232], reprinted by permission of the publisher, The Electrochemical Society, Inc.).

garded as a polysilazane with unique properties rather than as a variety of silicon nitride [240].

Silicon nitride films deposited by sputtering techniques also have frequently lower etch rates than Si_3N_4 if they are less dense, contain large quantities of gases, and/or nonstoichiometric.

A very extensive review of the literature on etching data of Si_3N_4 and silicon oxynitrides was prepared by Milek [230]. More recent results of etching studies have been reported on Si_3N_4 prepared by CVD [110, 111, 241–243], by sputtering techniques [244–246], and for anodic conversion of Si_3N_4 to SiO_2 [247].

The etch rate of silicon oxynitride films of the general formula $Si_xO_yN_z$ is strongly influenced by the deposition conditions and by film density and stoichiometry [230]. Typical etch rates for CVD $Si_xN_yO_z$ films (prepared from SiH_4, NH_3, and NO) in 48% HF are 300–500 Å/min, as compared with 30,000–50,000 Å/min for SiO_2 and 130–150 Å/min for Si_3N_4; in 85% H_3PO_4 at 180°C the etch rates are 10–100 Å/min for $Si_xN_yO_z$, 8–10 Å/min for SiO_2, and 60–100 Å/min for Si_3N_4 [248]. More recently, some additional data have been published [249], including data for glow-discharge deposited films [147].

Other nitrides for which etching data are available include CVD amorphous Ge_3N_4 films that rapidly dissolve in concentrated HF, HNO_3, and hot H_3PO_4 [250, 250a]. Aluminum nitride, BN, and GaN and have been included in Section III.B.4 on compound semiconductors.

B. Semiconductors

1. Elemental Semiconductors

Silicon is by far the most important elemental semiconductor because of its widespread use in modern microelectronic and photovoltaic devices. Another elemental semiconductor, germanium, has become of lesser importance. Because of their similarities many of the etching processes can be applied to both. The discussion of Si and Ge etching shall be kept very brief because of space limitations. A more detailed presentation has been published elsewhere [27a]. Specific details on etchant composition, etching conditions, etch rates, and applications are listed in the tables of Section IV.C.

Selenium is a group VI elemental semiconductor for which very little information on etching exists. However, all three forms of Se_8 are soluble in H_2SO_4. In addition, the amorphous modifications are soluble also in CS_2, the monoclinic form in HNO_3, and the hexagonal crystal form in $CHCl_3$.

The discussion of elemental semiconductor etching that follows is concerned exclusively with Si and Ge.

a. Isotropic Liquid Etching. Isotropic chemical etching of semiconductors in liquid reagents is the most widely used etching process for removal of work-damaged surfaces, for creating structures or planar surfaces in single-crystal slices, and for patterning deposited semiconductor films. For Si, etchants containing HF, HNO_3, and H_2O are most frequently used; Ge etchants based on HF, H_2O_2, and H_2O are typical. Extensive studies have been reported on the mechanism of Si etching in $HF–HNO_3$

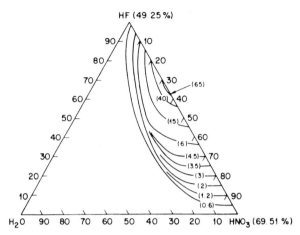

Fig. 7. Curves of constant rate of change of die thickness (mils per minute combined for two Si wafer surfaces) as a function of etchant composition, in the 49% HF–70% HNO₃ system (from Schwartz and Robbins [256], reprinted by permission of the publisher, The Electrochemical Society, Inc.).

[251, 252], in ternary mixtures of HF–HNO₃–H₂O, and in aqueous HF–HNO₃–CH₃COOH compositions [253–256]. In high-HF etchants the HNO₃ concentration determines the etch rate because oxidation is the rate-limiting step. In high-HNO₃ compositions, the etch rates are a function of only the HF because in this case dissolution is the rate-limiting process [253]. In Figs. 7 and 8 isoetch rate contours are shown for HF–

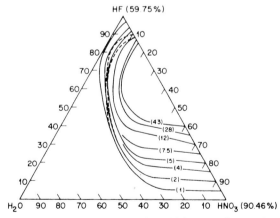

Fig. 8. Curves of constant rate of change of die thickness (mils per minute combined for two Si wafer surfaces) as a function of etchant composition in the 60% HF–90% HNO₃ system; the effect of added catalyst (NaNO₂) is shown as the dashed lines (from Schwartz and Robbins [256], reprinted by permission of the publisher, The Electrochemical Society, Inc.).

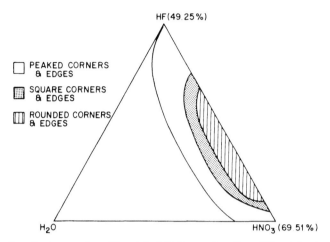

Fig. 9. Resultant geometry of the etched Si die as a function of the etchant composition in the 48% HF–70% HNO₃ system (from Schwartz and Robbins [256], reprinted by permission of the publisher, The Electrochemical Society, Inc.).

HNO₃–H₂O in normal and high concentration acids. The resulting geometry effects on initially rectangular (111)-plane parallelepipeds (n-type, 2Ω-cm dice) are indicated in Figs. 9 and 10 for the same etchants. The activation energies for both Si and Ge in HF–HNO₃–CH₃COOH etchant confirmed that the processes are diffusion controlled [257]. Silicon is also soluble, to a very small extent, in HF solutions and BHF [258–260]. Iso-

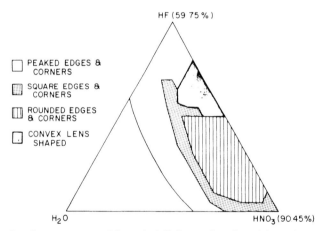

Fig. 10. Resultant geometry of the etched Si die as a function of the etchant composition in the 60% HF–90% HNO₃ system (from Schwartz and Robbins [256], reprinted by permission of the publisher, The Electrochemical Society, Inc.).

tropic liquid etching has been used for thinning of Ge and Si slices [261–266], for prepassivation surface cleanup [267], and for polishing [268–279]. Germanium etchants based on HNO_3-HF-H_2O or $HNO_3-HF-CH_3COOH$ are difficult to control, mainly because of variable induction periods [280–284]. The $HF-H_2O_2-H_2O$ etch system affords much better control [280, 284, 285]. Removal of thin layers Ge can be accomplished with 3% H_2O_2 at pH 3.8 [286]. Additional isotropic etchants [287–295] for Ge have been included in Table IV of Section IV.

b. Anisotropic Liquid Etching. In anisotropic or orientation-dependent etching, the etch rate varies within the principal crystallographic directions of the semiconductor single crystal. Orientation effects have been attributed to crystallographic properties, particularly the density of surface free bonds, the relative etch rate increasing with the number of available free bonds [8, 296–298].

Anisotropic liquid etchants for Si are usually alkaline solutions used at elevated temperature [21, 124, 298–312]. The essential feature for silicon technology of all these solutions is their up to 100 times higher etch rate in the ⟨100⟩ direction than in the ⟨111⟩ direction [303, 304]. For example, in the case of the water–ethylenediamine–pyrocatechol etchant, the etch rates of (100), (110), and (111) oriented Si are approximately 50:30:3 μm/hr, respectively [311]. Anisotropic etching of (100) Si through a patterned SiO_2 mask creates precise V grooves with the edges being (111) planes, at an angle of 54.7° from the (100) surface. Preferential etching allows fabrication of high-density monolithic integrated circuits [310, 313, 314], Si-on-sapphire integrated circuits [268, 305], and other Si devices that require structuring and patterning [311, 315–320]. Anisotropic etching of Ge for crystallographic studies has been reported [321].

c. Electrochemical and Selective Chemical Etching. Etching of semiconductors in liquid reagents by application of an external emf is used for preparing mirrorlike surfaces and creating very thin single-crystal films. Fundamental, theoretical, and practical aspects of semiconductor electrochemistry have been thoroughly treated in several books and reviews [30–37, 39–41].

Specific shapes can be imparted to Ge and Si crystals by controlled localized electrochemical etching [322–324]. Electropolishing of Ge and Si has been achieved by several techniques [325–328].

Selective electrochemical etching of single-crystal Si substrates having suitable epitaxial structures has been employed for preparing very thin Si crystals [269, 329–336]. *P*-type and heavily doped *n*-type Si can be dissolved anodically in dilute (1–5 *N*) HF at sufficiently low voltages, whereas *n*-type Si does not dissolve. Selective electrochemical thinning of

n^+-type Si substrates is also possible with alkali solutions in which the etch rates are strongly dependent on the electrode potentials [303, 337, 338].

Anodic oxidation of Si in electrolyte solutions based on organic media [339], followed by oxide dissolution, has been described for sectioning in the determination of Si diffusion profiles [340]. Objects in contact with the Si surface can either slow down or enhance the local etch rate considerably [341]. Substrate and etching conditions in the anodic dissolution of Si in aqueous HF can lead to brown layers, etch pits, and porous channels [269, 322, 328, 329, 331–333, 337, 342–344] caused by preferential etching and partial dissolution at localized sites [251, 331–333]. Single-crystal films of porous Si, formed purposely from n- and p-type Si by anodic reaction in concentrated HF [329, 344–346], are very similar or identical to these brown channeled layers.

Selective etching to dissolve Si of different dopant types and resistivities can also be achieved by chemical technique without use of external electrodes [263, 268, 304, 347–358], exemplified by Fig. 11 and the data presented in Table VI of Section IV.

d. Gas- and Vapor-Phase Etching. Gas- and vapor-phase etching are widely used for polishing of Si substrate wafers *in situ* prior to epitaxial crystal growth. The most successful reagent is sulfur hexafluoride, SF_6. It produces a smooth, mirrorlike surface when reacted in a dilution with H_2 at 950°C [359] or (more usually) above 1050°C, according to the overall reaction [360]:

$$4\ Si\ (s) + 2\ SF_6\ (g) \rightarrow SiS_2\ (s\ or\ l) + 3\ SiF_4\ (g).$$

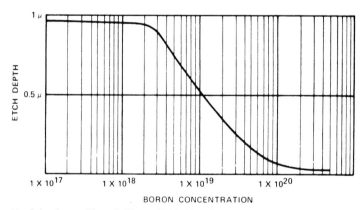

Fig. 11. Selective etching of silicon: Si ⟨100⟩ etch rate per minute versus boron concentration. Etchant system is KOH–H_2O–isopropyl alcohol at 80°C (from Kuhn and Rhee [348], reprinted by permission of the publisher, The Electrochemical Society, Inc.).

Since the free energy of the reaction is -706.81 kcal/mole at 1400°C [361], the etching proceeds spontaneously and irreversibly, producing volatile sulfides and fluorides of Si as the reaction products [360]. The etch rate as a function of temperature for 0.1% SF_6 is shown in Fig. 12, and as a function of SF_6 partial pressure in Fig. 13 [362]. The advantages of SF_6 over other reagents lie in the noncorrosiveness, nontoxicity, and low temperature needed, in addition to the excellent planarity attainable which makes the process applicable for wafer thinning.

Other vapor etchants used for polishing Si are HCl [363–370], HBr [365], HI [371, 372], Cl_2 [373], H_2S [374], HI–HF [375–376], and H_2O [377, 378].

In general, gas- and vapor-phase etching is not dependent on the resistivity level or type of the Si. Conditions can be selected in the HCl–H_2 etching system that result in anisotropic etching of various crystal planes at different rates; this has been exploited technologically in selective epitaxy [367, 369, 370], as noted in Section IV, Table VII.

Similar to Si, (111)-Ge can be polish etched with anhydrous HCl, HI–H_2, H_2S–H_2 [380]. Superior results are obtainable with H_2–H_2O vapor at 900°C which produces clean, structureless surfaces [380].

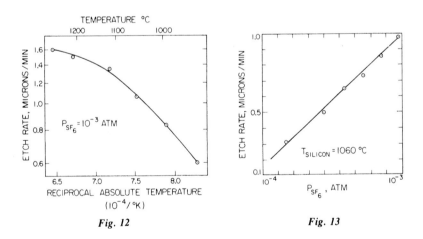

Fig. 12   *Fig. 13*

Fig. 12. Vapor phase etching of silicon: Si etch rate versus substrate temperature for a SF_6 pressure of 10^{-3} atm (from Stinson *et al.* [362], reprinted by permission of the publisher, The Electrochemical Society, Inc.).

Fig. 13. Vapor phase etching of silicon: Si etch rate versus SF_6 partial pressure at 1060°C (from Stinson *et al.* [362], reprinted by permission of the publisher, The Electrochemical Society, Inc.).

e. Chemical–Mechanical Polishing. Polishing by combined chemical and mechanical processes is usually the last step in preparing flat and specular wafers in silicon device manufacture. The generally preferred technique is the silica-sol (Syton [381]) method [382, 383]. The medium consists of a colloidal suspension of silica gel in aqueous NaOH solution of controlled pH and is dispensed on the polishing pad of a rotating polishing machine. Silicon removal proceeds by oxidation of the surface by water in the presence of the alkali ions and continuous dissolution of the surface oxide, aided by the silica gel which serves as a mild abrasive [383–386]. The process can also be used for polishing Ge wafers, but H_2O_2 must be added to the dispersion to achieve a smooth surface finish [383].

Another process for Si employs an aqueous solution containing copper and fluoride ions [387, 388]. The Cu^{+2} ions are reduced by Si to metallic Cu, and Si is oxidized to Si^{+4}. The Cu deposit on the wafer surface is removed by a polishing cloth, while the oxidized Si dissolves as fluosilicate.

Mechanical polishing without use of abrasive particles can be combined with liquid chemical etching by moving the semiconductor wafer with uniform pressure on a polishing cloth soaked with the etchant liquid. This technique has been used for Ge [389] and GaAs [390, 391].

Several reviews are available on semiconductor slicing, lapping, and polishing, and the damage introduced by these operations [5, 382–385, 387, 392–394].

2. Compound Semiconductors

Several reviews are available on etching of compound semiconductors [2, 5, 7, 10, 13, 19–23, 43a]; a very recent survey [27a] was prepared in conjunction with this chapter to complement the tables in Section IV.D. We shall therefore only outline this section, and present the details later in Tables VIII–XIII.

a. Group IV Compound Semiconductors. The only group IV compound semiconductor of technical importance is silicon carbide (SiC). Etchants consist of molten alkalis or borax [395–397], Cl_2–O_2 at 900°C and above [396, 398], and H_2 above 1600°C [399] (Table VIII). Electrolytic etching in HF solution is specific for *p*-type SiC [397, 400].

b. Group III–V Compound Semiconductors. The single most important compound semiconductor is single-crystal gallium arsenide (GaAs). Etching reactions of this, as well as other compound semiconductors, are complicated because of crystallographic surface orientation effects. Chemical etching of GaAs (and other III–V and II–VI compound semiconductors) proceeds by oxidation-reduction-complexing reactions analogous, in principle, to the general mechanism for Si and Ge etching.

The most commonly employed etchants for GaAs are Br_2-CH_3OH [401-404], $NaOH-H_2O_2$ [405, 406], $H_2SO_4-H_2O_2-H_2O$ [406-408], and $NH_4OH-H_2O_2-H_2O$ [404, 409]. High-viscosity etchants such as glycerol-$HCl-HNO_3$ are preferred for chemical polishing [410]. Jet etching and/or slice rotation on a polishing pad have been used with several etchants for polishing and thinning [23, 266, 390, 391, 404, 411, 412].

Orientation-dependent etching characteristics of $8H_2O_2-1H_2O-1H_2SO_4$ were utilized for etching channels of various geometries for Gunn-effect logic circuits [413] and for superlattice structures deposited by molecular beam epitaxy [414]. Gallium arsenide double heterostructure lasers have been fabricated using a smoothly acting selective etchant consisting of $CH_3OH-H_3PO_4-H_2O_2$ [415]. Several additional etchants have been used for crystallographically preferential etching [403, 408, 409, 416-418]. Older but useful chemical solution etchants [402, 419-424] are also listed in Section IV, Table IX. It should also be noted that very thin (≤ 100 Å) carbon films are impervious to common etchants for III-V compounds and can therefore be used as excellent pattern-definition masks [425].

Etching of GaAs can also be accomplished by electrolytic techniques in alkali solutions [426-430], in $HCl-H_2O$ [431], or in HNO_3-H_2O [432].

Vapor-phase etching of GaAs substrates in preparation for epitaxy includes $H_2-H_2O-N_2$ at 1000°C [433], and anhydrous HCl in a H_2-As ambient above 870°C [434]. Since this latter process is critically dependent upon the purity of the HCl etchant, $AsCl_3$, which is available at very high purity, is an attractive alternate etchant [434a].

Redox solutions of suitable composition and pH can etch GaAs selectively with respect to $Ga_{1-x}Al_xAs$, or vice versa [22, 435]. The same behavior is displayed by GaP-InGaP and GaP-GaAlAs heterostructures. Other redox systems etch selectively with respect to the dopant type in these materials. Aqueous redox solutions that are stable in both acid and alkaline solutions include I_2-KI, $K_3Fe(CN)_6-K_4Fe(CN)_6$, and $C_6H_4O_2-C_4H_6O_2$ (quinone-hydroquinone). Redox etchants that are stable only in the acid pH range are $FeCl_3-FeCl_2$ and $Ce(SO_4)_2-Ce(NO_3)_3$.

Gallium phosphide can be solution etched in the redox systems just noted and in the etchants listed in Table X [20, 402, 436-446]. Several electrochemical etchants have also been included [266, 432, 443, 447, 448].

Etching characteristics of other III-V compound semiconductors for which data have been published are listed in Section IV, Table XI [8, 402, 418, 423, 449-465].

c. *Other Compound Semiconductors.* Etching characteristics for II-VI compound semiconductors [162, 466-478] are listed in Table XII; those for selected other compound and alloy semiconductors [479–501] are summarized in Section IV, Table XIII.

C. Conductors

Conducting materials may conveniently be divided into two categories: elemental metals and alloys. Many general references for metal etching are available [11, 14, 15, 24, 25, 29, 38, 42, 44, 502].

Electrochemical etching procedures have been devised for virtually every conductor known [29, 503]. The material to be etched is made the anode of an electrochemical cell; the etch rate and surface finish are determined by the cell voltage, current density, and cathode material, as well as by the etching medium and other factors involved in chemical etching processes. The method of potential scan for etching metals [58] involves determining the potential range for oxidation using electrochemical methods, and then substituting a chemical reagent of suitable oxidizing potential for the applied voltage.

In the semiconductor device industry, the metal most often involved in pattern delineation processes is Al, and accordingly, many etchants for Al have been reported [14, 46, 57, 58, 504, 505].

An assortment of etchants is available for other important elemental metals: Cu and Ni [14, 15, 506–509]; Au and Ag [14, 15, 58, 510–512]; Pd and Pt [14, 507, 513, 514]; Rh [514]; Cr, Mo, and W [14, 15, 51, 507, 515–521]; Pb and Sn [15, 29, 513]; Nb and Ta [14, 29, 162, 513, 522, 523]; Ti, Zr, and Hf [14, 15, 29, 507, 524]; Mn [469, 513]; Co [29]; and Fe [29, 469].

Details of etching procedures for these and other elemental metals are listed in Section IV, Table XIV; etchants for several important alloys appear in Table XV.

D. Miscellaneous Materials

Included in this category are materials which may be required to be etched in various thin film processes, but which do not readily fall into one of the categories already discussed: insulator, semiconductor, or conductor.

The majority of entries in Section IV, Table XVI are oxides of various metals. Theory and procedures for oxide film dissolution have been reviewed [525]. Iron oxides [526–528] are a particularly important example, as are the garnet materials [529–531], generally mixed oxides of rare-earth elements.

A few nonmetallic elements have been included in the miscellaneous materials category, and a few other materials of known interest, mostly silicides or carbides complete the list.

IV. TABLES OF ETCHANTS AND ETCHING CONDITIONS

A. Guide to the Use of Tables

In this Section we have compiled and referenced specific etchants and etching conditions used for numerous technically important thin film materials and substrates. The crystallographic orientation and the dopant type and concentration, or electrical resistivity of semiconductor materials have been defined where critical. The average etch rate or recommended period of etching is also noted if the information was available. Typical applications are indicated in the last column.

The compositions of concentrated aqueous reagents noted in the tables are given in the accompanying tabulation.

Reagent	Wt %	Reagent	Wt %
HCl	37	H_2O_2	30
HF	49	N_2H_2 (hydrazine)	64
H_2SO_4	98	NH_4OH	29 (as NH_3)
H_3PO_4	85	NaOH	97 (min.)
$HClO_4$	70	KOH	85 (approx.)
HNO_3	70	Na_2O_2 (sodium peroxide)	93
CH_3COOH	99		

Additional definitions are

EDTA stands for ethylenediaminetetraacetic acid.
Aqua regia consists of 3 parts by volume of HCl plus 1 part by volume of HNO_3.
BHF stands for buffered HF prepared by mixing NH_4F, H_2O, and HF at appropriate ratios, i.e., 6 vol NH_4F (40%) plus 1 vol HF.
The abbreviation Bé stands for Baumé, designation of solution density.
Reagent ratios cited are on a volume basis unless otherwise specified.
All etching temperatures refer to degrees centigrade and are room temperature, unless otherwise noted.
The liquid diluent is deionized and/or distilled H_2O unless otherwise indicated.
Gaseous reagents are pure, concentrated gases unless otherwise noted.
The abbreviation SCE stands for saturated calomel electrode.

The tables (pp. 435–480) have been arranged according to the following outline:

1. Insulators and Dielectrics

> Table I—General Survey
> Table II—Selected Bulk Silicate Glasses

2. Elemental Semiconductors

> Tables III–VII—Silicon and Germanium

3. Compound Semiconductors

> Table VIII—Silicon Carbide
> Table IX—Gallium Arsenide
> Table X—Gallium Phosphide
> Table XI—Other Group III–V Compounds
> Table XII—Group II–VI Compounds
> Table XIII—Other Selected Compound Semiconductors

4. Conductors

> Table XIV—Elemental Metals
> Table XV—Metal Alloys and Superconductors

5. Miscellaneous Materials

> Table XVI—Miscellaneous Materials.

B. Insulators and Dielectrics

Table I

General Survey of Insulators and Dielectrics

Material	Modification, structure	Formation[a] or anneal temp (°C)	HF	BHF	H_3PO_4	H_2SO_4	NaOH (~30%)	Other etchants, fig. no.
SiO_2	Amorphous	<500	c	d	e,f	g	e,f	HF vapor (Figs. 3–4)
	Amorphous or crystalline	>500, thermal or bulk	d	h	e,f	g	e,f	HF vapor (Figs. 3–6)
SiO	Amorphous	<200, evap.	g	g	g	g	g	HF + HNO_3; NH_4F + KOH[f]
TiO_2	Amorphous	150–300	c	d	f,h	h,i	h,j	
	Crystalline	1000	e	g	e,f	e,f	e,j	
Ta_2O_5	Amorphous	≤500	d	h	g,j	g,j	g,j	KOH + H_2O_2 + H_2O[f]
	Crystalline	>500	g	g	g,j	g,j	g	KOH + H_2O_2 + H_2O[f]
ZrO_2	Monoclinic	800–1000	h,j	g	f,h	f,j	g	
Al_2O_3	Amorphous	<500	c	d	h,i	g	g	H_3PO_4 + CrO_3 + H_2O
	Crystalline	1000	g	g	f,h	g	g	
	Sapphire	>1000, fusion	g	g	g	g	g	H_3PO_4 + H_2SO_4 (f); gas phase (f); molten salts (f)
$MgAl_2O_4$	Spinel	>1000, fusion	g	g	g	g	g	H_3PO_4 + H_2SO_4(f); gas phase (f)
PSG	Amorphous	300–1100	c	d	e,f	g	e,f	(Fig. 3)
BSG	Amorphous	300–900	c	e	e,f	g	e,f	(Fig. 4)
AsSG	Amorphous	300–1100	d	h	e,f,j	g,j	e,f,j	(Fig. 5)
AlSG	Amorphous	300–1000	d	h	f,g,j	g	e,f,j	
PbSG	Amorphous	300–800	c	d	f,h,j	g,j	f,h,j	HNO_3 + H_2O (high Pb)

Etchants, etch rates, and conditions[b]

(Continued)

435

Table I (*Continued*)

Material	Modification, structure	Formation[a] or anneal temp (°C)	Etchants, etch rates, and conditions[b]					Other etchants, fig. no.
			HF	BHF	H_3PO_4	H_2SO_4	NaOH (~30%)	
Multisilicates	Amorphous	>1000, fusion	d	h	e, f	g	e, f	(Table II)
GaAs-oxides	Amorphous	<300, plasma	g	g	g	g	g	HCl (f)
	Amorphous	600, anodic	h	g	h, f	j	g	
$Si_xO_yN_z$	Amorphous	700–1000	d	h	e	g	g	HCl (f)
Si_3N_4	Amorphous	700–1000	h	e	h	g	g	(Fig. 6)
$Si_xN_yH_z$	Amorphous	250, plasma	c	h	d	g	g	
Ge_3N_4	Amorphous	400–600	c	d	f, h	e	e	HNO_3
HfO_2	Monoclinic	400–550	h	e	e, f	f, g	f, g	
Nb_2O_5	Rhombic	>1000, bulk	h, i	g, j	j	e, j	h	
GeO_2	Hexagonal	>1000, bulk	c	g	h	h	h	H_2O^f, HCl
	Tetragonal	>1000, bulk	g	g	j	j	e, f	

[a] Deposition method: CVD unless otherwise indicated.
[b] Etching temperature: cold unless otherwise indicated.
[c] Etch rate: very high.
[d] Etch rate: high.
[e] Etch rate: low.
[f] Etching temperature: hot.
[g] Etch rate: nil.
[h] Etch rate: medium.
[i] Etching temperature: warm.
[j] Etch rate: best guess.

Table II

Selected Bulk Silicate Glasses[a,b]

		Etch rate at 25°C (μm/min)		
Glass code	Glass type composition	HF–H$_2$O[c]	HF–HNO$_3$–H$_2$O[d]	BHF[e]
H115[f]	Zinc borosilicate	39.1	28.6	0.080
C7720[g]	Lead borosilicate	9.7	10.0	0.09
C7059[g]	Barium aluminosilicate	9.0	4.7	0.16
C1715[g]	Calcium aluminosilicate	7.5	3.0	0.22
C9741[g]	Aluminoborosilicate	6.0	5.4	0.13
C1717[g]	Calcium aluminosilicate	5.6	1.8	
C0211[g]	Alkali zinc borosilicate	5.5	2.4	
C1720[g]	Calc.magn. aluminoborosilicate	4.3	2.6	
C167OH[f,g]	Barium aluminoborosilicate	3.2	2.7	0.6
C7070[g]	Borosilicate	3.2	1.9	0.040
C7740[g]	Borosilicate	1.9	1.9	0.063
SiO$_2$[f]	CVD (450°C) silica film	1.80	0.44	0.48
C7052[g]	Aluminoborosilicate	1.7	1.8	
SiO$_2$[f]	Thermally grown silica film	0.30	0.11	0.11
SiO$_2$	Fused natural silica	0.29	0.15	0.12

[a] Data from Kern (79).
[c] 1 vol HF 49% + 1 vol H$_2$O.
[d] 20 vol HF 49% + 14 vol HNO$_3$ 70% + 66 vol H$_2$O.
[e] 454 g NH$_4$F (crystal) + 654 ml H$_2$O + 163 ml HF 49%.
[f] Experimental sample.

[b] According to metal oxide components of ≥5 wt %. Lapped and polished disks unless otherwise stated; immersed vertically in etch solution with light agitation; silica included for comparison. Listed in order of decreasing HF etch rates.
[g] Corning Glass Works.

C. Elemental Semiconductors

Table III

Isotropic Liquid Etching: Si

No.	Etchant	Substrate, etching conditions	Application, etch rate, remarks	Refs.
1	HF, HNO$_3$, H$_2$O, or CH$_3$COOH	(111) Si, n-type 2 Ω-cm; 25.0° bath, stirring, sample in agitated Teflon basket	Ternary diagrams of isoetch rate contours and resultant geometry versus composition of etchants; allows selection of optimal cond. for any appl. (see Figs. 7–10)	[256, 253, 254]
2	HF, HNO$_3$, H$_2$O, or CH$_3$COOH	(111), (100), (110) Si, n-type, 3 Ω-cm; Pt beaker, 0–50°, Pt-mesh sample basket, agitation	Graphs of reaction rates versus 1/T, for several compositions; E_a ranging from 4 to 20 kcal/mol	[255]
3	HF, HNO$_3$ (a) 98 HNO$_3$, 2 HF (b) 98 HNO$_3$, 2 HF (c) 98 HNO$_3$, 2 HF (d) 91 HNO$_3$, 9 HF (e) 84 HNO$_3$, 16 HF (f) 75 HNO$_3$, 25 HF (g) 66 HNO$_3$, 34 HF	(111) Si, n-type 0.05–8 Ω-cm, p-type 12–78 Ω-cm, stirring and sample rotation, 30° 0 r/min 45 r/min 88 r/min 88 r/min 88 r/min 88 r/min 88 r/min	General etching; no difference in etch rates for n-, p-, n-p-type 0.25 μm/min ⎫ 0.60 μm/min ⎬ rotation effects 0.92 μm/min ⎭ 5.0 μm/min best control with 10 μm/min good results using ~20 μm/min <65% HF etchants 50 μm/min	[252]
4	900 ml HNO$_3$, 95 ml HF, 5 ml CH$_3$COOH, 14 g NaClO$_2$	Si wafers; 0.5 liter/min CO$_2$ bubble stream in illumin. Teflon app.; float-mounted	For large-area wafer thinning to <25 μm; 15 μm/min, uniform ±2.5 μm, independent of orient., or resistivity blemish-free	[264]
5	15HNO$_3$, 5HF, 3CH$_3$COOH	n- and p-type Si; planetary jet polishing appl. with reciprocating nozzle at ~950 cycles/min, 4 cm³/sec flow rate	For polishing and thinning n- and p-type Si; nonpreferential, smooth surfaces to <20 μm thickness; ~0.5 μm/sec	[266]

438

	Composition	Material / Conditions	Results	Ref.
6	$10HNO_3$ (65%), $1HF$ (50%)	Si wafers >200-μm thick; 4 liter/min CO_2 bubble stream in illumin. Teflon appl. wafer rotates 8 r/min, on/off 2:1	For small-area wafer thinning to 3 μm; (12–15 μm/min, uniform to <0.3 μm	[265]
7	One of the following: (a) $5HNO_3$, $3HF$, $3CH_3COOH$ (b) 17 ml ethylene diamine, 8 ml H_2O, 3 g pyrocatechol (Ref. 311)	(100) Si, 0.25-mm thick wafers For bulk thinning, followed by (b) For final thinning	For wafer thinning to 2 μm ~25 μm/min to ~5 μm thickness ~0.8 μm/min to ~2 μm thickness	[263]
8	$745HNO_3$ (65%), $105HF$ (40%), $75CH_3COOH$ (96%), $75HClO_4$ (70%)	Si power device wafers (pnp); etch for 10 sec	For prepassivation surface etching 1700 Å/10 sec; 18 MΩ-cm H_2O rinse	[267]
9	One of the following: (a) 95 ml HNO_3 (65%), 5 ml HF (40%), 1.0 g $NaNO_2$ (b) 100 HNO_3 (65%), 40 H_2O, $6HF$ (40%)	n- and p-type Si except high conc. B doped poly Si epi Si (100), n and p doped bulk Si (100), low doped poly Si epi Si (100), n and p doped bulk Si (100), low doped	For pattern etching with SiO_2 mask 3 μm/min; smooth pattern edges 4 μm/min; nonlinear pattern edges 4 μm/min; nonlinear pattern edges 0.8 μm/min; smooth pattern edges 0.6 μm/min; nonlinear pattern edges 0.5 μm/min; nonlinear pattern edges	[268]
10	50 ml HF, 50 ml CH_3COOH, 200 mg $KMnO_4$ (fresh)	(111), (100), epi Si; n, $p < 2 \times 10^{16}$; 18°	For epi Si etching ~0.2 μm/min	[269]
11	(a) HF (b) $4H_2O$, $1HF$ (c) 96 ml H_2O, 2 ml HF, 4.3 g NaF	(111) Si, n-type 2 Ω-cm; 25°	For surface etching 0.30 Å/min 0.47 Å/min 0.78 Å/min	[258]
12	108 ml HF and 350 g NH_4F per 1000 ml	Single-crystal Si n-type 0.2–0.6 Ω-cm p-type 0.4 Ω-cm p-type 15 Ω-cm	For epi Si etching 0.43 ± 0.08 Å/min 0.45 ± 0.08 Å/min 0.23 ± 0.1 Å/min	[259]

(Continued)

TABLE III (Continued)

No.	Etchant	Substrate, etching conditions	Application, etch rate, remarks	Refs.
13	$15HNO_3$, $5CH_3COOH$, $2HF$ (planar etch)	Si	For general etching	[270]
14	110 ml CH_3COOH, 100 ml HNO_3, 50 ml HF, 3 g I_2 (iodine etch)	Si	For general etching	[9]
15	$30HNO_3$, $20HF$, $1Na_2HPO_4$ (2%)	Si	For general etching; produces superior surface finish	[271]
16	$30HNO_3$, $25CH_3COOH$, $20HF$, $1Na_2HPO_4$ (2%)	Si	For polishing	[271]
17	$9HNO_3$, $1HF$ (white etch)	Si; 15 sec	For polishing	[272]
18	$14CH_3COOH$, $10HF$, $5HNO_3$	Si; 0.5–3 min	For polishing	[273]
19	$5HNO_3$, $3HF$, $3CH_3COOH$	Si; 2–3 min	For slow polishing (slightly preferential for crystal defects)	[274]
20	$5HNO_3$, $3HF$, $3CH_3COOH$, $0.06Br_2$; (CP-4)	(111), (100) Si; 2–3 min	For fast polishing; 25 µm/min; ("chemical polish No. 4")	[274]
21	10 ml H_2O_2 (33%), 3.7 g NH_4F	Si	For polishing; 1 µm/14 min	[275]
22	1000 ml H_2O, 100 g NH_4F, 2 ml H_2O_2	Si	For pattern etching; low degree of undercutting of photoresist mask because of nearly neutral pH	[276]
23	$95HNO_3$, $5HF$	Si wafer attached to Teflon holder; rotating	For wafer thinning	[277, 278]
24	$9HNO_3$, $1HF$	Si; jet technique	For wafer thinning microscopy	[261]
25	NaOH (4%), add NaOCl (40%), until no H_2 evolu. on Si	Si; float wafer specimen; approx. 80°	For thinning slices for electron microscopy	[279]

Table IV

Isotropic Liquid Etching: Ge

No.	Etchant	Substrate, etching conditions	Application, etch rate, remarks	Refs.
1	50 wt % HF, 50 wt % H_2O_2	(100), (110), (111) Ge, n-type, 3 Ω-cm preetched; solution used within 0.5–4 hr, 25°, agitated (100) (111) (110)	For general etching, and polishing; orientation dependent; triaxial composition versus etch rate given 11.2 $\mu m/min$ 21.1 $\mu m/min$ 39.3 $\mu m/min$	[285]
2	175 ml H_2O_2 (3–4%), 25 ml 0.2 M KH_2PO_4 containing 12 ml H_3PO_4 per liter	(111), (100), (201) Ge, n-type, high resistivity; magnetic stirrer. Etchant composition yields 2.7–3% H_2O_2 of pH 3.8	For controlled fractional-μm removal; etch rate insensitive to concentration and orientation at pH 3.8, 0.02 $\mu m/min$; sensitive to pH (min. at pH 4)	[286]
3	5H_2O, 1H_2O_2	(111) Ge; 26°	For general etching	[287]
4	4H_2O, 1HF, 1H_2O_2 (No. 2 etch; superoxol etch)	(100), (111) Ge; 1–3 min	For general etching; slow to attack polished surfaces; may develop etch figures	[287, 288, 289]

(Continued)

441

Table IV *(Continued)*

No.	Etchant	Substrate, etching conditions	Application, etch rate, remarks	Refs.
5	100 ml H_2O_2 (10 vol %), 8 g NaOH	Ge; 70° Freshly prepared After 1 hr	For controlled etching 5 μm/min 1.25 μm/min	[290]
6	5HNO_3, 3HF, 3CH_3COOH with 0.06 Br_2 (CP-4)	(100), (111) Ge; 1.5 min for general etching, $\geq$2 min for polishing	For general etching and polishing; slightly preferential for crystal defects	[288, 291, 292]
7	5HNO_3, 3HF, 3CH_3COOH (CP-4A, CP-6, CP-8)	Ge; 23–70°; 2–3 min	For slow polishing; much slower than CP-4 at 23°	[288, 293]
8	11 ml CH_3COOH with 30 mg I_2 dissolved, 10 ml HNO_3, 5 ml HF (iodine etch A)	(100), (111) Ge; 4 min	For general etching and polishing; better than CP-4 for (100) Ge	[294]
9	1NaOCl (10%), 10H_2O	(100), (111) Ge; 40°, 40 min or as required for thinning	For general etching and thinning	[295]
10	9HF, 1HNO_3	Ge; jet technique	For small-area thinning for electron microscopy	[261]
11	NaOCl, H_2O	Ge; warm, float specimen	For thinning slices for electron microscopy	[279]

442

Table V

Anisotropic Liquid Etching: Si

No.	Etchant	Substrate, etching conditions	Application, etch rate, remarks	Refs.
1	KOH solutions	(100) epi Si, n-type, 0.01–10 Ω-cm; 60°	For patterning epi Si on sapphire; considerably lower etch rates with p-type	[305]
	(a) KOH (4 N)		Si, 60 Å/sec	
	(b) 6KOH (4 N), 1 isopropanol		27 Å/sec diffusion controlled for $\leq 4 N$ KOH	
	(c) 6KOH (6 N), 1 isopropanol		35 Å/sec activation controlled for $\geq 5 N$ KOH	
2	50 ml H_2O, 15 g KOH, 15 ml isopropanol	P doped $\leq 1.5 \times 10^{20}/cm^3$ Si B doped $<5 \times 10^{17}/cm^3$ Si	For patterning epi Si on sapphire or spinel. B doping $>5 \times 10^{17}/cm^3$ lowers etch rate	[268]
		(100) epi Si	0.9 μm/min	
		(100 bulk Si	1.1 μm/min	
		poly Si	0.7 μm/min	
3	100KOH (2 M), 25 isopropanol	(100) Si; 44–45° 2 Ω-ohm p-type	For preparing planar recessed Si structures ~10 Å/sec	[313]
4	KOH solutions; hydrazine	(100) Si	For texturizing and V grooving solar cells for V-groove etching (no texturizing) for diffuse-reflectivity texturizing	[318]
	(a) KOH (3–50%)	70–90°, SiO_2 masked		
	(b) 60 hydrazine, 40H_2O	110°, 10 min, unmasked		
5	Hydrazine–H_2O, equimolar	Si bulk crystal; 120° (100) Si	For structural etching in presence of Al; SiO_2; 200 μm/hr (111) minimum. (211) maximum etch rate	[296]

(Continued)

443

Table V (*Continued*)

No.	Etchant	Substrate, etching conditions	Application, etch rate, remarks	Refs.
6	Hydrazine–H_2O (a) 65 hydrazine, $35H_2O$ (b) 80 hydrazine, $20H_2O$	(100) Si, 3–5 Ω-cm; 100°, reflux	For shaping Si bulk and films; higher etch rate in<100> direction than along <111> 1.6 μm/min; V-groove patterns 0.7 μm/min; flat-bottom patterns	[310]
7	17 ml ethylene diamine, 8 ml H_2O, 3 g pyrocatechol	n-, p-types; 0.001–100 Ω-cm resist.; N_2 bubbling, reflux, 110° + ±1° (100) (110) (111) SiO_2	For shaping of Si films 50 μm/hr 30 μm/hr 3 μm/hr ~200 Å/hr, suitable as etch mask	[311]
8	Tetramethylammonium hydroxide, or trimethyl-2-hydroxyethyl ammonium hydroxide (0.5 wt %)	(100), (111) Si; 80–90° (100) n type (100) p type (111) n type Thermal SiO_2 CVD SiO_2	For alkali-free Si etching 3600 Å/min 2300 Å/min 163 Å/min 3 Å/min, suitable as etch mask 7 Å/min, suitable as etch mask	[312]
9	4 M NH_4F–1 M, $Cu(NO_3)_2$	n-type Si, 10–100 Ω-cm; 22° (100) (110) (111)	For very high resolution patterning; anisotropic displacement etching 0.185 μm/min 0.117 μm/min 0.012 μm/min	[298]
10	100 g KOH in 100 ml H_2O	(110) Si, SiO_2 masked; boiling (110) moat etching	For vertical deep etching of moats in (110) Si 50 μm/6 min	[301]
11	63.3 wt % H_2O, 23.4 wt % KOH, 13.3 wt % isopropanol	(111), (100) Si bulk (111), doped with As, P, Sb, B (100), doped with As, P, Sb	For structural etching, independent of resistivity 0.97 μm/min 0.04 μm/min (but not for B doped)	[304]

Table VI

Electrochemical and Selective Chemical Etching: Si

No.	Etchant	Substrate, etching conditions	Application, results, remarks	Refs.
1	5% HF	Epi Si, bulk Si; Pt cath., ~150 mA/cm^2, ~6 V	Selective substrates dissolve leaving thin epi Si films (0.5–20 μm thick, 10 cm^2)	[329]
		n-epi Si >0.3 Ω-cm on n^+ bulk Si <0.015 Ω-cm	n^+ dissolves leaving n epi	
		p-epi Si >0.5 Ω-cm on n-epi on n^+ bulk Si <0.015 Ω-cm	n^+ dissolves anodically, n epi is etched off chemically leaving p epi	
2	5% HF	(111), (100) Si, n, p type; Pt gauze cath., Pt wire to n^+ Si, 10 V, 18°, 5 cm sep., darkness	Preferential etching and thinning	[269]
		N_D >3 × 10^{18}/cm^3 (0.01 Ω-cm)	complete dissolution	
		N_D 3 × 10^{18} − 2 × 10^{16}/cm^3 (0.01–0.3 Ω-cm)	partial dissolution decreases with N_D forming brown, porous structure.	
		N_D ~2 × 10^{16}/cm^3 (>0.3 Ω-cm)	no etching	
		N_A >5 × 10^{15}/cm^3 (>3 Ω-cm)	dissolution increases with N_A conc.	
3	5 wt % HF	(111), (100) Si, n^+-type, 75–175 μm thick best; Pt-mesh cath., 2 Pt wires to n^+ Si, 1 cm sep., darkness, controlled rate of Si immersion	Polishing and thinning of n^+ Si; thinning of n^+n leaving n-epi layer	[330]
		(111), 0.0012 Ω-cm: 115 mA/cm^2, 4.3 V		
		(111), 0.0038 Ω-cm: 80 mA/cm^2, 5.2 V		
		(111), 0.013 Ω-cm: 70 mA/cm^2, 5.8 V		
		(100), 0.0035 Ω-cm: 125 mA/cm^2, 5.3 V		
		(100), 0.011 Ω-cm: 135 mA/cm^2, 6.3 V		
4	1HF, 1H$_2$SO$_4$, 5H$_2$O: freshly mixed	p^+ bulk Si, n-epi Si, N_D 2 × 10^{16}/cm^3; Pt cath., 0.5 V, agitation, no darkness needed: 80mA/cm^2 at 25°, 200 mA/cm^2 at 65°	Preferential etching of p^+ Si without formation of brown film or rough surface	[333]

(Continued)

Table VI (*Continued*)

No.	Etchant	Substrate, etching conditions	Application, results, remarks	Refs.
5	5 wt % HF (2.5 N)	(111), (100) Si; Pt-mesh cath., 2 cm sep., 20°, N_2 bubbling, darkness n^+ 0.001–0.01 Ω-cm: 65–150 mA/cm², 3.5–6 V p 0.01–1 Ω-cm: 120–160 mA/cm², 4 V	Polishing and thinning n^+, p Si optimum polish thinning optimum polish thinning	[328]
6	5% HF	(111), (110), (100) Si, n^+, $N_D > 2 \times 10^{18}$ /cm³; conditions as in Ref. 269 ~130 mA/cm², 10 V, $N_D > 2 \times 10^{18}$ /cm³; <130 mA/cm², <10V, $N_D > 2 \times 10^{18}$ /cm³; ~130 mA/cm², 10 V, N_D 2×10^{16} /cm³	Polishing n^+ Si bright finish results brown layer forms brown, pitted surface results	[331]
7	Ethylene glycol with 0.04 N KNO₃, 2.5% H₂O, 1–2 g/liter Al(NO₃)₃·9H₂O	n- and p-type Si; illumination, 2–8 mA/cm² 90–280 V, 15–20°, wafer holder, back of high resist, Si metallized $N_D \leq 10^{20}$/cm³, and $N_A \leq 3 \times 10^{18}$/cm³	Analytical sectioning. Anodization followed by oxide stripping n^+, p Si anodization 2.2 Å/V (linear), indep. of dopant type and resistivity	[340]
8	63.3 wt % H₂O, 23.4 wt % KOH, 13.3 wt % isopropanol	(100) Si, B doped, no electrodes, 80° (100) doped with B, 10^{14}–10^{18}/cm³ (100) doped with B, 10^{19}/cm³ (100) doped with B, 10^{20}/cm³	For selective electrodeless etching 0.99–0.95 μm/min 0.63 μm/min 0 μm/min	[304]
9	KOH (4 N)	(100) epi Si on sapphire; 60° n type, 0.001–10 Ω-cm p type, 0.01–10 Ω-cm	For etching epi Si on sapphire 50 Å/sec 50 Å/sec	[305]

10	8CH$_3$COOH, 3HNO$_3$, 1HF	Single-crystal Si, n and p type doped with As, P, Sb or B; no electrodes for resistivities <0.01 Ω-cm for resistivities >0.068 Ω-cm for SiO$_2$ films	For selective electrodeless etching of n- and p-type Si 0.7–3 μm/min no etching occurs (same for Si$_3$N$_4$) 0.05 μm/min	[351, 355]
11	33CH$_3$COOH, 26HNO$_3$, 1HF	Poly Si; undoped, B doped (930°) undoped B doped, <0.01 Ω-cm	For poly-Si etching ~1500 Å/min <1000 Å/min	[353]
12	KOH solutions	(100) Si; bulk crystal n and p type	For anisotropic etching as a function of dopant type; R = surface roughness/ etch depth	[349]
	(a) 50 g KOH, 100 ml H$_2$O, 50 ml CH$_3$OH	85° p^+, 0.018 Ω-cm, 5 × 10^{18} B/cm^3 p^-, 35 Ω-cm, ~3 × 10^{14} B/cm^3 n^+, 0.01 Ω-cm, 4 × 10^{18} Sb/cm^3 n^-, 9.5 Ω-cm, 5 × 10^{14} P/cm^3	1.06 μm/min 1.38 μm/min 1.38 μm/min 1.38 μm/min $\left.\right\}$ $R = 0.04$	
	(b) 50 g KOH, 100 ml 2-propanol, 50 ml CH$_3$OH	80° p^+ p^-, n^+, n^- $\left.\right\}$ as in (a)	0.80 μm/min 1.04 μm/min $\left.\right\}$ $R = 0.17$	
	(c) 50 g KOH, 100 ml Butyl cellosolve, 50 ml CH$_3$OH	92° p^+ p^- n^+ n^- as in (a)	1.16 μm/min 1.53 μm/min 1.35 μm/min 1.60 μm/min $\left.\right\}$ $R = 0.14$	
	(d) Hydrazine	72° p^+ p^- n^+ n^- as in (a)	1.10 μm/min 1.30 μm/min 1.00 μm/min 1.23 μm/min $\left.\right\}$ $R = 0.27$	

Table VII

Gas and Vapor Phase Etching: Si, Ge

No.	Etchant	Substrate, etching conditions	Application, etch rate, remarks	Refs.
1	SF$_6$–H$_2$ (0.006–0.02 vol % SF$_6$)	(111) Si, p type, 0.004, 30 Ω-cm. B doped polished and BHF, etched; 950–1100°, 7.5–20 liter/min⁻¹ H$_2$ 950°C, 0.011% SF$_6$, 7.5 liter/min H$_2$ 1100°C, 0.02%, SF$_6$, 10 liter/min H$_2$	For *in situ* polishing prior to epitaxy; smooth, reflecting at 950–1100°C; rough at 900°C 0.235 μm/min 0.755 μm/min	[359]
2	SF$_6$–H$_2$ (10⁻⁴–10⁻³ atm SF$_6$)	(111) Si, polished; 1050–1100° quartz tube reactor, ~100 liter/min H$_2$ (~25 cm/sec linear velocity)^2 1050°C, 2 × 10⁻⁴ atm SF$_6$ 1050°C, 1 × 10⁻³ atm SF$_6$	For wafer thinning and *in situ* polishing prior to epitaxy; smooth etching at ~≥1050°; preferential ~850°C. SiO$_2$ masks severely undercut 0.2 μm/min 0.9 μm/min	[360]
3	SF$_6$–H$_2$ or He (10⁻⁴–10⁻³ atm SF$_6$)	(111) Si, p-type, 50 Ω-cm; 7.5 cm-diameter tube reactor, 1060–1100°, linear velocity ~100 cm/sec H$_2$, He 1060°, 2 × 10⁻⁴ atm SF$_6$ 1060°, 1 × 10⁻³ atm SF$_6$	For *in situ* polishing prior to epitaxy; smooth etching at ≥1060°C; preferential ≤1050°, SiO$_2$ attacked >1060° (see Figs. 12, 13) 0.2 μm/min 1.0 μm/min	[362]
4	HCl–H$_2$ (1.3–6 vol % HCl)	(111) Si, lapped or polished: 1180–1275°, quartz tube reactor 1275°, 2% HCl 1275°, 3% HCl 1275°, 5.6% HCl	For polishing with good control 1.5 μm/min 3.0 μm/min 8.0 μm/min	[363]
5	HCl–H$_2$ (3.5–5.4 mole % HCl)	Epi Si, n-type; quartz tube reactor, 1100–1350° 1125–1350°, 3.5 mole % HCl 1175–1325°, 5.4 mole % HCl	For etching in conjunction with epi Si from SiCl$_4$ 1.1 μm/min 2.3 μm/min	[364]

6	HCl–H₂ (1.5–5 mole % HCl)	(111) Si, chemically polished; tube reactor 6 liter/min H₂ (300 cm/min linear velocity), 1150–1300° 1280°, 1.4 mole % HCl 1280°, 3.0 mole % HCl 1280°, 5.0 mole % HCl	For polishing, independent of resistivity and dopant type 1.0 μm/min 2.6 μm/min 4.8 μm/min	[365]
7	HCl–H₂ (5% HCl)	(100), (110), (111) Si; 1150–1200° (111) Si, 1200° (110) Si, 1200° (100) Si, 1200°	For orientation-dependent etching in di-electric isolation 1.48 μm/min 3.0 μm/min 3.4 μm/min	[367]
8	HCl (1.5–20 mole %)	(100) Si, n-type, epi reactor, 1050–1250° 1050°, 20 mole % HCl 1100°, 20 mole % HCl 1200°, 20 mole % HCl 1250°, 20 mole % HCl	For anisotropic etching in epitaxy ~20 μm/min, strongly anisotropic (100) 20 μm/min 21 μm/min 25 μm/min, weakly anisotropic (100)	[370]
9	HBr–H₂ (1–6 mole % HBr)	(111) Si, chemical polished; tube reactor, 6 liter/min H₂ (300 cm/min), 1050–1300° 1260°, 2.0 mole % HBr 1260°, 3.0 mole % HBr 1260°, 5.0 mole % HBr	For polishing, superior to HCl no. 6, independent of resistivity and dopant type; only moderate temp. dependence 1.0 μm/min 1.8 μm/min 4.0 μm/min	[365]
10	HI–HF–He–H₂ (0.5–2.8 × 10⁻² atm HI, 10⁻³% HF, 80% He, 20% H₂)	(111), (100) Si polished, n and p-type, high and low resist.; 10 liter/min (50 cm/sec), 900° quartz tube reactor <1.2 × 10⁻² atm, HI >1.2 × 10⁻² atm, HI	For low-temperature polishing; no difference between n,p-types: low, high resistivity, (111), (100) 0.12 μm/min: smooth 0.46 μm/min: stepped atm HI	[375, 376]

(Continued)

449

Table VII (Continued)

No.	Etchant	Substrate, etching conditions	Application, etch rate, remarks	Refs.
11	Cl_2–He (0.2% Cl_2)	(111) Si, low resist. p-type, polished; quartz tube reactor. 1000–1100° 1000° 1040° 1100°	For rapid polishing; smooth finish between 1000–1100° and $\leq$1% Cl_2. Preferential <1000° 0.73 μm/min 0.87 μm/min 1.0 μm/min	[373]
12	H_2S–H_2, H_2O–H_2, HCl–H_2	(111)Si, polished; horiz. reactor, 100 liter/min H_2 (=25 cm/sec), $\geq$1100–1200°. 1200°, H_2S, 5.5 × 10⁻³ atm 1200°, H_2O, 1.3 × 10⁻⁴ atm 1200°, HCl, 1.0 × 10⁻² atm	For very rapid polishing. Faster and smoother than H_2O or HCl 15.1 μm/min 0.071 μm/min 0.183 μm/min	[374]
13	HCl–H_2 (15% HCl)	(111) Ge, n-, p-, p^+-type, precleaned quartz tube reactor, 820–830°; 11.4 liter/min HCl–H_2, 110 sec 12.2 liter/min HCl–H_2, 110 sec	For polish etching (111) Ge; etch rate indep. of temp. >800°; mirror bright, optically flat surface; (100) Ge dev. square pits 5 μm/110 sec 13 μm/110 sec	[379]
14	HI–H_2	(211) Ge, p-type 0.01 Ω-cm; preetched with NaOCl soln.; 36 mm i.d. quartz tube reactor, 911°, I_2 140 cm/min linear velocity 700 cm/min linear velocity	For polish etching (211) Ge; temperature most critical for smoothness 23 mg/5 min for 1.39 cm² 39 mg/5 min for 1.39 cm²	[371]
15	H_2–H_2O	(111), (110), (100) Ge, n- and p-type, preetched with I_2 etch; 2.5 and 5.5-cm i.d. quartz tube reactors, 900°, 26 Torr H_2O partial pressure, 4 liter/min H_2–H_2O, 30 min	For polish etching; clean, structureless surfaces; large excess H_2 (but not Ar) impedes etch rate; superior to H_2–H_2S under similar conditions	[380]

450

D. Compound Semiconductors

Table VIII

Group IV Compound Semiconductors: SiC

No.	Etchant	Substrate, etching conditions	Applications, etch rate, remarks	Refs.
1	KOH or NaOH	SiC; fusion at >600°; 900° for 2 min	General etching	[395]
2	Na_2O_2	SiC; fusion between 350° and 900° 500° 900°	General etching; rapid etching at the higher temperature 0.1 mg/cm²-min 1 mg/cm²-min	[395]
3	$Na_2O \cdot 2B_2O_3 \cdot 10H_2O$ (borax)	SiC; fusion at 1000° for 2 min	General etching; excess borax removed with NaOH soln.	[395]
4	H_2	SiC, hexagonal; horiz. quartz tube furnace, gas-phase, 2.5 liter/min (8.5 cm/sec linear velocity) 1600° 1650° 1700° 1750°	Preparing smooth SiC crystal surfaces; nonpreferential etch Face A: 0.3 μm/min; Face B: 0.25 μm/min 1.5 μm/min; 0.8 μm/min 2 μm/min; 0.8 μm/min 4 μm/min; 2 μm/min	[399]
5	26% O_2–6% Cl_2 in Ar	β-SiC crystal: 900° Solution-grown crystals Epi crystals, undoped	Pattern etching β-SiC; similar to α-SiC, (111) face smooth but with etch pits: ($\bar{1}\bar{1}\bar{1}$)[396] face no pits; thermal oxide etch mask 0.3–0.5 μm/min 0.02 μm/min	

451

Table IX

Group III–V Compound Semiconductors: GaAs

No.	Etchant	Substrate, etching conditions	Application, etch rate, remarks	Refs.
1	$4H_2SO_4$, $1H_2O_2$, $1H_2O$	(100) GaAs; 50°	Fast etching, ~3 μm/min	[406]
2	1NaOH (1 M); $1H_2O_2$ (0.76 M)	(100) GaAs; 30°	Slow etching, 0.2 μm/min	[406]
3	$2Br_2$, $98CH_3OH$	(100), ($\bar{1}\bar{1}\bar{1}$)GaAs, Cr doped; rotating slices, jet nozzle	Jet polishing, nonpreferential, ~8 μm/min; smooth, flat	[266]
4	$700H_2O_2$, $1NH_4OH$ (29.5%)	(100), (111), ($\bar{1}\bar{1}\bar{1}$)GaAs; rotating slices on polishing pad	Planar polishing, nonpreferential, ~18 μm/hr	[404]
5	$20H_2O_2$, 1NaOCl	{100}, {111}A, {111}B GaAs; freely rotating slices on polishing pad	Planar polishing, nonpreferential, ~25 μm/3 hr	[390, 412]
5A	$3H_2SO_4$, $1H_2O_2$ (33%), $1H_2O$	GaAs; 60°, polishing pad	Planar polishing, ~1 μm/5 min	[412]
6	8 glycerol, 1HCl, $1HNO_3$	{111} GaAs, 0.13 Ω-cm, n-type	High polishing, 0.37 mg/cm²-min	[410]
7	$8H_2O_2$, $1H_2SO_4$, $1H_2O$	(100) GaAs, epi	Structural etching, SiO_2 mask, 8 μm/min, lateral dist. depends on mask alignment	[410]
8	99 wt % CH_3OH, 1 wt % Br_2	{110}, {111}B, {100}, {111}A GaAs	Preferential structural etching, etch rate {110} $\gtrsim$ {111}B $\gtrsim$ {100} $\gg$ {111}A	[403]

9	973H$_2$O, 20NH$_4$OH, 7H$_2$O$_2$	$\langle 111 \rangle B$, $\langle 100 \rangle$, $\langle 111 \rangle A$ GaAs	Selective removal through SiO$_2$ masks, flat profiles, reduced undercutting	[409]
		$\langle 111 \rangle B$	0.20 μm/min	
		$\langle 100 \rangle$	0.12 μm/min	
		$\langle 111 \rangle A$	0.037 μm/min	
10	10 citric acid (50 wt % aq. sol.), 1H$_2$O$_2$	(111)B, (100), (111)A GaAs	Preferential etching through photoresist masks, flat bottomed holes, no attack of resist. Etch rates (111)B > (100) > (111)A	[418]
11	3CH$_3$OH, 1H$_3$PO$_4$, 1H$_2$O$_2$	[110], [100], Ga [111] GaAs	Preferential structural etching, ~2 μm/min, except Ga[111] reduced twofold	[415]
12	1–20Br$_2$, 99–80CH$_3$OH	GaAs (for solution or pad etching)	Polishing	[402, 412]
13	3HNO$_3$, 2H$_2$O, 1HF	GaAs	Rapid polish etching	[419]
14	2HCl, 2H$_2$O, 1HNO$_3$	($\bar{1}\bar{1}\bar{1}$) GaAs, 10 min	General etching of ($\bar{1}\bar{1}\bar{1}$) plane	[420]
15	5NaOH (5%), 1H$_2$O$_2$	GaAs, 5 min	Fast etching, 10–15 μm/min	[421]
16	8–12AgNO$_3$ (1%), 5HNO$_3$, 1HF	(111), ($\bar{1}\bar{1}\bar{1}$) GaAs	Etching both (111) and ($\bar{1}\bar{1}\bar{1}$) planes	[419]
17	75H$_2$O, 20H$_2$SO$_4$, 5H$_2$O$_2$	GaAs	Polish etching	[423]
18	40HCl, 4H$_2$O$_2$, 1H$_2$O	GaAs, jet etching, 20°	Thinning specimens for electron microscopy	[422]

(Continued)

Table IX (*Continued*)

No.	Etchant	Substrate, etching conditions	Application, etch rate, remarks	Refs.
19	$25HClO_4$, $75CH_3COOH$	GaAs electrolytic, gently flowing from an orifice above sample at 42 V	Thinning specimens for electron microscopy	[424]
20	10–40% KOH or NaOH	GaAs, p type and heavily doped n type, electrolytic, 1–5 A/cm^2	Electropolishing to mirror-smooth surfaces	[428]
21	10% KOH	(100), (110), (111) GaAs. n type, electrolytic, flowing 10% KOH	Anodic dissolution	[426]
22	3 M NaOH,	⟨100⟩ GaAs, p type, spray electrolytic, 100 mA/cm^2	Selective removal of p-type substrate leaving n-type epi GaAs or $GaAs_{1-x}P_x$	[429]
23	0.025 M NaOH–0.001 M EDTA	(100), (111) Ga, (111) As GaAs n type; electrolytic, illumination	Electropolishing	[430]
24	HCl–H_2O	GaAs, n type, electrolytic	Controlled thinning	[431]
25	0.01–1 N HNO_3	(100) GaAs, n type; electrolytic, 10–20 mA/cm^2, 2–3 V	Controlled electroetching	[432]
26	H_2, AsH_3, HCl: (900, 3, 2 cm^3/min)	(100), ($1\bar{1}\bar{1}$) GaAs, Te, Zn, Si, Cr doped; vapor phase, 900°	Substrate polishing prior to epitaxy. 7–11 μm/min, nonpreferential, specular	[434]
27	$100CH_3OH$, $1Br_2$	GaAs	General etching; 8 μm/hr	
28	$5H_2SO_4$, $1H_2O$	GaAs	Polishing; 25 μm/hr	
29	$70H_2O$. $20H_2O_2$, 10 formic acid	GaAs	Surface cleanup	
30	$95CH_3OH$, $5Br_2$	(100) GaAs, n-type; CVD SiO_2 as etch mask	Preferential etching of {332} Ga plane	[416]

Table X

Group III–V Compound Semiconductors: GaP

No.	Etchant	Substrate, etching conditions	Application, etch rate, remarks	Refs.
1	1–20% Br$_2$, 99–80% CH$_3$OH	GaP: (solution or Pellon cloth techniques)	General etching and polishing	[402]
2	1% Br, 99% CH$_3$OH	GaP	Polishing, highest-quality surface; ~0.25 μm/min	[20]
3	2HCl, 1HNO$_3$, 1H$_2$O	{111} GaP; hot	Polishing {111} surface	[438]
4	2HCl, 2H$_2$O, 1HNO$_3$	P {111} GaP	Polishing P {111} surface	[439]
5	2HCl, 2H$_2$O, 1HNO$_3$	(111), ($\bar{1}\bar{1}\bar{1}$), (100) GaP; 60°, 1–2 min	Polishing	[442]
6	Aqua regia	($\bar{1}\bar{1}\bar{1}$) GaP	Groove and pattern etching; SiO$_2$ mask	[441]
7	2HCl, 2H$_2$SO$_4$, 2H$_2$O, 1HNO$_3$	P {111} GaP: 5 min cold, then 50°; or 10 sec etching on Pellon cloth	Polishing P {111} surface; Ga {111} face pitting; etch rate depends on Te carrier conc.	[440]
8	3H$_2$SO$_4$, 1H$_2$O$_2$ (33%), 1H$_2$O	GaP: 60°, 5 min	Surface etching for saw damage removal; 1 μm/5 min etches p-type preferentially	[443, 444]
9	1 CH$_3$OH sat. with Br$_2$, 1 H$_3$PO$_4$, freshly mixed	⟨111⟩ GaP; chem.-mech. technique wafer rotates face down	Work damage removal, prior to no. 10 etch	[436, 437]
10	5H$_2$SO$_4$, 1H$_2$O$_2$, 1H$_2$O	⟨111⟩ GaP; 80°, 5 min	Substrate preparation for epi growth, after no. 9 and before no. 11 etch; 0.6 μm/min optimal	[437]
11	Etchant no. 9	⟨111⟩ GaP; 50°	Substrate final etch for epi growth; immediately after no. 10 etch; 1.5 μm/min optimal	[437]

(Continued)

455

Table X (*Continued*)

No.	Etchant	Substrate, etching conditions	Application, etch rate, remarks	Refs.
12	1.0 M K$_3$Fe(CN)$_6$–0.5 M KOH	(100), (111) GaP, n- and p-type: 60–95°, stirring	Polishing and mesa etching; SiO$_2$ or Ti mask	[445]
		P(111), n-type, 95°	260 μm/hr, polished	
		Ga(111), n-type, 80°	115 μm/hr, smooth textured	
		P(100), n-type, 80°	210 μm/hr, polished	
13	Cl$_2$–H$_2$O, sat. soln.	(111), (100) GaP, p epi on n substrate	Polishing; ~0.5 μm/min	[446]
14	H$_3$PO$_4$	($\bar{1}\bar{1}\bar{1}$), (111), (100) GaP; 150–200°C	Groove etching: Au mask, V grooves result	[442]
15	1HF, 1H$_2$O$_2$	GaP, p type on n substrate	Preferential etching of p-type GaP on n type	[443]
16	7NaOCl (5.25 wt %), 1HCl	GaP, n type on p substrate; electrolytic n-type anode; Pt cathode	Preferential etching of n-type	[443]
17	16H$_2$O, 2NaOCl, 2HCl	{111}GaP; P($\bar{1}\bar{1}\bar{1}$) by chemical jet etching, Ga(111) by jet electroetching	Polishing; flat, smooth surfaces	[266]
18	Cl$_2$, CH$_3$OH (concentration not specified)	Ga(111), P(111) GaP; electrolytic jet	Localized thinning of specimens for transmission electron microscopy	[447]
19	3 N NaOH	Ga(111), P($\bar{1}\bar{1}\bar{1}$) GaP; 20°, electrolytic; Pt cathode	Anodic dissolution and selective etching	[448]
20	0.1 N HNO$_3$	($\bar{1}\bar{1}\bar{1}$) GaP, n-type; electrolytic; 20 mA/cm^2	Anodic dissolution; 1250 Å/300 sec	[432]
21	2HNO$_3$, 1HCl	GaP; 1–2 min	Polishing of some orientations	[437a]
22	Cl$_2$, CH$_3$OH; sat. soln.	GaP; immerse sample while Cl$_2$ bubbles through sat. soln., ≥20 min	Polishing	[402]

Table XI

Other Group III–V Compound Semiconductors

Material	No.	Etchant	Conditions	Application, etch rate, remarks	Refs.
AlN	1	10% NaOH or hot H_3PO_4	30–80° (NaOH)	General etching	[449]
BN	2	H_3PO_4 or H_2O_2 (3%)	Hot	General etching of pyrolytic films	[450, 450a]
BP	3	10% NaOH	Electrolytic; in dark, 0.1–10 A/cm²	Electrochem. etching; p-type etches >100x faster than n-type	[451]
AlSb	4	(a) 1HF, $1H_2O_2$, $1H_2O$ (b) 1HCl, $1HNO_3$	1 min (a), followed by 2 sec (b)	Etching ($1\bar{1}\bar{1}$) face	[452]
GaN	5	50% NaOH	5–90°	General etching	[453]
	6	0.1 N NaOH	Electrolytic jet etching	Selective etching of single-crystal, vapor-transport grown films	[454]
	7	H_3PO_4	Hot	Etches but develops etch figures	[455]
GaSb	8	99–80% CH_3OH, 1–20% Br_2	—	General polishing	[402]
	9	$9HNO_3$, 1HF	1–5 min	General polishing	[402]
	10	$2HNO_3$, 1HF, $1CH_3COOH$	15 sec	Polishing; develops pits on (111) face	[452]
	11	1HF, $1HNO_3$, $1H_2O$	—	Etching ($1\bar{1}\bar{1}$) face	[457]
	12	10 citric acid (50 wt %), $1H_2O_2$	p-type; 55°; photoresist masks	Pattern etching (100) face; 10 Å/sec	[418]

(Continued)

Table XI (*Continued*)

Material	No.	Etchant	Conditions	Application, etch rate, remarks	Refs.
InAs	13	Etchant no. 8	—	Polishing (111) and ($\bar{1}\bar{1}\bar{1}$) faces	[460]
	14	5HNO$_3$, 3HF, 3CH$_3$COOH, 0.06 Br$_2$ (CP-4)	—	General etching	[458]
	15	75HNO$_3$, 15HF, 15CH$_3$COOH, 0.06 Br$_2$	55°	Etching ($\bar{1}\bar{1}\bar{1}$) face; etch pits on (111) face	[458]
	16	HCl	75°	General etching; 5 mg/cm^2 min	[457, 459]
	17	0.4 M Fe^{3+}–6NHCl	—	General etching	[452]
	18	99.6 ml CH$_3$COOH, 0.4 g Br$_2$	—	General polishing	[423]
InP	19	99CH$_3$OH, 1Br$_2$, or 90CH$_3$OH, 10Br$_2$	—	General polishing	[402, 461]
	20	1HCl, 1HNO$_3$	—	Etching (100) face; hillocks on ($\bar{1}\bar{1}\bar{1}$) face	[461]
InSb	21	I$_2$, CH$_3$OH (concentration not specified)	—	General polishing	[402]
	22	Etchant no. 17	—	General etching	[452]
	23	1HF, 1HNO$_3$	2–5 sec	Polishing ($\bar{1}\bar{1}\bar{1}$) and (110) faces; no etching on (111) or (100) faces	[423, 465]
	24	5HF, 5HNO$_3$, 2H$_2$O	20 sec	Etching (100) and (110) faces	[8]
	25	5HNO$_3$, 3HF, 3CH$_3$COOH (CP-4A)	5–30 sec	General polishing	[463, 464]
	26	1H$_2$O, 1CH$_3$COOH, 1 CP-4	1 min	Etching ($\bar{1}\bar{1}\bar{1}$), ($\bar{1}\bar{1}\bar{2}$) faces, etc.	[463]
	27	2HF, 1HNO$_3$, 1CH$_3$COOH	5–30 sec	Polishing; etch pits on (111) face	[452]
	28	4H$_2$O, 1HF, 1H$_2$O$_2$	5–10 sec	Etching ($\bar{1}\bar{1}\bar{1}$) face	[457]
	29	Etchant no. 12	38°; photoresist masks	Pattern etching (100) face; 60 Å/sec	[418]

Table XII

Group II–VI Compound Semiconductors

Material	No.	Etchant	Conditions	Application, etch rate, remarks	Refs.
BeO	1	10% KOH	—	General etching	[162, 468]
	2	HCl	120°	General etching	[467]
	3	H_3PO_4–H_2SO_4	boiling	General etching, at low rate	[467, 468]
CdO	4	Mineral acids or NH_4^+ salts	—	General etching	[469]
CdS	5	100 ml H_2O, 1 ml H_2SO_4, 0.08 g Cr_2O_3	80°, 10 min	Etching of [10$\bar{1}$0] face	[471]
	6	16 N H_2SO_4–0.5 M $K_2Cr_2O_7$	95°, 5–10 min	General etching and polishing; etch pits on ($\bar{1}\bar{1}$) face	[472]
	7	300 ml precip. silica, 90 ml HNO_3, 10 g $AlCl_3$/liter H_2O	Chem.-mech. polishing, Poromeric disk, 240 r/min, 370 g/cm²	Polishing (0001) face	[475]
	8	70H_2O, 30HCl	Rotating on Pellon cloth	Polishing Cd, S, and prismatic faces	[473]
	9	1000 ml H_2O, 13.7 g KCl, 0.5 ml HCl	pH 2.5, Politex Pix pad, 58 r/min 0.25–0.40 kg/cm²	Polishing Cd face	[476]
	10	1000 ml H_2O, 13.3 g KCl, 16 ml HCl	pH 1.0; same as no. 9	Polishing S face	[476]
CdSe	11	30HNO_3, 20H_2SO_4, 10CH_3COOH, 0.1HCl	40°, 8 sec	General etching; thick films forming are soluble in H_2SO_4	[472]
CdTe	12	3HF, 2H_2O_2, 1H_2O	—	Polishing (111) and ($\bar{1}\bar{1}$) faces; etch pits on ($\bar{1}\bar{1}$) face	[472]

(Continued)

Table XII (*Continued*)

Material	No.	Etchant	Conditions	Application, etch rate, remarks	Refs.
	13	$2HNO_3$, $2HCl$, $1H_2O$	—	Polishing	[477]
	14	20 ml H_2O, 10 ml HNO_3, 4 g $K_2Cr_2O_7$	—	Polishing to mirrorlike surface	[477]
	15	99.5 CH_3OH, $0.5Br_2$	Strip deposit formed with CS_2, 45 min	General etching	[478]
CdTe (0.05)–HgTe (0.95)	16	$6HNO_3$, $1HCl$	Rinse in $1HCl$–$1CH_3OH$	Polishing	[479]
HgSe	17	$50HNO_3$, $20H_2SO_4$ (18 N), $10CH_3COOH$, $1HCl$	40°, 10–15 min	Polishing	[472]
HgTe	18	$6HNO_3$, $1HCl$, $1H_2O$	10–15 min	Polishing	[472]
ZnO	19	Mineral acids or alkali or NH_4Cl solns.	—	General etching	[469]
ZnS	20	Etchant no. 6	95°, 10 min	Polishing to high polish; etch pits on (111) face	[472]
ZnSe	21	Etchant no. 15	Same as no. 15	General etching; etch pits on (111) face	[478]
ZnTe	22	$4HF$, $3HNO_3$	Strip film formed with HCl, then H_2O	General etching and polishing	[472]
	23	Etchant no. 15	Same as no. 15		[478]

Table XIII

Other Selected Compound Semiconductors

Material	No.	Etchant	Conditions	Application, etch rate, remarks	Refs.
Ag_2Se	1	$5H_2SO_4$, $1H_2O_2$	$50°$, 5 min; rinse in EDTA solution, then H_2O	Polishing of some orientations	[479]
	2	$2KOH$ (sat.), 2 ethylene glycol, $1H_2O_2$	$80°$, 2 min following damage removal with etchant no. 1	Polishing	[479]
Ag_2Te	3	$3NH_4OH$, $2H_2O_2$	Remove film by brushing under water	Polishing of some orientations	[479]
Bi_2Se_3	4	$1H_2O$, $1HCl$	Damaged layers following much polishing	Polishing and removal of work-damaged layers	[470]
	5	$2HNO_3$, $1HCl$	May be diluted with H_2O	Cleaning and etching	[470, 481]
Bi_2Te_3	6	Etchant No. 5	—	Cleaning and etching	[470, 481]
$Hg_{1-x}Cd_xTe$	7	$80CH_3OH$, $20Br_2$	Rinse with CH_3OH	Removal of surface	[482, 483]
In_2O_3 Sn doped	8	H_2SO_4	$50–60°$; Cr film etch mask	Pattern etching; $\sim 0.5\ \mu m/hr$	[484]
	9	HCl		Pattern etching	[485]
	10	$1.0\ M$ oxalic acid, aq. soln.	$50°$; Shipley AZ-1350 etch mask	Pattern etching of annealed films	[486]
$In_2O_3 : SnO_2$ (4 : 1)	11	55% HI	Immerse sample vertically; Shipley AZ-1350H mask postbaked 90 min, $120°$	Precision pattern etching; 25 Å/sec	[487]
	12	$2HCl$, $1H_2O$, Zn powder	Paint Zn powder–H_2O slurry on substrate, immerse horiz. into $HCl–H_2O$; $20°$; mask as in no. 10	Rapid etching; 150 Å/sec	[487]
In_2Te_3	13	$19CH_3COOH$ sat. with citric acid, $1Br_2$	Finish by flooding with CH_3COOH sat. with citric acid	Polishing	[488]
$In_2Te_{3-x}Sb_x$ alloys	14	4 citric acid soln. sat. with Br_2, $3HNO_3$, $1HF$	—	Polishing	[488]

(Continued)

461

Table XIII (Continued)

Material	No.	Etchant	Conditions	Applications, etch rate, remarks	Refs.
PbS	15	30HCl, 10HNO$_3$, 1CH$_3$COOH	50°, few min, then rinse with 10% CH$_3$COOH	Polishing	[489]
	16	HNO$_3$	70°	Rapid etching	[489]
PbSe	17	5KOH (45%), 5 ethylene glycol, 1H$_2$O$_2$	Electrolytic; add H$_2$O$_2$ during etching to maintain rate; remove stains with 50% CH$_3$COOH	Thinning specimens for electron transmission microscopy	[490]
	18	Etchant no. 17	40°, 3 min	Polishing	[480]
PbTe	19	45 ml H$_2$O, 35 ml glycerol, 20 ml C$_2$H$_5$OH, 20 g KOH	Electrolytic; 4–6 V, 0.2 A/cm^2	Thinning specimens for electron transmission microscopy	[493]
	20	Etchant no. 19	Electrolytic; 10 V; rinse in C$_2$H$_5$OH	Polishing	[492] [494]
Pb$_{1-x}$Sn$_x$Se	21	10 ethylene glycol, 10KOH (sat. aq. soln. at 25°), 1H$_2$O$_2$	Felt covered wheel saturated with etchant	Polishing	[495] [496]
Pb$_{1-x}$Sn$_x$Te	22	Etchant no. 19	Electrolytic; 10 V, rinse in C$_2$H$_5$OH	Polishing	[492] [494]
	23	95HBr, 5Br$_2$	1–2 min, rinse many times with C$_2$H$_5$OH followed with slight etching with no. 22	Polishing; faster than etchant no. 22	[494]
SnO$_2$	24	3–10H$_2$O, 1HCl	Electrolytic, 5–40 mA/cm^2	Pattern precision etching; SiO$_2$ mask used; 1600 Å/min at 20 mA/cm^2	[498]
	25	HCl, Zn powder	Zn powder in photoresist in Ref. 500	Pattern etching	[497, 499, 500]
SnO$_2$ Sb doped	26	HCl, Zn powder	—	Pattern etching	[501]

462

E. Conductors

Table XIV

Elemental Metals

	Etching conditions	Etch rate or etch time	Remarks	Refs.
		Aluminum		
1	$4H_3PO_4$, $4CH_3COOH$, $1HNO_3$, $1H_2O$	350 Å/min	Polishing etch; contact to noble metals is possible without increases in etch rate and undercutting	[57, 504*]
2	75 g Na_2CO_3, 35 g $Na_3PO_4 \cdot 12H_2O$, 16 g $K_3Fe(CN)_6$, 0.5 liter H_2O	1300 Å/min	Polishing etch; contact to noble metals is possible without increase in etch rate and undercutting	[57]
3	$16-19H_3PO_4$, $1HNO_3$, $0-4H_2O$; $40°$; stirring	1500–2500 Å/min	Gas evolution occurs	[14]
4	$1HClO_4$, $1(CH_3CO)_2O$	3 μm/min	—	[29]
5	$74.1H_3PO_4$, $18.5H_2O$, $7.3HNO_3$; $50°$	9000 Å/min	—	[46*, 505]
6	$1HCl$, $4H_2O$: $80°$	—	For fine line etching; photoresist masks can be used	[507*, 532]
7	20% NaOH; $60-90°$	—	Photoresist mask can be used	[14, 507]
8	$FeCl_3$; $12-36°Bé$	—	—	[15, 513]
9	$10HCl$, $1HNO_3$, $9H_2O$; $49°$	25–50 μm/min	—	[15]
10	0.1 M $K_2B_4O_7$, 0.51 M KOH, 0.6 M $K_3Fe(CN)_6$, pH 13.6	1 μm/min	No H_2 is evolved	[58]
11	1 Electro Glo 100 (Electro Glo Co., Chicago), $3H_3PO_4$; $79°$, $7-10$ V, 0.06 A/cm², Pb cathode	—	Electrochemical polish; excellent polishing occurs	[533]
12	Other electrochemical etches	—	—	[29]

(Continued)

Table XIV (*Continued*)

Etching conditions	Etch rate or etch time	Remarks	Refs.
Antimony			
1 Aqua regia or hot H_2SO_4	—	—	[469]
2 5% $AgNO_3$	—	—	[534]
3 1 g $FeCl_3$, 3 ml HCl, 12 ml H_2O	—	—	[534]
Arsenic			
1 HNO_3	—	—	[469]
Beryllium			
1 5 g H_2SO_4, 75 g H_3PO_4 (98%), 7 g CrO_3, 13 ml H_2O; 49–50°	$\mu m/min$	Passive film formed removable by immersion for 15–30 sec in 10% H_2SO_4	[29, 468*]
2 HCl	—	—	[535]
3 1% H_2SO_4; preset voltage, 30 V	≥ 240 Å/min	Electrochemical etch	[468]
4 1% KOH: preset voltage, 50 V	≥ 60 Å/min	Electrochemical etch	[468]
5 Other electrochemical etches	—	—	[29, 536]
Bismuth			
1 Hot H_2SO_4	—	—	[469]
2 5% $AgNO_3$	—	—	[534]
3 49 ml sat. KI soln., 1 ml HCl; 5–7 V, 20 A/dm², stainless steel or carbon cathode	Periods of 30 sec	Electrochemical etch; allow brown film formed to dissolve between etching periods	[29]

464

Cadmium

	Composition	Rate/Time	Remarks	Ref.
1	3 fuming HNO_3 (type not specified), $1H_2O$	Periods of 5–10 sec	Etching periods are followed by rinsing in a rapid stream of H_2O	[29]
2	$7CH_3COOH$, 3 fuming HNO_3 (type not specified)	30 sec–1 min	—	[29]
3	$1 g I_2$, 3 g KI. 10 ml H_2O	—	—	[534]
4	Electrochemical etches and polishes	—	—	[29, 537, 538]

Chromium[b]

	Composition	Rate/Time	Remarks	Ref.
1	1 vol (1 g NaOH, 2 ml H_2O), 3 vol (1 g $K_3[Fe(CN)_6$, 3 ml H_2O)	250–1000 Å/min	Does not require depassivation; for etching Cr masks and vacuum-deposited Cr in general; photoresist mask can be used	[14*, 15, 507, 510*]
2	1 g $Ce(SO_4)_2 \cdot 2(NH_4)_2SO_4 \cdot 2H_2O$, 5 ml HNO_3, 25 ml H_2O; 28°	85 Å/min	Oxide coating etches at 200 Å/min	[519]
3	164.5 g $Ce(SO_4)_2 \cdot 2(NH_4)_2SO_4 \cdot 2H_2O$, 43 ml $HClO_4$, H_2O to make 1 liter; 25–50°	—	Recommended for precision patterning of Cr masks on glass substrates	[520]
4	$2FeCl_3$, 42°Bé, $1HCl$; 80°	—	Good for plated Cr; photoresist mask can be used	[15, 507]
5	HCl	>1500 Å/min	Must depassivate with Zn rod	[14, 516]
6	454 g $AlCl_3 \cdot 6H_2O$. 135 g $ZnCl_2$, 30 ml H_3PO_4, 400 ml H_2O	—	—	[14]
7	9 sat. $Ce(SO_4)_2$ soln.. $1HNO_3$	800 Å/min	—	[14, 517]
8	Dilute HCl or HNO_3	—	Photoresist mask can be used	[15, 507, 513, 517]
9	$1HClO_4$, $20CH_3COOH$; 45–50 V, 15–20 A/dm², stainless steel cathode	5–10 sec	Electrochemical polish	[29]

(Continued)

Table XIV (Continued)

	Etching conditions	Etch rate or etch time	Remarks	Refs.
	Cobalt			
1	1HCl, 1C$_2$H$_5$OH: 8–9 V, 250 A/dm^2, stainless steel cathode	0.5–1.5 min	Electrochemical polish: bluish-green anodic film is soluble in H$_2$O; gives polished surface with slight grain boundary delineation	[29]
2	H$_3$PO$_4$ (98%), 1–1.5 V, 1–2 A/dm^2, Co cathode	5–10 min	Electrochemical polish: solid black film forms which is removed by wiping with cotton wool	[29]
3	Other electrochemical etches	—	—	[539, 540]
	Copper			
1	FeCl$_3$, 42°Bé; 49°	50 μm/min	Use more dilute solutions for slower etching	[14*, 15, 508* 509*]
2	20–30% H$_2$SO$_4$, 10–20% CrO$_3$ or K$_2$Cr$_2$O$_7$; 49°	37 μm/min	—	[14*, 15, 509*]
3	1 g (NH$_4$)$_2$S$_2$O$_8$, 3 ml H$_2$O: 32–49°	25 μm/min	Addition of 5 ppm Hg as HgCl$_2$ activates etchant at lower temperatures	[15, 513*, 541*–543*]
4	5HNO$_3$, 5CH$_3$COOH, 2H$_2$SO$_4$; dilute with H$_2$O as desired	—	Etches Cu and Cu-based alloys at same rate as Ni and Ni-based alloys	[506]
5	2 g KI, 1 g I$_2$, 4 ml H$_2$O; dip into etch, rinse, remove residue with Neutra-clean (Shipley Co., Newton, Mass.)	—	Rapid etch, but undercutting is limited by formation of an insoluble Cu compound at line edges	[14]
6	4HNO$_3$, 11H$_3$PO$_4$ (98%), 5CH$_3$COOH: 60–70°	1–2 min	Polishing etch	[29]
7	2H$_3$PO$_4$ (98%), 1H$_2$O: 1.5–2 V, 6–8 A/dm^2, Cu cathode	15–30 min	Electrochemical polish	[4*, 29, 544*, 545*]
8	Other electrochemical etches	—	—	[546]

Gallium

1	Mineral acids or alkali solutions	—	Must be processed below melting point	[469]

Gold

1	4 g KI, 1 g I_2, 40 ml H_2O	0.5–1 μm/min	Better control of line edges than in more concentrated solution; solution is opaque, so removal from solution to observe end point is required	[14, 15, 511]
2	3HCl, 1HNO$_3$; 32–38°	25–50 μm/min	—	[15]
3	NaCN, H_2O_2, H_2O mixtures (unspecified composition)	—	—	[14, 15]
4	0.5 g I_2, 2 g NH_4I, 10 ml H_2O, 15 ml C_2H_5OH; 20°	700 Å/min	Converts the surface of an underlying Ag layer to the iodide, thus preventing undercutting	[510]
5	0.4 M $K_3Fe(CN)_6$, 0.2 M KCN, 0.1 M KOH	600 Å/min	Fresh solution must be used; no attack on Pd is observed.	[58]
6	100 ml H_2O, 0.5–10 ml HCl, 10–30 g NaCl; 4–5 V, 0.5–2 A/dm², Mo cathode; 20–40°	—	Electrochemical etch which retains bright surface; only small amount Cl_2 evolved	[547]
7	Other electrochemical etches	—	—	[29, 548]

Hafnium

1	1–2% HF	—	Negative photoresist can be used as mask	[14, 549]
2	5–6HClO$_4$, 100CH$_3$COOH; 18 V, stainless steel cathode. Use several successive dips with continued agitation	5–10 sec dips	Electrochemical polish	[29]
3	See Zirconium etch no. 1	—	—	[29]

(Continued)

467

Table XIV (*Continued*)

Etching conditions	Etch rate or etch time	Remarks	Refs.
Indium			
1 Mineral acids	—	—	[469]
2 $1HNO_3$, $3CH_3OH$; 40–50 V, 30 A/dm², stainless steel cathode; cool bath	1–2 min	Electrochemical polish	[29]
Iron			
1 $3HNO_3$, $7HCl$, $30H_2O$; 60–70°	2–3 min	Dense brown viscous layer forms on surface; layer is soluble in solution	[29]
2 ~10% $KAl(SO_4)_2 \cdot 12H_2O$	—	Slowly soluble	[469]
3 3 liter 10% HNO_3, 0.3 m³/hr O_3 injected; 30°	30 μm/min	Etches iron plate smoothly; O_3 removes passive film	[550]
4· $1HClO_4$, $20CH_3COOH$; 45–60 V, 40–80 A/dm², stainless steel cathode	15–30 sec	Electrochemical polish; solid film sometimes forms on the surface during washing; removable with dil. HF	[29]
5 Other electrochemical polishes	—	—	[29]
Lead			
1 $FeCl_3$ 36–42°Bé; 43–54°	—	—	[15, 513]
2 $9FeCl_3$ 42°Bé, $1HCl$ 20°Bé; 43–49°	—	—	[15]
3 $1H_2O_2$, $4CH_3COOH$	Periods of 5–10 sec	Alternate with immersion in a soln. of 10 g molybdic acid and 140 ml NH_4OH in 240 ml H_2O to which 60 ml HNO_3 is finally added	[29]
4 $1HNO_3$, $19H_2O$	—	—	[534]
5 $35HClO_4$, $63(CH_3CO)_2O$, $20H_2O$; <30°, 50–70 V, 9–12 A/dm², Pb or Cu cathode	5–10 min	Electrochemical polish	[29]
6 Other electrochemical polishes	—	—	[29]

Lithium

1	CH_3OH	30 sec	Surface achieves high luster and is smooth	[551]

Magnesium

1	240–400 g $FeCl_3$, 330 ml HCl, H_2O to make 1 liter	—	—	[469]
2	1–3HNO_3, 17–19H_2O	—	—	[507, 513, 534]
3	3 fuming HNO_3 (type not specified), 1H_2O; wash immediately after removal from solution	Periods of 3 sec	Reaction reaches almost explosive violence after $\sim$1 min; if allowed to continue, it ceases after several minutes leaving a polished surface	[29]
4	7H_3PO_4 (98%), 13C_2H_5OH; 1–2 V, 0.5 A/dm², stainless steel or Ni cathode	0.25 μm/min	Electrochemical polish; there is considerable initial gassing; shake anode to remove adhering bubbles	[29]
5	Other electrochemical etches	—	—	[29]

Manganese

1	1HCl, 1H_2O	—	—	[513]
2	Other dilute mineral acids	—	—	[469]

Mercury

1	Slightly diluted HNO_3	—	—	[469]

Molybdenum

1	5H_3PO_4, 3HNO_3, 2H_2O	—	Polishing etch; contact to noble metals possible without increase in etch rate and undercutting	[57]
2	1H_2SO_4, 1HNO_3, 1–5H_2O; 25–54°	$\sim$12 μm/min (25°) $\sim$25 μm/min (54°)	Photoresist mask can be used at 25°	[15, 507, 518]

(Continued)

Table XIV (Continued)

	Etching conditions	Etch rate or etch time	Remarks	Refs.
	Molybdenum (Continued)			
3	200 g $K_3[Fe(CN)_6]$, 20 g NaOH, 3–3.5 g sodium oxalate, add H_2O to make 1 liter	~1 μm/min	Also usable as electrochem. etch using stainless steel cathode at 6 V; photoresist masks applicable	[14, 15, 51*, 507*, 552]
4	$38H_3PO_4$, $15HNO_3$, $30CH_3COOH$, $75H_2O$	0.5 μm/min	Photoresist mask can be used	[518]
5	$1H_2SO_4$, $7CH_3OH$, 80–120 A/dm², stainless steel cathode; no agitation	1 min	Electrochemical polish	[29]
6	100 ml H_3PO_4, 20 ml H_2SO_4, 40 ml H_2O, 0.25 g MoO_3; 70°, 8 V, 0.6–0.9 A/cm², stainless steel, graphite or Pt cathode; stirring	9.4 μm/min	Electrochemical polish; superior surface finish	[553]
7	Other etches	—	—	[29, 38, 57, 554–556]
	Neptunium			
1	HCl	—	—	[557]
	Nickel			
1	$5HNO_3$, $5CH_3COOH$, $2H_2SO_4$, H_2O as desired	—	Etches Ni and Ni-based alloys at same rate as Cu and Cu-based alloys	[506]
2	$1HNO_3$, $1HCl$, $3H_2O$	—	Spray etching recommended	[507]
3	$FeCl_3$ 42–49°Bé; 43–54°	12–25 μm/min	Photoresist mask may be used	[15, 507*, 513*]
4	$3HNO_3$, $1H_2SO_4$, $1H_3PO_4$ (98%), $5CH_3COOH$: 85–95°	0.5–1 min	Gives very good polish	[29]
5	$1HNO_3$, $1H_2O$; or $9H_3PO_4$, $1HNO_3$; or $90H_3PO_4$, $15HNO_3$, $4HCl$, $1H_2O$	—	Polishing etchants for electroless Ni–P; very little undercutting despite contact to noble metals	[57]
6	$10H_2SO_4$, $10H_2O_2$, H_3PO_4, 5–7 $NiSO_4$ 30%; 50°	1–2 μm/min	For electroless Ni films; no undercutting; similar composition useful for electroless Cu	[558]

No.	Solution composition	Time/Rate	Description	Ref.
7	$57H_2SO_4$, $43H_2O$: 0.5 A/cm²	2 min, ~4 μm/min	Electrochemical polish	[559, 560]
8	Other electrochemical polishes	—	—	[29]

Niobium

No.	Solution composition	Time/Rate	Description	Ref.
1	2 Lactic acid, $1H_2SO_4$, 1HF: 15–20 V, Pt cathode, stirring	5–10 min	Electrochemical polish	[29]
2	7HF, $7HNO_3$, $26H_2O$; 49°, 12–20 V, 20–34 A/dm², Pt cathode	—	Electrochemical etch	[469]
3	$9H_2SO_4$, 1HF: 35–45°, 50 V, 2 A/dm², Pt or carbon cathode	5–10 min	Electrochem. etch; temp. rises during use; cool bath	[29]

Osmium

No.	Solution composition	Time/Rate	Description	Ref.
1	Aqua regia	—	—	[557]

Palladium

No.	Solution composition	Time/Rate	Description	Ref.
1	1HCl, $10HNO_3$, $10CH_3COOH$	1000 Å/min	—	[561]
2	Aqua regia	—	—	[513]

Platinum

No.	Solution composition	Time/Rate	Description	Ref.
1	$8H_2O$, 7HCl, $1HNO_3$; 85°	400–500 Å/min	—	[562, 563*]
2	Aqua regia; precede etching by 30 sec immersion in HF	—	Etch times limited because photoresist mask is destroyed	[14, 507]
3	3 M HCl; −0.3–+1.4 V versus SCE, modified triangular waveform ~600 Hz; magnetic stirring	~1000 Å/min	Electrochem. etch: good resolution: either pos. or neg. photoresists usable	[514]

Plutonium

No.	Solution composition	Time/Rate	Description	Ref.
1	$1H_3PO_4$ (98%), 1 diethylene glycol; 5 V, 5 A/dm², stainless steel cathode	5–10 min	Electrochemical polish	[29]

(Continued)

Table XIV (*Continued*)

Etching conditions	Etch rate or etch time	Remarks	Refs.
Polonium			
1 Dilute mineral acids	—	—	[557]
Potassium			
1 C_2H_5OH or $(CH_3)_2CHOH$	3 sec (ethanol) or 10 sec (isopropanol)	Brilliant, smooth surface obtained	[551]
Praesiodymium			
1 Mineral acids	—	—	[557]
Rhenium			
1 Dilute HNO_3	—	—	[557]
Rhodium			
1 $3\,M$ HCl, $-0.3 - +1.4$ V versus SCE, modified triangular waveform ~600 Hz; magnetic stirring	—	Electrochem. etch; 1 μm line spacings of Rh films over 5000 Å Ti on Si clearly resolved	[514]
Ruthenium			
1 Fused alkalis	—	—	[557]
Samarium			
1 Mineral acids	—	—	[557]
Selenium			
1 H_2SO_4	—	See also text under elemental semiconductors	[469]

Silver				
1	11 g Fe(NO$_3$)$_3$, 9 ml H$_2$O; 44–49°	20 μm/min	Photoresist mask can be used	[14*, 15, 507*]
2	5–9 HNO$_3$, 1–5 H$_2$O; 39–49°	12–25 μm/min	—	[14*, 15, 517*]
3	35 g AgCN, 37 g KCN, 38 g K$_2$CO$_3$, 100 ml H$_2$O; 2.5–3.0 V, 1 A/cm^2, Ag cathode; slow stirring	10 min	Electrochem. etch; best polishing in region of voltage and current instability	[29]
4	3HNO$_3$, 19H$_2$O; 2 V, stainless steel cathode	—	Electrochemical etch	[507]
5	KI–I$_2$ etches listed for Cu and Ag	0.3–1 μm/sec	Immersion followed by H$_2$O rinse	[14]
6	4CH$_3$OH, 1NH$_4$OH, 1H$_2$O$_2$	60 Å/sec	Useful for pattern etching with photoresist mask; rinse quickly after etching	[512]
Sodium				
1	CH$_3$(CH$_2$)$_7$CH$_2$OH (nonyl alcohol)	30 sec	Brilliant, smooth surface	[551]
Strontium				
1	Liquid ammonia	—	—	[557]
Tantalum				
1	9NaOH or KOH (30%), 1H$_2$O$_2$; heat alkali to 90°; then add H$_2$O$_2$	1000–2000 Å/min	Metal (e.g., Au) mask must be used; very little under-cutting; etches Ta$_2$O$_5$ and TaN at same rate as Ta	[162, 163]
2	5H$_2$SO$_4$, 2HNO$_3$, 2HF	5–20 sec	Polishing etch	[29, 522, 523]
3	1–2HNO$_3$, 1HF, 1–2H$_2$O	—	H$_2$O may be omitted for faster etch, especially if film contains oxygen and resists etching; faster etch reduces resist attack	[14, 513]
4	9H$_2$SO$_4$, 1HF; 35–45°, 50 V, 2 A/dm^2, Pt or carbon cathode	5–10 min	Electrochemical etch; temp. of solution rises during use and it may be necessary to cool bath	[29]

Table XIV (*Continued*)

	Etching conditions	Etch rate or etch time	Remarks	Refs.
	Tellurium			
1	$2HNO_3$, $3H_2O$	—	—	[513]
2	240 g $(NH_4)_2S_2O_8$, bring to 1 liter with H_2O	—	—	[513]
	Terbium			
1	Mineral acids	—	—	[557]
	Thallium			
1	HNO_3 or H_2SO_4	—	—	[557]
	Thorium			
1	$14CH_3COOH$, $4HClO_4$, $1H_2O$; $\sim10°$, 60 A/dm², stainless steel cathode	7–12 sec	Electrochemical polish	[29]
	Tin			
1	$FeCl_3$ 36–42°Bé; 32–54°	—	—	[15, 513]
2	$1HNO_3$, $49C_2H_5OH$	—	—	[534]
3	$HClO_4$; <35°, 50–60 V, 40 A/dm², Al cathode; hold piece vertically and rotate at 50–100 r/min	420 $\mu m/min$	Electrochemical polish; on large surfaces. orange peel effect may occur	[29]
4	Other electrochemical polishes	—	—	[29]
	Titanium			
1	$9H_2O$, $1HF$; 32°	12 $\mu m/min$	Photoresist mask may be used	[14, 15, 507]
2	$7H_2O$, $2HNO_3$, $1HF$; 32°	18 $\mu m/min$	Photoresist mask may be used	[14, 15, 507]

474

No.	Solution composition and conditions	Rate	Comments	References
3	180 ml C_2H_5OH, 20 ml n-butyl alcohol, 12 g $AlCl_3$, 56 g $ZnCl_2$; 30–50 V, 12 A/dm², stainless steel cathode	—	Electrochemical polish	[469]
4	3$HClO_4$, 50CH_3COOH; 20°, 30 V, 30–40 A/dm², Ti 2 min cathode		Electrochem. etch; anode to cathode distance about 3 cm	[29]
5	Other electrochemical etches	—		[29, 564, 565]
6	See Zirconium etch no. 1	—		[29]

Tungsten

No.	Solution composition and conditions	Rate	Comments	References
1	34 g KH_2PO_4, 13.4 g KOH, 33 g $K_3Fe(CN)_6$, H_2O to make 1 liter	~1600 Å/min	Photoresist mask may be used; high resolution (1–2 μm) can be achieved	[515]
2	5% KOH, 5% $K_3Fe(CN)_6$, 1% surfactant; ~23°, 0.2 A/cm², Pt cathode	~2.3 μm/min	Electrochem. etch; photoresist mask can be used; good pattern resolution	[51]
3	5–10% NaOH; 6 V, 3–6 A/dm², stainless steel cathode	—	Electrochem. etch; rotation of anode or agitation of electrolyte with N_2 is necessary	[29, 513, 566*]
4	Other etches	—		[51, 58, 515, 555, 567]

Uranium

No.	Solution composition and conditions	Rate	Comments	References
1	1–2$HClO_4$, 20CH_3COOH: 50–60 V, 5 A/dm², stainless steel cathode	1.9 μm/min	Electrochemical polish	[29]
2	Other electrochemical polishes	—		[29]

Vanadium

No.	Solution composition and conditions	Rate	Comments	References
1	1–2$HClO_4$, 18–19CH_3COOH; <35°, 50–60 V, 16–24 A/dm², stainless steel cathode	1–2 min	Electrochemical polish; a second polishing of about the same duration in a fresh soln. sometimes necessary	[29]

Ytterbium

No.	Solution composition and conditions	Rate	Comments	References
1	Mineral acids	—		[557]

(Continued)

Table XIV (Continued)

	Etching conditions	Etch rate or etch time	Remarks	Refs.
	Yttrium			
1	Dilute mineral acids or hot KOH solutions	—	—	[557]
	Zinc			
1	$2-3HNO_3$, $17-18H_2O$; $38-49°$	$25\ \mu m/min$	—	[15, 513*]
2	40 g CrO_3, 3 g Na_2SO_4, 10 ml HNO_3, 190 ml H_2O	$7\ \mu m/min$	Dense layer formed during treatment is soluble in water	[29]
3	1 g CrO_3, 5 ml H_2O; 60 V, $250-350$ A/dm²; Pt, Ni, or Zn cathode	$40-45$ sec	Electrochem. polish: tendency for a passive film formation	[29]
4	$20-45\%$ KOH; $0-50°$: interrupted dc or sine wave method	—	Electrochem. etch; Zn is amalgamated first by dipping in a soln. of 50 g/liter $HgCl_2$ for 30 sec	[568]
5	Other electrochemical etches	—	—	[29, 569–571]
	Zirconium			
1	$45HNO_3$, $8-10HF$, $45H_2O$ or H_2O_2; swab for $5-10$ sec, rinse in running H_2O	$5-10$ sec	Brownish-yellow vapor is evolved; similar solution can be used for Ti and Hf	[29]
2	$2HClO_4$, $7CH_3COOH$, 4 ethylene glycol; $30-50$ V, >100 A/dm², stainless steel cathode	$20-30$ sec	Electrochemical polish	[29]
3	Other electrochemical etches	—	—	[29]
4	Chemical polish etch	—	—	[572]

[a] Starred (*) reference numbers refer to secondary numbers.

[b] In acids, depassivation of Cr must be induced by (1) physical contact with electropositive metals (Al wire, Zn rod or pellets); (2) Cr^{+3} ions in aqueous solution; or (3) application of a cathodic potential. Then Cr dissolves rapidly in nearly all mineral acids.

476

Table XV

Metal Alloys and Superconductors

Etching conditions	Etch rate or etch time	Remarks	Refs.
Inconel			
1 FeCl₃ 36–49°Bé; 43–54°	—	—	[15, 513]
Invar			
1 5CH₃COOH, 2HNO₃, 2HCl, 1H₂O	—	—	[469]
2 FeCl₃ 36–42°Bé	—	Photoresist mask can be used	[507]
Kovar[a]			
1 400CH₃COOH, 200HNO₃, 75H₂O, 9HCl; rinse in running H₂O after etching	15–20 sec	Polishing etch; heavy oxide is first removed in 6 M HCl	[469]
Monel			
1 5HNO₃, 5CH₃COOH, 2H₂SO₄; dilute with H₂O as desired.	—	—	[506]

(Continued)

477

Table XV *(Continued)*

Etching conditions	Etch rate or etch time	Remarks	Refs.
Nichrome[b]			
1 $FeCl_3$, 36°Bé: 43°	—	Photoresist mask may be used	[507, 513]
2 $1HNO_3$, 1HCl, $3H_2O$	—	Photoresist mask may be used	[507]
3 4HCl, $1H_2O$	—	—	[513]
4 $7H_3PO_4$, $1H_2SO_4$, $2H_2O$: 11 V, Cu cathode.	—	Electrochemical polish	[469]
Permalloy			
1 $3.9\,M\ H_2SO_4$, $1.12\,M\ H_2O_2$, $0.4–4\,M$ HF	4 μm/min	Edges can be beveled using a Ti overcoat	[54]
NbSn			
1 $5H_2SO_4$, $4HNO_3$, 1HF: 12 V, graphite electrode	—	Electrochem. polish: rinse immediately in H_2O	[469]

[a] Trade name of Westinghouse Electric Corp.
[b] Trade name of Driver-Harris Co.

F. Miscellaneous Materials

Table XVI

Miscellaneous Materials

	Etching conditions	Etch rate or etch time	Remarks	Refs.
P	C_2H_5OH (abs)	—	Red phosphorus	[557]
	CS_2; C_6H_6; alkalis; ether; $CHCl_3$; toluene	—	Yellow (white) phosphorus	[557]
S	CS_2 or toluene	—	—	[557]
Ag_2O	$1NH_4OH$, $4H_2O$	—	—	[469]
	10% KCN	—	—	[469]
CrO_3	2 g $Ce(NH_4)_2(SO_4)_3 \cdot 2H_2O$, 10 ml HNO_3, 50 ml H_2O; 28°	200 Å/min	Cr itself etches at 85 Å/min	[519]
Cr_2O_3	164.5 g $Ce(NH_4)_2(NO_3)_6$, 90 ml HNO_3, H_2O to make 1 liter.	—	Used for removing Cr_2O_3 contamination from gold surfaces	[573]
CuO	$1HCl$, $2H_2O$; hot	—	Little loss of Cu	[469]
FeO or	Dilute HF	—	Little or no loss of Fe	[469]
Fe_2O_3	10% ammonium citrate; warm	—	No loss of Fe	[469]
	$1HCl$, $1H_2O$	—	Used for pattern etching	[526]
MoO_3	$9NH_4OH$. $1H_2O_2$; rinse in running H_2O and dry.	—	—	[469]
	25% KOH, rinse with H_2O, dip in 6 M HCl, rinse with H_2O, then CH_3OH and dry.	—	Electrochemical polish	[469]
PbO	HNO_3, followed by H_2O rinse, then CH_3OH rinse, and dry.	—	Do not expose piece to air during treatment	[469]
WO_3	20% Na_2O_2; hot	—	—	[469]

(*Continued*)

479

Table XVI *(Continued)*

	Etching conditions	Etch rate or etch time	Remarks	Refs.
Garnets	H_3PO_4; 150–180°	0.05–0.9 μm/min	Bubble domain collapse field, H_0, can be easily tailored by etching	[529, 530]
CuSi	1 N NaCl; I–V characteristics depend on % Si.	—	Electrochemical etch; black powder forms	[574]
CoSi	1 N NaCl or 1 N Na_2SO_4; I–V characteristics depend on % Si.	—	Electrochemical etch; black powder or glassy film forms	[574]
CrSi	$60H_3PO_4$, $5HNO_3$, 1HF	600–900 Å/min	Also effective for pure Si	[575]
CrSiO	1HCl, $1H_2O$; or $1H_2SO_4$, $3H_2O$; 50–70°	—	With depassivation of Cr; gas bubbles cling to surfaces; irregular line edges.	[14]
	2 g $K_3[Fe(CN)_6]$, 1 g NaOH, 10 ml H_2O; 50–60°	1000 Å/min	Used on Cr-20 at. % SiO films; good edge definition.	[14]
NiSi	1 N NaCl or 1 N Na_2SO_4; I–V characteristics depend on % Si.	—	Electrochemical etch; black powder or glassy film forms.	[574]
PtSi	Aqua regia	—	Dissolves more rapidly than Pt if protective SiO_2 film is removed.	[563, 576]
	4(sat. I_2/CH_3COOH), $3HNO_3$, 1HF	—	—	[577]
SIPOS[a]	$6H_2O_2$, 1HF, $10NH_4F$ (40%)	2000 Å/min	Rate refers to 20 at. % oxygen films; undoped Si is unaffected.	[578]
WC	2 g NaOH, 2 g sodium tungstate, 100 ml H_2O; 21 V, 400 A/dm^2, Cu or carbon electrode	2 min	Electrochemical polish; used for W–C–Co alloys.	[29]

[a] Semiinsulating-polycrystalline-silicon (SIPOS) films are chemically vapor-deposited polycrystalline-silicon doped with oxygen or nitrogen atoms.

V. SUMMARY AND CONCLUSIONS

Fundamental principles of chemical etching reactions, processes, and techniques have been briefly outlined. Considerations in pattern delineation etching of thin films have been discussed, and the importance of surface contamination before and after etching processes has been noted. Chemical etching of specific materials of technical importance has been discussed and referenced. Extensive tables summarizing etchants, conditions, and applications have been presented for insulators, dielectrics, semiconductors, conductors, and miscellaneous materials to facilitate the selection of suitable systems. Most of the information has been retrieved from the literature which contains widely dispersed data. It is obviously impossible to include all etchants for all materials in a brief summary, but an attempt has been made to present significant and representative information covering the literature through 1977.

ACKNOWLEDGMENTS

We wish to thank C. E. Tracy for his help in tabulating the silicon etching data, R. D. Vibronek for carrying out many of the glass etching measurements, and G. L. Schnable for offering many suggestions and for critically reviewing the manuscript.

REFERENCES

1. J. W. Faust, Jr., in "Solid State Physics, Part A: Preparation, Structure, Mechanical and Thermal Properties" (K. Lark-Horovitz and V. A. Johnson, eds.), Methods of Experimental Physics, Vol. 6, Ch. 2.8. Academic Press, New York, 1959.
2. J. W. Faust, Jr., in "The Surface Chemistry of Metals and Semiconductors" (H. C. Gatos, ed.), pp. 151–173. Wiley, New York, 1960.
3. N. Hackerman, in The Surface Chemistry of Metals and Semiconductors" (H. C. Gatos, ed.), pp. 313–325. Wiley, New York, 1960.
4. P. Lacombe, in "The Surface Chemistry of Metals and Semiconductors" (H. C. Gatos, ed.), pp. 244–284. Wiley, New York, 1960.
5. P. J. Holmes, in "The Electrochemistry of Semiconductors" (P. J. Holmes, ed.), Ch. 8. Academic Press, New York, 1962.
6. B. A. Irving, in "The Electrochemistry of Semiconductors" (P. J. Holmes, ed.), pp. 256–288. Academic Press, New York, 1962.
7. J. W. Faust, in "Semiconducting Compounds" (R. K. Willardson and H. L. Goering, eds.), Vol. 1, pp. 445–468. Reinhold, New York, 1962.
8. H. C. Gatos and M. C. Lavine, *Prog. Semicond.* **9,** 1–46 (1965).
9. "Integrated Circuit Silicon Device Technology; X-Chemical Metallurgical Properties of Silicon, "ASD-TDR-63-316, Vol. X, AD 625, 985. Research Triangle Inst., Research Triangle Park, North Carolina, (1965).
10. M. Aven and J. S. Prener "The Physics and Chemistry of II-VI Compounds," pp. 141, 155, 733. North-Holland, Publ., Amsterdam, 1967.
11. C. V. King, in "The Surface Chemistry of Metals and Semiconductors" (H. C. Gatos, ed.), pp. 357–380. Wiley, New York, 1960.

12. S. K. Ghandhi, "The Theory and Practice of Microelectronics," Ch. 7. Wiley, New York, 1968.
13. C. D. Dobson, in "Gallium Arsenide Lasers" (C. H. Gooch, ed.), pp. 193–222. Wiley (Interscience), New York, 1969.
14. R. Glang and L. V. Gregor, in "Handbook of Thin Film Technology" (L. I. Maissel and R. Glang, eds.), Ch. 7. McGraw-Hill, New York, 1970.
15. R. J. Ryan, E. B. Davidson, and H. O. Hook, in "Handbook of Materials and Processes For Electronics" (C. A. Harper, ed.), Ch. 14. McGraw-Hill, New York, 1970.
16. W. A. Pliskin and S. J. Zanin, in "Handbook of Thin-Film Technology" (L. I. Maissel and R. Glang, eds.), Ch. 11. McGraw-Hill, New York, 1970.
17. P. F. Kane and G. B. Larrabee, "Characterization of Semiconductor Materials." McGraw-Hill, New York, 1970.
18. H. F. Wolf, "Semiconductors," pp. 130–136. Wiley (Interscience), New York, 1971.
19. T. C. Harman and I. Melngailis, Appl. Solid State Sci. 4, 1–94 (1974).
20. B. Tuck, J. Mater. Sci. 10, 321 (1975).
21. W. R. Runyan, "Semiconductor Measurements and Instrumentation," Chs. 1, 2, 7, and 9. McGraw-Hill, New York, 1975.
22. R. Tijburg, Phys. Technol. September, p. 202 (1976).
23. D. J. Stirland and B. W. Straughan, Thin Solid Films 31, 139 (1976).
24. M. J. Pryor and R. W. Staehle, in "Treatise on Solid State Chemistry" (N. B. Hannay, ed.), Vol. 4, Ch. 9, Plenum, New York, 1976.
25. "Metals Reference Book" (C. J. Smithells, ed.), 5th Ed. Butterworth, London, 1976.
26. "Etching for Pattern Definition" (H. G. Hughes and M. J. Rand, eds.). Electrochem. Soc., Princeton, New Jersey, 1976.
27. W. Kern, ref. 26, pp. 1–18d.
27a. W. Kern, RCA Rev. 39, 278 (1978).
28. J. S. Judge, ref. 26, pp. 19–36.
29. W. Tegert, "The Electrolytic and Chemical Polishing of Metals," 2nd Ed. Pergamon, Oxford, 1959.
30. J. F. Dewald, in "Semiconductors" (N. B. Hannay, ed.), pp. 727–752. Reinhold, New York, 1959.
31. "The Surface Chemistry of Metals and Semiconductors" (H. C. Gatos, ed.). Wiley, New York, 1960.
32. "The Electrochemistry of Semiconductors" (P. J. Holmes, ed.). Academic Press, New York, 1962.
33. J. I. Pankove, ref. 32, Ch. 7.
34. E. A. Efimov and I. G. Erusalimchik, "Electrochemistry of Germanium and Silicon" (A. Peiperl, ed. and transl.) Sigma Press, Washington, D.C., 1963.
35. P. J. Boddy, J. Electroanal. Chem. 10, 199 (1965).
36. V. A. Myamlin and Y. V. Pleskov, "Electrochemistry of Semiconductors." Plenum, New York, 1967.
37. H. Gerischer, in "Electrochemistry," Part A (W. Jost, ed.), Physical Chemistry, Vol. 9, pp. 463–542. Academic Press, New York 1970.
38. P. V. Shchigolev, "Electrolytic and Chemical Polishing of Metals." Freund Publ. House, Holon, Israel, 1970.
39. Y. V. Pleskov, Prog. Surf. Membr. Sci. 7, 57–93 (1973).
40. A. K. Vijh, "Electrochemistry of Metals and Semiconductors." Dekker, New York, 1973.
41. G. L. Schnable and P. F. Schmidt, J. Electrochem. Soc. 123, 310C (1976).
42. "Corrosion" (L. L. Shreir, ed.), 2nd Ed. Newnes-Butterworth, London, 1976.

43. A. H. Agajanian, *Solid State Technol.* **16**(12), 73 (1973); **18**(4), 61 (1975); **20**(1), 36 (1977).
43a. C. W. Wilmsen and S. Szpak, *Thin Solid Films* **47**, 17 (1977).
44. L. Romankiw, ref. 26, pp. 161–193.
45. P. R. Camp, *J. Electrochem. Soc.* **102**, 586 (1955).
46. W. Kern, J. L. Vossen, and G. L. Schnable, *Annu. Proc. Reliab. Phys., 11th* p. 214 (1973).
47. B. A. Irving, ref. 32, Ch. 6.
48. J. L. Vossen, G. L. Schnable, and W. Kern, *J. Vac. Sci. Technol.* **11**, 60 (1974).
49. V. F. Dorfman, *Sov. Microelectron.* **5** (2), 99 (1976).
50. W. S. DeForest, "Photoresist Materials and Processes." McGraw-Hill, New York, 1975.
51. W. Kern and J. M. Shaw, *J. Electrochem. Soc.* **118**, 1699 (1971).
52. C. A. Deckert and D. A. Peters, *Proc. Kodak Microelectron. Semin.*, Eastman Kodak Co., Rochester, New York, pp. 3–25 (1978).
53. C. A. Deckert, *in* "Adhesion Measurement of Thin Films, Thick Films, and Bulk Coatings" (K. L. Mittal, ed.), pp. 327–341, ASTM, Philadelphia, Pennsylvania, 1978.
54. J. J. Kelly and G. J. Koel, *J. Electrochem. Soc.* **125**, 860 (1978).
55. Anonymous, *Ind. Res.* **19** (2), 19 (1977).
56. S. Somekh, *J. Vac. Sci. Technol.* **13**, 1003 (1976).
57. J. J. Kelly and C. H. de Minjer, *J. Electrochem. Soc.* **122**, 931 (1975).
58. D. MacArthur, ref. 26, pp. 76–90.
59. P. J. Holmes and R. C. Newman, *Proc. Inst. Electr. Eng., Part B, Suppl.* **15**, 287 (1959).
60. J. W. Faust, Jr., *Natl. Bur. Stand. (U.S.), Spec. Publ.* No. 337, pp. 436–441 (1970).
61. W. Kern, *RCA Rev.* **31**, 207 (1970).
62. W. Kern, *RCA Rev.* **31**, 234 (1970).
63. W. Kern, *RCA Rev.* **32**, 64 (1971).
64. W. Kern, *Solid State Technol.* **15**(1), 34 (1972; **15**(2), 39 (1972).
65. W. Kern and D. Puotinen, *RCA Rev.* **31**, 187 (1970).
66. A. Mayer and D. A. Puotinen, *Natl. Bur. Stand.* pp. 431–435 *(U.S.), Spec. Publ. No.* 337, (1970).
67. R. C. Henderson, *J. Electrochem. Soc.* **119**, 772 (1972).
68. R. L. Meek, T. M. Buck, and C. F. Gibbon, *J. Electrochem. Soc.* **120**, 1241 (1973).
69. D. A. Kiewit, I. J. D'Haenens, and J. A. Roth, *J. Electrochem. Soc.* **121**, 310 (1974).
69a. D. A. Peters and C. A. Deckert, *Electrochem. Soc. Extend. Abstr.* **78-2**, 637 (1978).
70. P. Rai-Choudhury, *in* "Semiconductor Silicon 1973" (H. R. Huff and R. R. Burgess, eds.), pp. 243–257. Electrochem. Soc., Princeton, New Jersey, 1973.
71. D. R. Oswald, *J. Electrochem. Soc.* **123**, 531 (1976).
72. J. A. Amick, *Solid State Technol.* **19**(11), 47 (1976).
73. D. Tolliver, *Solid State Technol.* **18**(11), 33 (1975).
74. M. G. Yang, K. M. Koliwad, and G. E. McGuire, *J. Electrochem. Soc.* **122**, 675 (1975).
75. R. Seltzer, *Circuits Manuf.* **15**(11), 56 (1975).
76. R. A. Geckle, *Electron. Packag. Prod.* **15**(7), 127 (1975).
77. G. L. Schnable, W. Kern, and R. B. Comizzoli, *J. Electrochem. Soc.* **122**, 1092 (1975).
78. R. L. Meek, *J. Electrochem. Soc.* **121**, 172 (1974).
79. W. Kern, unpublished observations.
80. W. Kern, *Semicond. Prod. Solid State Technol.* **6**(10), 22 (1963); **6**(11), 23 (1963).
81. "Clean Surfaces: Their Preparation and Characterization for Interfacial Studies" (G. Goldfinger, ed.). Dekker, New York, 1970.
82. Ref. 15, pp. 44, 45, 54, 55, 59–66.

83. Ref. 50, Chs. 3 and 7.
84. L. Holland, "The Properties of Glass Surfaces," Ch. 5. Wiley, New York, 1964.
85. R. Brown, *in* "Handbook of Thin Film Technology" (L. I. Maissel and R. Glang, eds.), Ch. 6, pp. 37–42. McGraw-Hill, New York, 1970.
86. D. M. Mattox, "Surface Cleaning in Thin Film Technology," AVS Monogr. Am. Vac. Soc., New York, 1975.
87. R. R. Sowell, R. E. Cuthrell, D. M. Mattox, and R. D. Bland, *J. Vac. Sci. Technol.* **11**, 474 (1974).
88. J. R. Vig, C. F. Cook, N. Schmidtal, J. W. LeBus, and E. Hafner, "Surface Studies for Quartz Resonators," R&D Tech. Rep. ECOM- 4251. U.S. Army Electron. Command, Fort Monmouth, New Jersey (1974).
89. D. A. Bolon and C. O. Kunz, *Polym. Eng. Sci.* **12**, 109 (1972).
90. P. H. Holloway and D. W. Bushmire, *Annu. Proc. Reliab. Phys. 12th* p. 180 (1974).
91. R. C. Snogren, "Handbook of Surface Preparation." Palmerton, New York, 1974.
92. J. L. Vossen, *J. Appl. Phys.* **47**, 544 (1976).
93. L. Holland, *J. Vac. Sci. Technol.* **14**, 5 (1977).
94. R. B. Gillette, J. R. Hollahan, and G. L. Carlson, *J. Vac. Sci. Technol.* **7**, 534 (1970).
95. J. L. Vossen, J. J. O'Neill, Jr., K. M. Finlayson, and L. J. Royer, *RCA Rev.* **31**, 293 (1970).
96. C. C. Chang, P. Petroff, G. Quintana, and J. Sosniak, *Surf. Sci.* **38**, 341 (1973).
97. D. V. McCaughan and R. A. Kushner, *Thin Solid Films* **22**, 359 (1974).
98. G. J. Kominiak and J. E. Uhl, Rep. SAND 75–0455. Sandia Lab., Albuquerque, New Mexico (1975).
99. Ref. 81, Ch. 5.
100. T. Smith, *Surf. Sci.* **27**, 45 (1971).
101. A. M. Morgan and I. Dalins, *J. Vac. Sci. Technol.* **10**, 523 (1973).
102. R. W. Kirk, *in* "Techniques and Applications of Plasma Chemistry" (J. R. Hollahan and A. T. Bell, eds.), Ch. 9. Wiley, New York, 1974.
103. G. J. Kominiak and D. M. Mattox, Rep. SAND 75–6110. Sandia Lab., Albuquerque, New Mexico (1976); *Thin Solid Films* **40**, 141 (1977).
104. Plasma Etching and Deposition Technology; Program, Atlanta, Ga. Meet. *J. Electrochem. Soc.* **124**, 252C–324C (1977).
105. J. S. Judge, *J. Electrochem. Soc.* **118**, 1772 (1971).
106. E. F. Duffek and D. Pilling, *Electrochem. Soc. Extend. Abstr.* No. 111, p. 244, Spring Meeting (1965).
107. S. A. Harrell and J. R. Peoples, Jr., *Electrochem. Soc. Extend. Abstr.* No. 112, p. 247, Spring Meeting (1965).
108. C. C. Mai and J. C. Looney, *Semicond. Prod. Solid State Technol.* **9**(1), 19 (1966).
109. J. Lawrence, *Electrochem. Soc. Extend. Abstr.* **72-2**, 466 (1972).
110. V. Harrap, *in* "Semiconductor Silicon 1973" (H. R. Huff and R. R. Burgess, eds.), pp. 354–362. Electrochem. Soc., Princeton, New Jersey, 1973.
111. R. Herring and J. B. Price, *Electrochem. Soc. Extend. Abstr.* **73–2**, 410 (1973).
112. D. S. Herman, M. A. Schuster, and H. G. Oehler, *Electrochem. Soc. Extend. Abstr.* **71–1**, 167 (1971).
113. J. J. Gajda, *Annu. Proc. Reliab. Phys., 12th* p. 30 (1974).
114. P. J. Holmes and J. E. Snell, *Microelectron. Reliab.* **5**, 337 (1966).
115. T. W. O'Keeffe and R. M. Hardy, *Solid-State Electron.* **11**, 261 (1968).
116. R. A. Moline, R. R. Buckley, S. E. Haszko, and A. U. MacRae, *IEEE Trans. Electron Devices* **ED-20**, 840 (1973).
117. G. Bell and J. Hoepfner, ref. 26, pp. 47–55.

118. W. A. Pliskin, *Thin Solid Films* **2**, 1 (1968).
119. L. A. Murray and N. Goldsmith, *J. Electrochem. Soc.* **113**, 1297 (1966).
120. B. Swaroop, *in* "Thin Film Dielectrics" (F. Vratny, ed.), p. 407. Electrochem. Soc., New York, 1969.
121. W. A. Pliskin and R. P. Gnall, *J. Electrochem. Soc.* **111**, 872 (1964).
122. D. M. Brown, W. E. Engeler, M. Garfinkel, and F. K. Heumann, *J. Electrochem. Soc.* **114**, 730 (1967).
123. W. Kern, *RCA Rev.* **29**, 557 (1968).
124. W. A. Pliskin and H. S. Lehman, *J. Electrochem. Soc.* **112**, 103 (1965).
125. D. H. Grantham and J. Swindal, *Int. Microelectron. Symp.* p. 118. Int. Soc. Hybrid Microelectron. Montgomery, Alabama (1975).
126. T. Kubota, *Jpn. J. Appl. Phys.* **11**, 1413 (1972).
127. P. C. Huang and P. M. Schaible, *Electrochem. Soc. Extend. Abstr.* **76–2**, 754 (1976).
128. M. J. Rand, *J. Electrochem. Soc.* **114**, 274 (1967).
129. S. Krongelb, *Electrochem. Technol.* **6**, 251 (1968).
130. W. Kern and J. P. White, *RCA Rev.* **31**, 771 (1970).
131. Y. Avigal, I. Beinglass, and M. Schieber, *J. Electrochem. Soc.* **121**, 1103 (1974).
132. N. Goldsmith and W. Kern, *RCA Rev.* **28**, 153 (1967).
133. B. E. Deal, P. J. Fleming, and P. L. Castro, *J. Electrochem. Soc.* **115**, 300 (1968).
134. M. L. Barry, *in* "Chemical Vapor Deposition" (J. M. Blocher, Jr. and J. C. Withers, eds.), pp. 595–617. Electrochem. Soc., New York, 1970.
135. E. L. MacKenna, *Proc. Semicond./IC Proc. Prod. Conf.* pp. 71–83. Ind. Sci. Conf. Manage., Chicago, Illinois (1971).
136. L. Hall, *J. Electrochem. Soc.* **118**, 1506 (1971).
137. W. Kern, *RCA Rev.* **37**, 78 (1976).
138. W. Kern, *RCA Rev.* **37**, 55 (1976).
139. M. L. Hammond and G. M. Bowers, *Trans. Metall. Soc. AIME* **242**, 546 (1968).
140. T. L. Chu, J. R. Szedon, and G. A. Gruber, *Trans. Metall. Soc. AIME* **242**, 532 (1968).
141. W. J. Kroll, Jr., R. L. Titus, and J. B. Wagner, Jr., *J. Electrochem. Soc.* **122**, 573 (1975).
142. A. K. Gaind, G. K. Ackermann, V. J. Lucarini, and R. L. Bratter, *J. Electrochem. Soc.* **122**, 573 (1975).
143. J. Kraitchman and J. Oroshnik, *J. Electrochem. Soc.* **114**, 405 (1967).
144. H. C. Lam and Y. W. Lam, *Thin Solid Films* **41**, 43 (1977).
145. J. Kraitchman, *J. Appl. Phys.* **38**, 4323 (1967).
146. J. H. Alexander, R. J. Joyce, and H. F. Sterling, *in* "Thin Film Dielectrics" (F. Vratny, ed.), pp. 146–208. Electrochem. Soc., New York, 1969.
147. Ref. 102, pp. 362–374.
147a. A. G. Revesz and K. H. Zaininger, *RCA Rev.* **29**, 22 (1968); T. L. Chu, *J. Vac. Sci. Technol.* **6**, 25 (1969).
148. J. Dey, M. Lundgren, and S. Harrel, *Kodak Photoresist Semin. Proc.* Vol. 2, p. 4 (P-192-B). Eastman Kodak Co., Rochester, New York (1968).
149. H. F. Wolf, ref. 18, p. 372.
150. W. A. Pliskin and R. P. Esch, ref. 26, pp. 37–46.
151. L. E. Katz and W. C. Erdman, *J. Electrochem. Soc.* **123**, 1249 (1976).
152. S. C. H. Lin and I. J. Pugacz-Muraszkiewicz, *J. Appl. Phys.* **43**, 119 (1972).
153. I. J. Pugacz-Muraszkiewicz and B. R. Hammond, *J. Vac. Sci. Technol.* **14**, 49 (1977).
154. E. T. Fitzgibbons, K. J. Sladek, and W. H. Hartwig, *J. Electrochem. Soc.* **119**, 735 (1972).
154a. M. Yokozawa, H. Iwasa, and I. Teramoto, *Jpn. J. Appl. Phys.* **7**, 96 (1968).

154b. D. R. Harbison and H. L. Taylor, *in* "Thin Film Dielectrics" (F. Vratny, ed.), pp. 254–278. Electrochem. Soc., New York, 1969.

155. M. Balog, M. Schieber, S. Patai, and M. Michman, *J. Cryst. Growth* **17**, 298 (1972).

156. Y. W. Hsueh and H. C. Lin, *Annu. Rep., Conf. Electr. Insul. Dielectr. Phenom.* p. 515 (1974).

157. E. Kaplan, M. Balog, and D. Frohman-Bentchkowsky, *J. Electrochem Soc.* **123**, 1570 (1976).

158. W. H. Knausenberger and R. N. Tauber, *J. Electrochem. Soc.* **120**, 927 (1973).

159. J.P.S. Pringle, *J. Electrochem. Soc.* **119**, 482 (1972).

160. P. W. Wyatt, *J. Electrochem. Soc.* **122**, 1660 (1975).

161. B. H. Hill, *J. Electrochem. Soc.* **115**, 668 (1969).

162. J. Grossman and D. S. Herman, *J. Electrochem. Soc.* **116**, 674 (1969).

163. H. M. Day, A. Christou, W. H. Weisenberger, and J. K. Hervonen, *J. Electrochem. Soc.* **122**, 769 (1975).

164. R. N. Tauber, A. C. Dumbri, and R. E. Caffrey, *J. Electrochem. Soc.* **118**, 747 (1971).

165. J. A. Aboaf, *J. Electrochem. Soc.* **114**, 948 (1967).

166. N. Hashimoto, Y. Koga, and E. Yamada, *in* "Thin Film Dielectrics" (F. Vratny, ed.), pp. 327–337. Electrochem. Soc., New York, 1969.

167. Y. Koga, M. Matsushita, K. Kobayashi, Y. Nakaido, and S. Toyoshima, *in* "Thin Film Dielectrics" (F. Vratny, ed.), pp. 355–377. Electrochem. Soc., New York, 1969.

168. M. T. Duffy and W. Kern, *RCA Rev.* **31**, 754 (1970).

169. M. Mutoh, Y. Mizokami, H. Matsui, S. Hagiwara, and M. Ino, *J. Electrochem. Soc.* **122**, 987 (1975).

170. H. Katto and Y. Koga, *J. Electrochem. Soc.* **118**, 1619 (1971).

171. A. J. Learn, *J. Appl. Phys.* **44**, 1251 (1973).

172. K. Iida, *J. Electrochem. Soc.* **124**, 614 (1977); M. Hirayama and K. Shohno, *J. Electrochem. Soc.* **122**, 1671 (1975).

173. E. Ferrieu and B. Pruniaux, *J. Electrochem. Soc.* **116**, 1008 (1969).

174. H. Harada, S. Satoh, and M. Yoshida, *IEEE Trans. Reliab.* **R-25**, 290 (1976).

175. R. G. Frieser, *J. Electrochem. Soc.* **113**, 357 (1966).

176. T. N. Kennedy, *Electron. Packag. Prod.* **14** (12), 136 (1974).

177. R. S. Nowicki, *J. Vac. Sci. Technol.* **14**, 127 (1977).

178. S. K. Tung and R. E. Caffrey, *J. Electrochem. Soc.* **114**, 257C, Abstr. RNP-24 (1967).

179. V. Y. Doo and P. J. Tsang, *Electrochem. Soc. Extend. Abstr.* No. 16, p. 33, Spring Meeting (1969).

180. P. J. Tsang, R. M. Anderson, and S. Cvikevich, *J. Electrochem. Soc.* **123**, 57 (1976).

181. K. M. Schlesier, J. M. Shaw, and C. W. Benyon, Jr., *RCA Rev.* **37**, 358 (1976).

182. G. C. Schwartz and V. Platter, *J. Electrochem. Soc.* **122**, 1508 (1975).

183. A. Reisman, M. Berkenblit, J. Cuomo, and S. A. Chan, *J. Electrochem. Soc.* **118**, 1653 (1971).

184. M. F. Ehman, *J. Electrochem. Soc.* **121**, 1240 (1974).

185. H. M. Manasevit, *J. Electrochem. Soc.* **121**, 293 (1974).

186. J. M. Green, *J. Electrochem. Soc.* **119**, 1765 (1972).

187. M. Safdar, G. H. Frischat, and H. Salge, *J. Am. Ceram. Soc.* **57**, 106 (1974).

188. B. Siesmayer, R. Heimann, and W. Franke, *J. Cryst. Growth* **28**, 157 (1975).

189. I. I. Baram, *Zh. Prikl. Khim.* **38**, 2181 (1965).

189a. M. Balog, M. Schieber, M. Michman, and S. Patai, *Thin Solid Films* **41**, 247 (1977).

190. W. Kern and R. C. Heim, *Electrochem. Soc. Extend. Abstr.* No. 92, p. 234, Spring Meeting (1968).

191. W. Kern and R. C. Heim, *J. Electrochem. Soc.* **117**, 562 (1970).

192. W. Kern, G. L. Schnable, and A. W. Fisher, *RCA Rev.* **37**, 3 (1976).
193. J. M. Eldridge and P. Balk, *Trans. Metall. Soc. AIME* **242**, 539 (1968).
194. E. H. Snow and B. E. Deal, *J. Electrochem. Soc.* **113**, 263 (1966).
195. W. Kern, *Electrochem. Soc. Extend. Abstr.* **76-1**, 119 (1976).
196. A. S. Tenney and M. Ghezzo, *J. Electrochem. Soc.* **120**, 1091 (1973).
197. S. Nishimatsu and T. Tokuyama, *Electrochem. Soc. Extend. Abstr.* No. 170, p. 24, Fall Meeting (1967).
198. T. Tokuyama, T. Miyazaki, and M. Horiuchi, *in* "Thin Film Dielectrics" (F. Vratny, ed.), pp. 297–326. Electrochem. Soc., New York, 1969.
199. W. Kern and R. C. Heim, *J. Electrochem. Soc.* **117**, 568 (1970).
200. K. Jinno, H. Kinoshita, and Y. Matsumoto, *J. Electrochem. Soc.* **124**, 1258 (1977).
201. K. Chow and L. G. Garrison, *J. Electrochem. Soc.* **124**, 1133 (1977).
202. P. F. Schmidt, W. van Gelder, and J. Drobek, *J. Electrochem. Soc.* **115**, 79 (1968).
203. P. Balk and J. M. Eldridge, *Proc. IEEE* **57**, 1558 (1969).
204. K. Sugawara, T. Yoshimi, and H. Sakai, *in* "Chemical Vapor Deposition," Fifth International Conference (J. M. Blocher, Jr. and H. E. Hinterman, eds.), pp. 407–412. Electrochem. Soc., Princeton, New Jersey, 1975.
205. W. Kern and A. W. Fisher, *RCA Rev.* **31**, 715 (1970).
206. A. H. El-Hoshy, *J. Electrochem. Soc.* **117**, 1583 (1970).
207. D. M. Brown and R. P. Kennicott, *J. Electrochem. Soc.* **118**, 293 (1971).
208. L. Rankel Plauger, *J. Electrochem. Soc.* **120**, 1428 (1973).
209. F. N. Schwettmann, R. J. Dexter, and D. F. Cole, *J. Electrochem. Soc.* **120**, 1566 (1973).
210. R. O. Schwenker, *J. Electrochem. Soc.* **118**, 313 (1971).
211. S. S. Chang, U.S. Patent 3,784,424 (1974).
212. W. Kern, *J. Electrochem. Soc.* **116**, 251C (1969).
213. J. Wong, *J. Electrochem. Soc.* **119**, 1071 (1972).
214. M. Ghezzo and D. M. Brown, *J. Electrochem. Soc*, **120**, 110 (1973).
215. R. B. Fair, *J. Electrochem. Soc.* **119**, 1389 (1972).
216. F. C. Eversteijn, *Philips Res. Rep.* **21**, 379 (1966).
217. S. K. Tung and R. E. Caffrey, *J. Electrochem. Soc.* **117**, 91 (1970).
218. P. J. Tsang, R. M. Anderson, and S. Crikevich, *J. Electrochem. Soc.* **123**, 57 (1976).
219. S. M. Spitzer, B. Schwartz, and G. D. Weigle, *J. Electrochem. Soc.* **122**, 397 (1975).
220. T. Sugano and Y. Mori, *J. Electrochem. Soc.* **121**, 113 (1974).
221. G. W. Morey, "The Properties of Glass," 2nd Ed., Ch. 4. Reinhold, New York, 1954.
222. L. Holland, "The Properties of Glass Surfaces," Chs. 3 and 5. Wiley, New York, 1964.
223. T. Yoshida and M. Koyama, U.S. Patent 3,839,113 (1974).
224. M. Dumesnil and R. Hewitt, *J. Electrochem. Soc.* **117**, 100 (1970).
225. F. M. Ernsberger, *J. Am. Ceram. Soc.* **42**, 375 (1959).
226. W. A. Pliskin, *J. Electrochem. Soc.* **114**, 620 (1967).
227. D. M. Mattox and G. J. Kominiak, *J. Electrochem. Soc.* **120**, 1535 (1973).
228. W. A. Pliskin, P. D. Davidse, H. S. Lehman, and L. I. Maissel, *IBM J. Res. Dev.* **11**, 461 (1967).
229. H. N. Farrer and F. J. C. Rossotti, *J. Inorg. Nucl. Chem.* **26**, 1959 (1964).
230. J. T. Milek, "Silicon Nitride for Microelectronic Applications, Part 1—Preparation and Properties," pp. 1–118. IFI/Plenum, New York, 1971.
231. C. A. Deckert, *J. Electrochem. Soc.* **124**, 320 (1978).
232. W. van Gelder and V. E. Hauser, *J. Electrochem. Soc.* **144**, 869 (1967).
233. D. C. Miller, *J. Electrochem. Soc.* **120**, 1771 (1973).
234. C. A. Deckert, unpublished observations (1975).

235. W. Kern and R. S. Rosler, *J. Vac. Sci. Technol.* **14**, 1082 (1977).
236. E. A. Taft, *J. Electrochem. Soc.* **118**, 1341 (1971).
237. R. Gereth and W. Scherber, *J. Electrochem. Soc.* **119**, 1248 (1972).
238. Y. Kuwano, *Jpn. J. Appl. Phys.* **8**, 876 (1969).
239. M. J. Rand, *Electrochem. Soc. Extend. Abstr.* **77-2**, 419 (1977).
240. W. A. Lanford and M. J. Rand, *Electrochem. Soc. Extend. Abstr.* **77-2**, 421 (1977).
241. V. D. Wohlheiter and R. A. Whitner, *J. Electrochem. Soc.* **119**, 945 (1972).
242. H. J. Stein, *J. Electron. Mater.* **5**, 161 (1976).
243. V. D. Wohlheiter, *J. Electrochem. Soc.* **122**, 1736 (1975).
244. G. J. Kominiak, *J. Electrochem. Soc.* **122**, 1272 (1975).
245. C. J. Mogab, P. M. Petroff, and T. T. Sheng, *J. Electrochem. Soc.* **122**, 815 (1975).
246. A. W. Stephens, J. L. Vossen, and W. Kern, *J. Electrochem. Soc.* **123**, 303 (1976).
247. C. J. Dell'Oca, *J. Electrochem. Soc.* **120**, 1225 (1973).
248. D. M. Brown, P. V. Gray, F. K. Herrmann, H. R. Philipp, and E. A. Taft, *J. Electrochem. Soc.* **155**, 311 (1968).
249. M. J. Rand and J. F. Roberts, *J. Electrochem. Soc.* **120**, 446 (1973).
250. H. Nagai and T. Niimi, *J. Electrochem. Soc.* **115**, 671 (1968).
250a. T. Yashiro, *J. Electrochem. Soc.* **119**, 780 (1972).
251. D. R. Turner, *J. Electrochem. Soc.* **107**, 810 (1960).
252. D. L. Klein and D. J. D'Stefan, *J. Electrochem. Soc.* **109**, 37 (1962).
253. H. Robbins and B. Schwartz, *J. Electrochem. Soc.* **106**, 505, 1020 (1959).
254. H. Robbins and B. Schwartz, *J. Electrochem. Soc.* **107**, 108 (1960).
255. B. Schwartz and H. Robbins, *J. Electrochem. Soc.* **108**, 365 (1961).
256. B. Schwartz and H. Robbins, *J. Electrochem. Soc.* **123**, 1903 (1976).
257. A. F. Bogenschütz, W. Krusemark, K.-H. Löcherer, and W. Mussinger, *J. Electrochem. Soc.* **114**, 970 (1967).
258. S. M. Hu and D. R. Kerr, *J. Electrochem. Soc.* **114**, 414 (1967).
259. W. Hoffmeister, *Int. J. Appl. Radiat. Isot.* **2**, 139 (1969).
260. E. Cave, RCA Solid State Div., personal communication (1972).
261. G. R. Booker and R. Stickler, *Br. J. Appl. Phys.* **13**, 446 (1962).
262. G. Das and N. A. O'Neil, *IBM J. Res. Dev.* **18**, 76 (1974).
263. C. J. Schmidt, P. V. Lenzo, and E. G. Spencer, *J. Appl. Phys.* **48**, 4080 (1975).
264. A. I. Stoller, R. F. Speers, and S. Opresko, *RCA Rev.* **31**, 265 (1970).
265. J. Freyer, *J. Electrochem. Soc.* **122**, 1238 (1975).
266. B. A. Unvala, D. B. Holt, and A. San, *J. Electrochem. Soc.* **119**, 318 (1972).
267. R. E. Blaha and W. R. Fahrner, *J. Electrochem. Soc.* **123**, 515 (1976).
268. C. Raetzel, S. Schild, and H. Schlötterer, *Electrochem. Soc. Extend. Abstr.* **74-2**, 336 (1974).
269. M. J. J. Theunissen, J. A. Apples, and W. H. C. G. Verkuylen, *J. Electrochem. Soc.* **117**, 959 (1970).
270. Ref. 21, p. 199, Table 7.3.
271. R. R. Stead, U.S. Patent 2,973,253 (1961).
272. Ref. 5, p. 370.
273. "Book of ASTM Standards." ASTM, Philadelphia, Pennsylvania, 1967.
274. P. J. Holmes, *Proc. Inst. Elect. Eng., Part B. Suppl.* **17**, 861 (1959).
275. Telefunken, AG, Brit. Patent 962,335 (1964).
276. M. Chappey and P. Meritet, Fr. Patent 1,266,612 (1961).
277. R. J. Jacodine, *J. Appl. Phys.* **36**, 2811 (1965).
278. M. V. Sullivan and R. M. Finne, *Electrochem. Soc. Meet. Abstr.* No. 156, Fall Meeting (1960).
279. B. A. Irving, *Br. J. Appl. Phys.* **12**, 92 (1961).

280. B. Schwartz and H. Robbins, *J. Electrochem. Soc.* **111**, 196 (1964).
281. B. A. Irving, *J. Electrochem. Soc.* **109**, 120 (1962).
282. D. R. Turner, ref. 32, Ch. 4.
283. J. W. Faust, Jr., *in* "Reactivity of Solids" (J. W. Mitchell, R. C. De Vries, R. W. Roberts, and P. Cannon, eds.), pp. 337–343. Wiley, New York, 1969.
284. M. F. Ehman, J. W. Faust, Jr., and W. B. White, *J. Electrochem. Soc.* **118**, 1443 (1971).
285. B. Schwartz, *J. Electrochem. Soc.* **114**, 285 (1967).
286. W. Primak, R. Kampwirth, and Y. Dayal, *J. Electrochem. Soc.* **114**, 88 (1967).
287. P. R. Camp, *J. Electrochem. Soc.* **102**, 586 (1955).
288. J. P. McKelvey and R. L. Longini, *J. Appl. Phys.* **25**, 634 (1954).
289. B. W. Batterman, *J. Appl. Phys.* **28**, 1236 (1957).
290. Ref. 32, p. 369.
291. F. L. Vogel, W. G. Pfann, H. E. Corey, and E. E. Thomas, *Phys. Rev.* **90**, 489 (1953).
292. R. D. Heidenreich, U.S. Patent 2,619,414 (1952).
293. Ref. 32, p. 368.
294. P. Wang, *Sylvania Technol.* **11**, 50 (1958).
295. G. A. Geach, B. A. Irving, and R. Phillips, *Research (London)* **10**, 411 (1957).
296. W. W. Harvey and H. C. Gatos, *J. Electrochem. Soc.* **105**, 654 (1958).
297. H. C. Gatos and M. C. Lavine, *J. Electrochem. Soc.* **107**, 433 (1960).
298. W. K. Zwicker and S. K. Kurtz, "Semiconductor Silicon 1973" (H. R. Huff and R. R. Burgess, eds.), pp. 315–326. Electrochemical Soc., Princeton, New Jersey, 1973.
299. I. J. Pugacz-Muraszkiewicz and B. R. Hammond, *J. Vac. Sci. Technol.* **14**, 49 (1977).
300. H. A. Waggener, R. C. Kragness, and A. L. Tyler, *Int. Electron Devices Meet.* Abstr. No. 11.1, p. 68 (1967).
301. A. I. Stoller, *RCA Rev.* **31**, 271 (1970).
302. H. A. Waggener, R. C. Kragness, and A. L. Tyler, *Electronics* **40**(23), 274 (1967).
303. H. A. Waggener and J. V. Dalton, *Electrochem. Soc. Extend. Abstr.* **72-2**, 587 (1972).
304. J. B. Price, *in* "Semiconductor Silicon 1973" (H. R. Huff and R. R. Burgess, eds.), pp. 339–353. Electrochem. Soc., Princeton, New Jersey, 1973.
305. D. P. Clemens, *Electrochem. Soc. Extend. Abstr.* **73-2**, 407 (1973).
306. D. F. Weirrauch, *J. Appl. Phys.* **46**, 1478 (1975).
307. K. E. Bean, R. L. Yeakley, and T. K. Powell, *Electrochem. Soc. Extend. Abstr.* **74-1**, 68 (1974).
308. M. J. Declercq, J. P. DeMoor, and J. P. Lambert, *Electrochem. Soc. Extend Abstr.* **75-2**, 446 (1975).
309. D. B. Lee, *J. Appl. Phys.* **40**, 4569 (1969).
310. M. J. Declercq, L. Gerzberg, and J. D. Meindl, *J. Electrochem. Soc.* **122**, 545 (1975).
311. R. M. Finne and D. L. Klein, *J. Electrochem. Soc.* **114**, 965 (1967).
312. M. Asano, T. Cho, and H. Muraoka, *Electrochem. Soc. Extend. Abstr.* **76-2**, 911 (1976).
313. I. Bassou, H. N. Yu, and V. Maniscalco, *J. Electrochem. Soc.* **123**, 1729 (1976).
314. K. E. Bean and W. R. Runyan, *J. Electrochem. Soc.* **124**, 56 (1977).
315. T. R. Payne and H. R. Plumlee, *IEEE J. Solid-State Circuits* **SC-8**, 71 (1973).
316. K. E. Bean and J. R. Lawson, *IEEE J. Solid-State Circuits* **SC-9**, 111 (1974).
317. W. Tsang and S. Wang, *J. Appl. Phys.* **46**, 2164 (1975).
318. C. R. Baraona and H. W. Brandhorst, *IEEE Photovoltaic Spec. Conf. Proc.*, Scottsdale, AZ pp. 44–48 (1975).
319. F. Restrepo and C. E. Backus, *IEEE Trans. Electron Devices* **ED-23**, 1195 (1976).
320. R. K. Smeltzer, *J. Electrochem. Soc.* **122**, 1666 (1975).

321. R. C. Ellis, Jr., *J. Appl. Phys.* **25**, 1497 (1954); **28**, 1068 (1957).
322. A. Uhlir, *Bell Syst. Tech. J.* **35**, 333 (1956).
323. G. L. Schnable and W. M. Lilker, *Electrochem. Technol.* **1**, 203 (1963).
324. Ref. 33, pp. 297–308.
325. G. R. Booker and R. Stickler, *J. Electrochem. Soc.* **109**, 1167 (1962).
326. M. V. Sullivan, D. L. Klein, R. M. Finne, L. A. Pompliano, and G. A. Kolb, *J. Electrochem. Soc.* **110**, 412 (1963).
327. C. E. Hallas, *Solid State Technol.* **14**(1), 30 (1971).
328. R. L. Meek, *J. Electrochem. Soc.* **118**, 437 (1971).
329. H. J. A. van Dijk and J. de Jonge, *J. Electrochem. Soc.* **117**, 553 (1970).
330. R. L. Meek, *J. Electrochem. Soc.* **118**, 1240 (1971).
331. M. J. J. Theunissen, *J. Electrochem. Soc.* **119**, 351 (1972).
332. M. J. Hill, *J. Electrochem. Soc.* **120**, 142 (1973).
333. C. P. Wen and K. P. Weller, *J. Electrochem. Soc.* **119**, 547 (1972).
334. T. I. Kamins, *Proc. IEEE* **60**, 915 (1972).
335. T. I. Kamins, *J. Electrochem. Soc.* **121**, 286 (1974).
336. T. I. Kamins, *Solid-State Electron.* **17**, 667 (1974).
337. H. A. Waggener, *Bell Syst. Tech. J.* **49**, 473 (1970).
338. H. A. Waggener and J. V. Dalton, *Electrochem. Soc. Extend. Abstr.* **70-2**, 450 (1970).
339. A. Manara, A. Ostidich, G. Pedroli, and G. Restelli, *Thin Solid Films* **8**, 359 (1971).
340. H. D. Barber, H. B. Lo, and J. E. Jones, *J. Electrochem. Soc.* **123**, 1404 (1976).
341. J. Sanada, K. Furuno, K. Shima, and T. Momoi, *Jpn. J. Appl. Phys.* **16**, 299 (1977).
342. R. Memming and G. Schwandt, *Surf. Sci.* **4**, 109 (1966).
343. P. H. Bellin and W. K. Zwicker, *J. Appl. Phys.* **42**, 1216 (1971).
344. Y. Watanabe, Y. Arita, T. Yokoyama, and Y. Igaraschi, *J. Electrochem. Soc.* **122**, 1351 (1975).
345. Y. Arita and Y. Sunohara, *J. Electrochem. Soc.* **124**, 285 (1977).
346. Y. Yashiro, K. Saito, and T. Suzuki, *Electrochem. Soc. Extend. Abstr.* **74-2**, 351 (1974).
347. J. B. Price and W. C. Roman, *Electrochem. Soc. Extend. Abstr.* **72-2**, 584 (1972).
348. G. L. Kuhn and C. J. Rhee, *J. Electrochem. Soc.* **120**, 1563 (1973).
349. J. P. White, RCA Solid State Division, unpublished observations (1974).
350. A. Bohg, *J. Electrochem. Soc.* **118**, 401 (1971).
351. H. Muraoka, T. Ohhashi, and Y. Sumitomo, *in* "Semiconductor Silicon 1973" (H. R. Huff and R. R. Burgess, eds.), pp. 327–338. Electrochem. Soc., Princeton, New Jersey, 1973.
352. Y. Yasuda and T. Moriya, *in* "Semiconductor Silicon 1973" (H. R. Huff and R. R. Burgess eds.), pp. 271–284. Electrochem. Soc., Princeton, New Jersey, 1973.
353. R. E. Chappelow and P. T. Lin. *J. Electrochem. Soc.* **123**, 913 (1976).
354. J. Lawrence, Can. Patent 903,650 (1972).
355. Y. Sumitomo, K. Niwa, H. Sawazaki, and K. Sakai, *in* "Semiconductor Silicon 1973" (H. R. Huff and R. R. Burgess, eds.), pp. 893–904. Electrochem. Soc., Princeton, New Jersey, 1973.
356. Y. Sumitomo, T. Yasui, H. Nakatsuka, T. Oohashi, H. Tsutsumi, and H. Muraoka, *Electrochem. Soc. Extend. Abstr.* **72-1**, 74 (1972).
357. K. E. Peterson, *Appl. Phys. Lett.* **31**, 521 (1977).
358. H. Guckel, S. Larsen, M. G. Lagally, G. Moore, J. B. Miller, and J. D. Wiley, *Appl. Phys. Lett.* **31**, 618 (1977).
359. Y. S. Chiang and G. Looney, RCA Laboratories, unpublished observations (1970).
360. P. Rai-Choudhury, *J. Electrochem. Soc.* **118**, 266 (1971).

361. S. E. Mayer and D. E. Shea, *J. Electrochem. Soc.* **111**, 550 (1964).
362. L. J. Stinson, J. A. Howard, and R. C. Neville, *J. Electrochem. Soc.* **123**, 551 (1976).
363. G. A. Lang and T. Stavish, *RCA Rev.* **24**, 488 (1963).
364. W. H. Shepherd, *J. Electrochem. Soc.* **112**, 988 (1965).
365. L. V. Gregor, P. Balk, and F. J. Campagna, *IBM J. Res. Dev.* **9**, 327 (1965).
366. W. Runyan, "Silicon Semiconductor Technology," pp. 72–73. McGraw-Hill, New York, 1965.
367. K. E. Bean and P. S. Gleim, *Proc. IEEE* **57**, 1469 (1969).
368. K. Sugawara, Y. Nakazawa, and Y. Sugita, *Electrochem. Technol.* **6**, 295 (1968).
369. K. Sugawara, *J. Electrochem. Soc.* **118**, 110 (1971).
370. M. Druminski and R. Gessner, *J. Cryst. Growth* **31**, 312 (1975).
371. A. Reisman and M. Berkenblit, *J. Electrochem. Soc.* **112**, 812 (1965).
372. L. D. Dyer, *AIChE J.* **18**, 728 (1972).
373. J. P. Dismukes and R. Ulmer, *J. Electrochem. Soc.* **118**, 634 (1971).
374. P. Rai-Choudhury and A. J. Noreika, *J. Electrochem. Soc.* **116**, 539 (1969).
375. J. P. Dismukes and E. R. Levin, in "Chemical Processing of Microelectronic Materials" (L. D. Locker and J. Childress, eds.), p. 135. Am. Inst. Chem. Eng., New York, 1970.
376. E. R. Levin, J. P. Dismukes, and M. D. Coutts, *J. Electrochem. Soc.* **118**, 1171 (1971).
377. T. L. Chu, G. A. Gruber, and R. Stickler, *J. Electrochem. Soc.* **113**, 156 (1966).
378. W. G. Oldham and H. Holmstrom, *J. Electrochem. Soc.* **114**, 381 (1967).
379. J. A. Amick, E. A. Roth, and H. Gossenberger, *RCA Rev.* **24**, 473 (1963).
380. T. L. Chu and R. W. Kelm, *J. Electrochem. Soc.* **116**, 1261 (1969).
381. R. J. Walsh and A. H. Herzog, U.S. Patent 3,170,273 (1963).
382. E. Mendel, *Semicond. Prod. Solid State Technol.* **10**, 27 (1967).
383. T. M. Buck and R. L. Meek, *Nat. Bur. Stand. (U.S.), Spec. Publ.* No. 337, pp. 419–430 (1970).
384. J. T. Law, *Solid State Technol.* **14**(1), 25 (1971).
385. R. B. Herring, *Solid State Technol.* **19**(5), 37 (1976).
386. E. L. Kern, G. L. Gill, and P. Rioux, in "Semiconductor Silicon 1977" (H. R. Huff and E. Sirtl, eds.), pp. 182–186. Electrochem. Soc., Princeton, New Jersey, 1977.
387. E. Mendel and K. Yang, *Proc. IEEE* **57**, 1476 (1969).
388. L. H. Blake and E. Mendel, *Solid State Technol.* **13**(1), 42 (1970).
389. A. Reisman and R. Rohr, *J. Electrochem. Soc.* **111**, 1425 (1964).
390. V. L. Rideout, *J. Electrochem. Soc.* **119**, 1778 (1972).
391. H. Hartnagel and B. L. Weiss, *J. Mater, Sci.* **8**, 1061 (1973).
392. T. M. Buck, in "The Surface Chemistry of Metals and Semiconductors" (H. C. Gatos, ed.), pp. 107–135. Wiley, New York, 1960.
393. R. B. Soper, *Nat. Bur. Stand. (U.S.), Spec. Publ.* No 337, pp. 412–418 (1970).
394. A. C. Bonora, in "Semiconductor Silicon 1977" (H. R. Huff and E. Sirtl, eds.), pp. 154–169. Electrochem. Soc., Princeton, New Jersey, 1977.
395. J. W. Faust, in "Silicon Carbide" (J. R. O'Connor and J. Smiltens, eds.), p. 403. Pergamon, Oxford, 1960.
396. R. W. Bartlett and M. Barlow, *J. Electrochem. Soc.* **117**, 1436 (1970).
397. W. V. Muench and I. Pfaffeneder, *Thin Solid Films* **31**, 39 (1976).
398. R. C. Smith, *Electrochem. Soc. Extend. Abstr.* No. 15, Fall Meeting (1963).
399. T. L. Chu and R. B. Campbell, *J. Electrochem. Soc.* **112**, 955 (1965).
400. R. W. Brander and A. L. Boughey, *Br. J. Appl. Phys.* **18**, 905 (1967).
401. M. V. Sullivan and G. A. Kolb, *J. Electrochem. Soc.* **110**, 585 (1963).
402. C. S. Fuller and H. W. Allison, *J. Electrochem. Soc.* **109**, 880 (1962).

403. Y. Tarui, Y. Komiya, and Y. Harada, *J. Electrochem. Soc.* **118**, 118 (1971).
404. J. C. Dyment and G. A. Rozgonyi, *J. Electrochem. Soc.* **118**, 1346 (1971).
405. D. W. Shaw, *J. Electrochem. Soc.* **113**, 958 (1966).
406. I. Shiota, K. Motoya, T. Ohmi, N. Miyamoto, and J. Nishizawa, *J. Electrochem. Soc.* **124**, 155 (1977).
407. M. V. Sullivan and L. A. Pompliano, *J. Electrochem. Soc.* **108**, 60C (1961).
408. S. Iida and K. Ito, *J. Electrochem. Soc.* **118**, 768 (1971).
409. J. J. Gannon and C. J. Nuese, *J. Electrochem. Soc.* **121**, 1215 (1974).
410. R. D. Packard, *J. Electrochem. Soc.* **112**, 871 (1965).
411. R. W. Bicknell, *J. Phys. D* **6**, 1991 (1973).
412. E. W. Jensen, *Solid State Technol.* **16**(8), 49 (1973).
413. O. Wada, S. Yanagisawa, and H. Takanashi, *J. Electrochem. Soc.* **123**, 1546 (1976).
414. W. T. Tsang and A. Y. Cho, *Appl. Phys. Lett.* **30**, 293 (1977).
415. J. L. Merz and R. A. Logan, *J. Appl. Phys.* **47**, 3503 (1976).
416. L. A. Koszi and D. L. Rode, *J. Electrochem. Soc.* **122**, 1676 (1975).
417. J. G. Grabmaier and C. B. Watson, *Phys. Status Solidi* **32**, K13 (1969).
418. M. Otsubo, T. Oda, H. Kumabe, and H. Miki, *J. Electrochem. Soc.* **123**. 676 (1976).
419. J. L. Richards and A. J. Crocker, *J. Appl. Phys.* **31**, 611 (1960).
420. J. G. White and W. C. Roth, *J. Appl. Phys.* **30**, 946 (1959).
421. D. N. Nasledov, A. Y. Patrakova, and B. V. Tsatvenkov, *Zh. Tekh. Fiz.* **28**, 779 (1958).
422. E. Biedermann and K. Brack, *J. Electrochem. Soc.* **113**, 1088 (1966).
423. Ref. 21, p. 199, Table 7.4.
424. E. S. Meieran, *J. Appl. Phys.* **36**, 2544 (1965).
425. G. H. Olson and V. S. Ban, *Appl. Phys. Lett.* **28**, 734 (1976).
426. T. Ambridge, C. R. Elliot, and M. M. Faktor, *J. Appl. Electrochem.* **3**, 1 (1973); **4**, 135 (1974).
427. M. M. Faktor, D. G. Fiddyment, and M. R. Taylor, *J. Electrochem. Soc.* **122**, 1566 (1975).
428. Y. V. Pleskov, *Dokl. Akad. Nauk SSSR* **143**, 1399 (1962).
429. C. J. Nuese and J. J. Gannon, *J. Electrochem. Soc.* **117**, 1094 (1970).
430. A. Yamamoto and S. Yano, *J. Electrochem. Soc.* **122**, 260 (1975).
431. D. L. Rode, B. Schwartz, and J. V. DiLorenzo, *Solid-State Electron.* **17**, 1119 (1974).
432. B. Schwartz, F. Ermanis, and B. H. Brastad, *J. Electrochem. Soc.* **123**, 1089 (1976).
433. C. Lin, L. Chow, and K. Miller, *J. Electrochem. Soc.* **117**, 407 (1970).
434. R. Bhat, D. J. Baliga, and S. K. Ghandhi, *J. Electrochem. Soc.* **121**, 1378 (1975).
434a. R. Bhat and S. K. Ghandhi, *J. Electrochem. Soc.* **124**, 1447 (1977).
435. R. P. Tijburg and T. van Dongen, *J. Electrochem. Soc.* **123**, 687 (1976).
436. W. G. Oldham, *Electrochem. Technol.* **3**, 57 (1965).
437. A. G. Sigai, C. J. Nuese, R. E. Enstrom, and T. Zameroski, *J. Electrochem. Soc.* **120**, 947 (1973).
437a. Ref. 8, p. 38.
438. M. I. Val'kovskaya and Y. S. Boyarskaya, *Sov. Phys.—Solid State* **8**, 1976 (1967).
439. R. H. Saul, *J. Electrochem. Soc.* **115**, 1185 (1968).
440. E. Hájková and R. Fremunt, *Phys. Status Solidi A* **10**, K35 (1972).
441. N. E. Schumaker, M. Kuhn, and R. A. Furnage, *IEEE Trans. Electron Devices* **ED-18**, 627 (1972).
442. T. Uragaki, H. Yamanaka, and M. Inoue, *J. Electrochem. Soc.* **123**, 580 (1976).
443. W. H. Hackett, Jr., T. E. McGahan, R. W. Dixon, and G. W. Kammlott, *J. Electrochem. Soc.* **119**, 973 (1972).

444. N. E. Schumaker and G. S. Rozgonyi, *J. Electrochem. Soc.* **119**, 1233 (1972).
445. L. Rankel Plauger, *J. Electrochem. Soc.* **121**, 455 (1974).
446. A. Milch, *J. Electrochem. Soc.* **123**, 1256 (1976).
447. B. D. Chase, D. B. Holt, and B. A. Unvala, *J. Electrochem. Soc.* **119**, 310 (1972).
448. R. L. Meek and N. E. Schumaker, *J. Electrochem. Soc.* **119**, 1148 (1972).
449. T. L. Chu and R. W. Kelm, Jr., *J. Electrochem. Soc.* **122**, 995 (1975).
450. M. J. Rand and J. F. Roberts, *J. Electrochem. Soc.* **115**, 423 (1968).
450a. M. Hirayama and K. Shomo, *J. Electrochem. Soc.* **122**, 1671 (1975).
451. T. L. Chu, M. Gill, and S. S. Chu, *J. Electrochem. Soc.* **123**, 259 (1976).
452. H. C. Gatos and M. C. Lavine, *J. Electrochem. Soc.* **107**, 427 (1960).
453. T. L. Chu, *J. Electrochem. Soc.* **118**, 1200 (1971).
454. J. I. Pankove, *J. Electrochem. Soc.* **119**, 1118 (1972).
455. A. Shintani and S. Minagawa, *J. Electrochem. Soc.* **123**, 706 (1976).
456. A. T. Churchman, G. A. Geach, and J. Winton, *Proc. R. Soc., Ser. A* **238**, 194 (1956).
457. J. W. Faust and A. Sagar, *J. Appl. Phys.* **31**, 331 (1960).
458. E. P. Warekois and P. H. Metzger, *J. Appl. Phys.* **30**, 960 (1959).
459. L. Bernstein, *J. Electrochem. Soc.* **109**, 270 (1962).
460. B. L. Sharma, *Solid-State Electron.* **9**, 728 (1966).
461. B. Tuck and A. J. Baker, *J. Mater. Sci.* **8**, 1559 (1973).
462. V. Wrick, G. J. Scilla, L. F. Eastman, R. L. Henry, and E. M. Swiggard, *Electron. Lett.* **12**, 394 (1976).
463. J. W. Allen, *Philos. Mag.* **2**, 1455 (1957).
464. J. F. Dewald, *J. Electrochem. Soc.* **104**, 244 (1957).
465. J. D. Venables and R. M. Broudy, *J. Appl. Phys.* **29**, 1025 (1958); see also personal communication, cited in D. B. Holt, *J. Appl. Phys.* **31**, 2231 (1960).
466. A. Reisman, M. Berkenblit, J. Cuomo, and S. A. Chan, *J. Electrochem. Soc.* **118**, 1653 (1971).
467. M. F. Ehman, *J. Electrochem. Soc.* **121**, 1240 (1974).
468. M. T. Shehata and R. Kelly, *J. Electrochem. Soc.* **122**, 1359 (1975).
469. E. A. James, RCA Laboratories, personal communication (1963).
470. J. W. Faust, *J. Electrochem. Soc.* **105**(12) 252C (1958).
471. J. Woods, *Br. J. Appl. Phys.* **11**, 296 (1960).
472. E. P. Warekois, M. C. Lavine, A. N. Mariano, and H. C. Gatos, *J. Appl. Phys.* **33**, 690 (1962).
473. M. V. Sullivan and W. R. Bracht, *J. Electrochem. Soc.* **114**, 295 (1967).
474. W. H. Strehlow, *J. Appl. Phys.* **40**, 2928 (1969).
475. V. Y. Pickhardt and D. L. Smith, *J. Electrochem. Soc.* **121**, 1067 (1974).
476. A. A. Pritchard and S. Wagner, *J. Electrochem. Soc.* **124**, 961 (1977).
477. M. Inoue, I. Teramoto, and S. Takayanagi, *J. Appl. Phys.* **33**, 2578 (1962).
478. A. Sagar, W. Lehman, and J. W. Faust, *J. Appl. Phys.* **39**, 5336 (1968).
479. Ref. 8, p. 41.
480. Ref. 8, p. 42.
481. I. Teramoto and S. Takayanagi, *J. Appl. Phys.* **32**, 119 (1961).
482. S. G. Parker and J. E. Pinnell, *J. Electrochem. Soc.* **118**, 1868 (1971).
483. Ref. 19, p. 43.
484. J. C. C. Fan and F. Bachner, *J. Electrochem. Soc.* **122**, 1719 (1975).
485. J. Kane, W. Kern, and H. P. Schweizer, *Thin Solid Films* **29**, 155 (1975).
486. J. A. Thornton and V. L. Hedgcoth, *J. Vac. Sci. Technol.* **13**, 117 (1976).
487. G. Bradshaw and A. J. Hughes, *Thin Solid Films* **33**, L5 (1976).
488. Ref. 6, pp. 256, 261–262.

489. R. F. Brebrick and W. W. Scanlon, *J. Chem. Phys.* **27**, 607 (1957).
490. H. Abrams and R. N. Tauber, *J. Electrochem. Soc.* **116**, 103 (1969).
491. B. B. Houston and M. K. Norr, *J. Appl Phys.* **31**, 615 (1960).
492. M. K. Norr, *J. Electrochem. Soc.* **109**, 433 (1962).
493. E. Levine and R. N. Tauber, *J. Electrochem. Soc.* **115**, 107 (1968).
494. Ref. 19, p. 35.
495. A. R. Calawa, T. C. Harman, M. Finn, and P. Youtz, *Trans. AIME* **242**, 374 (1968).
496. Ref. 19, pp. 38–39.
497. R. Muto and S. Furuuchi, *Asahi Garasu Kenkyu Hokoku* **23**, 27 (1973).
498. B. J. Baliga and S. K. Ghandhi, *J. Electrochem. Soc.* **124**, 1059 (1977).
499. H. B. Bullinger, U.S. Patent 3,615,465 (1971).
500. H. Kusakawa, T. Yata, and W. Fujinaga, Jpn. Patent 75 85,895 (1975).
501. J. Kane, W. Kern, and H. P. Schweizer, *J. Electrochem. Soc.* **123**, 270 (1976).
502. D. A. Vermilyea, *Annu. Rev. Mater. Sci.* **1**, 373 (1971).
503. R. Weiner, *Metalloberfläche* **27**, 441 (1973).
504. T. Agatsuma, A. Kikuchi, K. Nabada, and A. Tomozawa, *J. Electrochem. Soc.* **122**, 825 (1975).
505. W. Kern and R. B. Comizzoli, *J. Vac. Sci. Technol.* **14**, 32 (1977).
506. H. K. Johnston and T. L. Larson, U.S. Patent 3,702,273 (1972).
507. "Applications Data for Kodak Photosensitive Resists," Pamphlet P-91. Eastman Kodak Co., Rochester, New York (1966).
508. W. N. Greer, Plating (*East Orange, N.J.*) **48**, 1095 (1961).
509. J. R. Sayers and J. Smit, *Plating* (*East Orange, N.J.*) **48**, 789 (1961).
510. K. Naraoka and M. Maeda, Jpn. Patent 73 08,706 (1973).
511. M. C. Zyetz and A. M. Despres, *Am. Vac. Soc. Symp., 13th, Extend. Abstr.* p. 169 (1966).
512. F. Okamato, *Jpn. J. Appl. Phys.* **13**, 383 (1974).
513. Dynachem Tech. Data, "Etchants for Metals and Thin Films," Lithoplate, Covina, California.
514. R. F. Frankenthal and D. H. Eaton, *J. Electrochem. Soc.* **123**, 703 (1976).
515. T. A. Shankoff and E. A. Chandross, *J. Electrochem. Soc.* **122**, 294 (1975).
516. A. Rogel, *Rev. Sci. Instrum.* **37**, 1416 (1966).
517. F. Woitsch, *Solid State Technol.* **11**(1), 29 (1968).
518. D. M. Brown, W. R. Cady, J. W. Sprague, and P. J. Salvagni, *IEEE Trans. Electron Devices* **ED-18**, 931 (1971).
519. A. R. Janus, *J. Electrochem. Soc.* **119**, 392 (1972).
520. Anonymous, "Incidential Intelligence About Kodak Resists," Vol. 7, No 1, p. 4. Eastman Kodak Co., Rochester, New York (1969).
521. G. S. Kelsey, *J. Electrochem. Soc.* **124**, 927 (1977).
522. Y. H. Choo and O. F. Devereux. *J. Electrochem. Soc.* **123**, 1868 (1976).
523. T. Takamura and H. Kihara-Morishita, *J. Electrochem. Soc.* **122**, 386 (1975).
524. C. W. Halsted and M. W. Haller, *Electrochem. Soc. Extend. Abstr.* **76-2**, 748 (1976).
525. J. W. Diggle, *in* "Oxides and Oxide Films" (J. W. Diggle, ed.), Vol. 2, Ch. 4. Dekker, New York, 1973.
526. J. B. MacChesney, P. B. O'Connor, and M. V. Sullivan, *J. Electrochem. Soc.* **118**, 776 (1971).
527. W. Sinclair, D. L. Rousseau, and J. J. Stancavish, *J. Electrochem. Soc.* **121**, 925 (1974).
528. G. W. Kammlott and W. Sinclair, *J. Electrochem. Soc.* **121**, 929 (1974).

529. S. J. Licht, *J. Electron. Mater.* **4**, 757 (1975).
530. T. Kasai, *Jpn. J. Appl. Phys.* **14**, 1421 (1975).
531. D. C. Miller, *J. Electrochem. Soc.* **120**, 1771 (1973).
532. I. Haller, M. Hatzakis, and R. Srinivasan, *IBM J. Res. Dev.* **12**, 251 (1968).
533. M. Zamin, P. Mayer, and M. K. Murthy, *J. Electrochem. Soc.* **123**, 1377 (1976).
534. "Lange's Handbook of Chemistry" (N. A. Lange, ed.), 10th Ed. McGraw-Hill, New York, 1967.
535. J. B. Mooney and J. O. McCaldin, *J. Electrochem. Soc.* **124**, 625 (1977).
536. J. W. Johnson, S. C. Chen, J. S. Chang, and W. J. James, *Corros. Sci.* **17**, 813 (1977).
537. Y. Okinaka, *J. Electrochem. Soc.* **117**, 289 (1970).
538. R. D. Armstrong, *Corros. Sci.* **11**, 693 (1971).
539. R. D. Cowling and A. C. Riddiford, *Electrochim. Acta* **14**, 981 (1969).
540. W. K. Behl and J. E. Toni, *J. Electroanal. Chem.* **31**, 63 (1971).
541. Becco Bulletin No. 97. Becco Chem. Div., Buffalo, New York.
542. Becco Bulletin No. 99. Becco Chem. Div., Buffalo, New York.
543. Becco Bulletin No. 102. Becco Chem. Div., Buffalo, New York.
544. R. J. Schaefer and J. A. Blodgett, *J. Electrochem. Soc.* **123**, 1701 (1976).
545. P. Jacquet, *Rev. Metall.* (*Paris*) **42**, 133 (1945).
546. D. Landolt, R. H. Muller, and C. W. Tobias, *J. Electrochem. Soc.* **118**, 40 (1971).
547. K. Umeda, Jpn. Patent 75 01,351 (1975).
548. R. P. Frankenthal and D. E. Thompson, *J. Electrochem. Soc.* **123**, 799 (1976).
549. F. Huber, W. Witt, and I. H. Pratt, *Proc., Electron. Components Conf.* p. 66 (1967).
550. Y. Ohno, T. Matsuoka, and K. Miyaji, Jpn. Patent 74 11,3737 (1974).
551. R. N. Castellano and P. H. Schmidt, *J. Electrochem. Soc.* **118**, 653 (1971).
552. Ref. 14, pp. 7–38.
553. M. Zamin, P. Mayer, and M. K. Murthy, *J. Electrochem. Soc.* **124**, 1558 (1977).
554. L. A. Colom and H. A. Levine, U.S. Patent 3,639,185 (1975).
555. G. D. Barnett and A. Miller, U.S. Patent 3,232,803 (1966).
556. R. de Bernardy and L. F. Donaghey, *Electrochem. Soc. Extend. Abstr.* **76-2**, 648 (1976).
557. "Handbook of Chemistry and Physics" (R. C. Weast and S. M. Selby, eds.), 48th Ed. Chem. Rubber Publ. Co., Cleveland, Ohio, 1967–1968.
558. K. Baba, Jpn. Patent 74 47,225 (1974).
559. M. J. Graham, G. I. Sproule, D. Caplan, and M. Cohen, *J. Electrochem. Soc.* **119**, 883 (1972).
560. B. MacDougall and M. Cohen, *J. Electrochem. Soc.* **123**, 191 (1976).
561. M. S. Shivaraman and C. M. Svensson, *J. Electrochem. Soc.* **123**, 1258 (1976).
562. M. J. Rand, *J. Electrochem. Soc.* **122**, 811 (1975).
563. M. J. Rand and J. F. Roberts, *Appl. Phys. Lett.* **24**, 49 (1974).
564. E. J. Kelly, *Proc. Int. Congr. Met. Corros., 5th, Tokyo, 1972* p. 137 (1974).
565. E. J. Kelly, *J. Electrochem. Soc.* **123**, 162 (1976).
566. J. W. Johnson and C. L. Wu, *J. Electrochem. Soc.* **118**, 1909 (1971).
567. G. S. Kelsey, *J. Electrochem. Soc.* **124**, 814 (1977).
568. T. P. Dirkse, D. DeWit, and R. Shoemaker, *J. Electrochem. Soc.* **115**, 442 (1968).
569. M. Eisenberg, H. F. Bauman, and D. M. Brettner, *J. Electrochem. Soc.* **108**, 909 (1961).
570. R. D. Armstrong and G. M. Bulman, *J. Electroanal. Chem.* **25**, 121 (1970).
571. R. W. Powers and M. W. Breiter, *J. Electrochem. Soc.* **116**, 719 (1969).
572. F. M. Cain, "Zirconium and its Alloys," Am. Soc. Met., Columbus, Ohio, 1953.

573. P. H. Holloway and R. L. Long, Jr., *IEEE Trans. Parts, Hybrids, Packag.* **PHP-11**, 83 (1975).
574. A. T. Kuhn, H. Shalaby, and D. W. Wakeman, *Corros. Sci.* **17**, 833 (1977).
575. R. K. Waits, *Trans. Metall. Soc. AIME* **242**, 490 (1968).
576. M. J. Rand and J. F. Roberts, *Electrochem. So. Extend. Abst.* **73-2**, 432 (1973).
577. D. Shinoda, *Electrochem. Soc. Extend. Abstr.* No. 502, p. 462, Fall Meeting (1968).
578. T. Matsushita, T. Aoki, T. Otsu, H. Yamoto, H. Hayashi, M. Okayama, and Y. Kawana, *Jpn. J. Appl. Phys.* **15** (Suppl. 15-1), 35 (1976).

V-2

Plasma-Assisted Etching Techniques for Pattern Delineation

C. M. MELLIAR-SMITH AND C. J. MOGAB

Bell Telephone Laboratories
Murray Hill, New Jersey

I. INTRODUCTION

Recent advances in silicon integrated circuit technology, and pattern delineation processes in particular, have placed an increasing emphasis on thin film etching techniques. Improved lithographic techniques have permitted a reduction in feature size in photo and electron sensitive resists into a range where conventional liquid phase chemical etching is no longer a viable means of pattern transfer. Accordingly, there has been a growing emphasis on the use of gas phase plasma-assisted etching methods, which, as we will discuss, have an inherently better resolution

497

capability. Under the generic title "plasma-assisted etching" we include ion milling, sputter etching, reactive ion etching, and plasma etching. Although all of these techniques make use of plasma activated species, the mechanisms involved in the etching processes differ. Ion etching (ion milling and sputter etching) is a physical process while plasma etching occurs primarily by chemical reaction. There is, in fact, a gray area between these two extremes which we will refer to as reactive ion etching (sometimes called "reactive sputter etching"), in which both physical and chemical effects contribute to etching.

Ion etching can be most graphically described as an atomic scale version of sand blasting. Ions are extracted from a gaseous plasma (usually an inert gas such as argon) and accelerated to the substrate where the surface is eroded by momentum transfer. In ion milling, the ions are generated in a plasma remote from the substrates and subsequently accelerated towards them, while in sputter etching the substrates are an integral part of the cathode of a parallel plate discharge.

Plasma etching employs a glow discharge to generate active species such as atoms or free radicals from a relatively inert molecular gas. The active species diffuse to the substrate where they react with the surface to produce volatile products. Several different reactor configurations are used. In most cases, plasma etching is carried out using a higher pressure discharge ($\sim 10-100$ times greater pressure) than is normal for ion etching methods and etching occurs predominantly by direct chemical reaction. In fact, the samples can be shielded from the plasma in order to eliminate ion bombardment effects, but this is not always done. For ion etching using an inert gas, the physical process has been well established (see Chapters II-1 and II-5). However, it has been demonstrated that the addition of reactive gases (reactive ion etching) to the ion source enhances the physical etch rate and also introduces reactive chemical etching as well. Indeed, in the parallel plate configuration, the substrates are loaded onto the active electrode where ion bombardment, and hence physical erosion contributes to the etching process.

Plasma-assisted etching techniques have received very concentrated development in the past few years, but have been known for much longer. Sputter erosion was first observed in 1852 by Grove [1] and deliberately applied to etching a substrate in 1954 by Castaing and Laborie [1a], while the effectiveness of methyl radicals to etch thin lead films was demonstrated at the turn of the century [1b]. Recently several review articles have been published on ion beam milling [2-7], sputter etching [8, 9], and plasma etching [10-13]. It is not the purpose of this chapter to duplicate this work; rather we have tried to describe the mechanism of the processes in sufficient detail that their application and optimization for any etching problem might be facilitated.

II. PHYSICAL AND CHEMICAL PHENOMENA IN LOW PRESSURE PLASMAS

A. Introduction

A plasma can be loosely defined as an assembly of positively and negatively charged particles which is electrically neutral on a macroscopic scale. Plasmas can be formed in gases, liquids, or solids. In particular, low pressure gas discharges conform to this definition and the terms discharge and plasma are increasingly used interchangeably.

Plasma-assisted pattern transfer techniques employ partially ionized gases composed of ions, electrons, and neutrals produced by low pressure ($\sim 10^{-4}-10$ Torr) electric discharges. Various methods can be used to generate a gaseous discharge, but the same processes are basic to its initiation and maintenance.

A discharge is initiated when an electron is released by some means (e.g., photoionization or field emission) in a gas to which an electric field is applied. The electron is accelerated by the field thereby gaining kinetic energy. It subsequently loses kinetic energy in each collision with a gas molecule (or atom), but initially, the losses amount to only a small fraction of its total energy since these collisions will necessarily be elastic; the energy being too low to excite or ionize a molecule. The average fraction of the electron energy lost in an elastic collision with a gas molecule is $\sim 2m/M$, where m is the electron mass and M is the mass of the molecule; thus, typically only $\sim 10^{-5}$ of the kinetic energy is lost per elastic collision. Meanwhile, the electron continues to gain energy between collisions enabling it, ultimately, to ionize a gas molecule through an inelastic collision which is a much more efficient means of energy transfer. (All but a fraction $\sim m/M$ of the initial kinetic energy is converted to potential energy of the struck particle.) Ionization, in turn, frees another electron and a cascade rapidly occurs resulting in a gas ionized throughout its volume, provided that a sufficient voltage, the breakdown potential, has been applied. Electrons created in ionizing collisions and by secondary processes (see Chapter II-1) are removed from the plasma by drift and diffusion to the boundaries, by recombination with positive ions and, in certain gases, by attachment to neutrals to form negative ions. The discharge reaches a self-sustained steady state when the generation and loss processes balance each other.

Inelastic collisions can also produce molecular fragmentation and excited neutral particles. The electronically excited states of molecules and atoms account for much of the luminous glow of the gaseous discharge while the fragments are often highly reactive atoms and radicals.

In the bulk of the discharge, the field of each charged particle is screened by other charged particles in its vicinity. The screening distance is called the Debye length and depends on the mean electron energy and electron density. Ordinarily the Debye length is much smaller than the dimensions of the discharge which can then properly be termed a plasma.

Pattern generation usually involves thermally sensitive materials which further restricts considerations to low temperature gaseous discharges (cold plasmas). These plasmas are characterized by electron densities in the range $10^9 - 10^{12}$ cm^{-3}, a Debye length of $\sim 10^{-2}$ cm, an average electron energy of several electron volts corresponding to an electron temperature of $\sim 10^4 - 10^5$ °K, and a gas temperature near that of the ambient. The electron temperature can thus be much higher than the gas temperature.

This remarkable state of affairs exists because of the inefficient transfer of kinetic energy during elastic collisions between electrons and gas molecules. Accordingly, the cold plasma provides a unique environment of highly energetic particles "immersed" in a "cool" matrix. In the final analysis, it is the electrons that are responsible for this environment since they convert electrical energy to potential energy of activated gaseous species which, in turn, can be utilized to produce etching by chemical and/or physical interactions with solid surfaces.

B. Methods for Plasma Production

Various techniques have been devised to produce low pressure discharges as discussed in Chapter II-1. Those methods primarily used for etching will be mentioned. Direct current glow discharges can be used as ion sources for ion milling and for sputter etching of conductive materials on conductive substrates. However, rf discharges are generally employed for plasma-assisted etching for several reasons: (i) Electrons can pick up sufficient energy during their oscillations in an rf field to cause ionization. The discharge can thus be sustained independent of the yield of secondary electrons from the walls and electrodes; in fact, the electrodes can be placed outside the discharge as is commonly done in so-called electrodeless discharges. (ii) The efficiency of ionizing collisions is also enhanced by the electron oscillations so that rf discharges can be operated at pressures as low as ~ 1 mTorr. This is useful in sputter etching, where it is undesirable to have sputtered material reflected back to the etched surface as a result of collisions with gas molecules. The yield of atoms and radicals in molecular gas discharges is also increased by the use of rf excitation as compared to a dc discharge of the same field strength and pressure. This has definite advantages for etching processes relying on chemical reactions. (iii) Electrodes within the discharge can be covered with insulating material. This permits sputter etching of insulators and also eliminates problems due to the build up of insulating material on metal electrodes, which can occur in reactive gases such as those employed in plasma etching.

C. Chemical Reactions in Molecular Gas Discharges

The high energy electrons in a low pressure discharge produce nonequilibrium steady-state conditions which can transform a normally inert molecular gas such as CF_4 into a highly reactive medium. Essentially all the components of the plasma are potential reactants. Examples of reactions occurring between ions and molecules, electrons and molecules, ions and ions, and electrons and ions can be found in the literature of plasma chemistry [14–16].

Discharge chemistry is important to plasma and reactive ion etching which involve reactions between species formed in the plasma and the material being etched. The breadth and complexity of reactions occurring in molecular gas plasmas defies a comprehensive discussion and, in any case, the current state of understanding of the chemistry underlying plasma etching is rather rudimentary. Accordingly, only a brief discussion will be given of some of the reactions known to be important in molecular gas discharges.

The steady-state constitution of any discharge is determined by the relative rates of production and loss of the various species present. In general, more than one reaction can contribute to the production and loss of each species which quickly adds to the overally complexity. However, a single production or loss mechanism may be dominant.

1. Production of Ions, Atoms, and Radicals

Ionization in molecular gas discharges normally occurs predominantly by electron impact. Reactions of this type include:

$$O_2 + e \rightarrow O_2^+ \ + 2e \qquad \text{ionization} \tag{1}$$
$$CF_4 + e \rightarrow CF_3^+ + F + 2e \qquad \text{dissociative ionization} \tag{2}$$

Ionization potentials of atoms and molecules typically range from 8 to 20 eV, which is well above the mean electron energy in a discharge used for plasma etching. Thus, only electrons in the high energy tail of the electron energy distribution take part in these reactions.* Electron impact disso-

---

* Stepwise ionization, in which a molecule is first excited to a metastable state and subsequently ionized, is possible and can occur by successive impacts with electrons having energies less than the ionization potential; however, this is not usually a dominant mechanism.

ciation which requires less energetic electrons also occurs with high probability. Thus, reactions such as

$$e + O_2 \rightarrow 2O + e$$
$$\rightarrow O + O^-$$ (3)

$$e + CF_4 \rightarrow CF_3 + F + e$$
$$\rightarrow CF_3 + F^-$$ (4)

$$e + C_2F_6 \rightarrow 2CF_3 + e$$ (5)

are a major source of atoms, free radicals, and negative ions in the discharge. Reaction (3) is important in photoresist stripping with oxygen plasmas and reactions (4) and (5) for plasma etching Si, SiO_2, and Si_3N_4. Atomic oxygen reacts readily with organic materials to form CO, CO_2, and H_2O [17–20] while atomic F is thought to be the active species for etching silicon and silicon compounds in CF_4 discharges [21–25]. A major problem in plasma etching contact windows for silicon integrated circuit fabrication is that F atoms etch Si much faster than SiO_2. Typically, the etch rate ratio Si/SiO_2 is greater than 10:1 when F atoms are the etchant. Under these conditions, it is virtually impossible to prevent degradation of the silicon substrate during window etching, particularly if shallow junctions are present. However, gases such as C_2F_6, C_3F_8, and CHF_3 can be used to achieve etch rate ratios $(Si/SiO_2) \lesssim 1:10$ which is satisfactory for window etching. The favorable etch rate ratio has been attributed to the low F atom concentration in these plasmas and the ability of CF_3 radicals to react preferentially with SiO_2 [22, 26]. Existing evidence is consistent with this hypothesis although CF_3 radicals have not, as yet, been confirmed as the active species experimentally.

Electron impact is also important in the production of excited states of atoms and molecules. Electronically excited states can be produced in atoms while molecules can have rotationally and vibrationally excited states as well. The dissociation of a molecule such as occurs in reactions (3)–(5) must be preceded by the formation of an electronically excited state which rapidly ($\sim 10^{-13}$ sec) dissociates. Excited species can also decay by ordinary dipole radiation which requires $\sim 10^{-9}$ sec. Radiation from electronically excited species accounts for the uv-visible emission of the discharge; the spectroscopic analysis of this radiation can be very useful in determining the species present in the plasma [19, 20, 24, 25, 27]. However, the relatively short lifetimes of excited species against radiative decay may preclude their participation in additional reactions. An exception to this is found in excited *metastable* species which cannot return to their ground states by direct radiative transitions; metastables can participate in reactions because they have relatively long lifetimes [28].

One additional process involving electron-molecule reactions is of interest. Many of the gases used in plasma etching contain one or more halogens (especially F, Cl, and Br) or oxygen. Atoms such as these, which lack one or two electrons in their outer shell, can capture an electron in the discharge to form stable negative ions; that is, they have a positive electron affinity. The formation of negative ions can reduce the conductivity of a given ionized gas because the overall effect is to replace a mobile electron with a heavier and therefore less mobile ion. This can sometimes change the behavior of the discharge and cause instabilities which lead to nonuniform etching. These effects are particularly prevalent in halocarbon gases containing Br atoms, two or more Cl atoms, or hydrogen and a halogen. Uniformity can usually be obtained by additions of suitable diluents or by choosing appropriate power and pressure conditions.

2. Loss Mechanisms

In molecular gas plasmas, loss of electrons occurs by diffusion and drift, dissociative recombination (e.g., $e + O_2^+ \rightarrow 2O$), and dissociative attachment (e.g., $e + CF_4 \rightarrow CF_3 + F^-$). Other loss mechanisms exist but at the electron densities pertinent to plasma etching they are not usually as rapid as these. Electron losses are balanced, of course, through electron production by ionization as already discussed.

Ions are similarly lost by diffusion and by recombination with electrons. Owing to the overall neutrality maintained by the plasma, electrons and positive ions diffuse at equivalent rates determined by the ambipolar diffusion coefficient, despite the great disparity in their intrinsic mobilities. This results from the incipient local electric field which is established as the more mobile electrons attempt to move away from the ions. The field adjusts so as to retard the electrons while speeding up the ions. The formation of negative ions increases the ambipolar diffusion coefficient resulting in more rapid electron loss.

Neutral atoms and radicals are lost primarily by recombination; either by homogeneous gas phase reactions or through surface reactions on the walls of the container. The contribution of gas phase versus surface recombination losses depends on many factors such as the condition and material of the surfaces present, the probability of two and three body gas phase reactions, and the surface area to volume ratio of the discharge. The recombination of F atoms in discharges of F_2 and Ar, for example, cannot occur by a two body homogeneous gas phase reaction $(F + F \rightarrow F_2)$ because the single molecule formed is unable to conserve both energy and momentum. Thus, gas phase recombination requires a third body such as an Ar atom $(F + F + Ar \rightarrow F_2 + Ar)$. This makes the recombination rate sensitive to the pressure at which the discharge is operated. On the other hand, a surface can also act as a third body independent of the pressure (F + F $\xrightarrow[\text{wall}]{}$ F$_2$). The rate of such reactions is strongly dependent on the catalytic properties of the surface. The probability of F atom recombination on Teflon®* is small ($\leq 7 \times 10^{-5}$) [29, 30], while for copper it is approximately three orders of magnitude higher [31]. Surfaces which are good catalysts for recombination can often be passivated by the addition of small amounts of impurities to the gas phase resulting in reduced losses. For example, it has been reported that the addition of 0.01–0.05% of N$_2$, N$_2$O, or NO to very pure O$_2$ increases the yield of oxygen atoms significantly in a microwave discharge [32, 33]. The addition of small amounts of water vapor has a similar effect.

The reactor surfaces exposed to the plasma can play a major role in determining etch rates. One manifestation of the importance of exposed surfaces is frequent need to "condition" a plasma reactor, by some means developed empirically, in order to achieve optimal etch rates, etch rate control, reproducibility or, in some instances, any etching at all [18, 34].

The balance of production and loss processes determines the steady-state concentration of species in the plasma as well as their average lifetimes. Once a nonequilibrium species has escaped the discharge, however, its lifetime is determined solely by the rates of various loss mechanisms. This lifetime is important in determining the etch rate for downstream or shielded reactions where the material to be etched is placed outside the plasma boundaries. To maximize etch rates in systems of this kind, it is necessary that the materials of construction have surfaces with a low recombination probability and reactivity for the active species. Etching reactions with species which can recombine in homogeneous two body collisions are not favorable for this type of system because the lifetimes of such species are usually small, owing to the high probability of two body collisions. Thus, for example, CF$_3$ radicals can recombine to form C$_2$F$_6$ molecules in a two body process because the large C$_2$F$_6$ molecule has sufficient internal degrees of freedom to conserve energy and momentum. The etching of contact windows in SiO$_2$ films is reportedly not a viable process "downstream" of a C$_2$F$_6$, C$_3$F$_8$, or CHF$_3$ discharge [26]. In contrast, F atoms created in a CF$_4$–O$_2$ plasma have a low recombination rate with respect to gas phase reactions and thus in the presence of appropriate surfaces can be transported considerable distances outside the plasma [23].

*® Registered trademark, E. I. Du Pont de Nemours and Co., Wilmington, Delaware.

3. An Example—The CF_4–O_2 Discharge

Discharges in molecular gases containing F can be used to etch Si, SiO_2, and Si_3N_4 Tetrafluoromethane (CF_4) is a particularly useful gas for this application because it is relatively inert in the absence of a discharge. The addition of O_2 to a CF_4 discharge has a remarkable effect on the etch rates of these materials. As the amount of O_2 is increased, the etch rate first rises sharply, then reaches a maximum and finally decreases again. Optical spectroscopic studies of the CF_4–O_2 plasma [24] indicate that the concentration of atomic fluorine in the discharge has a similar dependence on the O_2 content of the feed gas; this is illustrated in Fig. 1 which shows the etch rate of Si and the intensity of emission of atomic fluorine plotted against the percentage of O_2 in the feed gas.

Mass and infrared spectroscopic studies of the reactor effluent indicate that CO, CO_2, and COF_2 are all stable end products of the plasma [23, 25, 35]. The amount of CF_4 converted to products is shown in Fig. 2.

It is thought that oxygen reacts with CF_3 radicals generated by reaction (1) to produce F atoms and the oxidation products noted above [25].

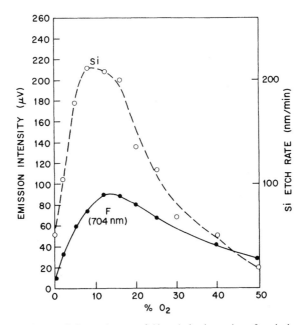

Fig. 1. Dependence of the etch rate of Si and the intensity of emission from excited F atoms on the O_2 concentration in a CF_4–O_2 discharge (courtesy of Harshbarger *et al.* [24]).

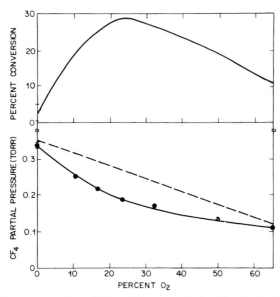

Fig. 2. The lower curve shows CF$_4$ partial pressure in the effluent of a CF$_4$–O$_2$ discharge run at 200 W, 0.3 Torr. The upper curve indicates the percentage of CF$_4$ converted to products. ---, feed gas; ●, effluent.

In addition, the oxygen may be effective in reducing the carbon contamination of surfaces which occurs due to impact dissociation (e.g., of CF$_3^+$). This adsorbed carbon would consume F atoms in the absence of oxygen [35]. Fluorine atoms are active in etching and thus the etch rates are strongly dependent on oxygen content. For etching silicon, the presence of oxygen also influences the rate of etching by F atoms. It has been observed that oxygen, probably as atomic oxygen, competes with F atoms for active surface sites, thereby, inhibiting etching. Consequently, the maximum etch rate for Si occurs at a lower O$_2$ concentration than that which produces a maximum F atom concentration as can be seen in Fig. 1. Oxygen adsorption does not occur to an appreciable extent on the SiO$_2$ surface so that the maximum etch rate for SiO$_2$ coincides with the maximum F atom concentration [25].

III. THE DESIGN OF ETCHING EQUIPMENT

The design of ion etching equipment has been covered in Chapters II-1 and II-5 and will not be reviewed further here.

A. Plasma Etching Systems

Commercial equipment for plasma etching has been developed mainly for application to silicon integrated circuit fabrication. It is designed for batch loading of silicon wafers and for the most part is highly automated although some smaller less automated equipment is available. The essential components of an etching reactor include a vacuum chamber and vacuum pump, a power supply to energize the discharge, and gas handling equipment to monitor and control the flow of gas to the reactor.

Two general types of reactor design have been employed, both of which depend on rf power to sustain a discharge for reasons mentioned in Section II.B. The first type consists of a chamber with external electrodes into which the wafers are placed vertically in a suitable carrier (Fig. 3). The chamber is usually cylindrical and the rf power may be applied to plates affixed to the cylinder or to a coil wound about it. The second type of apparatus has internal electrodes, usually a simple parallel plate (planar) arrangement, and the wafers are loaded horizontally on the grounded electrode for plasma etching or on the active rf electrode for reactive ion etching.

Both kinds of systems have similar requirements for vacuum production, admission and control of gas flows, rf power generation, and monitor and control instrumentation. These aspects will be considered after a discussion of the differences in reactor design.

1. Tube-Type Reactors

The earliest commercial reactors were initially developed for stripping of organic photoresists with an oxygen plasma [36]. These reactors consist of a quartz tube into which a boat containing the wafers in a vertical array is placed. The system is evacuated, partially backfilled with oxygen using

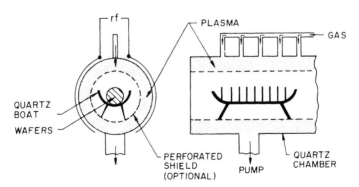

Fig. 3. Tube-type reactor.

dynamic pumping, and a discharge is ignited and maintained until all traces of photoresist are removed. For most resist stripping operations, the materials exposed to the discharge, other than the photoresist, are relatively unaffected by an oxygen plasma, and there is no pattern to transfer. This makes plasma stripping a noncritical process and so there is little concern about the uniformity of stripping; that is, the variation of stripping rate from wafer to wafer or across a single wafer.

With no stringent requirements on uniformity, the tube-type reactor is a natural choice for resist stripping since the vertical loading arrangement allows for the highest density of material per unit volume of reactor space. Uniformity is not totally neglected; most of the commercial systems employ gas inlet manifolds arranged so as to provide uniform *gas flow* over the wafers. This is not always sufficient, however, as a uniform flow of feed gas does not ensure a uniform concentration of plasma-activated species. As an example of this, Fig. 4 shows the variation of emission from

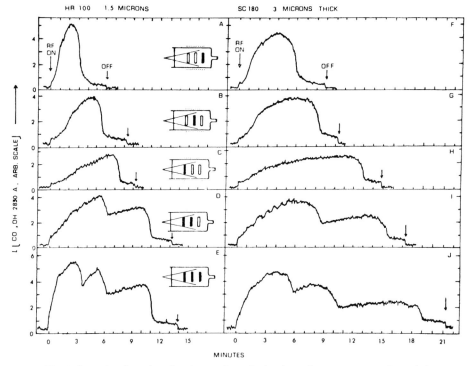

Fig. 4. Intensity-time plots for photoresist stripping in a tube-type reactor. Occupied positions in the wafer carrier are shown as filled rectangles in the inset (courtesy of Degenkolb *et al.* [19]).

electronically excited CO product molecules during resist stripping of wafers in a tube-type reactor [19]. The disappearance of emission from CO signifies the completion of resist removal and can be used as a convenient means of end point detection [19, 37]. Notice that the wafers, which initially were uniformly coated with resist, strip at different rates; faster at the rear of the reactor and more slowly at the front.

The variability observed in tube-type reactors arises partly from the nonuniformity of the plasma which tends to be doughnut shaped, partly from the gas flow pattern, and partly from temperature nonuniformity. The lack of temperautre control is a major shortcoming since stripping and etching rates are generally temperature sensitive (see Section IV.B.5).

The tube-type reactor used for resist stripping was initially applied to plasma etching simply by changing the feed gas. Uniformity then became a major consideration since selectivity was generally poorer and fidelity of pattern transfer was required. In order to enhance uniformity while retaining material handling capacity, the so-called "etch tunnel" was introduced [38]. The etch tunnel is a perforated metal cylinder (Fig. 3), placed inside and concentric with the reactor tube, which confines the plasma to the annular space between the reactor walls and the tunnel surface. With the wafers inside this Faraday cage, etching occurs "downstream" (outside) of the plasma proper by diffusion of active neutral species through the perforations. These tunnels have been cited by their manufacturers as producing excellent uniformity while shielding the wafers from much of the uv radiation from the plasma and reducing wafer temperature. As mentioned in Section II, such an arrangement requires relatively long lived active species for the attainment of practical etch rates and is somewhat restrictive because of this.

2. Parallel-Plate Reactors

In this type of reactor the wafers are placed flat on the ground electrode. A disadvantage of this arrangement, compared to tube-type reactors, is the smaller batch capacity per unit volume.

This is amply compensated by several advantages. First, the wafers can be heated or cooled simply, for example, by the use of heating elements in the ground electrode or by circulating fluid through a hollow electrode. Substantial heat transfer occurs by convection at pressures exceeding ~ 0.1 Torr. This greatly relaxes the requirements for good thermal contact between the wafer and its temperature controlled support, in contrast to rf sputter etching and ion milling where the working pressure is much lower. Secondly, in principle, the concentration of active species can be made uniform throughout the bulk of the plasma which is unperturbed by the wafers.

One means of doing this, the radial flow technique, was originated to enhance uniformity in reactive plasma deposition [39]. In this system (Fig. 5), the flow velocity profile is used to offset the electron density distribution so that a molecule resides in lower electron density regions longer than in high density regions of the plasma. Since activation ultimately can be traced to electron–molecule collisions, this implies a uniform probability of activation throughout the plasma provided that the flow rate, electrode spacing, and power density are properly adjusted. Moreover, this type of flow pattern tends to compensate for gas-phase depletion effects. An additional advantage of planar systems is that the wafer back is automatically shielded from the discharge. This is particularly useful when the back of the wafer is a high etch rate material (Section IV.B.2).

Other designs for planar reactors are similar to sputtering systems in that the gases are admitted to the chamber with no attempt to direct the flow, while in at least one commercial system, the gas enters the reactor at the center of the rf active electrode. Whatever the design particulars, the materials of construction are important in this type of system. The electrodes should be inert to the etchant and should have a low recombination probability for the active species. By way of example, aluminum is an excellent material for use with CF_4 plasmas because an inert nonvolatile aluminum fluoride layer is quickly established which passivates the electrode surfaces and discourages the recombination of fluorine atoms.

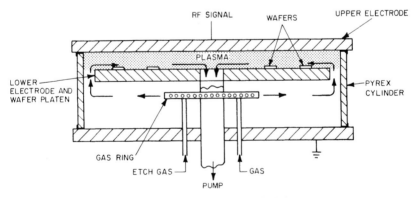

Fig. 5. Radial flow reactor (parallel plate).

3. Gas Handling

Gas flow rates in plasma etching systems typically are in the range ~ 50–500 sccm, with the upper limit dictated by the working pressure and the speed of the pump employed to evacuate the chamber. Gases can be admitted to the reactor through metering valves after passing through a flow measuring device such as a rotameter or mass flow meter. It is often important to control the flow rate and the pressure in the reactor independently; this requires a throttling valve on the pump which makes flow measurement highly desirable for reproducibility.

4. Vacuum System and Pressure Measurement

Generally, only a rough vacuum is needed for successful plasma etching. However, it is still advisable to follow good vacuum practice in making seals and connections as well as in general cleanliness. Both real and virtual leaks should be avoided, and trapping of the pump to prevent oil backstreaming into the reactor is useful. Organic contaminants are not usually a problem in oxygen containing plasmas but they can cause difficulties such as polymer formation in oxygen-free discharges.

Adsorbed water vapor can also be deleterious. Strongly bound water cannot be pumped out of an unbaked chamber but will readily desorb when a plasma is ignited. This can decrease etch rates and in particularly bad cases, may suppress etching totally. At the very least, reproducibility will be affected. In CF_4 plasmas, for example, the presence of water vapor results in scavenging of F atoms by H atoms formed from fragmentation of H_2O molecules.

Two stage oil-sealed rotary pumps are commonly used with pumping speeds ranging from ~ 10 to 50 ft^3/min. Corrosive and toxic gases can be formed in the discharge (e.g., CO and COF_2 in CF_4-O_2 plasmas), so good ventilation of the exhausted gases is mandatory even when using relatively harmless feed gases. Pump maintenance is important for the same reason and frequent oil changes are often needed. In some cases it is necessary to provide a cold trap between pump and reactor so that corrosive constituents of the effluent can be condensed before reaching the pump.

Most plasma etching processes permit a reasonable latitude in pressure; however, with the variety of gases often employed, it is desirable to have pressure measuring instrumentation which is independent of gas composition. Commercially available capacitance manometers are an excellent choice for pressure measurement as they are entirely independent of gas composition, mechanically robust, corrosion resistant, and quite accurate.

5. RF Power Generation

Power requirements are mostly specific to particular etch processes and reactor geometry. Low power density (~ 0.1 W/cm^3)* usually suffices for attaining acceptably high etch rates and isotropic etching (Section V.B.1) of materials such as polycrystalline Si. Certain applications, such as etching of contact windows in phosphosilicate glass films, require much higher power densities (~ 1 W/cm^3) to achieve adequate selectivity and rate. Radio frequency power supplies are commercially available which can provide a maximum output of 1 kW or more, while being capable of delivering very low power with high stability. An impedance matching network is usually required to match the output impedance of the generator (typically 50 Ω) to the plasma impedance in order to ensure maximum useful power transfer (see Chapter II-1).

Some means of monitoring the power delivered to the matching network is desirable for reproducibility and to protect the power supply against extended operation at high reflected power levels. In-line watt meters are convenient for this. The true power delivered to the plasma depends on many factors, such as the presence of ground loops and shunt paths and the quality factor of the matching network. The specification of power or power density should therefore be considered only as a relative measure pertaining to a particular combination of gas, reactor, and matching network.

The operating frequency is usually 13.56 MHz, but frequencies ranging from 300 kHz to microwave have been used. To date, there has been no definitive published work on the effect of frequency on etching characteristics.

IV. PROCESS PARAMETERS

A. Factors Controlling Ion Etching Rates

The rate of ion etching is, to a first approximation, related to the product of two factors—the impinging ion flux density and the sputtering yield. The former factor is related to the ion source design and has been described in Chapter II-5. This section deals with the physical factors that affect the sputtering yield which is defined as the number of atoms ejected from the substrate per incident ion. The major factors involved include the composition of the substrate and the ion beam, the ion acceleration

---

* Refers to the volume of a tube-type reactor or the volume of the region between the plates of a parallel-plate reactor.

voltage, and the angle of incidence of the beam onto the substrate. Most of these factors have been studied from an experimental point of view. However, recently several workers [40], and in particular Sigmund [41], have correlated these data with the development of a theoretical understanding of the processes involved in ion etching. A detailed review of this work is beyond the scope of this chapter and readers interested in this aspect are referred to the original literature.

1. Elemental Composition of Target

Figure 6a shows the sputtering yield as a function of target atomic number for a variety of elements [42]. There is a strong periodicity for the transition metals with the sputtering rate related to the degree of filling of the 3d, 4d, and 5d electron shells. Under the conditions shown in Fig. 6a, carbon has the lowest sputtering yield of less than 0.1 atoms/ion while silver has the highest at 2.7 atoms/ion. Morgulis and Tischenko [43] showed empirically that the sputtering rate of many elements was correlated with the inverse of the heat of sublimation. The correlation can be seen by comparing Figs. 6a and b.

The ion etching rate, as opposed to the sputtering yield, has been measured for a wide variety of materials by many workers involved in ion etching for pattern delineation. A partial listing of rates, materials, and conditions is given in Table I. It should be noted, however, that although test results have been normalized for current density variations in the experiments, they only represent net etching rates and they will show the experimental scatter which seems to be typical of etching rates from different literature sources.

Care should be taken when applying these measured etch rates to a pattern delineation process involving rf sputter etching. The rates shown in Tables I and II, for the most part, were measured using a uniform unmasked film of the target material in contrast to a more typical pattern delineation system where alternating areas of etched film and masking material can be very closely spaced. Under these heterogeneous conditions, Vossen and co-workers [8, 49] and Maissel et al. [50] have observed significant interaction between these adjacent materials, which can radically alter their relative ion etch rates compared to those measured from a homogeneous surface.

It has been established that during rf sputtering that up to 30% of the sputtered material can be redeposited onto the cathode [51], the amount being related to the gas pressure, target potential, and magnetic field strength [52]. Redeposition can be due to resputtering from the chamber walls and anode by energetic neutrals and negative ions present in the

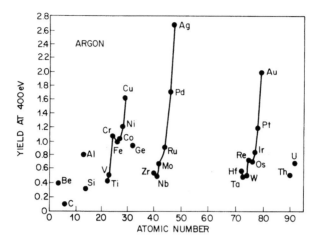

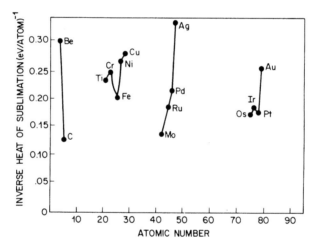

Fig. 6. (a) Sputtering yield as a function of substrate atomic number using 400 V argon ion bombardment [42]. (b) Inverse heat of sublimation as a function of element atomic number (plotted from data in Spencer and Schmidt [2]).

plasma [51, 53, 54] or by thermal reemission [55]. (These effects are less severe in an ion beam etching system and the relative etching rates are less affected by redeposition from the gas phase.) The effect of redeposition is to coat the surface being sputter etched with a thin (1–10 monolayers) film which is made up of a mixture of materials from the mask, the substrate, and the cathode table. In consequence, there is a uniform etch rate for all the materials and no preferential etching of either the mask or

Table I

Ion Etching Rates for Various Materials

	Rate (Å/min)	Ion energy (keV)	References
Metals			
Aluminum	300–700	0.5	[4, 6, 44]
	450–750	1	[45, 46]
Gold	1050–1500	0.5	[6]
	1600–2150	1	[6, 45]
Tungsten	180	0.5	[4]
Tantalum	150–330	0.5	[4, 44]
Titanium	200	0.5	[44]
	200	1	[45]
Molybdenum	230	0.5	[4]
	400	1	[45]
Copper	450	0.5	[46]
Chromium	200–400	1	[45, 46]
Zirconium	320	1	[45]
Silver	2000	1	[45]
Manganese	270	1	[46]
Vanadium	220	1	[46]
Permalloy	330–450	0.5	[4, 44]
Iron	320	1	[45]
Niobium	300	1	[45]
Dielectrics			
SiO_2	280–420	0.5	[4, 6, 46]
	380–670	1	[6, 45]
Al_2O_3	83	0.5	[4]
	130	1	[45]
$LiNbO_3$	640	1	[6]
Semiconductors			
Silicon	215–500	0.5	[4]
	360–750	1	[45, 46]
GaAs	650	0.5	[4]
	2600	1	[45]
Resist materials			
AZ1350 (Shipley)	200–420	0.5	[4, 44]
	600	1	[45]
Riston 14 (du Pont)	250	0.5	[4]
KTFR (Kodak)	390	1	[45]
Polymethylmethacrylate	840	1	[45]
COP[47]	860	0.5	[44]

Table II

Sputter-Etching Rates for Various Materials

	Rate (Å/min)	Power density (W/cm²)	References
Metals			
Aluminum	120–160	1.6	[9]
Gold	200–900	1.6	[9, 48]
Nichrome	100	1.6	[9]
Copper	200–350	1.6	[9]
Tungsten	70	1.6	[9]
Platinum	900	2.0	[48]
Nickel	500	2.0	[48]
Titanium	50	2.0	[48]
Dielectrics			
SiO_2	120	1.6	[9]
Si_3N_4	60	1.6	[9]
Al_2O_3	20–50	1.6	[9]
Resist Materials			
KTFR (Kodak)	70–300	1.6	[9]
AZ340 (Shipley)	70–300	1.6	[9]

the substrate. The relative composite etch rate is dependent on the mask and underlying substrate composition, the rates for a wide variety of which have been measured by Vossen and Davidson [49]. It is of interest to note that for rf sputter etching in chemically reactive gases such as Ar/O_2 or Ar/halocarbon mixtures, this nonpreferential etch rate has not been detected and wide differences have been observed between masking and substrate materials. Coburn *et al.* [35] have recently demonstrated that carbon deposited by impact dissociation of halocarbon ions is very effective in reducing sputtering rates. The addition of small amounts (10^{-5} Torr) of molecular O_2 to the plasma, resulted in the removal of the carbon, restoring the sputtering rates to those expected of the material.

Redeposition during Ar sputter etching can be reduced by using a "capture anode" of the type shown in Fig. 12 of Chapter II-1, while contamination of the wafers by the cathode material can be suppressed by using a dark space cathode shield.

2. Ion Etching Gas

The sputtering yield is known to be dependent on the atomic weight of the incident ion [56] and ion etching rates have also been found to be markedly dependent on the addition of small quantities of reactive gas to the inert gas supply. This latter phenomenon falls in the category of reactive ion etching.

The effect of the atomic weight of the impinging ion is a purely physical phenomenon, related to the momentum transfer effect behind ion etching, with the sputtering yield showing a periodicity similar to that of atomic number (shown in Fig. 7) as was observed for the dependence of yield on target material. The maximum sputtering rate is reached for each row in the periodic table as the electron shells are filled, with the fortunate effect that the inert gas ions have the highest sputtering rates. Based on the data shown in Fig. 7, it might seem attractive to use Xe for the ion beam, but this is not so since the advantage of the heavy element is significantly reduced at the lower voltages generally used for ion etching where the inert gas ions all have similar sputtering yields [41], as shown in Fig. 8.

The addition of a chemically active gas to the ion plasma has the effect of changing the etching rate due to a chemical interaction between the substrate and the added gas. The majority of the work has been directed towards the additions of O_2 [46, 48, 57–60] and halocarbons [11, 61–73] to both ion beam and sputter etch processes.

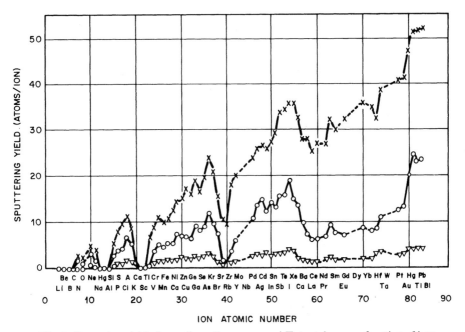

Fig. 7. Sputtering yield of: ×, silver; ○, copper; and ▽, tantalum as a function of bombarding ion atomic number (from Almen and Bruce [56]).

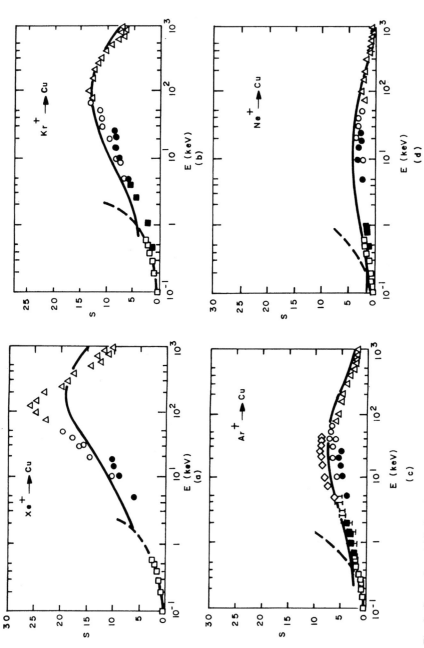

Fig. 8. Sputtering yields for silver bombarded with Xe, Kr, Ar, and Ne, showing a comparison of theoretical calculations by Sigmund [41] to experimental results of ○, Almén and Bruce [56]; ●, Guseva [56a]; ■ (in b), Keywell [56b]; □, Wehner *et al.* [56c]; ●, Yonts *et al.* [56d]; △, Dupp *et al.* [77]; ■ (in c,d), Weijsenfeld [56e]; I, Southern [56f]; ● (in d), Rol *et al.* [56g]).

The addition of O_2 reduces the sputtering yield of metals, such as Ti, Cr, and Al, which readily oxidize, but it has relatively little effect on more inert metals such as Au and Pt. In consequence, the relative sputtering rates of these two groups of materials can be adjusted by either using pure Ar to generate the ion beam or by adding varying amounts of O_2 to the Ar. The effect of O_2 partial pressure on the etching rate of various materials is shown in Fig. 9a for an ion beam system [46] and in Fig. 9b for an rf sput-

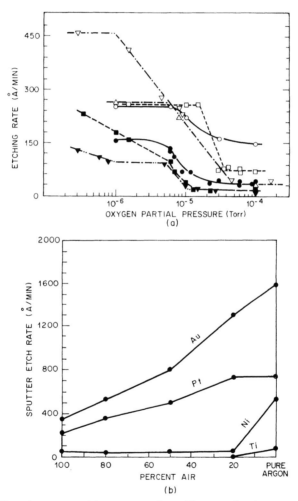

Fig. 9. Effect of oxygen partial pressure on the etching rate of various materials using (a) ion milling at 1 kV and 0.6 mA/cm² [46] and (b) rf sputter etching using 2 W/cm² [48]. (a) △, Al thin film; ▽, Al bulk: ■, Cr; ●, Mn: ▼, V; □, Si; ○, SiO₂. (b) Total pressure, 10 mTorr; power density, 2.0 W/cm² ((a) reprinted by permission of the publisher, Chapman & Hall).

ter etching system [48]. In both systems, the apparent effect is similar and is probably related to the formation of a chemisorbed layer of oxygen on the surface of active metals which can act as a "protective blanket" against the physical erosion of the ion beam. The effect of forming a thin discrete oxide layer on the surface is not clear, because the results of Cantagrel and Marchal [46] in Fig. 9a show that Si and SiO_2 etch at the same rate in pure Ar yet the addition of 10^{-4} Torr of O_2 reduces the Si etch rate to half that of oxide.

Argon/oxygen gas mixtures have generally been used to etch deep patterns into an inactive material using an active metal such as Cr as the etching mask. In this way, it is possible to obtain etching rate ratios between the mask material and the substrate which are higher than can be obtained by using a photoresist mask. The metal mask itself is often first patterned using a resist mask, prior to conversion from Ar ion etching to Ar/O_2 etching. Examples of this technique include the sputter etching of Au metallizations using Ni [48] and Ti [44] masks. It should be noted that a real leak in the vacuum station will have a similar effect as deliberately adding O_2, and as organic materials, such as photoresist, are attacked very quickly in an Ar/O_2 plasma, such leaks should be excluded from any ion etching system.

The use of a halocarbon plasma to increase the rf sputter etch rate of materials was first described by Hosokawa et al. [61]. Using a gas pressure of 20 mTorr and a power density of 1.3 W/cm^2, they observed the etch rates for various materials in pure Ar, halocarbon-12 (CCl_2F_2), and halocarbon-113 (CCl_2FCClF_2) and found the increased rates for Al and Si considerable. The etch rate for silicon in halocarbon-12 was further increased by a factor of 2 by the addition of 10% O_2. Beyond this value the rate dropped off again. Coburn et al. [35] have postulated that the reason for this is that the addition of a small amount of O_2 removes the carbon contamination produced by dissociative impact of halocarbon ions, while excessive O_2 addition results in substrate oxidation reducing the sputtering yield. In addition to the sputter etching, some plasma etching appears to be involved, because samples immersed in the plasma were etched, but by only a factor of $\frac{1}{6}$ of those placed on the cathode, and consequently subjected to more energetic ion bombardment. The rates of reactive ion etching of various materials are shown in Table III.

Reactive ion etching of Si and SiO_2 films using CF_4 and CHF_3 has been studied extensively [65–73] because of its importance to silicon integrated circuit technology. The use of reactive ion etching as opposed to sputter etching has the advantages of increased etch rate, etch rate selectivity, and sharp step profiles (see Section V). After Hosokawa's work, Heinecke [73] showed that the use of CHF_3, or the addition of H_2 to CF_4, re-

Table III

Reactive Ion Etching Rates of Various Materials in CF_4 and CHF_3

	p = 10 mTorr			p = 30 mTorr		
	Ar	CF_4	CHF_3	Ar	CF_4	CHF_3
Si	10	200	20	40	250	20
W	10	115	15	45	340	30
SiO_2	58	290	220	80	340	300
Si_3N_4	93	467	150	48	447	180
Shipley AZ 1350 H (on SiO_2)	245	487	40	280	480	50
Waycoat HR 100	187	180	~0	317	480	~0
Au	110	90	75	95	100	70
Cr	15	20	12	0	~0	10

[a] From Lehmann and Widmer [72].
[b] Etch rates in angstroms per minute at a power density of 0.5 W/cm².

sulted in a rapid increase of the etch rate of SiO_2 over that of Si, allowing the opening of contact windows in SiO_2 films over polysilicon. The etch rate ratio can be as high as 35 to 1 although the exact value is dependent on power density [66, 67, 69], flow rate [67, 68, 70], and pressure [70].

3. Ion Energy

The velocity of the incident ion has a marked effect on the sputtering yield. For ion beam systems, where the ion acceleration potential can be separated from the plasma generation potentials and readily measured, the effect of ion energy is easily perceived. For rf sputtering, where the surface can be bombarded by ions of variable energy, the effect of the substrate bias is less specific.

To achieve any sputtering at all, the ion energy must exceed a threshold voltage which is relatively low (less than 30 V for most materials [74]). Between this value and 100 V there is a very steep exponential rise in the sputtering yield, indicating that voltages above 100 V must be used for effective ion etching. The yield for Ni, Fe, and Ta under argon ion bombardment is shown in Fig. 10a for the energy range 100–600 eV; and from this it can be seen that above 300 V, the sputtering yield is not rising as rapidly as the ion energy, and consequently the etching efficiency (etching rate/input power) will be maximized if an accelerating potential of a few hundred volts is used. The rate of increase in sputtering yield with ion energy continues to fall above 1000 V, as shown in Fig. 10b [40], until at very high voltages, above 10^4 V, the sputtering yield starts to drop due to

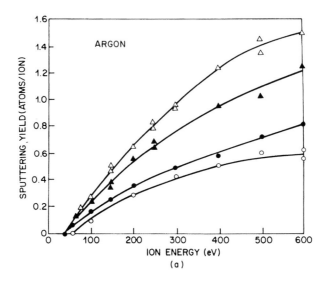

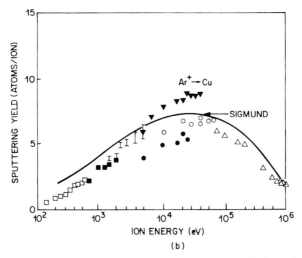

Fig. 10. Sputtering yield as a function of ion energy in low and high voltage ranges. (a) △, Ni; ▲, Fe; ●, Hf; ○, Ta. (b) □, Wehner *et al.* [56c]; ▼, Yonts, *et al.* [56d]; ●, Guseva [56a]; ○, Almén and Bruce [56]; ▲, Dupp *et al.* [77]; ■, Weijsenfeld [56e]; I, Southern *et al.* [56f] (data from (a) Laegried and Wehner [42] and (b) MacDonald [40, Fig. 5]).

implantation effects. The use of heavier bombarding ions reduces the degree of implantation and the turn-over voltage will occur at higher values for Xe as was shown in Fig. 8.

4. Angle of Incidence

Due to the momentum transfer mechanism of ion etching, the sputtering yield is dependent on angle of incidence of the beam. This effect has been studied using inert gas ions at both high (>20 keV) [75–79] and low [4, 5, 44, 80] accelerating voltages, using 200 eV Hg ions [81], and theoretically by Sigmund [41]. The results for low voltage ion etching are shown in Fig. 11 both as sputtering yield and etching rate. The difference in angular dependence for these two quantities is due to the cosine dependence of the ion flux which is usually referred to with respect to a plane perpendicular to the ion beam and hence is decreased at the substrate surface as the angle of incidence increases. For Au, the angular dependence of the sputtering yield is sufficiently modest that the cosine ion flux dependence is paramount and the etch rate falls off with angle of incidence. In contrast, for photoresist the reverse is true and the strong angular dependence (Fig. 11) can have a marked effect on the profile of an etched line when photoresist is used as the etch mask.

This effect is shown schematically in Fig. 12. The top edge of a photoresist feature is usually slightly rounded due to exposure and development conditions and this rounded portion will be subject to bombardment at various angles of incidence, ranging from perpendicular at the top ($\theta = 0°$) to grazing at the walls ($\theta = 90°$). In consequence, because photoresist mills fastest at some angle θ between 45° and 90°, a facet will form at this corner. This facet will continue to propagate until it intersects the film-resist interface, at which point the linewidth will start to decrease and a new facet will be milled into the thin film being etched. The resultant linewidth shrinkage can occur even when much of the photoresist remains and, generally, much thicker resist is required than would be expected from simple sputtering rate calculations. For example, Smith [82] has shown that to mill a groove in $LiNbO_3$, a polymer film 3 to 5 times thicker than the desired groove depth is required even though the $LiNbO_3$ mills at twice the rate of the resist film.

An additional problem that can arise due to an increase in etch rate with angle of incidence is the accelerated erosion of the photoresist mask when it passes over steps in the substrate. In extreme cases the resist mask can be completely removed and the underlying material eroded, while on flat portions of the wafer the photoresist is still an effective mask.

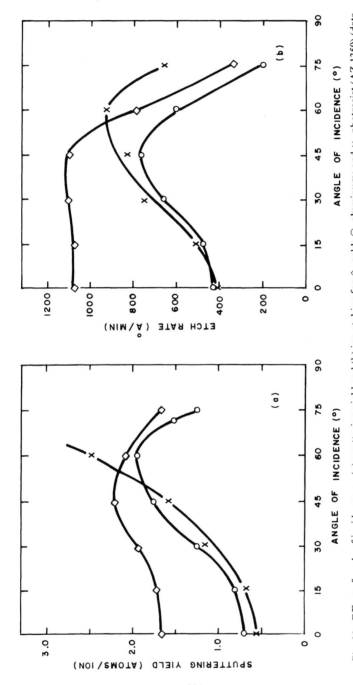

Fig. 11. Effect of angle of incidence on (a) sputtering yield and (b) ion etching for ◇, gold; ○, aluminum; and ×, photresist (AZ 1350) (data from Gloersen [4] and Somekh [44, 57]).

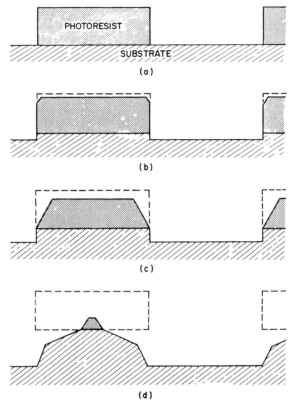

Fig. 12. Facet formation in photoresist due to ion bombardment. (a) The photoresist cross section prior to ion beam etching. (b) The onset of facet formation in the photoresist. (c) The photoresist facet intersects the original substrate surface plane. (d) Further etching exposes the underlying film causing a new facet to propagate into the substrate pattern (from Smith [82]).

B. Factors Controlling Plasma Etch Rates

Plasma etch rates depend on a number of factors such as feed gas composition and pressure, rf power, flow rate, surface temperature, and the surface area of etchable material. Other factors such as the materials of construction exposed to the plasma or the presence of residual gases, particularly water vapor, can be important, too, for reasons already discussed.

Unfortunately, but understandably, no unified treatment of plasma etching is available to provide a basis for quantitative prediction or interpretation of the effects of variations of the parameters subject to direct control. Consequently, only generalized guidelines can be given regarding their influence on etch rates.

1. Gas Composition

The two main criteria for selection of an etching gas are rather obvious. First, the gas must contain one or more constituents which, in principle, could form a volatile compound with the material to be etched. Secondly, the volatile compound expected should be relatively stable, that is it should have a reasonably large heat of formation. The latter condition, apart from providing a thermodynamic driving force for the etching reaction, ensures that the volatile compound, once formed, will not be decomposed easily to its constituents by the plasma. For example, the heat of formation of SiF_4 is 386 kcal/mole (exothermic) at room temperature. Silicon tetrafluoride is relatively stable in a plasma and Si is readily etched in many fluorine containing gases. In contrast, the heat of formation of SiH_4 is small (7.3 [83] to -14.8 [84] kcal/mole at room temperature); SiH_4 is readily decomposed in a plasma to form solid deposits (SiH_x) and Si does not etch at practical rates in a low temperature hydrogen plasma.

Incidentally, the term "volatile" is used in a relative sense. In the ideal plasma etching process, the reaction product is literally pumped away and is found in the gaseous effluent from the discharge. However, it often happens, particularly in the etching of metals, that the compound formed is volatile enough to escape from the surface being etched but simply redeposits in some form elsewhere in the reactor [85]. With the small quantities of material usually etched, there may be no visible evidence of this but the reactor can retain a "memory" of previous materials etched, nonetheless. The etching of silicon compounds in fluorine containing gases, fortunately, is an exception to this as SiF_4 is normally the end product and has ample volatility for complete removal.

Halides and carbonyls are numerous among the simple inorganic compounds which are volatile in the temperature range of interest for pattern generation. Experience has shown that carbonyls are not easily formed by plasma reactions with solids, probably because of surface kinetic limitations. It is not surprising then that essentially all gases reported to produce successful etching contain one or more halogens. Halocarbons have been used extensively in plasma etching because many of them are easy to handle, noncorrosive, and relatively nontoxic sources of halogens. When subjected to a discharge, these gases can be fragmented to produce reactive radicals for etching. Table IV lists some representative halocarbons and some materials etchable in discharges containing them along with the principal fragments observed by mass spectrographic analysis

Table IV

Some Halocarbon Gases for Plasma Etching

Etching gas	Etchable materials	Principal fragments and relative intensities from mass spectra					
CF_4	Si, SiO_2, Si_3N_4, Ti, Mo, Ta, W	CF_3 100	CF_2 12	F 7	CF 5		
C_2F_6	SiO_2, Si_3N_4	CF_3 100	C_2F_5 41	CF 10	CF_2 10	C 2	F 1
$CClF_3$	Au, Ti	CF_3 100	$CClF_2$ 21	CF_2 10	Cl 7	CF 4	
$CBrF_3$	Ti, Pt	CF_3 100	$CBrF_3$ 15	$CBrF_2$ 14	Br 5		
C_3F_8	SiO_2, Si_3N_4	CF_3 100	CF 29	C_3F_7 25	CF_2 9	C_2F_5 9	C_2F_4 7
CCl_4	Al, Cr_2O_3, Cr	CCl_3 100	Cl 41	CCl 41	CCl_2 29		

when these gases are irradiated with 70 eV electrons. Although this is well above the mean energy of electrons in a typical low temperature plasma, the mass spectra are still useful in providing an indication of the species which can be present in a discharge.

The addition of O_2 to many of the halocarbons and some other halogen containing gases often has beneficial effects on etching. As indicated in Section II.C.4, several percent O_2 markedly increases the concentration of atomic F in CF_4–O_2 discharges, and this in turn increases the etch rate for Si and silicon compounds as shown, for example, in Figs. 1 and 13. Similarly, the addition of O_2 to a CF_3Cl discharge can increase the concentration of both Cl and F atoms as shown in Fig. 14. In contrast, there is little effect on the Cl atom concentration in a CF_2Cl_2–O_2 discharge, or on the Br atom concentration in CF_3Br–O_2 plasmas [85].

Additions of O_2 can also be beneficial in controlling or eliminating polymer deposition which occurs under high throughput high power conditions in many of the halocarbon discharges, particularly with nonperfluorinated gases. These polymers result from the reactions of radicals to form high molecular weight condensables.

Mixtures of Cl_2 and O_2 have been used in plasma etching of Cr films for photolithographic mask fabrication [86, 87]. Aluminum etching can be carried out in gas discharges producing atomic Cl but free from F-containing constituents. Aluminum usually has an oxide surface film which must

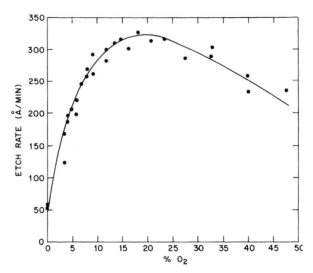

Fig. 13. Dependence of the etch rate of SiO_2 on the feed O_2 concentration in a CF_4–O_2 discharge. 200 W, 0.35 Torr, 150 sccm, 100°C.

be removed before the Cl atoms can react with the bulk material. Carbon tetrachloride and BCl_3 are reported to be useful in plasma etching Al because they can reduce the surface oxide and simultaneously provide Cl atoms for reaction with the bulk [12, 88, 89]. Fluorine-containing gases are not useful, presumably because a nonvolatile aluminum fluoride film which is resistive to chlorine attack is formed rapidly.

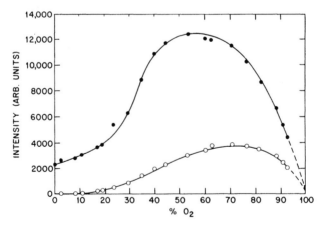

Fig. 14. Variation of emission intensity from excited F and Cl atoms in a CF_3Cl–O_2 discharge. 300 W; 0.35 Torr; 150 sccm; ●, Cl atom emission (725.7 mm); ○, F atom emission (703.7 nm).

The addition of H_2 (or H_2O vapor) to various halocarbon discharges can reduce the concentration of free halogen atoms and may also encourage polymer deposition. Again, depending on the application and the ability to control these effects, the presence of hydrogen may be desirable.

It is fair to say that almost all the gas mixtures presently used in pattern generation have been derived more or less empirically given that one or more constituents of the mixture might form a volatile compound with the material to be etched. In time, this empirical approach coupled with analytical techniques, which are beginning to be used to determine the species present in the plasma, will, hopefully, provide the information needed to put gas selection on a more rational basis.

2. Surface Area Effects

For certain combinations of etch gas and material to be etched, the etch rate depends on the amount of etchable surface exposed to the plasma [90]. This phenomenon, referred to as the loading effect, is a direct result of depletion of the plasma activated etching species, via a rapid etching reaction, being the dominant loss mechanism. A simple phenomenological model of the loading effect, which assumes the etch rate to be first order in the concentration of active species, indicates that the etch rate (R) is related to the surface area (A) of etchable material according to

$$R = \beta\tau G/(1 + K\beta\tau A), \qquad (6)$$

where G is the volume generation rate of active species, β the reaction rate constant, τ the lifetime of the active species in the absence of etchable material, and K is a constant for a given material and reactor geometry [90]. If identical wafers are being etched, as in integrated circuit processing, Eq. (6) can be written in terms of the number of wafers N

$$R = \beta\tau G/(1 + K\beta\tau A_w N), \qquad (7)$$

where A_w is the surface area of etchable material per wafer. The maximum possible etch rate R_0 can be determined from Eq. (7) by setting $A_w N = 0$. Equation (7) shows that the reciprocal etch rate (or R_0/R) is proportional to N as demonstrated in Fig. 15, for example. This linear relation allows the etch rate for any load size to be determined from two measurements on different loads.

The extent to which loading occurs depends on the factor $\beta\tau$ as well as the load size $(A_w N)$, as can be seen from Eq. (7). When $K\beta\tau A_w N \ll 1$, no noticeable change in etch rate will result from changing the load. Conversely, if $K\beta\tau A_w N \gtrsim 1$, the etch rate will increase as the surface area decreases. This has important consequences for control of linewidth and

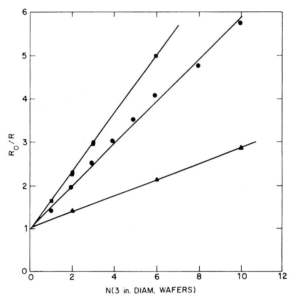

Fig. 15. Variation of etch rate with wafer load. The Si–N (plasma deposited) and Si were etched with a CF₄–8% O₂ plasma in a radial flow reactor. The photoresist was stripped in O₂ in a tube reactor. ■, SiN, 55°C; ●, Si, 100°C; ▲, negative photoresist (tube reactor).

feature size. To illustrate this, consider batch etching of polycrystalline Si for silicon gate metal–oxide-semiconductor (MOS) integrated circuit fabrication using a CF_4–O_2 plasma (Fig. 15) with 10 wafers present.

Notice that at the start of etching the etch rate corresponds to the value for $N = 10$, but as the polycrystalline Si clears, the etch rate increases until, when all wafers are just etched, the etch rate is approximately R_0. (The edge area of the etch features will be small in comparison to the initial surface area exposed so $A_w N \approx 0$ at the completion of etching). Consequently, any overetching will take place at a much higher rate than the nominal value applicable at the start of etching and an accelerated lateral etching will result on clearing. The seriousness of this effect can be seen if one considers the etching of 2 μm lines. If the initial rate is say 0.2 μm/min, then the overetch (lateral) rate will be ≈ 1.2 μm/min, which means that very little overretching is tolerable. In practice, however, some overetching is necessary, not only because of thickness nonuniformity, but even for ideally uniform films because the wafers will not etch uniformly since the loading effect is localized. Loading causes etching to be faster at wafer and mask edges and can lead to faster etching of small isolated features compared to large less-isolated features.

One way to minimize this consequence of loading is to utilize the rapid change in concentration of active species in the plasma near the completion of etching in order to detect the end point. This has been done for etching of polycrystalline Si and silicon nitride in CF_4-O_2 plasmas by monitoring the optical emission from electronically excited fluorine atoms, usually at 703.7 nm [24, 27]. Of course, if the film to be etched is not initially uniform in thickness, across a wafer or from wafer to wafer, end-point detection will be compromised and some overetching will be needed, or else a tedious process of stopping to withdraw finished wafers while returning unfinished ones for additional etching must be followed. The combination of good thickness uniformity and high sensitivity end-point detection may well prove to be adequate for very fine feature pattern transfer on a production scale; however, end-point detection has yet to be proven in this application.

An alternative approach is to etch under conditions where loading is insignificant. For practical batch sizes this implies a reduction in $\beta\tau$ (Eq. (7)) such that $K\beta\tau A_w N \ll 1$. Generally this requires a change in gas composition so that either a different active species prevails or the same active species with a substantially reduced lifetime is used. Notice that although $\beta\tau$ also appears in the numerator of Eq. (7), the etch rate need not decrease in proportion to $\beta\tau$ because G, the generation rate of active species, which depends mainly on power density and pressure, can be increased. On the other hand, since little loading occurs, end-point detection based on measurement of the *reactant* concentration is not feasible. Ideally, the end point should be detected by monitoring the volatile etching *product*. Product detection has been reported for "downstream" plasma etching of polycrystalline Si [91] and for plasma stripping of photoresist [19, 20, 27].

3. Pressure and Power

For a given gas, both the pressure and power, or more precisely the power density, affect the electron energy distribution which, in turn, determines the generation rate of active species. Pressure and power density are independent parameters. However, in practice their effects are not always easy to separate because it is usually the total power applied to the reactor, rather than the power density, which is measured. Of course, the nominal power density can always be defined by dividing the applied power by the reactor volume; however, the true plasma volume can be different than the geometric volume and can depend on pressure. In the parallel plate reactor of Fig. 5, for example, as the pressure is reduced, at fixed power, a point will be reached where the plasma tends to "spill" over the ground electrode and fill some of the volume beneath it; the power density thus decreases at fixed power. Similarly, in a tube-type reactor, the doughnut shaped plasma tends to fill more of the container volume as pressure is reduced.

Unfortunately, for practical reasons, almost all of the plasma etching rates reported in the literature are referenced to total applied power for a particular reactor. Consequently, data showing etch rate as a function of pressure (power) at a fixed total power (pressure) are difficult to interpret in terms of pressure (power) effects alone. Generally etch rates increase monotonically with applied power. Figure 16 is typical of data reported for the dependence of etch rate on power.

As pressure is increased (at constant power density), the collision rate will increase but the average electron energy will eventually decrease simply because the existing field will have less time to accelerate an electron between collisions. Since the generation rate for the active species de-

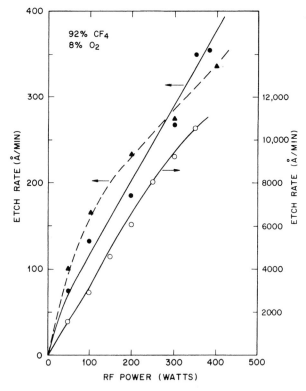

Fig. 16. Variation of etch rate with rf power, 92% CF_4–8% O_2 plasma; ●, thermal oxide, 0.5 Torr; ▲, CVD Si_3N_4, 0.3 Torr; ○, silicon, 0.42 Torr (2 in. wafer).

pends on the electron energy distribution, which determines the number of electrons in an appropriate energy range for generating a particular active species or its precursor, the etch rate must eventually decrease with increasing pressure, at a fixed power density. Conversely, as the pressure is reduced the collision frequency will decrease and the concentration of active species will begin to drop at low enough pressure. These limitations fix the useful pressure range in the region from about 0.1 to several Torr.

It has been tacitly assumed in the preceding discussion that power and pressure mainly affect the generation rate of active species. However, it is possible that changes in these variables might also influence the lifetime of the active species, as well; for example, by causing disproportionate changes in the concentration of the active species and some other species with which they can recombine. A moment's reflection reveals the complexity of the situation and makes clear the difficulty attached to any attempt to make quantitative predictions about the power and pressure dependence of etch rate for even the simplest gases.

One additional feature of the pressure-power dependence is noteworthy. Plasma stability and uniformity are sometimes unsatisfactory for certain combinations of gas, power, and pressure. In such cases, the desire to optimize uniformity may strongly influence the choice of these parameters.

4. Flow Rate

Plasma etching is normally done under dynamic flow to replenish continually the reactant species and remove volatile products. The inlet flow rate of the feed gas and the reactor pressure determine the residence time of the average molecule in the plasma. Flow rate and pressure can be varied independently if a throttling valve is used to control the pumping speed. The steady-state pressure in the reactor will be

$$p = F \cdot 760/S \quad \text{Torr} \tag{8}$$

where F and S are the inlet flow rate (STP) and pumping speed, respectively, measured in the same units. The average residence time t_r is

$$t_r = Vp/760F, \tag{9}$$

where V is the reactor volume. Notice that at fixed pumping speed, the residence time is unchanged by changes in pressure brought about by changing the inlet flow rate. At fixed pressure, the residence time can be decreased to a minimum value by increasing the flow rate to the maximum allowable value, as determined by the maximum available pumping speed.

For typical conditions of say 0.5 Torr, 150 cm^3/min in a 10 liter chamber, $t_r \cong 2.5$ sec, while the minimum residence time at the same pressure and a pumping speed of 550 liter/min would be ~1 sec.

If there were no loss mechanisms for active species other than removal by pumping, the residence time would fix τ in Eq. (6). However, if the lifetime of an active species against other loss mechanisms is small compared to the residence time, the concentration of etchant will be determined mainly by these other loss processes. Under such conditions, the etch rate should be relatively insensitive to flow rate over much of the range of attainable flow rates. As the flow rate is lowered, at constant pressure, residual gases resulting from real and virtual leaks will eventually reach significant proportions and can affect etch rates according to the way in which they interact with the feed gas in the plasma [10].

5. Temperature

The temperature of the surface being etched is a significant variable, inasmuch as the chemical reaction producing etching occurs on this surface. The factor β in Eq. (6) is effectively a reaction rate constant to which the temperature dependence can be assigned formally. Nearly all published data exhibit an Arrhenius type temperature dependence for etch rate with activation energies lying in the range of 0.05–0.5 eV. Table V lists some etch gas and material combinations along with measured activation energies. The values of activation energy for SiO_2 and Si in CF_4–O_2 plasmas vary somewhat from one investigator to another. This is probably due, in part, to the difficulty in measuring accurately small activation energies in a narrow temperature range, and in the case of Si may also be dependent on different sample surface areas being used. Assuming that β has an Arrhenius temperature dependence, it can be seen from Eq. (6) that when loading is present, the logarithm of etch rate is not strictly linear in reciprocal temperature.

In any case, it can be said that generally etch rates increase sufficiently with temperature to require temperature control in order to achieve constant and reproducible rates. As indicated in Section III.C.2, the parallel plate reactor lends itself rather easily to temperature control. One method of accomplishing temperature control in tubular reactors is to preheat the wafers in an inert gas plasma, and subsequently to control the input power to the etching plasma through a feedback loop using temperature as the control variable [95]. The latter technique is only applicable above ambient temperature, of course. This is a drawback when thermally sensitive masks such as electron resists are used.

The rf power dissipated in the plasma ultimately appears as heat and is a major source of temperature rise in thermally isolated wafers. Addition-

Table V

Measured Thermal Activation Energy for Plasma Etching

Etching gas	Equipment	Material/activation energy (eV/mole)	Refs.
CF_4-5% O_2	Tube reactor	SiO_2/0.20	[12]
CF_4-10% O_2	Radial flow reactor	SiO_2/0.16; Si/0.11	[10]
CF_4-O_2	"Downstream"	SiO_2/0.18; Si/0.11	[23]
CF_4-8% O_2	Radial flow reactor	Si/0.10	[90]
CF_3Br-He-O_2	Radial flow reactor	Ti/0.20	[85]
CF_4	Tube reactor shielded	SiO_2/0.19; PSG/0.08–0.12[a]; BSG/0.12–0.15[b]; ASG/0.11–0.16[c]	[94]
O_2	Tube reactor shielded	Photoresist/0.5	[138]
O_2	Tube reactor	Photoresist/0.27	[92]
O_2	Tube reactor	Graphite/0.28	[93]

[a] PSG denotes phosphosilicate glass film.
[b] BSG denotes borosilicate glass film.
[c] ASG denotes arsenosilicate glass film.

ally, the heat of reaction can also be substantial. Etching of polycrystalline Si at a typical rate of 2000 Å/min in a CF_4–O_2 plasma, for example, produces heat at a rate of about 2 W per 3 in. (7.6 cm) diameter wafer for a 30% mask coverage. The consequences of poor heat sinking can be gauged by assuming that all the heat of reaction is absorbed by the wafer. This would produce a temperature rise of approximately 60°C in a 20 mil (0.5 mm) thick wafer for each 5000 Å of film removed.

6. Mask Selection

Plasma etching processes for pattern transfer generally provide wide latitude in the choice of masking material. Photoresists are by far the most often employed materials, although almost any material in which a pattern can be generated is useful, provided it has an acceptable rate of erosion for the etching conditions used.

Masks which are totally inert to the plasma can be prepared for critical applications but usually only at the expense of additional processing, as

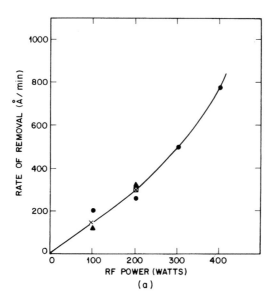

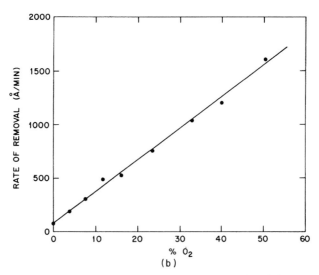

Fig. 17. Photoresist etch rate in $CF_4 - O_2$ plasma (a) as a function of rf power and (b) as a function of the feed gas O_2 concentration. (a) 92% CF_4-8% O_2 plasma; 0.3 Torr; 100 sccm; 40°C; ●, SC 180; ×, IC 3; ▲, HR 100; □, AZ 1350J (b) $CF_4 - O_2$ plasma, 200 W, 0.35 Torr, 40°C, AZ 1350J.

for example, in the use of Al films to mask against CF_4-O_2 plasmas. Normally, photoresists provide ample masking and, in fact, one of the attributes of plasma etching is its compatibility with photolithographic processing.

The rate of resist erosion is fixed by the same parameters that determine etch rates for other materials. However, unlike physical sputtering, it is almost always possible to choose conditions (gas composition, temperature, power, etc.) which produce adequate selectivity for the material to be etched in comparison to the photoresist mask.

By way of example, Figs. 17a and b illustrate resist erosion rates as a function of power and O_2 concentration for CF_4-O_2 plasmas. Note that there is very little difference between positive and negative working resists and that the rate increases rapidly with O_2 concentration.

In order that a masking layer be present through the completion of etching, it is necessary that the ratio of the film thickness to be etched to the initial mask thickness not exceed the ratio of etch rates, film to mask. Indeed, to allow some margin for nonuniform erosion of the mask, the ratio of initial thicknesses should not exceed about 80% of the etch rate ratio. The maximum mask thickness is generally dictated by the resolution required in the pattern to be replicated, so that as feature size is reduced a corresponding decrease in mask thickness is required. If the thickness of the material to be etched cannot be reduced accordingly, then the selectivity requirement becomes more stringent. At present, plasma etching processes which provide ample selectivity to produce features at the resolution limits of optical lithography, using commercially available photoresists, exist for essentially all of the materials to be patterned in silicon integrated circuit fabrication. Certain of these processes are still considered proprietary by their developers.

V. PATTERN EDGE PROFILES

The goal of any pattern etching process is the transfer of a pattern from a mask to an underlying film in order to replicate, exactly, the mask's features. But this objective establishes only a two-dimensional criterion for the quality of replication. To illustrate this point, consider the etched features shown in cross-section in Figure 18. All of these features have the same projected dimensions in the plane of the substrate, but each is clearly different from the others, by virtue of its profile. The profile provides an additional criterion to judge the quality of etching. Unlike the transfer of dimensional information from mask to film, no single ideal goal can be set for the edge profile. If the etched feature is to be covered with a

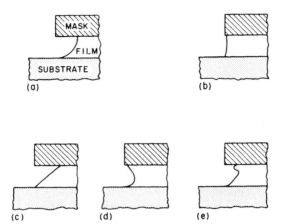

Fig. 18. Cross sections of several types of edge profiles. (a) isotropic, (b) anisotropic, (c) tapered, (d) reentrant, (e) breadloaf overhand overhang.

deposited film and conformal step coverage is desired, a profile such as Fig. 18c is preferable. If a conductor line is being defined, Fig. 18b will maximize the cross-sectional area. Similarly, if a feature is being etched in a chrome mask, Fig. 18b represents an ideal result.

The exact shape of the profile will generally have greater significance as the aspect ratio (height to width) of the feature increases. For film thicknesses in the range of a few tenths to 1 μm, this implies that the edge profile will become increasingly more important as feature size is reduced into the submicrometer range.

A. Ion Etching

A variety of mechanisms contribute towards the edge profile developed during ion etching. One of the most important of these, namely faceting, has already been described in the previous section and is shown in Fig. 12. For some applications, a tapered edge profile (Fig. 18c) may be desirable and a controlled faceting effect is considered advantageous. An example of the use of controlled faceting to produce a tapered wall is shown in Fig. 19. In this case, a Ti/Pt/Au metallization on a silicon integrated circuit was sputter etched using a Ni mask, the thickness of which was chosen so that faceting produced a tapered wall but without excessively narrowing the line.

The development of "trenches" at the base of etched steps is often observed after ion etching (Fig. 20a). This type of topography is due to several causes shown schematically in Fig. 21. The most common cause is

Fig. 19. Cross section scanning electron micrograph of Ti/Pt/Au stripe after rf sputter etching, demonstrating the resolution of the uniform gold process and the ability to produce sloped metal walls by controlled faceting (from Melliar-Smith [6]).

thought to be enhanced ion flux at the base of the step due to ion reflection off the side of the step. In addition, to this there will be some forward sputtering from the wall depositing material at the base of the wall. Whether a "trench" or "residual material" (effectively a reduced etch rate) is observed after milling will depend on which effect dominates—a factor controlled by wall slope angle. Redeposition from gas phase collisions in the plasma has also been considered partially responsible for the phenomena of trenching. The area close to the edge of a step will see a reduced solid angle (θ) for redeposition from above the sample, while further out a larger solid angle (θ') is apparent and hence more redeposition and a slower etch rate is observed.

Equally important is the direct redeposition of sputtered material onto the edge of a step. This subject has been covered from a theoretical point of view by many workers with respect to the generation of surface topography [96–107] and more specifically with respect to pattern delineation by others [3, 4, 80, 108]. This process is shown schematically in Fig. 22 for the formation of a redeposited veil on the side of a photoresist pattern. The veil is often resistant to liquid or plasma photoresist stripping and re-

EDGE PROFILE CONTROL

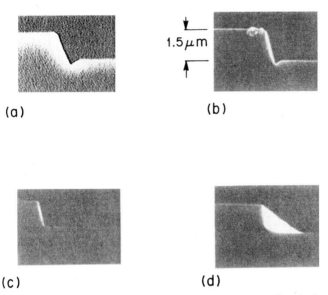

(a)   (b)

(c)   (d)

Fig. 20. Cross section scanning electron micrographs of edge profiles obtained by (a) rf sputter etching and (b), (c), (d) ion milling at various angles of incidence with the sample rotated. (b) $\theta = 15°$, (c) $\theta = 22.5°$, (d) $\theta = 30°$ (from Somekh [57].)

mains after resist removal, either in complete form or of reduced height if it is brittle and breaks away during substrate cleaning. Redeposition can cause special problems if a metallization is being etched and deposits build up along steps in the oxide. These conductive filaments can cause leakage paths between conductors, which in some metal-oxide-semicon-

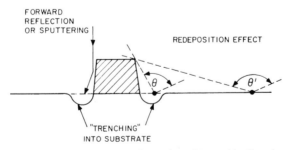

Fig. 21. Schematic diagram showing the formation of "trenching" at the side of patterns (from Melliar-Smith [6].)

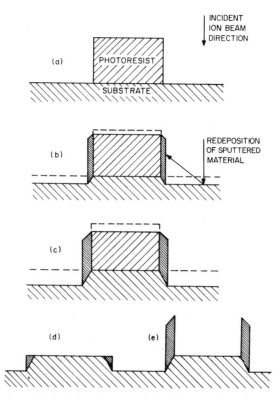

Fig. 22. Simple model depicting the stages in ion beam etching of grooves using a photoresist mask. (a) The photoresist cross section before etching. (b) and (c) As etching proceeds facets are formed and coating of the side walls by redeposition takes place. When the remaining photoresist is stripped after etching is completed, (d) the side walls may tear off, (e) or may remain, depending on how thin and brittle the walls are (from Glöersen [4]).

ductor (MOS) devices will be fatal if they have a resistance of less than 10^5 Ω. The most effective solution for redeposition is to minimize the thickness of the resist film and reduce as much as possible the slope of all steps present on the substrate.

In ion milling, unlike sputter etching, the substrates can be bombarded at an angle, and special edge profiles can be obtained as a result. An example is shown in Fig. 20. Figure 20a was rf sputter etched and Fig. 20b was ion milled at an angle of incidence of 15° with the sample rotated. Both edge profiles show trenching. The trenching can be reduced if the angle of incidence is increased to a value larger than 90° minus the slope of the side wall. This tilt angle then shadows the trench area from the primary beam for some portion of the rotation cycle and this compensates for the enhanced flux due to ion reflection (Fig. 20c). If the tilt angle is too great, the shadowing effect dominates and residual material is left at the base of the step (Fig. 20d).

Further specialized profiles can be achieved if a single-crystal material is ion bombarded at an angle. It has been known for some time [109] that different crystallographic planes etch at different rates and recently this effect has been utilized to produce blazed gratings in GaAs, using reactive ion etching (Ar + 0.3% Freon 12). Examples of the gratings are shown in Fig. 23.

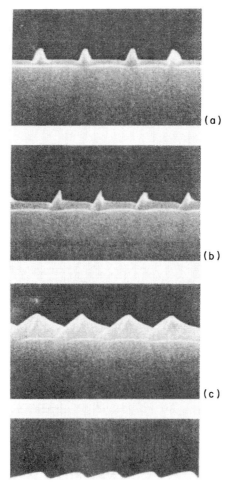

(a)

(b)

(c)

Fig. 23. Profiles due to ion milling of single crystal GaAs at various angles showing the development of various crystallographic faces (courtesy of R. D. Heidenreich and G. W. Kammlott).

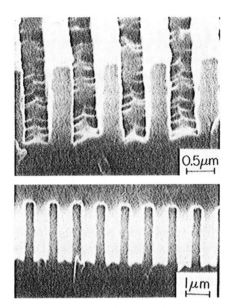

Fig. 24. Scanning electron micrographs of anisotropic reactive sputter etching of SiO_2 in a CHF_3 plasma; pressure, 7 mTorr; power density, 0.6 W/cm^2; de bias, -510 V; etch time, 68 min (from Lehmann and Widmer [72]).

Many of the problems described above can be obviated if reactive ion etching is used [72] and almost vertical edge profiles can be obtained as shown in Fig. 24. The reasons for this are related to the mechanism of reactive ion etching in which it is observed that the erosion process only occurs where the surface is struck by reactive ions. As these are directional, the etching is anisotropic. However because a chemical reaction between the substrate and the reactive ion also occurs to produce a volatile product, redeposition is not a factor. Lehmann and Widmer [72] also showed that reactive ion etching in CHF_3 produced and showed significantly less faceting of AZ1350 photoresists when compared to CF_4 reactive sputtering.

B. Plasma Etching

A popular misconception concerning plasma etching is that it produces no lateral etching under the mask edge (i.e., undercutting). If this were so, profiles such as Fig. 18b would be the rule rather than the exception. In fact, a wide range of profiles is possible though not always predictable or entirely reproducible. We consider two extremes of possible profiles and then discuss the development of actual profiles.

1. Isotropic Etching

In isotropic etching, every point on the exposed surface moves at the same rate normal to the local tangent to the surface at that point. The resulting edge profile for the just-etched feature is a quarter circle (in cross section) with the origin at the mask edge; overetching produces circular arcs centered at the mask edge with a radius proportional to the etch time as shown in Fig. 25 [40]. If the etched film is structurally and chemically homogeneous, if there is no mask lifting, and if the flux of active species is uniform over the entire profile throughout the etch cycle, then an isotropic profile should result. Figure 26 illustrates isotropic plasma etching of a deposited insulator film approximately 1-μm thick; a photoresist mask was used and the etching was carried out in a radial flow reactor using a CF_4–O_2 mixture.

Undercutting of the mask at the top surface of the etched feature is unavoidable with isotropic etching, although the dimensions at the base of the feature will replicate the mask in the "just-etched" condition. Overetching will cause undercutting at the lower surface as well. As indicated previously (see Section IV.B.2), undercutting can take place at the same rate as for vertical etching if no loading occurs, or at an accelerated rate with loading. In the latter case, gross undercutting and complete loss of dimensional control may result.

It is of interest to note that so-called negative undercutting has also been reported [111]. According to this report, under certain conditions the etched feature can protrude beyond the mask edge. No explanation or confirmation of this phenomenon has been given.

2. Anisotropic Etching

At the other extreme in edge profiles is the anisotropic profile which in the limiting case corresponds to Fig. 18b, where the etched feature has vertical walls and there is virtually no undercut of the mask. Such profiles

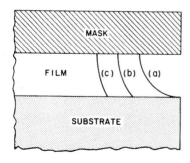

Fig. 25. Edge profiles for isotropic etching. (a) just etched, (b) 50% overetch, (c) 100% overetch.

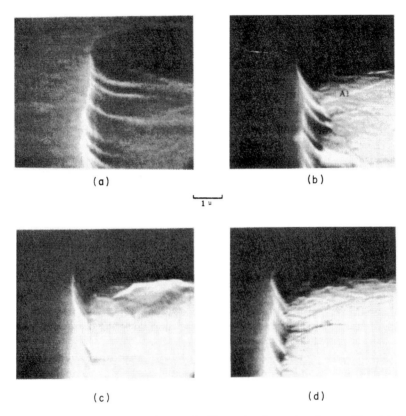

(a)   (b)

1 ᵘ

(c)   (d)

Fig. 26. Scanning electron micrographs illustrating isotropic edge profiles. (a) Underetch, (b) end point, (c) 50% overetch, (d) 100% overetch (courtesy of S. E. Haszko).

have been observed to result from plasma etching under certain conditions. The mechanism for the development of such a profile is not clear, but may be associated with charged particle bombardment coupled with chemical reaction, suggesting a process similar to reactive ion etching. The anisotropic profile can be generated under plasma conditions which are unlikely to produce physical sputtering of the etched film itself. However, it is possible that sputtering of weakly bound organic films could occur under the same conditions. Thus, ions accelerated across the ion sheath at the film-plasma interface may have energies high enough (several electron volts) to sputter and/or decompose polymer films formed from constituents of the plasma. The ion trajectories will be normal to the substrate (across the ion sheath), and thus polymer films may be removed faster than they form at the bottom of an etching feature, while on the side walls, which receive a lower ion flux, there may be net polymer deposition.

Anisotropic profiles have certain desirable properties for producing patterns with very fine features. In particular, the etched feature has exactly the mask dimensions. This is important when feature size is at the limit of resolution of the mask-making procedure for then it is not always possible to design the mask to accommodate changes in feature size which can occur during nonanisotropic etching.

Another type of anisotropic etching relates to differential rates for various crystallographic orientations. It has been observed [112] that single-crystal Si etches preferentially in CCl_4-O_2 plasmas, the etch rate for the (100) planes being about thirty times larger than the rate for (111) planes. Other halocarbon gases containing F or F and Cl do not preferentially etch Si.

3. Real Profiles and Their Development

In practice, etch profiles often lie somewhere between the extremes of isotropic and anisotropic etching. The development of an actual profile depends on the extent of overetch; the relative reactivity of mask, film, and substrate; the degree of adherence of the mask; the dimensions of the etched feature; and the pressure.

The mean free path corresponding to the pressures used in plasma etching ranges from ≈ 10 to $100~\mu m$, which is of the same order or larger than typical etch depths and mask openings. In consequence, an etching species which has entered a mask opening (see Fig. 27) can be assumed to impinge on the substrate without further collisions in the gas phase. If ran-

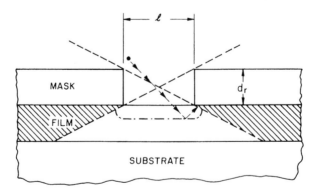

Fig. 27. Cross section of a masked film. The dashed lines define the solid angle corresponding to direct incidence of etching species. The cross-hatched areas of film can only be reached by "reflected" species.

dom motion of the active species prevails in the gas phase, as would be true for neutrals, the flux of primary particles lies within the solid angle subtended by the mask opening as shown in Fig. 27. The cross-hatched areas are protected from direct incidence, but can be reached by active species which are "reflected" from the etching surface as shown, for example, by the trajectory in Fig. 27. The "reflection coefficient" will depend on the reaction probability (fraction of incident species which react before desorbing) which, in turn, depends on the surface encountered (that is, mask, film, or substrate) by the primary particle.

Chemical reactions between solids and gases seldom have reaction probabilities near unity so it is likely that many reflections occur prior to an active species being consumed. In this case, all surfaces will receive a nearly uniform flux except when the ratio of mask thickness to feature size (d_r/l) is very large, then all incident particles arrive at near normal incidence. However, the latter case is unlikely to be of importance since there are fundamental limitations on pattern resolution which make $d_r >> l$ practically unattainable. The more likely situation is where the mask thickness is smaller than the mask opening $(d_r < l)$, which, in turn, is smaller than the mean free path. Under these conditions, it is argued [5, 113] that a high reaction probability (low reflection coefficient) for the film being etched will tend to produce reentrant angles (overhang) at the top of the profile, or near vertical walls for the just-etched condition. Conversely, a low reaction probability will tend to yield sloped walls or isotropic profiles. The effect of overetching depends on the reflection coefficient of the substrate. An inert substrate will cause undercutting to occur during overetching while a highly reactive substrate will suppress reflections of active species to areas under the mask.

Although these ideas are intuitively appealing, they have not, as yet, been demonstrated to be valid by direct experiment. Indeed, there is some evidence that for a rapidly etching material (polycrystalline Si) in a CF_4–O_2 plasma, the profile is nearly isotropic even with a comparatively inert substrate (SiO_2) and submicron feature size [114].

4. Control of Edge Profile

The truly anisotropic profile makes conformal step coverage with films deposited subsequent to patterning difficult to achieve. A prime example is the deposition of Al metallization over patterned phosphosilicate glass films for integrated circuit fabrication. High aspect ratio features tend to cause shadowing during film deposition which results in thinning of the conductor at etched steps in the insulator [115]. The thinned region inevitably leads to yield and reliability problems. Therefore, it is often desirable to produce a tapered profile similar to that shown in Fig. 18c. Many techniques have been devised to taper profiles in wet etching and some of these are applicable to plasma etching. In particular, the surface of the film to be etched can be altered to produce a faster etching region either by altering chemical composition during film growth, or by ion implantation of the surface region after growth, or by the use of a faster etching sacrificial surface layer. These techniques have been described fully in the literature on wet etching [116–118] (see Chapter V-1).

VI. ADVANTAGES AND DISADVANTAGES OF PLASMA-ASSISTED ETCHING

A new pattern delineation process is often judged or promulgated on its resolution capability and ion and plasma etching have been no exception. Indeed, under the correct circumstances, they have unparalleled ability to transfer resist patterns into thin films. However, resolution is by no means the only or sometimes even the major consideration in the successful utilization of these processes.

Plasma techniques have the advantage of being dry processes; the problems resulting from surface tension phenomena in liquid etching are eliminated. Resist adhesion is less critical as mask lifting due to capillarity is not possible. Also, there is no bubble formation to block etching and no problem with incomplete wetting. The disposal of toxic and corrosive liquids is obviated. In addition, because the etching species are generated by the electric discharge, etching terminates rapidly when the discharge is extinguished. Consequently, no etchant can be trapped in capillary channels to continue undesired etching after the nominal completion of the etching process. Nor are the problems associated with reactivity decay in liquid chemical etchants a significant factor.

Plasma and ion etching differ in that the former is a chemical process, the latter a physical process. To achieve good uniformity with the former process, care must be taken to ensure that surface residues (either from resist residues or from surface film formation on active metals) are not present on the surface to be etched. The lack of convective agitation, which can undercut and lift off residues in liquid etching, makes plasma etching rather sensitive to this problem, particularly where multilevel or alloy metallization films are involved [85]. The problem can be mitigated somewhat by using multiple plasma processes; introducing for example , a gas to clear the surface (oxygen to remove resist residues) followed by the gas used to etch the film. Ion etching, due to its physical attack, is less affected by thin surface films because these residues have a finite ion etch rate and are removed relatively rapidly.

Etch uniformity can be particularly important for high-resolution patterns where overetching to clean up slow etching residues will cause severe undercutting of the remainder of the pattern. In the event that some overetching is required to allow for ion flux variations across the beam, ion etching, due to its anisotropic etching rate, will also be less susceptible to undercutting of the resist.

It must be remembered, however, that this linewidth control will only be achieved if the resist pattern remains intact over the whole of the pattern being etched. The major problem is the faceting effect (Fig. 12), which can result in linewidth reduction very similar to that caused by undercutting, unless care is taken in designing the pattern delineation parameters correctly.

Probably the major disadvantage of the ion etching process is that it is not possible to obtain the chemical selectivity in etch rate between two materials that can be achieved with liquid etching. The relative ion etch rates for two films, even using reactive ion etching, rarely differ by more than a factor of 20 and often by much less. Thus, the successful utilization of ion etching usually requires that the layer beneath that being etched is, to some degree at least, sacrificial and some erosion of this layer should be acceptable. Plasma etching is much more selective although not generally as good as liquid etching. Plasma etching is, however, generally less harsh on the polymeric resist materials than either liquid or ion etching, and intermediate masking levels are therefore seldom needed.

The electronic damage introduced by plasma processing into metal-oxide-semiconductor devices (MOS) has been the subject of several studies, particularly by McCaughan and Kushner [119]. The operational characteristics of this type of device are very dependent on the purity and electronic structure of the gate oxide which can be degraded by inappropriate processing. The degree and type of damage are dependent both on the device structure being ion etched and the plasma environment to which it is exposed. It is well known that MOS devices can be degraded by ionizing radiation [120–123]. The various species to which they are exposed are listed in Table VI for rf sputter etching, ion beam milling, and plasma etching. All of the particles listed in Table VI can (if of sufficient energy) and do cause device degradation if they are permitted to impinge directly onto the gate oxide. Less susceptible are devices where the gate oxide is already covered by a polycrystalline Si or Al gate electrode, and least susceptible are the Ti/Pd/Au devices due to the higher absorption for x rays of the gold. Positive ions impinging directly onto the gate oxide can cause dielectric breakdown or produce mobile charge, fixed interface charge, or interface trapping states. Dielectric breakdown is particularly severe in that the damage cannot be annealed out [124] and that a dose of

Table VI

Particle Bombardment during Ion and Plasma Etching

Type of particle	rf Sputter etching	Ion beam milling	Plasma etching	
			Shielded	Unshielded
Electrons	Up to peak-to-peak rf voltage	≤25 eV (from neutralizer)	Insignificant	Up to peak-to-peak rf voltage ~100 V
Positive ions	Up to peak-to-peak rf voltage	Few eV less than ion accelerating potential	Insignificant	A few eV
Photons	Up to peak-to-peak rf voltage	≅tens of eV from plasma discharge and neutralizer electrons striking chamber walls	Up to peak-to-peak rf voltage	Up to peak-to-peak rf voltage ~100 V voltage ~100 V

the order of only 10^{15} ions/cm² is sufficient for its occurrence. The mobilization of otherwise immobile impurity atoms will also occur under these circumstances [125–127], giving rise to reliability problems if the devices are not annealed at around 900°C. The interface states are annealable at a lower temperature of 400°C. These appear to be caused by energetic photons [128] generated at the oxide surface by ion bombardment. Photons produced in the plasma will have a similar effect leading to both fixed charge and interface states, although some of the former may be attributable to impurity effects. Powell and Derbenwick [129] have shown that the threshold energy for photon damage is 8.8 eV.

Electron irradiation of MOS devices has been investigated in the 5–20 keV range [122, 130–132], and with more applicability to ion etching effects, by McCaughan and Kushner [119] for energies of 50–1250 eV. At doses of 10^{14}–10^{15} e/cm², increased interface state density and some reduction of the dielectric strength of the oxide was observed for electron energies above 100 eV. In consequence, ion beam and plasma systems are generally free of damage due to electron irradiation. When damage does occur, the interface states can generally be annealed out at less than 500°C; however, the dielectric strength degradation is more permanent [124]. Based on these studies, it is not generally recommended that a gate oxide be directly exposed to either rf sputter etching or ion beam milling. However, if the gate oxide is protected under the gate metallization, the devices are much less susceptible to damage. Under these circumstances, the major concern is photon damage because photons have a much greater penetration range in the gate metallization than electrons or ions, and are more likely to reach the gate oxide. Since the photon energy (and also the energy of the electrons which will cause x-ray fluorescence) is much lower in ion beam and plasma etching systems, MOS devices are generally free of damage if the gate metal itself is pattern delineated using these techniques [12, 88, 133], although exceptions to this have been noted [134].

The use of sputter etching however requires more care. The secondary electrons generated at the cathode by ion bombardment are accelerated across the full cathode fall potential and then drift across the plasma until they strike the anode, generating x rays with an energy up to the peak-to-peak rf voltage. In order to absorb these x rays, Ryden *et al.* [135] used an anode structure similar to that shown in Fig. 12 of Chapter II-1. For effective x-ray suppression, the capture anode depends on two factors; first, the secondary electrons leaving the cathode have trajectories which are distributed in a narrow angular cone perpendicular to the cathode and hence will penetrate deep into the anode tubes; and second, when the electrons strike the anode they generate x rays which have a large angular range and therefore are mostly absorbed in the tube walls.

The efficiency of the capture anode in preventing voltage shifts in MOS devices is shown in Fig. 28. Wafers processed using a conventional anode show a higher threshold voltage and a large standard deviation, in contrast to the capture anode product which is very similar to the control wafer processed by wet chemical means.

The extent of device contamination by impurities as a result of plasma etching is also of special interest, although there is little published information. In the case of photoresist stripping in an oxygen plasma, it is known that impurities present in the resist which do not form volatile oxides are left on the surface at the completion of stripping [136]. It has been shown that when Na is present in a resist being stripped from an oxidized Si wafer, the Na can be driven into the oxide by local fields [137]. The penetration is particularly severe in an unshielded plasma. Once incorporated in the oxide, the Na causes ion drift instability. However, various techniques can be used to getter or immobilize the incorporated sodium [138]. Moreover, stripping of resist from the sensitive gate oxide would not occur in actual practice for Si gate MOS fabrication where the gate oxide would be masked by a polycrystalline Si layer.

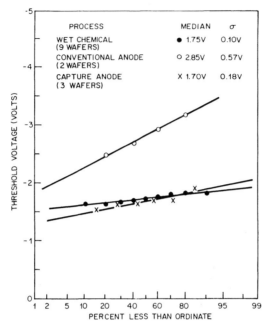

Fig. 28. Threshold voltage shift for MOS device produced by wet chemical rf sputter and rf sputter + capture anode processes (from Ryden *et al.* [135].)

Process	Median	σ
Wet chemical (9 wafers)	● 1.75 V	0.10 V
Conventional anode (2 wafers)	○ 2.85 V	0.57 V
Capture anode (3 wafers)	× 1.70 V	0.18 V

REFERENCES

1. W. R. Grove, *Philos. Trans. R. Soc. London* **142**, 87 (1852).
1a. R. Castaing and P. Laborie, *C.R. Acad. Sci.* **238**, 1885 (1954).
1b. M. Gomberg, *Ber. Dtsch. Ges. Chem.* **33**, 3150 (1900).
2. E. G. Spencer and P. H. Schmidt, *J. Vac. Sci. Technol.* **8**, S52 (1971).
3. H. I. Smith, *in* "Etching for Pattern Definition" (H. C. Hughes and M. J. Rand, eds.), p. 133. Electrochem. Soc., Princeton, New Jersey, 1976.
4. P. G. Glöersen, *J. Vac. Sci. Technol.* **12**, 28 (1975).
5. H. Dimigen and H. Luthji, *Philips Tech. Rev.* **35**, 199 (1975).
6. C. M. Melliar-Smith, *J. Vac. Sci. Technol.* **13**, 1008 (1976).
7. D. T. Hawkins, *J. Vac. Sci. Technol.* **12**, 1389 (1975).
8. J. L. Vossen and J. J. O'Neill, *RCA Rev.* **29**, 149 (1968).
9. G. N. Jackson, *Thin Solid Films* **5**, 209 (1970).
10. A. R. Reinberg, *in* "Etching for Pattern Definition" (H. G. Hughes and M. J. Rand, eds.), p. 91. Electrochem. Soc., Princeton, New Jersey, 1976.
11. J. A. Bondur, *J. Vac. Sci. Technol.* **13**, 1023 (1976).
12. R. G. Poulsen, *J. Vac. Sci. Technol.* **14**, 266 (1977).

13. L. Holland, *J. Vac. Sci. Technol.* **14**, 5 (1977).
14. "Chemical Reactions in Electric Discharges" (R. F. Gould, ed.), Advances in Chemistry Series, No. 80. Am. Chem. Soc. Publ., Washington, D.C., 1969.
15. "Techniques and Applications of Plasma Chemistry" (J. R. Hollahan and A. T. Bell, eds.). Wiley, New York, 1974.
16. F. K. McTaggart, "Plasma Chemistry in Electrical Discharges." Elsevier, Amsterdam, 1967.
17. S. M. Irving, *Solid State Technol.* **14**(6), 47 (1971).
18. J. F. Battey, *J. Electrochem. Soc.* **124**, 147 (1977).
19. E. O. Degenkolb, C. J. Mogab, M. R. Goldrick, and J. E. Griffiths, *Appl. Spectrosc.* **30**, 520 (1976).
20. E. O. Degenkolb and J. E. Griffiths, *Appl. Spectrosc.* **31**, 134 (1977).
21. H. Abe, Y. Sonobe, and T. Enomoto, *Jpn. J. Appl. Phys.* **12**, 154 (1973).
22. R. A. H. Heinecke, *Solid-State Electron.* **18**, 1146 (1975).
23. Y. Horiike and M. Shibagaki, *Jpn. J. Appl. Phys.* **15**(Suppl.), 13 (1976).
24. W. R. Harshbarger, T. A. Miller, P. Norton, and R. A. Porter, *Appl. Spectrosc.* **31**, 201 (1977).
25. C. J. Mogab, A. C. Adams, and D. L. Flamm, *J. Appl. Phys.* **49**, 3796 (1978).
26. R. A. H. Heinecke, *Solid-State Electron.* **19**, 1039 (1976).
27. R. G. Poulsen, *Electrochem. Soc. Extend. Abstr.* No. 77-1, p. 242 (1977).
28. E. E. Muschlitz, Jr., *Science* **159**, 599 (1968).
29. R. Foon and M. Kaufman, *Prog. React. Kinet.* **8**, 81 (1975).
30. P. S. Ganguli and M. Kaufman, *Chem. Phys. Lett.* **25**, 221 (1974).
31. P. C. Nordine and J. D. LeGrange, *AIAA J.* **14**, 644 (1976).
32. F. Kaufman and J. R. Kelso, *J. Chem. Phys.* **32**, 301 (1960).
33. F. Kaufman and J. R. Kelso, *Int. Combust. Symp., 8th* p. 230 (1960).
34. T. O. Headon and R. L. Burke, *Plasma Etch. Alum., Kodak Microelectron. Semin.*, *1977*.
35. J. W. Coburn, H. F. Winters, and T. J. Chuang, *J. Appl. Phys.* **48**, 3532 (1977).
36. S. M. Irving, *Electrochem. Soc. Extend. Abstr.* No. 67-2, p. 460 (1967).
37. E. O. Degenkolb and J. E. Griffiths, *Appl. Spectrosc.* **31**, 40 (1977).
38. R. Bersin and M. Singleton, U.S. Patent 3,879,597 (1976).
39. A. R. Reinberg, U.S. Patent 3,757,733 (1975).
40. R. J. MacDonald, *Adv. Phys.* **19**, 457 (1970).
41. P. Sigmund, *Phys. Rev.* **184**, 383 (1969).
42. N. Laegried and G. K. Wehner, *J. Appl. Phys.* **32**, 365 (1961).
43. N. D. Morgulis and V. D. Tischenko, *Bull. Acad. Sci. USSR, Phys. Ser.* **20**, 1082 (1957).
44. S. Somekh, Intel Corp., personal communication.
45. W. Laznovsky, *Res./Dev.* **26**(8), 47 (1975).
46. M. Cantagrel and M. Marchal, *J. Mater. Sci.* **8**, 171 (1973).
47. L. F. Thompson, E. D. Feit, and R. D. Heidenreich, *Polym. Eng. Sci.* **14**, 529 (1974).
48. E. F. Labuda, G. K. Herb, W. D. Ryder, L. B. Fritzinger, and J. M. Szaba, *Electrochem. Soc. Extend. Abstr.* No. 74-1, p. 195 (1974).
49. J. L. Vossen and E. B. Davidson, *J. Electrochem. Soc.* **119**, 1708 (1972).
50. L. I. Maissel, C. L. Standley, and L. V. Gregor, *IBM J. Res. Dev.* **16**, 67 (1972).
51. R. E. Jones, C. L. Standley, and L. I. Maissel, *J. Appl. Phys.* **38**, 4656 (1967).
52. J. L. Vossen, J. J. O'Neill, K. M. Finlayson, and L. J. Royer, *RCA Rev.* **31**, 293 (1971).
53. H. R. Koenig and L. I. Maissel, *IBM J. Res. Dev.* **14**, 168 (1970).
54. H. F. Winters and E. Kay, *J. Appl. Phys.* **38**, 3928 (1967).

55. L. I. Maissel, R. E. Jones, and C. L. Standley, *IBM J. Res. Dev.* **14**, 176 (1970).
56. O. Almén and G. Bruce, *Nucl. Instrum. Methods* **11**, 257 (1961); **11**, 279 (1961).
56a. M. I. Gruseva, *Sov. Phys.-Solid State* **1**, 1410 (1960).
56b. F. Keywell, *Phys. Rev.* **97**, 1611 (1955).
56c. G. K. Wehner, R. V. Stuart, and D. Rosenberg, General Mills Annu. Rep. Sputtering Yields, No. 2243 (1961) (unpublished).
56d. O. C. Yonts, C. E. Normand, and D. E. Harrison, *J. Appl. Phys.* **31**, 447 (1960).
56e. C. H. Weijsenfeld, *Philips Res. Rep., Suppl.* No. 2 (1967).
56f. A. L. Southern, W. R. Willis, and M. T. Robinson, *J. Appl. Phys.* **34**, 153 (1963).
56g. P. K. Rol, J. M. Fluit, and J. Kistemaker, *Physica* **26**, 1000 (1960).
57. S. Somekh, *J. Vac. Sci. Technol.* **13**, 1003 (1976).
58. M. Bernheim and G. Slodzian, *Int. J. Mass Spectrom. Ion Phys.* **12**, 93 (1973).
59. M. Cantagrel, *IEEE Trans. Electron Devices* **Ed-22**, 483 (1975).
60. M. Cantagrel, *J. Vac. Sci. Technol.* **12**, 1340 (1975).
61. N. Hosokawa, R. Matsuzaki, and T. Asmaki, *Jpn. J. Appl. Phys., Suppl.* **2**, 1 (1974).
62. G. C. Schwartz, L. B. Zielinski, and T. Schopen, *in* "Etching for Pattern Definition" (H. G. Hughes and M. J. Rand, eds.), p. 122. Electrochem. Soc., Princeton, New Jersey, 1976.
63. S. Somekh and H. C. Casey, Jr., *Appl. Opt.* **16**, 126 (1977).
64. S. Matsuo and Y. Takehara, *Jpn. J. Appl. Phys.* **16**, 175 (1977).
65. H. W. Lehmann and R. Widmer, *Appl. Phys. Lett.* **32**, 163 (1978).
66. J. A. Bondur, *Electrochem. Soc. Extend. Abstr.* No. 77-2, p. 371 (1977).
67. Y. Kurogi, M. Tajima, K. Mori, and K. Sugibuchi, *Electrochem. Soc. Extend. Abstr.* No. 77-2, p. 373 (1977).
68. L. M. Ephrath, *Electrochem. Soc. Extend. Abstr.* No. 77-2, p. 376 (1977).
69. M. Itoga, M. Inone, Y. Kitihara, and Y. Ban, *Electrochem. Soc. Extend. Abstr.* No. 77-2, p. 378 (1977).
70. G. C. Schwartz, L. B. Zielinski, and T. S. Schopen, *Electrochem. Soc. Extend. Abstr.* No. 77-2, p. 391 (1977).
71. H. F. Winters, J. W. Coburn, and E. Kay, *Electrochem. Soc. Extend. Abstr.* No. 77-2, p. 393 (1977).
72. H. W. Lehmann and R. Widmer, *J. Vac. Sci. Technol.* **15**, 319 (1978).
73. R. H. Heinecke, U.S. Patent 3,940,586 (1976).
74. R. V. Stuart and G. K. Wehner, *J. Appl. Phys.* **33**, 2345 (1962).
75. Y. A. Oyagi and S. Namba, *Opt. Acta* **23**, 701 (1976).
76. V. A. Molchanov, V. A. Snisar, and W. M. Chicerov, *Sov. Phys.—Dokl.* **14**, 43 (1969).
77. G. Dupp and A. Scharmann, *Z. Phys.* **192**, 284 (1966); **194**, 448 (1966).
78. K. B. Cheney, and E. T. Pitkin, *J. Appl. Phys.* **20**, 283 (1967).
79. N. Colombie, Thesis, Univ. of Toulouse, Toulouse, 1964.
80. R. E. Chapman, *J. Mater. Sci.* **12**, 1125 (1977).
81. G. Wehner, *J. Appl. Phys.* **30**, 1762 (1959).
82. H. I. Smith, *Proc. IEEE* **62**, 1361 (1974).
83. I. Barin and O. Knacke, "Thermochemical Properties of Inorganic Substances." Springer-Verlag, Berlin and New York, 1973.
84. "Selected Values of Thermodynamic Properties." *Natl. Bur. Stand. (U.S.), Circ. No.* 5000 (1952).
85. C. J. Mogab and T. A. Shankoff, *J. Electrochem. Soc.* **124**, 1766 (1977).
86. H. Abe, K. Nishioka, S. Tamura, and A. Nishimoto, *Jpn. J. Appl. Phys.* **15**(Suppl.), 25 (1976).
87. H. Abe, *Jpn. J. Appl. Phys.* **14**(Suppl.), 287 (1975).
88. D. B. Fraser, Bell Lab., personal communication (1977).

89. R. Reichelderfer, D. Vogel, and R. L. Bersin, *Electrochem. Soc. Extend. Abstr.* No. 77-2, p. 414 (1977).
90. C. J. Mogab, *J. Electrochem. Soc.* **124**, 1262 (1977).
91. Y. Horiike and M. Shibagaki, *Electrochem. Soc. Extend. Abstr.* No. 77-2, p. 243 (1977).
92. J. F. Battey, *J. Electrochem. Soc.* **124**, 437 (1977).
93. C. E. Gleit, *Adv. Chem. Ser.* No. 80, p. 232 (1969).
94. K. Jinno, H. Kinoshita, and Y. Matsumato, *J. Electrochem. Soc.* **124**, 1258 (1977).
95. R. G. Poulsen and M. Brochu, *in* "Etching for Pattern Definition" (H. G. Hughes and M. J. Rand, eds.), p. 111. Electrochem. Soc., Princeton, New Jersey, 1976.
96. A. G. D. Stewart and M. W. Thompson, *J. Mater. Sci.* **4**, 56 (1969).
97. M. J. Nobes, J. S. Colligon, and G. Carter, *J. Mater. Sci.* **4**, 730 (1969).
98. G. Carter, J. S. Colligon, and M. J. Nobes, *J. Mater. Sci.* **6**, 115 (1971).
99. A. R. Bayly, *J. Mater. Sci.* **7**, 404 (1972).
100. C. Catana, J. S. Colligon, and G. Carter, *J. Mater. Sci.* **7**, 467 (1972).
101. I. H. Wilson, *Radiat. Eff.* **18**, 95 (1973).
102. P. Sigmund, *J. Mater. Sci.* **8**, 1545 (1973).
103. G. Carter, J. S. Colligon, and M. J. Nobes, *J. Mater. Sci.* **8**, 1473 (1973); M. J. Witcomb, *J. Mater. Sci.* **9**, 551 (1974).
104. D. J. Barber, F. C. Frank, M. Moss, J. W. Steeds, and I. T. S. Tsong, *J. Mater. Sci.* **10**, 30 (1975).
105. T. Ishitani, M. Kato, and R. Shimizu, *J. Mater. Sci.* **9**, 505 (1974).
106. J. P. Ducommun, M. Cantagrel, and M. Marchal, *J. Mater. Sci.* **9**, 725 (1974).
107. B. B. Meckel, T. Nenadović, B. Perović, and A. Vlahov, *J. Mater. Sci.* **10**, 1188 (1975).
108. T. C. Tisone and P. D. Cruzan, *J. Vac. Sci. Technol.* **12**, 1058 (1975).
109. R. D. Heidenreich and G. W. Kammlott, Bell Lab., personal communication.
110. R. G. Brandes and R. H. Dudley, *J. Electrochem. Soc.* **120**, 140 (1973).
111. H. Abe, *Jpn. J. Appl. Phys.* **14**, 1825 (1975).
112. H. Kinoshita and K. Jinno, *Jpn. J. Appl. Phys.* **16**, 381 (1977).
113. A. Jacob, *Solid State Technol.* **19**(9), 70 (1976).
114. H. Komiya, H. Toyoda, T. Kato, and K. Inaba, *Jpn. J. Appl. Phys.* **15**(Suppl.), 19 (1976).
115. I. Blech, D. B. Fraser, and S. E. Haszko, *J. Vac. Sci. Technol.* **15**, 13 (1978).
116. W. Kern, J. L. Vossen, and G. L. Schnable, *Annu. Proc. Reliab. Phys.*, *11th* pp. 214–223 (1973).
117. L. H. Hall and D. L. Crosthwait, *Thin Solid Films* **9**, 447 (1972).
118. G. Bell and J. Hoepfner, *in* "Etching for Pattern Definition" (H. G. Hughes and M. J. Rand, eds.), p. 47. Electrochem. Soc., Princeton, New Jersey, 1976.
119. D. V. McCaughan and R. A. Kushner, *Proc. IEEE* **62**, 1236 (1974).
120. K. H. Zaininger, *Appl. Phys. Lett.* **8**, 140 (1966).
121. K. H. Zaininger and A. G. Holmes-Siedle, *RCA Rev.* **28**, 208 (1967).
122. E. H. Snow, A. S. Grove, and D. J. Fitzgerald, *Proc. IEEE* **55**, 1168 (1967).
123. J. P. Mitchell, *IEEE Trans. Electron Devices* **ED-14**, 764 (1967).
124. D. V. McCaughan and V. T. Murphy, *J. Appl. Phys.* **44**, 3182 (1973).
125. D. V. McCaughan and V. T. Murphy, *J. Appl. Phys.* **46**, 2008 (1973).
126. D. V. McCaughan and V. T. Murphy, *IEEE Trans. Nucl. Sci.* **NS-19**, 249 (1972).
127. D. V. McCaughan, R. A. Kushner, and V. T. Murphy, *Phys. Rev. Lett.* **30**, 614 (1973).
128. D. V. McCaughan, C. W. White, N. Tolk, and V. T. Murphy, *J. Appl. Phys.* to be published, (1978).
129. R. J. Powell and G. F. Derbenwick, *IEEE Trans. Nucl. Sci.* **NS-18**, 99, (1971).

130. M. Simmonds, *IEEE Trans. Electron Devices* **ED-15,** 966 (1968).
131. K. A. Pickar and L. R. Thibault, *J. Vac. Sci. Technol.* **10,** 1074 (1973).
132. J. P. Mitchell and D. K. Wilson, *Bell Syst. Tech. J.* **46,** 1 (1967).
133. T. P. Sosnowski. *Electrochem. Soc. Extend. Abstr.* No. 74-2, p. 162 (1974).
134. A. M. Voschenskov and J. L. Bartelt, *Electrochem. Soc. Extend. Abstr.* No. 75-2, p. 333 (1975).
135. W. D. Ryden, E. F. Labuda, and J. T. Clemens, *in* "Etching for Pattern Definition" (H. G. Hughes and M. J. Rand, eds.), p. 144. Electrochem. Soc., Princeton, New Jersey, 1976.
136. H. G. Hughes, W. L. Hunter, and K. Ritchie, *J. Electrochem. Soc.* **120,** 99 (1973).
137. E. P. G. T. Van DeVen and H. Kalter, *Electrochem. Soc. Extend. Abstr.* No. 76-1, p. 332 (1976).
138. H. Kalter and E. P. G. T. Van DeVen, *Electrochem. Soc. Extend. Abstr.* No. 76-1, p. 335 (1976).

Index

ISBN 0-12-728250-5

90293>

9 780127 282503
HARDCOVER BAR CODE